Fundamentals of
Multilevel Converters

Fundamentals of
Multilevel Converters

G. Tulasi Ram Das
Professor of Emeritus
JNTU Hyderabad.

BSP BS Publications
An Imprint of **BSP Books Pvt. Ltd.**
4-4-309/316, Giriraj Lane,
Sultan Bazar, Hyderabad - 500 095.

Fundamentals of Multilevel Converters

by **G. Tulasi Ram Das**

Published by:

BSP **BS Publications**
An Imprint of **BSP Books Pvt. Ltd.**
4-4-309/316, Giriraj Lane,
Sultan Bazar, Hyderabad - 500 095
Phone : 040 - 23445600, 23445688
e-mail : info@bspbooks.net
www.bspbooks.net

ISBN: 978-93-48734-70-9 (Hardback)

Dedicated to my family:

My wife Aruna

Son and daughter-in-law

Gaddam Abishek Raman Das and Deekshitha

Grand son

Gaddam Yuvaan Raman Das

Preface

This book is intended towards encouraging the graduate, under-graduate students and research scholars, and working engineers for understanding various topologies of multilevel inverters. In this book, few case studies are provided for practicing the simulation and analysis of multilevel inverters in wide variety of applications in the area of electrical engineering. Good number of exercise problems are also given in each chapter of multilevel inverters. Few pulse width modulation techniques suitable for multilevel inverters are presented for ready reference.

This book is organized into six chapters dealing with introduction of multilevel inverter, pulse width modulation techniques, diode-clamped multilevel inverters, flying-capacitor multilevel inverters, cascaded H-bridge multilevel inverters, and hybrid multilevel inverters. Few case studies are presented for diode-clamped, flying-capacitor, and cascaded H-bridge multilevel inverters for resistive and inductive loads. Advantages, disadvantages and application of multilevel inverters are also presented in every chapter.

Chapter 1 introduces the concept of multilevel inverter topologies at various levels, and given classification of multilevel inverters, Pulse width modulation Techniques with respect to multilevel inverters are presented in the chapter 2. Chapter 3 offers detailed operation and analysis of various topologies of diode-clamped multilevel inverters such as 3-level, 5-level, 7-level and m-level inverters, few case studies are presented by applying various modulation techniques for resistive and inductive loads. In chapter 4, various configurations of flying-capacitor based multilevel inverters are discussed elaborately for 3-level, 5-level and 7-level inverters and few case studies are included for better understanding and exercise problems are provided for practice through simulation. Chapter 5 discusses about the cascaded -H bridge inverters addressing symmetrical, asymmetrical and transistor-clamped topologies of multilevel inverter. Finally, Chapter 6 introduces hybrid multilevel inverters of various combinations and explained for better understanding and classification of advanced hybrid multilevel inverters for further research.

The orientation of the book is covered with variety of case studies by simulation using SIMULINK / MATLAB environment, and each chapter is concluded with summary and exercise problems for practice with learning objectives and learning outcomes. The readers can also practice these exercise problems with any available and suitable software tools. Each chapter is provided with theoretical Questions and Multiple Choice Questions with Answers.

This book is authored exclusively on multilevel inverters by covering all popular topologies used in many applications and suitable for under-graduate, post-graduate and research scholar program levels for learning, practicing and carryingout Dissertation work.

In recent years many publications are appeared in Journals and Conference Proceedings on Multilevel Inverters applied to Electrical Motor Drives, Electrical Vehicles, Power Systems such as FACTS, PQ, and Renewable Energy generation.

Prof. G. Tulasi Ram Das

Acknowledgement

I thank my research scholars who have completed their doctoral theses under my supervision and contributed the technical research papers published in reputed International and National Journals and conference proceedings. Some of their basic works relevant to the topics of this book and the content have been used including case studies in this book in various chapters.

I also acknowledge all the active researchers across the Globe whose papers available in public domain are referred and some of the contents been used in making this book to its present shape and cited those relevant papers in the reference.

I thank my father Late Sri G Janaki Das and mother Late Smt. G. Tara Bai and my brother Dr. G Jagmohan Das, for their blessings and constant encouragement and guidance in making this book possible. I also thank my Father-in-law, Late Sri. B Narayana for his continuous support.

I also profusely thank my family who have shown lot of tolerance and humility during the preparation of this book, which enabled me to continuously concentrate in successful completion, namely my wife Aruna, son Abishek Raman Das, daughter-in-law Deekshitha and grandson Master Yuvan Raman Das.

Prof. G. Tulasi Ram Das

Contents

Chapter 6: Introduction to Hybrid Multilevel Inverters

Nomenclature

MLI	Multi-Level Inverter
DCMLI	Diode-Clamped Multi-Level Inverter
FCMLI	Flying-Capacitor Multi-Level Inverter
NPC-MLI	Neutral Point Clamped Multi-Level Inverter
CHB-MLI	Cascaded H-Bridge Multi-Level Inverter
HMI	Hybrid Multi-level Inverter
PWM	Pulse Width Modulation
SPWM	Sinusoidal Pulse Width Modulation
THI-PWM	Third Harmonic Injected – Pulse Width Modulation
SHE-PWM	Selective Harmonic Elimination-Pulse Width Modulation
SVPWM	Space Vector Pulse Width Modulation
2-D SVM	Two-Dimensional Space Vector Modulation
3-D SVM	Three-Dimensional Space Vector Modulation
PS-PWM	Phase Shifted-Pulse Width Modulation
LS-PWM	Level Shifted-Pulse Width Modulation
CBPWM	Carrier Based Pulse Width Modulation
CBSVPWM	Carrier Based Space Vector Pulse Width Modulation
PD-CB-PWM	Phase Disposition-Carrier Based-Pulse Width Modulation
POD-CB-PWM	Phase Opposition Disposition-Carrier Based-Pulse Width Modulation
APOD-CB-PWM	Alternate Phase Opposite Disposition-Carrier Based-Pulse Width Modulation
CMLI	Cascaded Multi Level Inverter
CM	Common Mode
VSI	Voltage Source Inverter
VSC	Voltage Source Converter
SV	Small Vector
MV	Medium Vector
SSV	Switching State Vector
P	Prism
T	Tetrahedron
NZSV	Non-Zero State Vector

ZSV	Zero State Vector
THD	Total Harmonic Distortion
STATCOM	Static VAR Compensation
FACTS	Flexible AC Transmission System
SSSC	Static Synchronous Series Compensation
DVR	Dynamic Voltage Restorer
D-STATCOM	Distributed Static VAR Compensation
PQ	Power Quality
EMI	Electro-Magnetic Interference
EMC	Electro-Magnetic Compatibility
RFI	Radio Frequency Interference
NPC	Neutral Point Converter
SMC	Stacked Multi Cell
ANPC	Active Neutral Point Converter
MMI	Modular Multilevel Inverter
HB-NPC	H-Bridge-Neutral Point Converter
HMC	Hybrid Multilevel Converter
HEV	Hybrid Electric Vehicles
EV	Electric Vehicles
HVDC	High Voltage Direct Current
HV	High Voltage
AC	Alternating Current
BJT	Bipolar Junction Transistor
MOSFET	Metal Oxide Semiconductor Field Effect Transistor
IGBT	Insulated Gate Bipolar Transistor
GTO	Gate Turn-off Thyristor
IGCT	Insulated Gate Commutated Transistor
ETO	Emitter Turn-Off Transistor
MTO	MOS Turn-Off Transistor
MCT	MOS Controlled Transistor
SITH	Static Induction Thyristor
V	Voltage
V_{dc}	DC Voltage
V_{aN}, V_{bN}, V_{cN}	Pole Voltages
V_{an}, V_{bn}, V_{cn}	Phase Voltages
V_{ab}, V_{bc}, V_{ca}	Line Voltages

V_{ref}	Reference Voltage
I_a	Load Current
P	Active Power
Q	Reactive Power
S	Apparent Power
PF	Power Factor
T1	Output Terminal 1
T2	Output Terminal 2
Z, Z_1, Z_2	Mid-point Source Terminal, Terminal 1, Terminal 2
O	Mid-point Source Terminal

About the Author

G Tulasi Ram Das is a former professor of electrical and electronics engineering, University College of Engineering (Autonomous), Jawaharlal Nehru Technological University Hyderabad, Telangana State, India. He also served as academic administrator as Registrar, and Director, Academic and Planning, Jawaharlal Nehru Technological University Hyderabad, India, and as Vice-Chancellor, Jawaharlal Nehru Technological University Kakinada, Andhra Pradesh, India. Presently he is working as professor of Emeritus in Jawaharlal Nehru Technological University Hyderabad.

He holds doctoral degree from Indian Institute of Technology Madras (IITM), Master degree from Osmania University, Hyderabad, and Bachelor degree from Jawaharlal Nehru Technological University Hyderabad (JNTUH) in electrical and electronics engineering.

He also supervised 29 research scholars for the award of doctoral degrees, about 80 and more than 100 dissertation theses of post-graduate and under-graduate students respectively, in the areas of advanced Power Electronic Converters, Electrical Motor Drives, Flexible AC power Transmission Systems, Power Quality in Distribution Systems, Renewable Energy Sources, and Hybrid Electric Vehicles. He has contributed more than 170 Technical Research Papers published in the refereed International and National Journals and conference proceedings along with his scholars, during his 35 years of academic service in JNTU Hyderabad.

He is a Fellow of Institute of Engineers (India), Fellow of Institute of Electronics and Telecommunications Engineers (India), Member of Institute of Electrical and Electronics Engineers (IEEE), Member of Solar Energy Society of India (SESI), and Life Member of Indian Society of Technical Education (ISTE).

CHAPTER

1 Introduction to Multilevel Inverters

LEARNING OBJECTIVES

- ✓ *To understand and acquire knowledge about importance of multilevel inverters and its classification.*
- ✓ *To prepare learners to classify multilevel inverters.*
- ✓ *To prepare learners to classify basic pulse width modulation techniques.*
- ✓ *To recollect and remember the power semiconductor devices and their symbols for using them in multilevel inverters.*

1.1 Introduction

In recent years, multilevel inverters have received more attention in industrial applications, such as motor drives, Static VAR Compensators (STATCOMs), Flexible AC Transmission System (FACTS), high voltage direct current lines, electrical vehicle drives and renewable energy systems. The most attractive features of multilevel inverters are they can generate output voltages with extremely low distortion and lower dv/dt, increased power ratings, improved harmonic performance, and reduced electromagnetic interference (EMI) emission, draws input current with very low distortion that can be achieved with multiple dc levels that are synthesis of the output voltage waveform, and they generate smaller Common Mode (CM) voltage thus reducing the stress in the motor bearings. In addition, using sophisticated modulation methods, CM voltages can be eliminated. They can operate with a lower switching frequency.

There are numerous industrial applications requiring medium voltage and high-power converters. It is troublesome to achieve medium voltage and high power with directly connecting a single power semiconductor switch to the megawatt grid. For these reasons a new family of Multilevel Inverters has emerged as a solution for working with medium voltage and high-power ranges.

In this chapter the following topics are briefly discussed for Multilevel Inverters, such as various Inverter topologies, various pulse width modulation techniques, various types of power semiconductor devices and their applications.

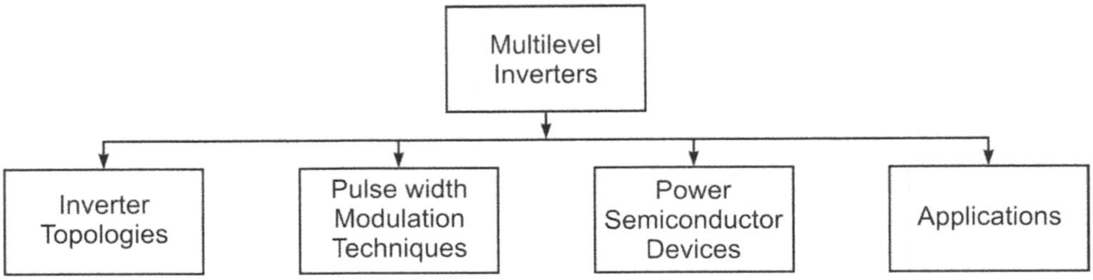

The general structure of the multilevel inverters is to synthesize a sinusoidal voltage from several small levels of voltages, typically obtained from capacitor voltage sources. The multilevel starts from three-level. The waveform obtained from a three-level inverter is a quasi-square wave output. As the number of levels increases, the synthesized output waveform adds more steps producing a staircase wave which approaches the sinusoidal wave with minimum harmonic distortion. Ultimately, a zero-harmonic distortion of the output wave can be achieved by an infinite number of levels. More levels also mean higher voltages can be achieved by connecting several switching devices in series, which inherently pose device voltage sharing problems. The voltage unbalance problem can be solved by replacing capacitor by a constant dc voltage source.

The multilevel inverters are used to minimize the output voltage distortion with improved fundamental voltage and also due to good power quality, low switching losses, high voltage capability and good electromagnetic compatibility (EMC).

The main disadvantages of multilevel inverters are that they require large number of switching devices at lower voltages and small voltage steps are supplied on the dc input side. The multilevel inverters provide more than two voltage levels and as the number of levels increases, the output total harmonic distortion decreases, resulting to sinusoidal waveform.

Table 1.1 Comparison of conventional two-level inverter and multilevel inverter.

Serial Nos.	Conventional 2-level Inverter	Multilevel Inverter
1.	Square wave output voltage.	Stair-case wave output voltage.
2.	Higher THD in output voltage.	Low THD in output voltage.
3.	High voltage levels are not possible.	High voltage levels are possible.
4.	Not applicable for High Voltage applications.	Applicable for High Voltage applications.
5.	More switching stresses on devices.	Reduced switching stresses on devices.
6.	Since dv/dt is high the EMI from system is high.	Since dv/dt is low the EMI from system is low.
7.	Higher switching frequency is used hence switching losses are high.	Low switching frequency can be used and hence reduction in switching losses.
8.	Power bus station control schemes are simple.	Control scheme becomes complex as number of levels increases.

Basic Multilevel Inverters:

A. ***Single-phase two-level inverter:*** Consider a basic single-phase two-level inverter with its load voltage waveform shown in Fig.1.1. for explaining the operation. When switches S_1 and S_2 are closed, then the load current flows from dc source, through switch S_1, load terminals '*a*' to '*b*', switch S_2, and back to the other terminal of dc source. The load current is positive when enters into load at terminal '*a*', and the positive voltage waveform is obtained across the load. Similarly, a negative voltage waveform across the load is obtained when conduction takes place from dc source, through switch S_3, load terminals '*b*' to '*a*', switch S_4, back to another terminal of dc source. The load current is negative when enters into load terminal '*b*'. Thus, a complete load voltage waveform (Phase) is obtained as shown in Fig. 1.1 (b) and load current wave shape depends on the type of load. The load phase voltage $V_o = V_{ab}$ obtained is of two-level, appears as $+V_{dc}$ and $-V_{dc}$.

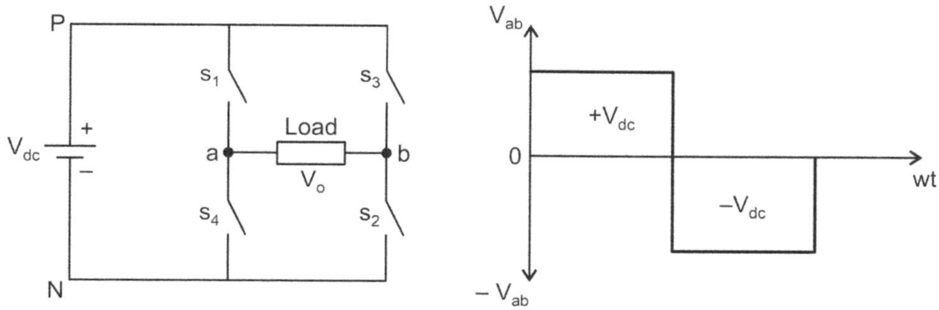

(a) (b)

Fig. 1.1 Single-phase two-level inverter (a) Basic Circuit, (b) Load voltage waveform.

B. ***Three-phase two-level inverter:*** Now, a three-phase two-level inverter with three-phase wye-connected load is considered as shown in Fig.1.2 to represent various load voltages normally come across in the multilevel inverters and their voltage waveform varies accordingly, as the levels are increased.

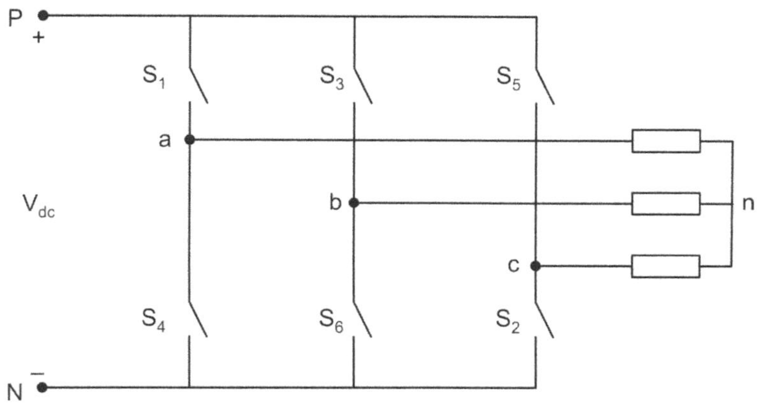

Fig. 1.2 Three-phase two-level inverter.

V_{aN}, V_{bN}, and V_{cN} are known as Pole voltages across load terminal and neutral point of the inverter circuit or second terminal of dc source 'N', respectively.

V_{an}, V_{bn}, and V_{cn} are known as Phase voltages across respective load terminal and neutral point of three-phase load 'n'.

V_{ab}, V_{bc}, and V_{ca} are known as Line voltages between phases of three-phase loads.

The typical various load voltage waveforms are shown in Fig. 1.5. – Fig. 1.7 for three level, five-level and seven-level outputs at a later-stage.

Three-phase inverter operation is explained with single-leg arrangement for easy understanding, with its load voltage waveform representing three-level output as shown in Fig. 1.3 (a) and (b). The inverter operation is explained as follows:

The inverter circuit consists of three-terminal dc sources, from which two-identical dc sources of value each V_{dc} are connected to terminals 'P' and 'N' and these terminals are separated by a mid-point terminal '*o*' and the load is connected between terminals '*a*' and '*o*'. The controlled power semiconductor switch 'S_w' is to be connected to three points of dc source terminals '1, '*o*', '2' each time and accordingly load voltage waveform varies. As the switch 'S_w' of the load terminal '*a*', when connects to the dc source terminal '1', then it produces a positive load voltage of $V_{ao} = +V_{dc}$, and similarly, when switch 'S_w' connects to the mid-point dc source terminal, then a zero load voltage appears as $V_{ao} = 0$. Finally, the switch 'S_w' when connects to dc source terminal '2', then it produces a negative load voltage, $V_{ao} = -V_{dc}$. The resultant load voltage waveform for V_{ao} is shown in Fig. 1.3 (b). Similarly, single-phase and three-phase bridge circuit topologies can be developed by adding one more leg and two more legs, respectively.

Observe the difference of load waveforms between two-level inverter and three-level inverter outputs, so that reader can easily identify the load voltage waveforms when they appear for various levels of inverters in various chapters of this book.

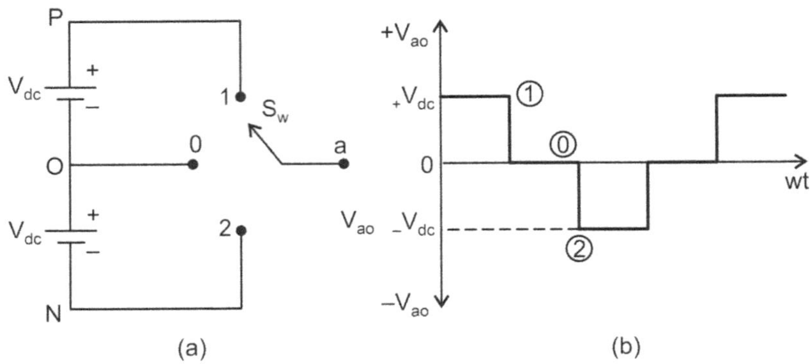

Fig. 1.3 Three-level Inverter (a) with single-leg and mid-point arrangements, and (b) three-level output voltage waveform.

From the Fig.1.3, it is noticed that the dc source voltage of the circuit is divided into two equal dc sources of each V_{dc}, separated by connecting mid-point common terminal as 'O'. Then the resultant three-level output voltage waveform is also divided into two parts, one representing positive and another representing negative separated by zero-period. In order to judge the three-level output voltage between load terminal '*a*' and mid-point common terminal 'O' as V_{aO}, in such a case, one has to count the waveform from positive (1) to negative (2) including zero-period (O), thus, get the three-level output voltage waveform by counting 1, O, and 2 waves together in Fig.1.3 (b).

C. ***Single-phase three-level bridge inverter:*** Consider a single-phase bridge circuit inverter for three-level output voltage waveform as shown in Fig.1.4 (a) and (b) respectively, for the explanation of the detailed operation. The circuit topology is having two dc sources as V_1 and V_2 with three source terminals indicated as 1, 2 and 3. Two switches are provided in the circuit as S_{wa} and S_{wb} connecting the load terminals '*a*' and '*b*' to three source terminals one at a time in proper sequence, respectively, in order to obtain a regular three-level output voltage (V_{ab}) waveform. Switch S_{wa} can be connected to dc source terminals 1, 2 and 3 to obtain positive output voltage across the load as $V_1 + V_2$, either V_1 or V_2 and zero voltage respectively. Similarly Switch S_{wb} can be connected to dc source terminals 3', 2' and 1' to obtain negative output voltage across the load as zero voltage, either V_1 or V_2 and $V_1 + V_2$, respectively.

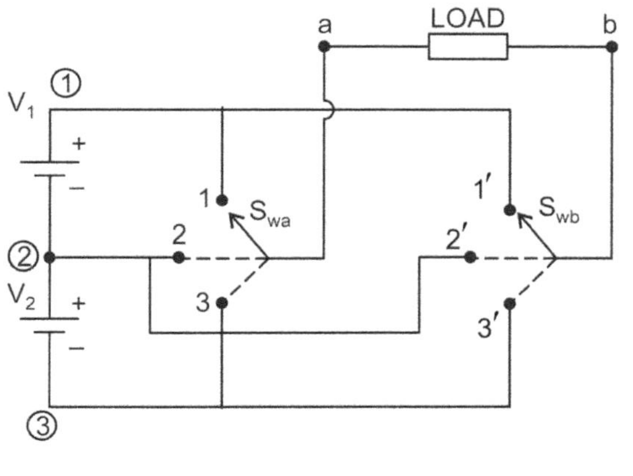

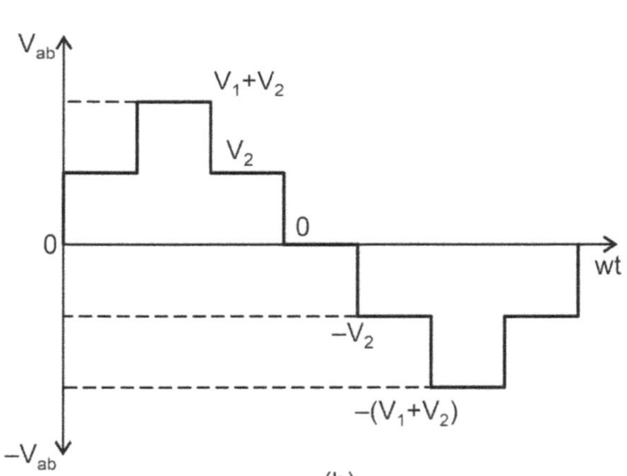

Fig. 1.4 Single-phase three-level bridge inverter (a) Basic single-phase bridge circuit inverter, (b) three-level output voltage (V_{ab}) waveform.

The operation of the above circuit topology is presented in six modes as follows and the switching sequence and respective output voltages are given in table 1.1.

Mode 1: To obtain $V_{ab} = +(V_1+V_2)$, the switch S_{wa} is to be connected to dc source terminal 1 and switch S_{wb} is to be connected to dc source terminal 3', respectively, and load current I_a flows from load terminals '*a*' to '*b*' is considered as positive current.

Mode 2: To obtain $V_{ab} = +V_2$, the switch S_{wa} is to be connected to dc source terminal 2 and switch S_{wb} is to be connected to dc source terminal 3', respectively, and load current I_a flows from load terminals '*a*' to '*b*' is considered as positive current.

Mode 3: To obtain $V_{ab} = 0$, the switch S_{wa} is to be connected to dc source terminal 3 and switch S_{wb} is to be connected to dc source terminal 3', respectively, and load current I_a is zero.

OR

Mode 4: To obtain $V_{ab} = 0$, the switch S_{wa} is to be connected to dc source terminal 1 and switch S_{wb} is to be connected to dc source terminal 1', respectively, and load current I_a is zero.

Mode 5: To obtain $V_{ab} = -V_2$, the switch S_{wa} is to be connected to dc source terminal 3 and switch S_{wb} is to be connected to dc source terminal 2', respectively, and load current I_a flows from load terminals '*b*' to '*a*' is considered as negative current.

Mode 6: To obtain $V_{ab} = -(V_1+V_2)$, the switch S_{wa} is to be connected to dc source terminal 3 and switch S_{wb} is to be connected to dc source terminal 1', respectively, and load current I_a flows from load terminals '*b*' to '*a*' is considered as negative current.

Table 1.2 Switching sequence and output voltages.

Modes	Switch S_{wa}	Switch S_{wb}	Output Voltage, V_{ab}	Load current, I_a
Mode 1	Terminal -1	Terminal -3'	$+ (V_1 + V_2)$	Positive
Mode 2	Terminal -2	Terminal -3'	$+ V_2$	Positive
Mode 3	Terminal -3	Terminal -3'	Zero Voltage	Zero
Mode 4	Terminal -1	Terminal -1'	Zero Voltage	Zero
Mode 5	Terminal -3	Terminal -2'	$- V_2$	Negative
Mode 6	Terminal -3	Terminal -1'	$- (V_1 + V_2)$	Negative

It is observed from the Fig.1.4 that, the dc source voltages V_1 and V_2 are not divided as it was done in Fig.1.3, and the output voltage in this case is measured between load terminals '*a*' and '*b*', unlike in Fig.1.3 where output voltage is measured between load terminal '*a*' and common mid-point terminal 'O'. The resultant output voltage is counted from only positive side of the waveform, for three-level, it is counted as $+(V_1+V_2)$, $+V_2$ and 0 together, but not counted both positive and negative sides of the waveform, if counted so, it will lead to five-level waveform, which is not correct.

Now it is appropriate to present various type of output voltages for three-phase bridge inverter shown in Fig.1.2 earlier for three-level. If the same three-phase bridge inverter is developed for five-level and seven-level and their typical (i) Line to mid-point Voltage, (Pole Voltage),V_{ao} (ii) Line to line Voltage, V_{ab} and (iii) Phase to neutral of star-connected load Voltage waveforms, (Phase Voltage) V_{an} are given in Fig. 1.5 (a), (b), (c) for three-level voltages and in Fig. 1.6 (a), (b), (c) for five-level voltages and Fig. 1.7 (a), (b), (c) for seven-level voltages, respectively, to observe the difference in representing various voltage waveforms and their respective waveshapes for better understanding and clarity.

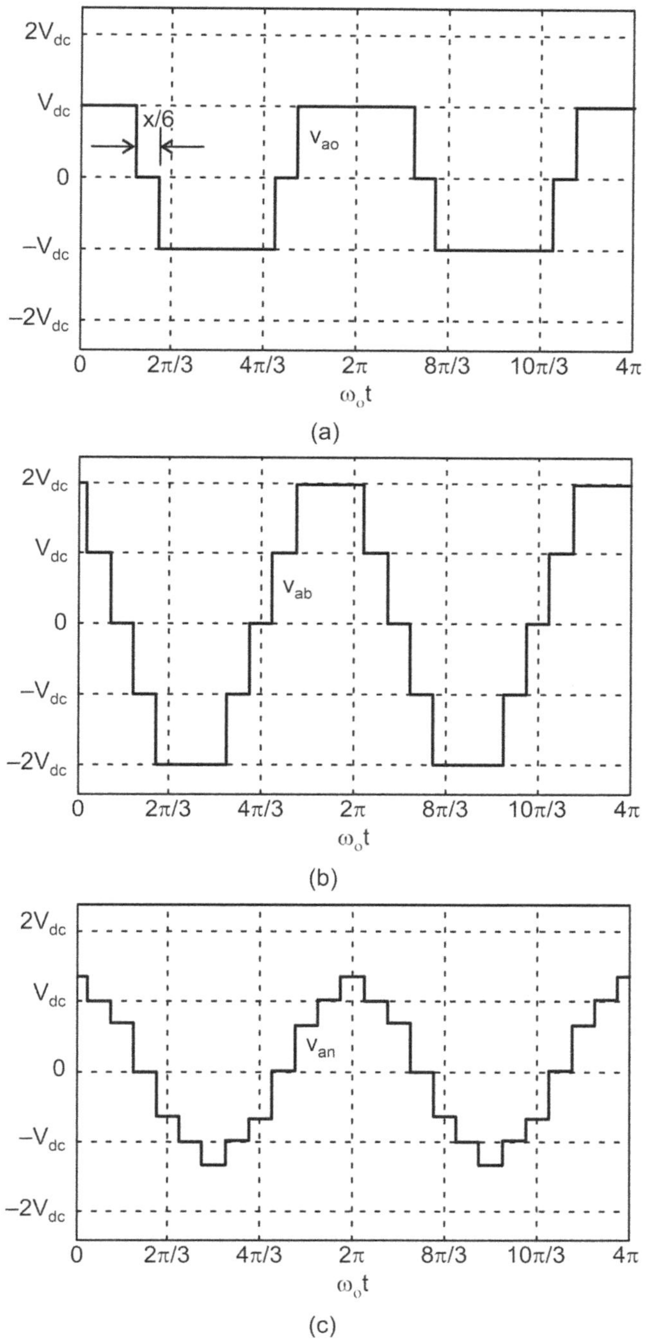

Fig. 1.5 Three-phase Three-level Inverter, typical (a) Pole Voltage, (line-to-midpoint), V_{ao}, (b) Line-to-line Voltage, V_{ab} and (c) line-to-neutral Voltage, V_{an}, (Phase Voltage) waveforms.

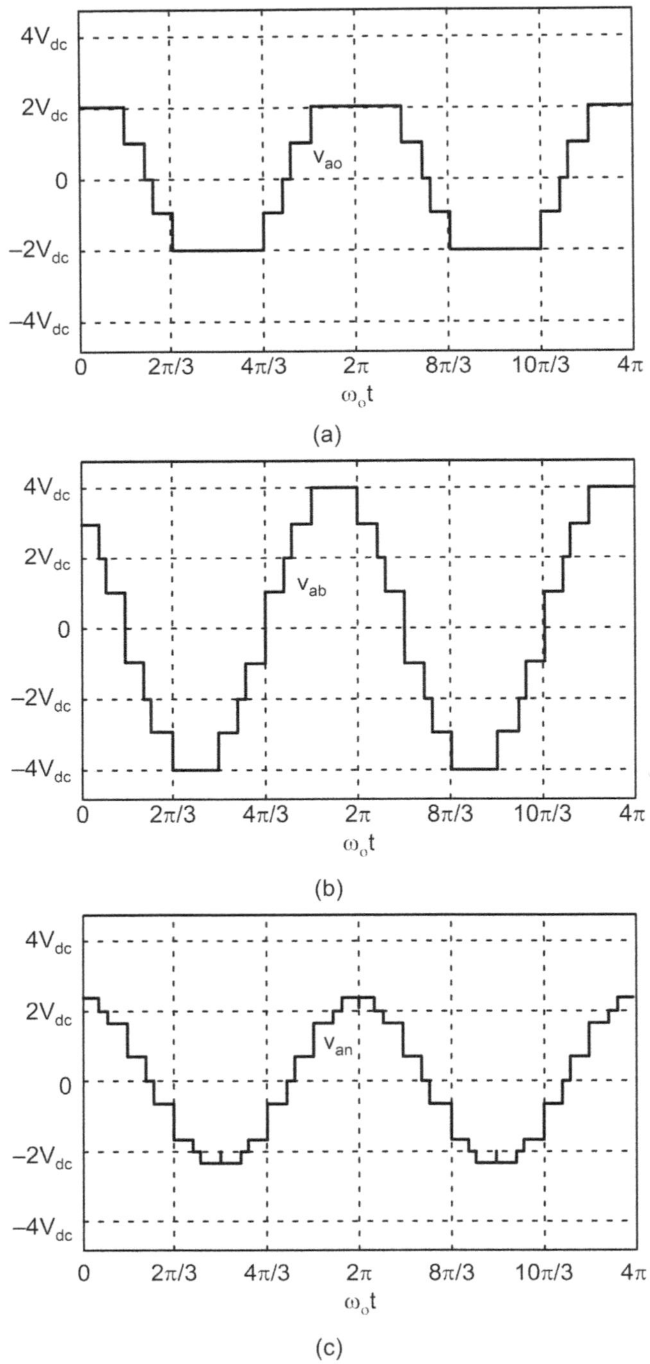

Fig. 1.6 Three-phase Five-level Inverter, typical (a) Pole Voltage, (line-to-midpoint), V_{ao}, (b) Line-to-line Voltage, V_{ab} and (c) line-to-neutral Voltage, V_{an}, (Phase Voltage) waveforms.

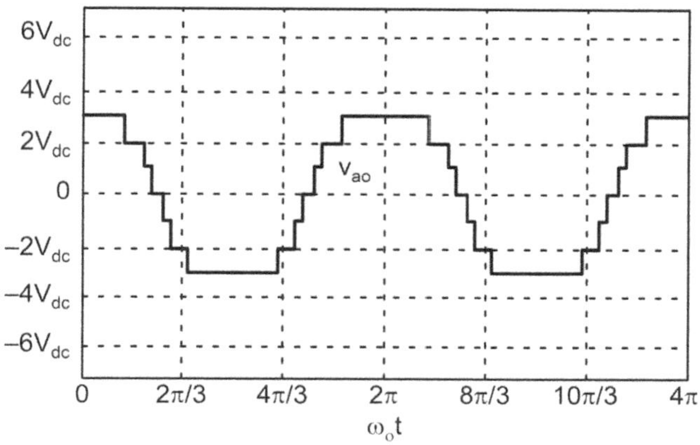

(a)

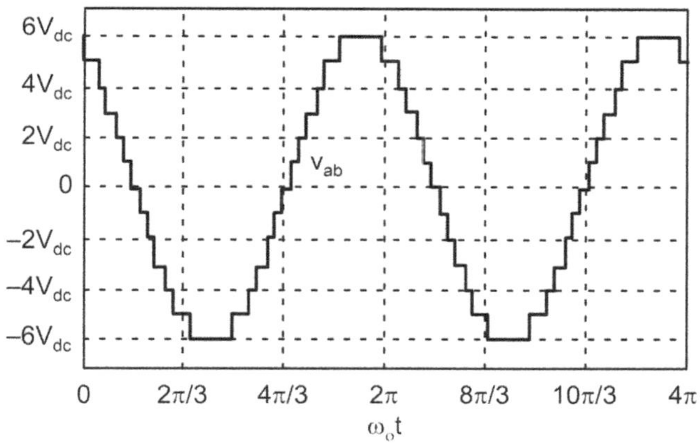

(b)

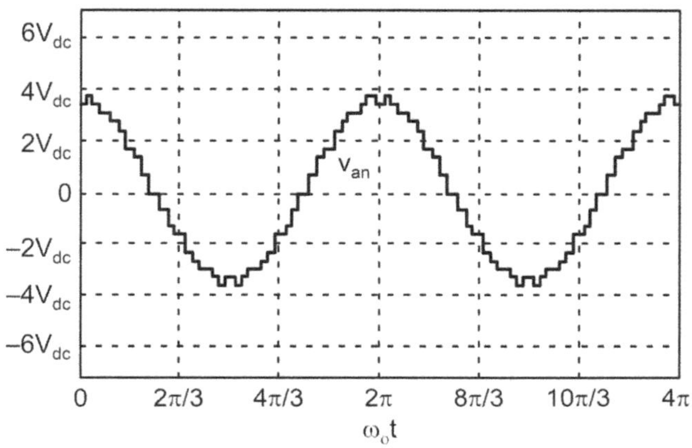

(c)

Fig. 1.7 Three-phase seven-level Inverter, typical (a) Pole Voltage, (line-to-midpoint), V_{ao}, (b) Line-to-line Voltage, V_{ab} and (c) line-to-neutral Voltage, V_{an}, (Phase Voltage) waveforms.

D. m-level generalized inverter: The m-level generalized inverter circuit is shown in Fig. 1.8, where the load terminal 'O' is connected through switch to various dc source terminals from V_1 to V_m. The output voltage V_o can be varied from V_1, V_2, . . ., $V_m/2$, $V_m/2$ -1, . . ., V_m -1, and V_m.

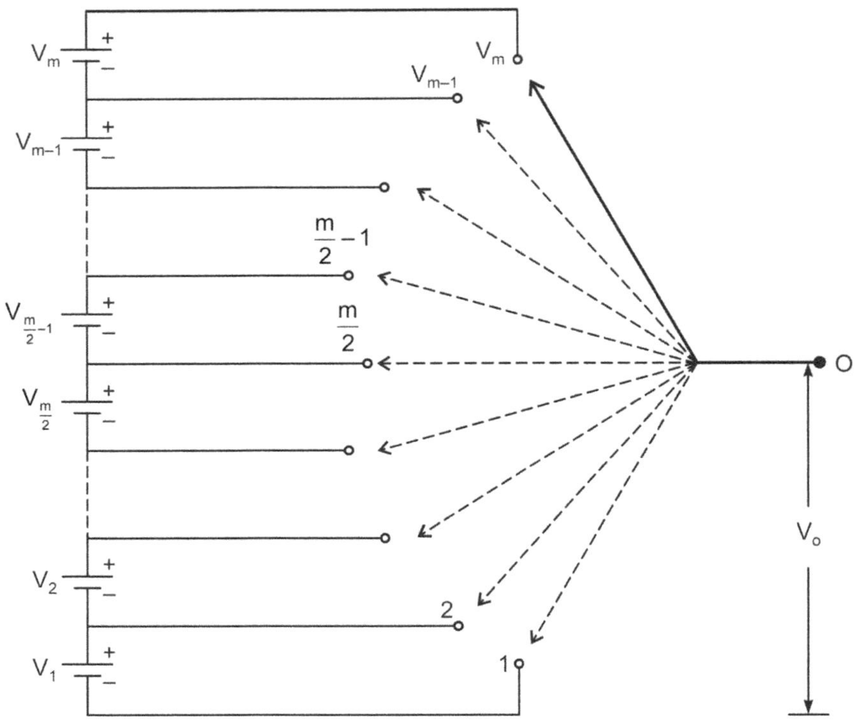

Fig. 1.8 m-level inverter.

1.2 Classification of Multilevel Inverters

The multilevel inverters are classified in various topologies based on voltage clamping facilities. They are

1. Diode-clamped multilevel inverters
2. Flying capacitor multilevel inverters
3. Cascaded H-bridge inverters
4. Hybrid Multilevel inverters

1. **Diode-clamped multilevel inverters:** In diode-clamped multilevel inverters the diodes are used as the clamping devices to clamp the d.c. bus voltage so as to achieve steps in the output voltage. It requires more clamping diodes.
2. **Flying capacitor multilevel inverters:** In flying capacitor multilevel inverters, the flying-capacitors are used to clamp the d.c. bus voltages to achieve steps in the output voltage. It requires more clamping capacitors.

3. **Cascade H-bridge inverters:** Several units of single phase full bridge inverters with separate d.c. source is connected in series to increase the number of levels in output voltage and acts as cascaded H-bridge multilevel inverters. Clamping diodes and clamping capacitors are not required.
4. **Hybrid Multilevel Inverters:** Hybrid topologives are also recently developed for making the quality of output voltage waveforms to which it requires higher number of different voltage magnitude dc sources. Generally, asymmetric dc source multilevel inverters generate higher output voltage steps with lower percentage of asymmetric. Total harmonic distortion compared with symmetric dc source multilevel inverters.

They can be used for single-phase, three-phase, five-phase and six-phase inverters of all the above three varieties. The multilevel inverters start from two-level, three-level, four-level, five-level, continues up to m-level inverters. The diode-clamped three-level inverters are popularly known as Neutral Point Clamped (NPC) inverters. Diode-clamped multilevel inverters or NPC inverters are presented in the chapter:3, in chapter:4, Flying capacitor multilevel inverters are discussed, Cascaded H-bridge inverters are explained in chapter:5, and in chapter:6 Hybrid multilevel inverters. The constructional details, basic operations, various levels, various topologies, advantages, disadvantages and applications are presented for all the four classified topologies.

1.3 Harmonic Distortions

Introduction: The basic two-level inverter can generate a square-wave output ac voltage of various frequencies across the load. The square-wave voltage consists of harmonic components of multiple frequencies, but the load demands pure sinusoidal waveform with zero harmonic component. The multilevel inverters also generate a non-sinusoidal voltage waveform and current waveform. When waveforms deviate from a sinewave shape, they contain harmonics. These current harmonics distort the voltage waveform and create distortion in the power system which can cause many problems. A power system can contain one or two different kinds of loads, a non-linear load or a linear load. Non-linear loads such as adjustable speed drives, power supplies for various electronic equipment are the origin of both current and voltage harmonics production in the distribution system leading to the power quality problem. In response to the power quality problem, IEEE 519 and IEC EN 61000-3 standards specify regulations governing harmonic compliance.

Some of the effects the harmonic can cause, to the equipment, to the installation, or both, are:

- Electronic equipment may be sensitive to the voltage distortion supplying it, due to higher voltage peaks, unexpected zero-crossing, affectation to protection circuits, etc.
- Digital circuits can be affected by misinterpretation of logical values in presence of harmonics.
- Reduced service life of components and equipment under continuous distorted supply voltage.
- Affectation of IT equipment such as memory losses, and turn offs.

- Uninterruptible Power Supplies may need to handle with high distorting loads, i.e., high current peaks may be over the range of the crest factor capacity of this UPS. In such case, the voltage distortion can even increase, if the inverter of the UPS is not capable enough.
- Added efficiency losses to the system composed by electrical installation and equipment.
- Unexpected resonances.
- Unwanted overload (or need to oversize) for transformers, and wirings.
- Malfunctions of motors and generators.
- Unwanted Circuit Breakers tripping or Fuses blowing.

As per the standards, IEEE 519: Recommended Practice and Requirements for Harmonic Control in Electric Power Systems, individual harmonic limit is 5%, Total harmonic distortion for low voltage is 8%. Some of the higher-order harmonics can be filtered out by using passive and active harmonic filters.

1.4 Classification of Modulation Techniques

The multilevel inverters are employed for facilitating variable output voltages at various frequencies or fixed frequency for a load or for a power system from a fixed dc source. The variation in output voltage is possible within the inverter by applying a suitable modulation technique for controlling the switching pattern of the inverter.

Pulse width modulation is a great method of controlling the amount of power delivered to a load without dissipating any wasted power. The main advantage of PWM is that power loss in the switching devices is very low. When a switch is off there is practically no current, and when it is on and power is being transferred to the load, there is almost no voltage drop across the switch.

Applications of Pulse Width Modulation (PWM): Techniques reduces the overall harmonic distortion THD in a load current.

- Pulse Width Modulation helps in voltage regulation and thus finds its use in controlling the speed of motors.

Advantages of Pulse Width Modulation (PWM): The advantages are:

- Pulse Width Modulation provides accuracy and quick response time.
- It provides high input Power Factor.
- Initial cost is low.
- PWM technique helps the motors to generate maximum torque even when they are running at lower speeds.

Disadvantages of Pulse Width Modulation (PWM): The disadvantages are:

- As the PWM frequency is high, switching losses is considerably high.
- It induces Radio Frequency Interference (RFI).

The modulation techniques are classified as follows:

1. Pulse width modulation (PWM) techniques:
   - Sinusoidal Pulse width modulation (SPWM) scheme.
   - Third Harmonic Injected Pulse width modulation (THI-PWM) scheme.
   - Selective Harmonic Elimination Pulse width modulation (SHE-PWM) scheme.
2. Carrier based pulse width modulation (CB-PWM) techniques:
   - Phase-shifted Carrier-based pulse width modulation (PS-CB-PWM) scheme.
   - Level-shifted Carrier-based pulse width modulation (LS-CB-PWM) schemes.
     - Phase Disposition carrier-based pulse width modulation (PD-CB-PWM) method.
     - Phase Opposition Disposition carrier-based pulse width modulation (POD-CB-PWM) method.
     - Alternate Phase Opposition Disposition carrier-based pulse width modulation (APOD-CB-PWM) method.
3. Space Vector modulation techniques:
   - Two-dimensional (2-D) algorithm scheme.
   - Three-dimensional (3-D) algorithm scheme.
   - Space vector carrier-based pulse width modulation (SV-CB-PWM) scheme.

The above all pulse width modulation techniques are addressed elaborately in chapter 2.

1.5 Classification of Power Semiconductor Switching Devices

The power semiconductor switching devices play an important role in controlling the magnitude and quality of the output voltage and load current, in coordination with various pulse width modulation schemes for multilevel inverters. There are wide variety of power semiconductor switching devices available in the literature and in the market. The selection of type of power semiconductor devices depends on various factors, such as, voltage and current ratings, power handling capability, thermal ratings, switching frequencies, turn-on and turn-off times, conduction and switching losses, voltage blocking capabilities, safe operating area (SOA), etc.

The classification of power semiconductor switching devices are given as follows.

1. Transistor Family Devices:
   - Power Bipolar Junction Transistor (BJT) devices.
   - Power MOSFET (Metal-Oxide Field-effect transisitor) devices.
   - Power IGBT (Insulated Gate Bipolar Transistor) devices.
2. Thyristor Family Devices:
   - Thyristor Power Devices.
   - Gate Turn-Off (GTO)-Thyristor Power Devices.

3. Transistor-Thyristor Family Devices:
   - Power IGCT (Insulated Gate Commutated Thyristor) Devices.
   - Power ETO (Emitter Turn-Off) Thyristor Devices.
   - Power MTO (MOS Turn-Off) Thyristor Devices.
   - Power MCT (MOS Controlled Thyristor) Devices.
   - Static Induction Thyristor (SITH) Devices.

Table. 1.3 Various Power Semiconductor Switching Devices and their symbols.

Sl.no.	Name of the Device	Symbols
1.	**Transistor Family Devices**	
1.1	BJT (Bipolar Junction Transistor)	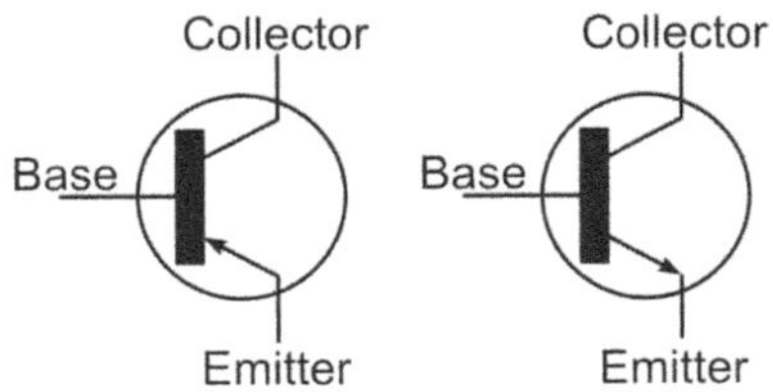 P-n-P n-P-n
1.2	MOSFET (Metal-Oxide Semiconductor Field Effect Transistor)	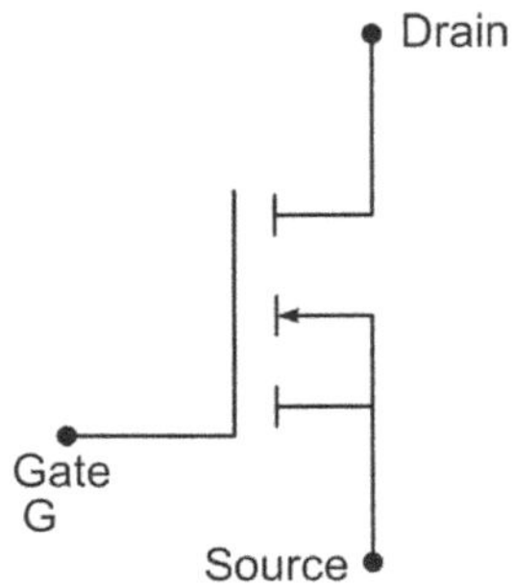 MOSFET
1.3	IGBT (Insulated Gate Bipolar Transistor)	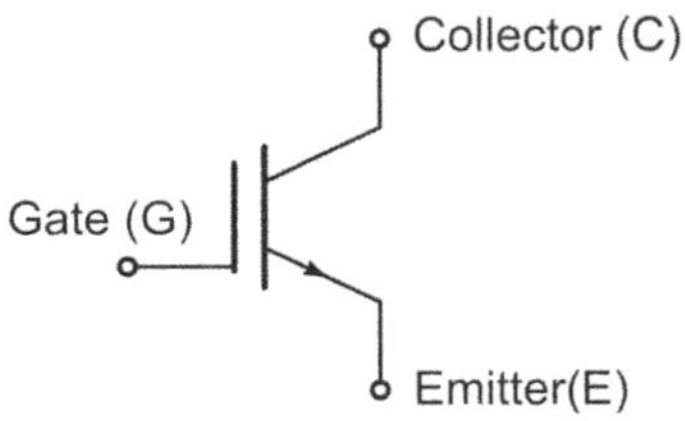
2.	Thyristor Family Devices	
2.1	Thyristor	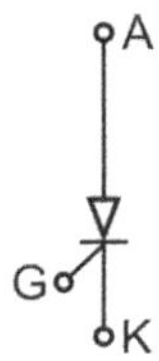

Contd...

Sl.no.	Name of the Device	Symbols
2.2	GTO-Thyristor (Gate-Turn-Off-Thyristor)	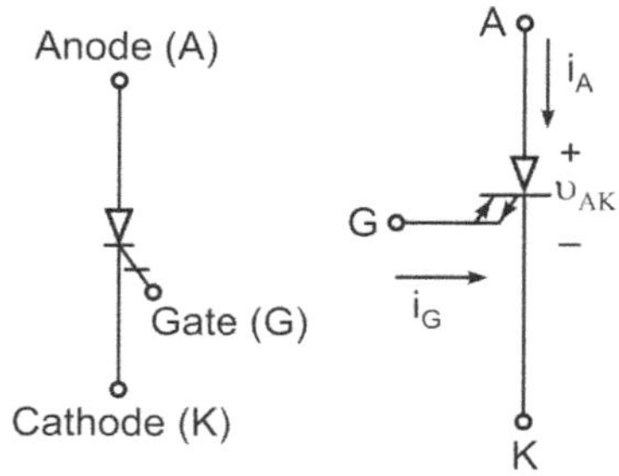
3.	Transistor-Thyristor Family Devices	
3.1	IGCT (Insulated Gate Commutated Transistor)	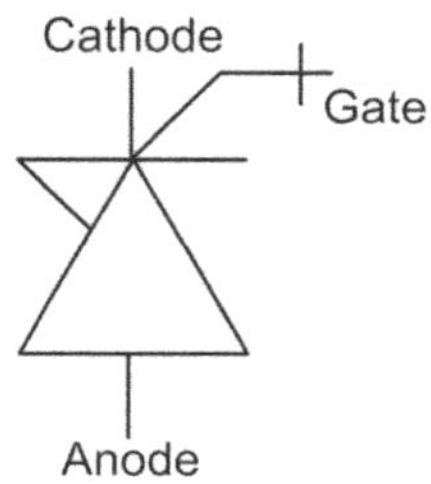
3.2	ETO – Thyristor (Emitter Turn-Off -Thyristor)	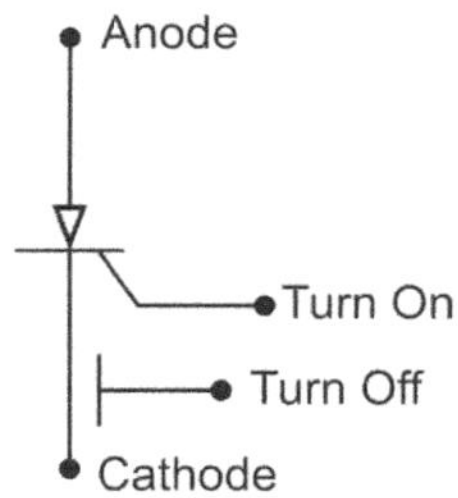
3.3	MTO – Thyristor (MOS-Turn-Off – Thyristor)	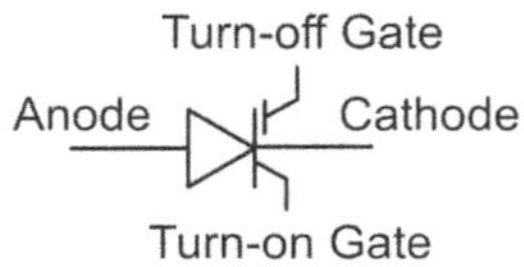
3.4	MCT – Thyristor (MOS-Controlled-Thyristor)	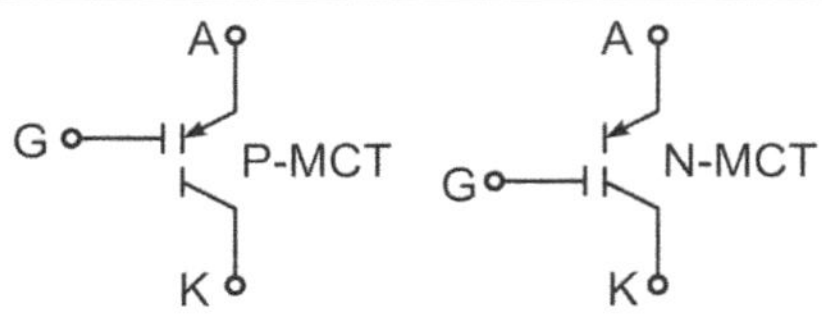
3.5	SITH – Thyristor (Static-Induction Thyristor)	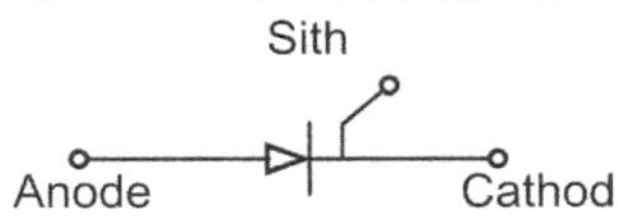 **Symbol of Sith**

1.6 Applications of Multilevel Inverters

The Multilevel Inverters are used in industrial motor drives, hybrid electric vehicles, locomotives drives, AC power transmission systems (FACTS) such as STATCOMs, SSSC, UPFC, AC distribution systems (PQ) such as DVR, D-STATCOM, UPQC, HVDC Transmission Systems such as converter and inverter stations, Standby power supplies and custom power devices, etc.

1.7 Summary

This chapter deals with the brief introduction of Multilevel inverters, their importance, basic circuits and operations, typical output voltage waveforms. The importance of Pulse width modulation for multilevel inverters and their classifications are also presented for clarity. The availability and suitability of power semiconductor switching devices along with their symbols are given for ready reference. The multilevel inverters are having advantages of producing 15% more output voltage with less harmonics by synthesizing compared to their conventional multi-pulse inverters. In recent years, the multilevel inverters are very widely used in many applications in the area of industrial drives, power transmission and distribution, and renewable energy systems.

Learning Outcomes

- To be able to acquire knowledge about importance of Multilevel inverters and to be able to give classification of various multilevel inverters. (CO1) :: (PO1)
- To be able to acquire knowledge about importance of pulse width modulation techniques and to be able to give classification of various schemes of pulse width modulations. (CO1):: (PO1)
- To be able to understand the power semiconductor devices, their symbols and names of the terminals and application of suitable device to multilevel inverters. (CO2) :: (PO1)

Questions

1.1 Classify multilevel inverters?

1.2 Discuss the features of multilevel inverter over conventional two-level inverter?

1.3 Explain the operation of basic single-phase two-level inverter with its output voltage?

1.4 Explain the operation of basic three-level inverter with single-leg arrangement and its output voltage?

1.5 Discuss the working of basic single phase three-level bridge inverter with its output voltage with the help of various modes?

1.6 Draw the pole voltage (line-to-midpoint), phase voltage (line-to-neutral), and line voltage(line-to-line) waveforms for three-phase three-level inverter?

1.7 Draw the pole voltage (line-to-midpoint), phase voltage (line-to-neutral), and line voltage(line-to-line) waveforms for three-phase Five-level inverter?

1.8 Draw and explain m-level generalized inverter basic diagram?

1.9 Discuss the effects of harmonics on the performance of the electrical equipment and installations?
1.10 Give advantages and disadvantages of pule width modulation?
1.11 Classify Sinusoidal Pulse width modulation techniques?
1.12 Classify Carrier based Pulse width modulation techniques?
1.13 Classify Space Vector Modulation techniques?
1.14 Name various power semiconductor devices?
1.15 Give applications of multilevel inverters?
1.16 Draw the symbols of Power BJT and Power IGBT with the name of the Terminals?
1.17 Draw the symbol of Power MOSFET with the name of the Terminals?
1.18 Draw the symbols of Power Thyristor and GTO-Thyristor with the name of the Terminals?
1.19 Draw the symbol of Power IGCT with the names of the terminals?
1.20 Draw the symbols of Power ETO, MTO and MCT with the names of the terminals?

Multiple Choice Questions

1.1 Identify the terminals of the power Bipolar Junction Transistor (BJT),
(a) Collector, Anode, Cathode (b) Emitter, Collector, Base
(c) Anode, Emitter, Base (d) Cathode, Collector, Emitter

1.2 Choose the appropriate terminals of the power IGBT
(a) Collector, Emitter, Base (b) Cathode, Anode, Gate
(c) Collector, Emitter, Gate (d) Collector, Anode, Base

1.3 Identify the terminals of the power MOSFET
(a) Anode, Cathode, Gate (b) Collector, Emitter, Base
(c) Source, Drain, Base (d) Source, Drain, Gate

1.4 Choose the appropriate terminals of the power Thyristor
(a) Anode, Collector, Gate (b) Anode, Cathode, Gate
(c) Emitter, Collector, Base (d) Gate, Anode, Emitter

1.5 Identify the terminals of the power GTO-Thyristor
(a) Cathode, Emitter, Gate
(b) Collector, Base, Cathode
(c) Anode, Cathode, bidirectional Gate
(d) Emitter, Base, Anode

1.6 Choose the appropriate terminals of the power IGCT
(a) Cathode, Anode, isolated Gate
(b) Cathode, Anode, bidirectional Gate
(c) Cathode, Anode, Base
(d) Collector, Emitter, Base

1.7 Identify the terminals of the power ETO
   (a) Anode, Cathode, Turn-on Gate, Turn-off Gate
   (b) Gate, Anode, Cathode
   (c) Base, Collector, Emitter
   (d) Collector, Cathode, Gate

1.8 Choose the appropriate terminals of the power MTO
   (a) Anode, Cathode, Gate1 for Turn-on, Gate2 for Turn-off
   (b) Base, Collector, Anode
   (c) Gate for turn-on, Gate for turn-off, Emitter, Collector
   (d) Anode, Cathode, Base for turn-on, Base for turn-off

1.9 Identify the terminals of the power MCT
   (a) Anode, Cathode, connected Gate
   (b) Anode, Cathode, Base
   (c) Anode, Cathode, isolated Gate
   (d) Collector, Emitter, Base

1.10 Choose the appropriate terminals of power SITH
   (a) Anode, Cathode, Base (b) Anode, Cathode, Gate
   (c) Emitter, collector, Base (d) Emitter, Collector, Gate

1.11 How many schemes of sinusoidal pulse width modulation for multilevel inverter are presented in this chapter
   (a) three (b) two
   (c) one (d) none

1.12 How many schemes of carrier-based pulse width modulation for multilevel inverters are discussed in this chapter
   (a) one (b) two
   (c) three (d) four

1.13 How many methods of level-shifted pulse width modulation for multilevel inverters are Presented
   (a) one (b) two
   (c) four (d) three

1.14 How many Space Vector Modulation schemes are available for multilevel inverters
   (a) three (b) two
   (c) one (d) none

Answers to Multiple Choice Questions

1.1	(b)	1.2	(c)	1.3	(d)	1.4	(b)	1.5	(c)
1.6	(b)	1.7	(a)	1.8	(a)	1.9	(c)	1.10	(b)
1.11	(a)	1.12	(b)	1.13	(d)	1.14	(a)		

CHAPTER

2 Modulation Techniques for Multilevel Inverters

LEARNING OBJECTIVES

- ✓ *To understand and acquire knowledge about importance of pulse width modulation techniques.*
- ✓ *To prepare learners to analyze various schemes of pulse width modulation techniques.*
- ✓ *To prepare learners to practice pulse width modulation schemes through case studies on basic multilevel inverters.*
- ✓ *To understand the suitability of pulse width modulation schemes for various multilevel inverters.*
- ✓ *To prepare learners to model the pulse width modulation schemes and simulate the performance of multilevel inverters.*

2.1 Introduction

The multilevel inverters are elaborately explained their functionality in the previous chapter, for three types of inverters, namely, Diode-clamped, Flying-capacitor and Cascaded H-bridge inverters of 3-level, 5-level and m-level. The industrial applications need three-phase multilevel inverters. The output voltages are obtained based on the number of switches working for one cycle. These switches are turned-on and -off by switching signals generated by adopting various modulation techniques. The most popular one is Pulse width modulation. The different switching techniques used for three-phase multilevel inverters are discussed in this chapter, and classification is shown in Fig.2.1.

The modulation techniques are presented for three-phase multilevel inverter as per the classification shown as in Fig.2.1 also given here.

Modulation Techniques for Multilevel Inverters

- Pulse Width Modulation Techniques
  - Sinusoidal Pulse Width Modulation Technique
  - Third Harmonic Injection PWM Technique
  - Selected Harmonic Elimination PWM Technique

- Multicarrier Pulse Width Modulation Technique
  - o Phase-shifted Carrier Pulse Width Modulation Technique
  - o Level-shifted Carrier Pulse Width Modulation Technique
    - In-Phase Disposition Carrier Pulse Width Modulation Technique (IPD)
    - Phase Opposite Disposition Carrier Pulse Width Modulation Technique (POD)
    - Alternate Phase Opposite Disposition Carrier Pulse Width Modulation Technique (APOD)
- Space Vector Pulse Width Modulation Technique
  - o 2D Algorithm
  - o 3D Algorithm
- Carrier Based Space Vector Pulse Width Modulation Technique

The Modulation Techniques are explained in detail as per above given sequence.

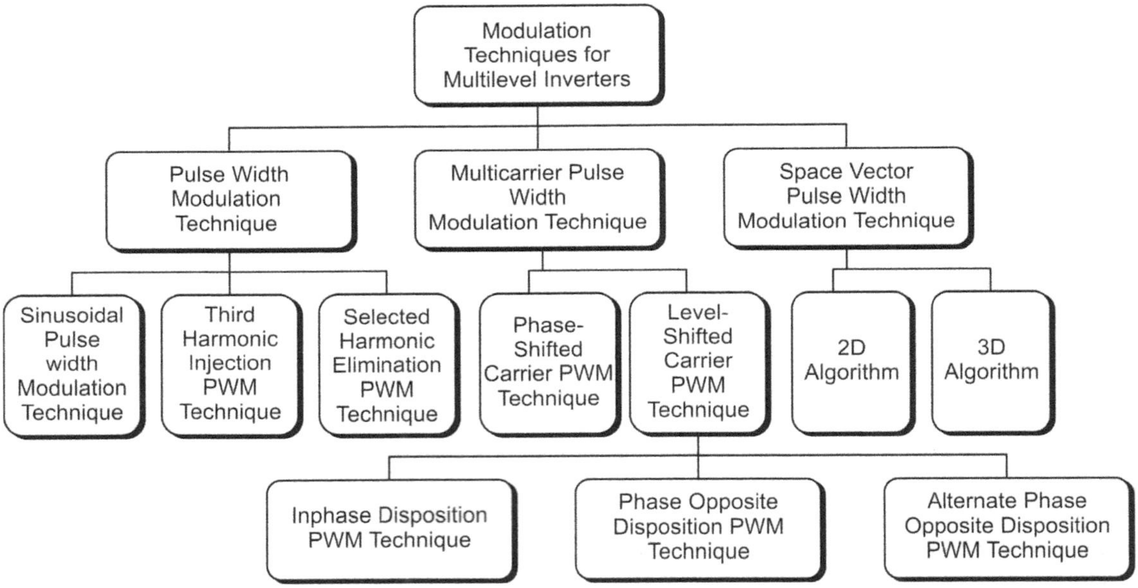

Fig. 2.1 Classification of various switching techniques used for three-phase multilevel inverters.

2.2 Pulse Width Modulation Technique

Pulse width modulation controls the output voltage by producing pulses of constant magnitude and variable duty cycle. The desired features of the modulation techniques are simple implementation, low Total Harmonic Distortion, lower switching losses and better DC link utilization. The PWM switching approach tackle the primary problems such as efficient dc bus utilization, less Total Harmonic Distortion in the output and take care of decrease in Electro Magnet Interference, switching loss, etc. All Pulse width modulated techniques have constant level pulses. However, the size or width of the pulses can be varied to reduce the harmonics in the inverter voltage.

The main objective of the PWM is to control the inverter output voltage and to reduce the harmonic content in the output voltage. These techniques are most efficient and they control the gating signals of the switching devices. The following are the some of the modulation methods which are most widely used.

- Sinusoidal Pulse Width Modulation (SPWM)
- Third Harmonic Injection Pulse Width Modulation (THIPWM)
- Selective Harmonic Elimination Pulse Width Modulation (SHE-PWM)
- Carrier Based Pulse Width Modulation (CBPWM)
- Space Vector Pulse Width Modulation (SVPWM)
- Carrier Based Space Vector Pulse Width Modulation (CBSVPWM)

2.2.1 Sinusoidal Pulse Width Modulation Technique

In this modulation technique, sinusoidal reference signal at 50Hz is compared with high frequency carrier signal (triangular waveform). Whenever the reference wave is higher than the carrier wave, three switches out of six are switched ON and the other three will remain in OFF.

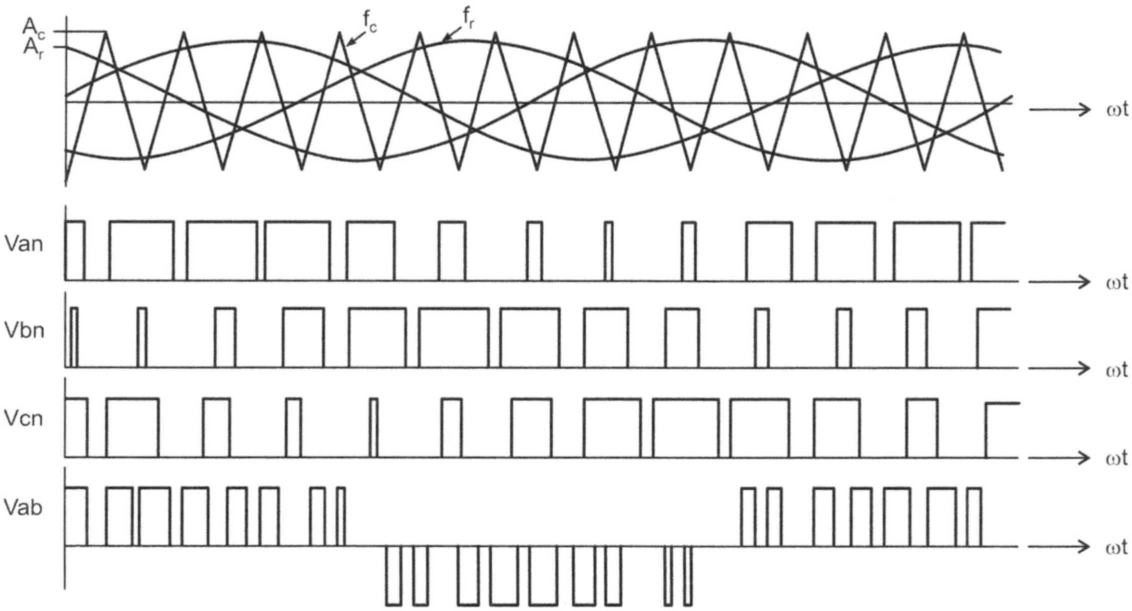

Fig. 2.2 Sinusoidal pulse width modulation technique.

position (in case of three phase two level inverter). Likewise, when the reference signal is lower than the triangular wave, then the switches which are in OFF position are switched ON and the ON state switches will go to OFF state. The modulation index depends on the magnitude of the reference signal.

The above Fig.2.2 shows SPWM for 2- level inverter. For multilevel inverter, the reference sine wave is compared with number of level shifted triangular waves for pulse creation i.e for m-level inverter, the reference is compared with (m-1) carrier signals, for example for FIVE level inverter, the reference wave is correlated with four triangular carrier waves for PWM generation.

2.2.2 Third Harmonic Injection Pulse Width Modulation Technique

The sinusoidal PWM is the simplest modulation scheme to understand but it is unable to fully utilize the available DC bus supply voltage. Due to this problem, the third-harmonic injection pulse-width modulation (THIPWM) technique was developed to improve the inverter performance. The sinusoidal PWM technique causes decrease in maximum achievable output voltage. By simply adding a third harmonic signal to each of the reference signals, it is possible to obtain a significant amplitude increase at the output voltage without loss of quality, as represented in Figure 2.3 (a). It is remarkable that the reference signal resulting from the addition of the third (V_3) and first harmonic (V_1) together is smaller in amplitude than the first harmonic. At the output, the obtained amplitude of the first harmonic is equal to the amplitude of the first harmonic reference. Note also that the third harmonic is not seen at the output voltage.

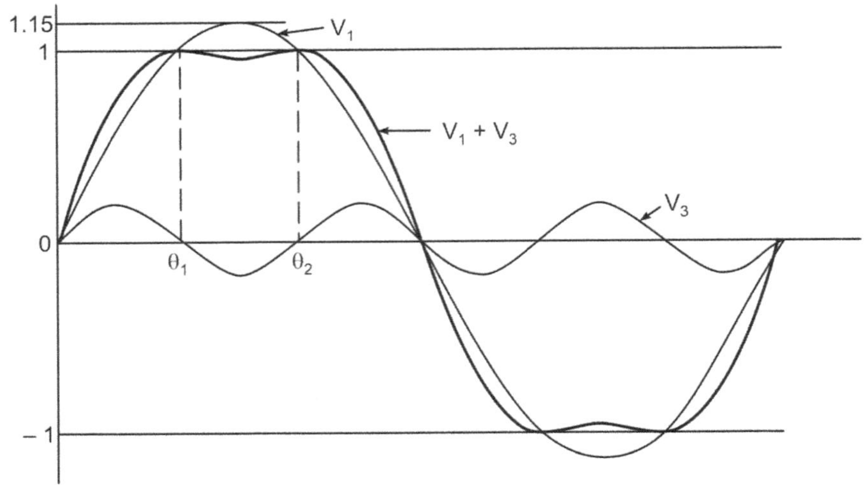

Fig. 2.3(a) Third Harmonic Injections to the Reference Signal.

On the other hand, the reference signal takes two maxima at t = π/3 and t = 2π/3 equal to 1. The first and third harmonic equations are given by

$$V_1 = V_{1max} \sin t \qquad .....(2.1)$$

$$V_3 = V_{3max} \sin 3t \qquad \text{.....(2.2)}$$

Therefore, when $t = \pi/3$, the first harmonic of the output voltage (line to neutral) takes the value $V_{bus}/2$. By substituting in Equation (E.2.1), we find

$$V_{bus}/2 = V_{1max} \sin (\pi/3) \qquad \text{.....(2.3)}$$

Consequently, the amplitude of the first harmonic yields

$$V_{1max} = V_{bus} / \sqrt{3} \qquad \text{.....(2.4)}$$

We see that the quality of the voltage waveform has not been significantly degraded. For each phase reference, the third harmonic injected is equal

$$V_{1max} \text{Sin} (wt) + V_{3max} \sin (3wt) \qquad \text{.....(2.5)}$$

$$V_{1max} \text{Sin} (wt-2\pi/3) + V_{3max} \sin (3wt) \qquad \text{.....(2.6)}$$

$$V_{1max} \text{Sin} (wt+2\pi/3) + V_{3max} \sin (3wt) \qquad \text{.....(2.7)}$$

***Increasing the Output Voltage*:** The addition of one-sixth of third harmonic component to the modulating waveform has the effect of reducing the peak value of the reference waveform by a factor of 0.866 without changing the amplitude of the fundamental. The third harmonic injected reference is a double peaked waveform and it is represented in Figure 2.3(b). The magnitude (peak) of reference is just 86.6% (at 60° and 120°).

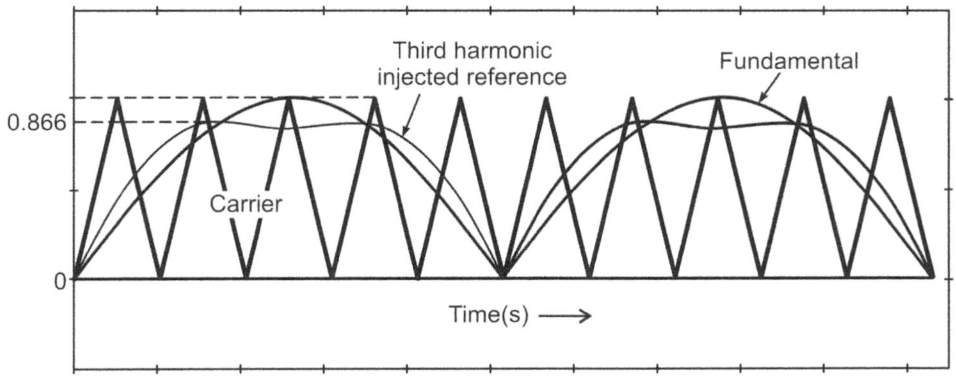

Fig. 2.3(b) Reference with One-Sixth of Third Harmonic component added.

This is similar to operating the PWM controller in lower modulation range while the aim of THIPWM is being fundamental enhancement. Hence, the reference should be boosted by multiplying a constant (K) (greater than unity), to utilize the dc source voltage effectively and get higher fundamental. Care must be taken while deciding the upper limit of K such that it should not cause any pulse dropping, which may introduce lower order harmonics and worsen the spectrum.

Thus, the modulating waveform becomes

$$y = K\,(\sin\theta + (1/6)\sin 3\theta) \qquad \text{where } \theta = \omega t \qquad \text{.....(2.8)}$$

Example E.2.1

Consider a three-phase two-level voltage source inverter whose output voltages are controlled by (a) Sinusoidal Pulse Width Modulation (SPWM) and (b) Third Harmonic Injected Pulse Width Modulation (THIPWM), the typical one phase to neutral voltage with both modulation schemes shown in Fig. E2.2 (a) and (b) respectively for the inverter given in Fig.E2.1.

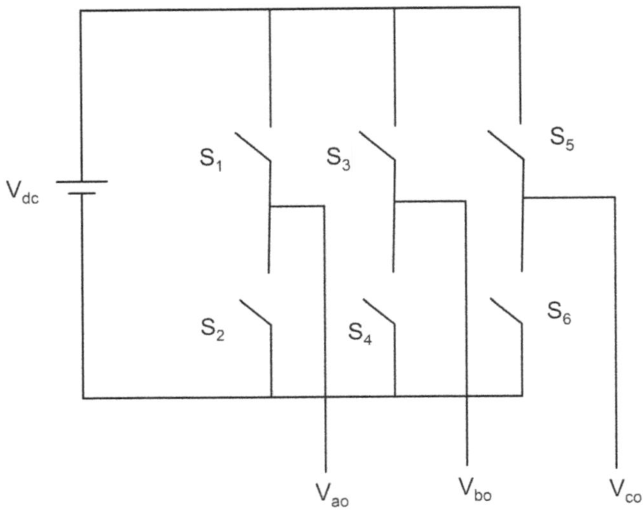

Fig. E.2.1 Three-phase two-level voltage source inverter.

For three phase two-level VSI, the sinusoidal PWM and Third Harmonic Injected PWM are applied to obtain the output voltages and presented the line to neutral voltage for one-phase load in Fig.E2.2 (a) and (b), to show the difference in the amplitude and closeness of the pulse widths of the output voltages.

There are few advantages in THIPWM method, when compared between THIPWM and SPWM. Utilization of DC voltage is more in THIPWM when compared to SPWM. Maximum output voltage of inverter is increased in THIPWM when compared to SPWM. From the figure E2.2 (a) and (b) harmonics presence in the inverter voltages are less in SPWM when compared to THIPWM.

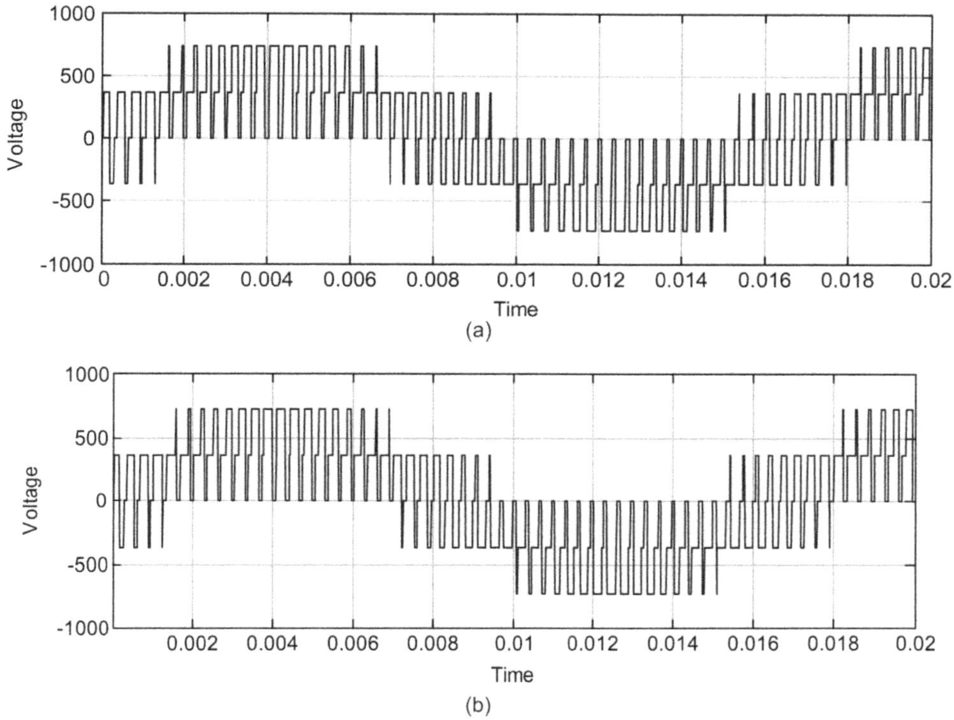

Fig. E.2.2 Line to Neutral Voltage (Van) of (a) SPWM and (b) THIPWM.

2.2.3 Selected Harmonic Elimination Pulse Width Modulation Technique

In medium- and high-power applications, it is not good practice to use high switching frequencies because the thermal losses could damage the switching devices. In recent year, multilevel inverters have received more attention because of their ability to generate high quality output waveforms with low switching frequency. Generally, the output voltage of the inverters must be sinusoidal. However, the waveforms of practical inverters are non- sinusoidal and contain certain harmonics. For low and medium power applications, square wave or quasi square wave may be acceptable but for high power applications, low distorted sinusoidal waveforms are required. By increasing number of levels in inverter the output voltage has more steps generating a staircase waveform, which has reduced harmonic distortion. There emerges need of multilevel inverter.

Concept of SHE Technique: Selective harmonic elimination PWM technique was introduced by Husmukh and Richard in 1973. The idea of this method is the basic square wave output is "chopped" a number of times. The chopping time is related to a set of switching angles of inverter switches, which are obtained by proper off-line calculations. By appropriate distribution of the switching angles to turn the inverter bridge switches on and off, the output waveform of the inverter is controlled and reaches the goal of eliminating lower order harmonics. In contrast to carrier based PWM methods, in which switching instants are determined by direct comparison of reference and carrier waveforms, in selective harmonic elimination method the exact moments of switching instants are calculated according to the desired fundamental component and the harmonic components to be eliminated. Because of complexity of equations to be solved to find switching instants, the number of switching angles is normally kept low to make the calculations simple. This also has the advantage of lowering the converter switching losses and switching instants are found by offline calculations. According to Hasmukh and Richard (1974), Chiasson (2004) and Sirisukprasert et al (2002), the selective harmonic elimination PWM technique is one of the optimal PWM techniques. It effectively reduces the harmonics content of inverter output waveform and generate higher quality spectrum through elimination of specific lower order harmonics. Therefore, it has been applied in power electronic controllers extensively and many related techniques have been proposed in recent year. The basic idea is to set up the notches at the specially designated sites of PWM waveform and the inverter alters directions many times per half cycle to control the inverter's output waveform appropriately.

Features of SHE PWM technique: The SHE PWM technique is used to synthesize an output waveform of both a half-bridge and a full-bridge inverter. Let us consider a full-bridge or H-bridge voltage source inverter, which comprises four switches and one dc source as depicted in figure 2.4. Three different voltage levels of an output waveform such as positive, negative and zero can be obtained. Figure 2.5 shows a generalized three-level SHE PWM waveform, which is synthesized by using the inverter circuit shown in Fig 2.4. The output waveform is chopped N times per quarter. Each switch is, therefore, switched N times per cycle to generate such waveform.

The H-bridge inverter as discussed above has three states of operation. Consider the generalized three-level SHE PWM waveform shown in Fig. 2.5 Let N be the number of switching angles per quarter-cycle. The output waveform is assumed to be odd quarter wave symmetry, whose amplitude equals E.

The Fourier series of the SHE PWM Waveform: Because of odd quarter-wave symmetry, the dc component and the even harmonics are equal to zero. The Fourier series of the three levels SHE PWM is as follows:

$$V_{out}(wt) = \sum_{n=1}^{\infty} a_n \sin(nwt) \qquad \text{.....(2.9)}$$

Where $$a_n = (4E/n\pi) \sum_{K=1}^{N} (-1)^{k+1} \cos(n\,\alpha_k)\ , \text{ for odd } n \qquad \text{.....(2.10)}$$

N = number of switching angles per quarter

α_k = switching angles satisfying the following condition

$$\alpha_1 < \alpha_2 < \ldots < \alpha_N < \pi/2 \qquad \text{.....(2.11)}$$

where E = amplitude of dc source and n = harmonic order

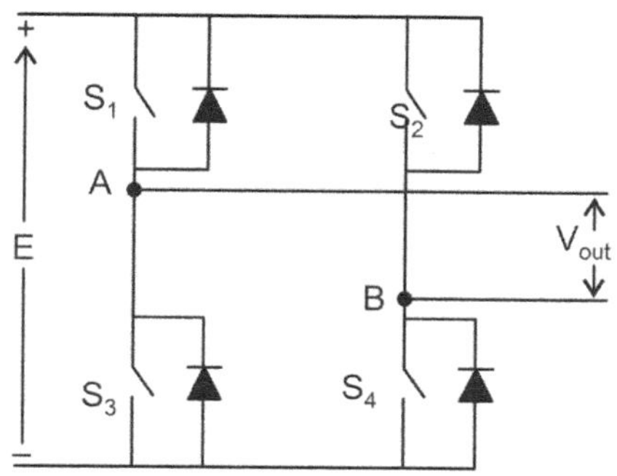

Fig. 2.4 Full Bridge Voltage Source Inverter.

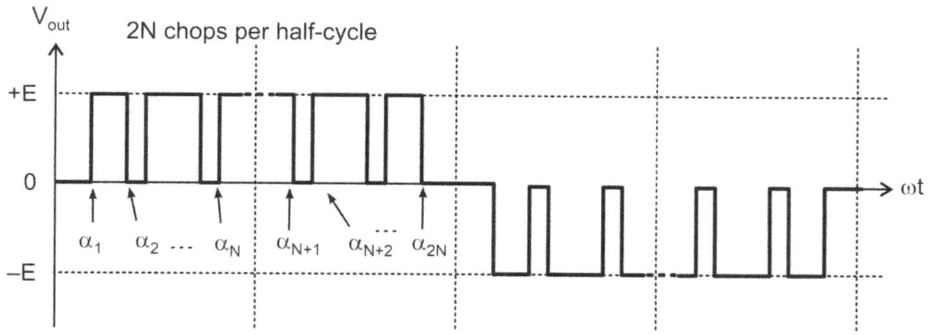

Fig. 2.5 Generalized three-level SHE PWM Waveform.

From equation (2.11), the non-linear system of equations for SHE PWM waveform can be written as follows

$$\cos(\alpha_1) - \cos(\alpha_2) + \ldots\ldots\ldots \pm \cos(\alpha_N) = (\pi/4)\,M \qquad \ldots..(2.12a)$$

$$\cos(3\alpha_1) - \cos(3\alpha_2) + \ldots\ldots\ldots \pm \cos(3\alpha_N) = (3\pi/4E)\,h_3 \qquad \ldots..(2.12b)$$

$$\cos(5\alpha_1) - \cos(5\alpha_2) + \ldots\ldots\ldots \pm \cos(5\alpha_N) = (5\pi/4E)\,h_5 \qquad \ldots..(2.12c)$$

$$\cos(N\alpha_1) - \cos(N\alpha_2) + \ldots\ldots\ldots \pm \cos(N\alpha_N) = (N\pi/4E)\,h_N \qquad \ldots..(2.12n)$$

Where M = modulation index and is given by $M=h_1/E$

The relation between the fundamental voltage and the maximum obtainable voltage is given by modulation index. The modulation index is defined as the ratio of the fundamental output voltage to the maximum obtainable fundamental voltage. The maximum fundamental voltage is obtained when all the switching angles are zero.

From equations (2.12a) to (2.12n), the cosine terms are negative with even values of N and positive with odd values of N. To control the amplitude of the fundamental component, the modulation index in (2.12a), M, is given. According to the above nonlinear system, N-1 surplus harmonic can be eliminated from the output waveform by setting equations (2.12b) to (2.12n) to zero. Basically, the lowest odd harmonic components need to be eliminated from a single-phase system. In a three-phase system, the lowest non-triplen harmonic components are to be eliminated. Generally, all triplen-harmonics in line-to-line voltage will be eliminated by 120 electrical degree phase shift characteristic.

In general equations (2.12a) to (2.12n) can be written as

$$F(\alpha) = B(M_I) \qquad .....(2.13)$$

The equations (2.12a) to (2.12n) represent a system of N transcendental equations, known as selective harmonic elimination (SHE) equations in terms of N unknowns $\alpha_1, \alpha_2, \alpha_3, \ldots \alpha_N$. For given values of M_I (from 0 to 1) it is required to get complete and all possible solutions of (2.12) when they exist with minimum computational burden and complexity.

Example E.2.2

Let us consider a three level five switching angle SHE PWM waveform as shown in fig (E.2.3) to solve the SHE PWM switching angles. In this example a single-phase system is assumed and modulation index of output voltage, M is 0.85.

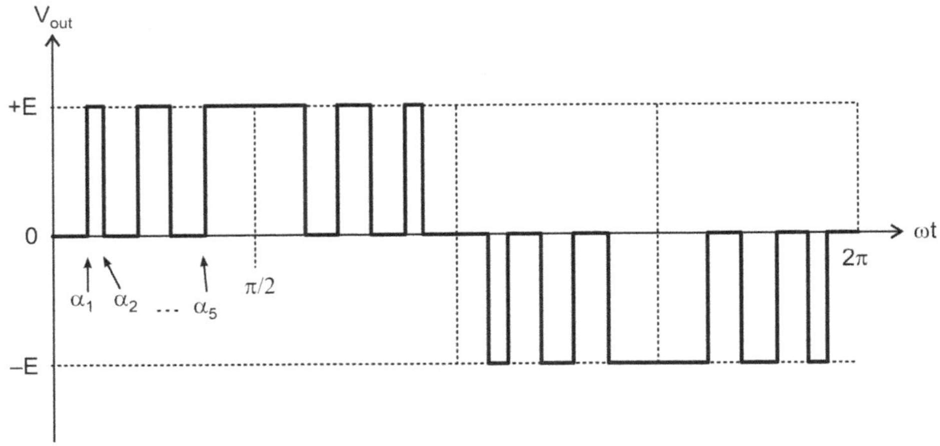

Fig. E.2.3 Output voltage waveform of a three-level five-angle SHE PWM.

From the output waveform as shown in figure E.2.3 five unknowns, α_1 to α_5, need to be solved. Because of a single-phase system, the lowest four odd harmonics, i.e., the 3rd, 5th, 7th, and 9th will be eliminated. From equations (2.12a) to (2.12n), five nonlinear equations will, therefore, be set up as follows:

$$\cos(\alpha_1) - \cos(\alpha_2) + \cos(\alpha_3) - \cos(\alpha_4) + \cos(\alpha_5) = 0.85\,(\pi/4) \qquad .....(E.2.2a)$$

$$\cos(3\alpha_1) - \cos(3\alpha_2) + \cos(3\alpha_3) - \cos(3\alpha_4) + \cos(3\alpha_5) = 0 \qquad .....(E.2.2b)$$

$$\cos(5\alpha_1) - \cos(5\alpha_2) + \cos(5\alpha_3) - \cos(5\alpha_4) + \cos(5\alpha_5) = 0 \qquad .....(E.2.2c)$$

$$\cos(7\alpha_1) - \cos(7\alpha_2) + \cos(7\alpha_3) - \cos(7\alpha_4) + \cos(7\alpha_5) = 0 \qquad .....(E.2.2d)$$

$$\cos(9\alpha_1) - \cos(9\alpha_2) + \cos(9\alpha_3) - \cos(9\alpha_4) + \cos(9\alpha_5) = 0 \qquad .....(E.2.2e)$$

Example E.2.3

Cascade Multilevel Inverter (CMLI) is one of the most important topologies in the family of multilevel and multi-pulse inverters. It requires least number of components when compare to diode-clamped and flying capacitors type multilevel inverters and no specially designed transformer is needed as compared to multi-pulse inverter. It has modular structure with simple switching strategy and occupies less space. The CMLI consists of a number of H-bridge inverter units with separate dc source for each unit and is connected in cascade or series as shown in Fig. E.2.4. Each H-bridge can produce three different voltage levels: $+V_{dc}$, 0, and $-V_{dc}$ by connecting the dc source to ac output side by different combinations of the four switches S_1, S_2, S_3, and S_4. The ac output of each H-bridge is connected in series such that the synthesized output voltage waveform is the sum of all of the individual H-bridge outputs.

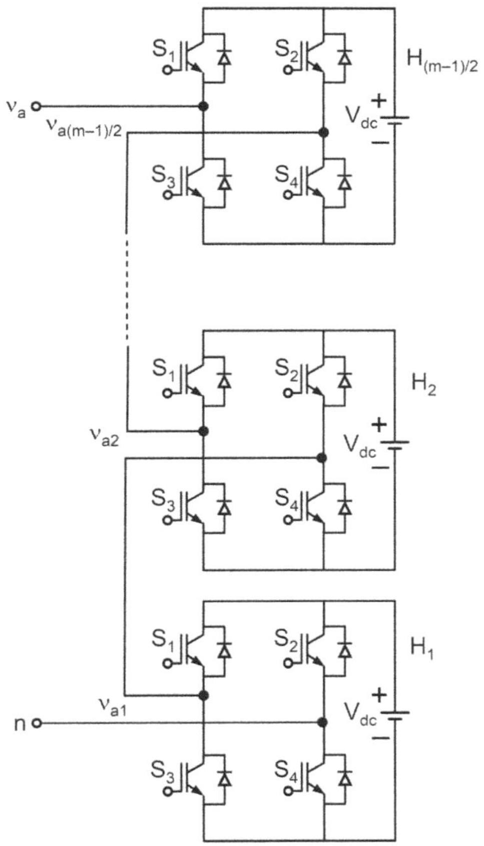

Fig. E.2.4 Configuration of single-phase cascade multilevel inverter.

By connecting the sufficient number of H-bridges in cascade and using proper modulation scheme, a nearly sinusoidal output voltage waveform can be synthesized. The number of levels in the output phase voltage is 2s+1, where s is the number of H-bridges used per phase. Fig. E.2.5 shows an 11-level output phase voltage waveform using five H-bridges. The magnitude of the ac output phase voltage is given by $v_{an} = v_{a1} + v_{a2} + v_{a3} + v_{a4} + v_{a5}$

In general, the Fourier series expansion of the staircase output voltage waveform as shown in Fig. E2.5 is given by

$$Van(\omega t) = \Sigma k = 1,3,5\ldots\ (4V_{dc}/k\pi)\ [\cos(k\,\alpha_1) + \cos(k\,\alpha_2) + \ldots + \cos(k\,\alpha_s)]\ \sin(k\omega t) \qquad \ldots..(2.14)$$

where s is the number of H-bridges connected in cascade per phase. For a given desired fundamental peak voltage V_1, it is required to determine the switching angles such that $0 \le \alpha_1 < \alpha_2 < \ldots < \alpha_s \le \pi/2$ and some predominant lower order harmonics of phase voltage are zero. Among 's' number of switching angles, generally one switching angle is used for fundamental voltage selection and the remaining (s-1) switching angles are used to eliminate certain predominating lower order harmonics. In three-phase power system, triplen harmonics are cancel out automatically in line-to-line voltage as a result only non-triplen odd harmonics are present in line to- line voltages.

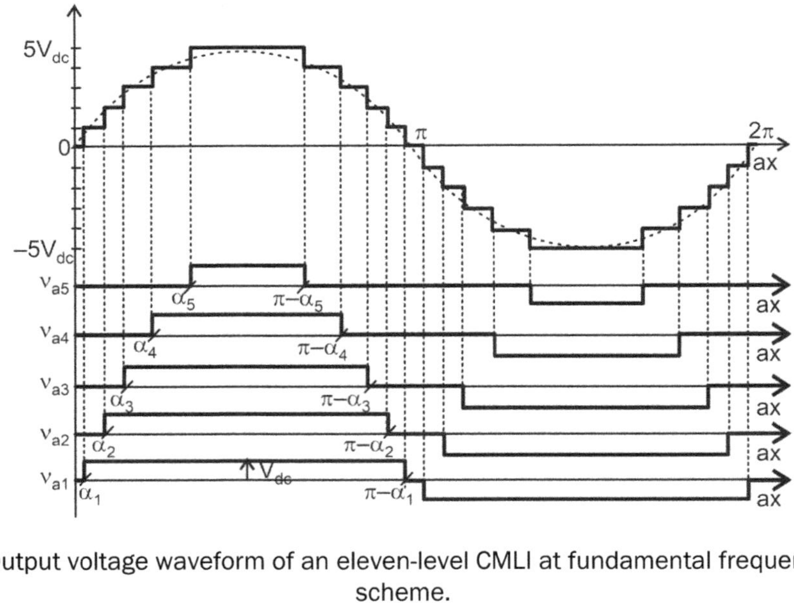

Fig. E.2.5 Output voltage waveform of an eleven-level CMLI at fundamental frequency switching scheme.

From (2.14), the expression for the fundamental voltage in terms of switching angles is given by

$$(4V_{dc}/\pi)\ [\cos(\alpha_1) + \cos(\alpha_2) + \ldots + \cos(\alpha_s)] = V_1 \qquad \ldots..(2.15)$$

The relation between the fundamental voltage and the maximum obtainable voltage is given by modulation index. The maximum fundamental voltage is obtained when all the switching angles are zero i.e. $V_{1max} = 4sV_{dc}/\pi$. Therefore, the expression for M_1 is

$$M_l = \pi V_l / 4sV_{dc} \quad .....(2.16)$$

For an 11-level cascade inverter, there are five H-bridges per phase i.e. s = 5 or five degrees of freedom are available; one degree of freedom is used to control the magnitude of the fundamental voltage and the remaining four degrees of freedom are used to eliminate 5th, 7th, 11th, and 13th order harmonic components as they dominate the total harmonic distortion. The above stated conditions can be written in the following way by combining (2.14) and (2.16);

$$\begin{aligned}
&\text{Cos}\,(\alpha_1) + \cos(\alpha_2) + \ldots + \cos(\alpha_s) = 5M_l \\
&\text{Cos}\,(5\,\alpha_1) + \cos(5\,\alpha_2) + \ldots + \cos(5\,\alpha_s) = 0 \\
&\text{Cos}\,(7\,\alpha_1) + \cos(7\,\alpha_2) + \ldots + \cos(7\,\alpha_s) = 0 \\
&\text{Cos}\,(11\,\alpha_1) + \cos(11\,\alpha_2) + \ldots + \cos(11\,\alpha_s) = 0 \\
&\text{Cos}\,(13\,\alpha_1) + \cos(13\,\alpha_2) + \ldots + \cos(13\,\alpha_s) = 0 \quad .....(2.17)
\end{aligned}$$

To solve these switching angles a number of methods are proposed in the literature to get all possible solutions for the non-linear SHE equations (2.13) or (2.17). Once the solution sets are obtained, the solution set having least THD is chosen.

Methods to solve SHE equations: The methods most commonly used to solve these equations are classified as follows as shown in Table 2.1

Table 2.1 Classification of Methods to solve non-linear SHE equations.

Sl. No.	Different Methods	Different algorithms
1.	Numerical methods	Conventional Newton-Raphson (NR) algorithm [4]
		Sequential homotopy [5]
2.	Algebraic methods	Resultant elimination theory [6,7]
		Groebner bases with symmetric polynomials [8]
3.	Evolutionary-algorithm-based optimization methods	Modified species-based particle swarm optimization [10,11]
		Ant-colony-based algorithm [12]
		Artificial neural network real-time angle generation [13]
		Colonial competitive algorithm [14]
		Bee algorithm [15]
		Universal method based on equal area criteria [16]

In numerical techniques, for fast and sure convergence, a good initial guess is necessary and finding multiple solutions is difficult and involves complicated calculations. Metaheuristic-based optimization methods may not converge on an exact solution of the equations but rather minimize an objective function and give an approximate solution with a particular initial guess. The solution to this method is highly dependent on the formulation of the objective function and search space. Also, the solution depends on the parameter assumed in the algorithm and requires expert programming for getting reliable solutions.

The limitations of numerical and metaheuristic methods can be overcome by algebraic methods that can provide all the solutions at a particular modulation index. However, because of the involvement of a high degree of polynomials, it is complicated and takes a lot of time to evaluate all possible solutions.

Recently a univariate-equation based SHE PWM method was proposed which can be implemented for a large number of inverter levels using general purpose micro controllers with reduced mathematical burden.

Merits of SHE PWM technique: Uses only 50% switching pulse which leads to less stress on switching devices and less switching loss.

- Quality of inverter output voltage and current waveform are improved.
- Increased fundamental voltage component.
- Optimization of output spectrum and switching angles distribution.
- Harmonic content can be flexibly redistributed and high-quality output waveforms are obtained.

De-merits:

- Implementation is complicated one
- Switching angles are very difficult to find out.

Applications: Where low switching frequency inverters are used (mainly multilevel inverters)

- High power electric drives for all AC motors
- High voltage direct current transmission systems
- Flexible ac transmission systems
- Interconnection of distributed generation units to a grid
- Power quality improvement

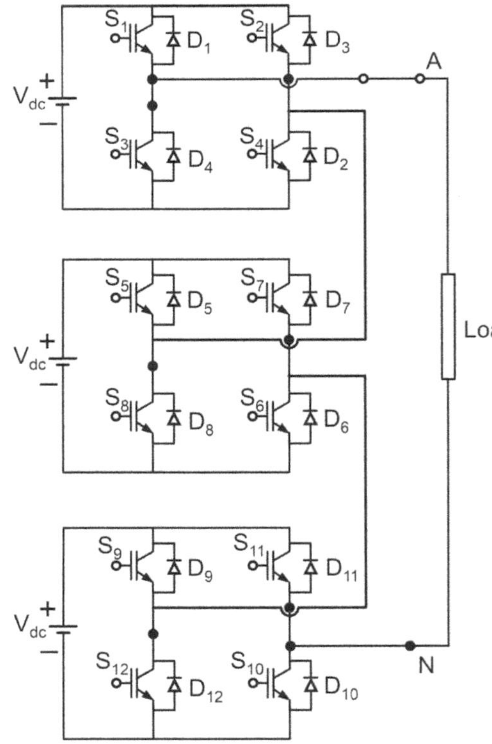

Fig. 2.6 The three H-bridge cells cascaded Inverter.

2.3 Multicarrier Pulse Width Modulation Technique

In recent years, carrier-based pulse width modulation techniques gained importance because of simplicity in expansion to higher levels. This carrier-based technology can be easily extended to higher inverter levels and can be applied to sinusoidal and space vector modulation schemes. The carrier-based Sinusoidal PWM schemes are explained in this section

2.3 and the carrier-based Space Vector Modulation scheme is explained in the section 2.6 at the end of this chapter.

Carrier based PWM schemes are classified as:

1. Phase-Shifted PWM (PS-PWM)
2. Level shifted PWM (LS-PWM)

2.3.1 Phase-shifted Carrier Pulse Width Modulation Technique

Phase shifted PWM (PS-PWM) is a natural extension of traditional PWM techniques. This modulation shifts the phase of each carrier in a proper angle to reduce the harmonic content of the output voltage. The phase shift is introduced between the carrier signals of contiguous cells, producing a phase-shifted switching pattern between them. In this way, when connected together, a stepped multilevel waveform is originated. It has been demonstrated that the lowest distortion can be achieved when the phase shifts between carriers is 180º/k, where k is number of power cell as shown in Fig. 2.8. This strategy leads to the cancellation of all carrier and associated sideband harmonics up to the kth carrier group.

The harmonics for the multilevel PWM techniques can be expressed in general form as double summation Fourier series as given by (2.18) below

$$\begin{aligned} F(t) = \frac{A_{00}}{2} &+ \sum_{n=1}^{\infty}\{A_{0n}.\cos(n\omega_0 t) + B_{0n}\sin(n\omega_0 t)\} \\ &+ \sum_{m=1}^{\infty}\{A_{m0}\cos(m\omega_c t) + B_{m0}\sin(m\omega_c t)\} \\ &+ \sum_{m=1}^{\infty}\sum_{\substack{n=-\infty \\ n\neq 0}}^{\infty}\{A_{mn}\cos(m\omega_c t + n\omega_0 t)\} + B_{mn}\sin(m\omega_c t + n\omega_0 t) \quad \text{.....(2.18)} \end{aligned}$$

For PS-PWM, the Fourier analysis of voltage is given by (2.19). By observing (2), we can see that for PS-PWM, the first set of harmonics is centered about 2kth multiple of carrier frequency.

$$\begin{aligned} V_{PS\_AN}(t) = kMV_{dc}\cos(\omega_0 t) &+ \frac{2V_{dc}}{\pi} \\ &\times \sum_{m=1}^{\infty}\sum_{\substack{n=-\infty \\ n=0}}^{\infty}\left\{\begin{array}{c}\frac{1}{m}J_{2n-1}(km\pi M)\cos((km+n+1\pi \\ \cos(2km\omega_0 t + (2n-1)\omega_0 t\end{array}\right\} \quad \text{.....(2.19)} \end{aligned}$$

In this technique, the carriers have same equal amplitude, frequency and are phase shifted by 90º from one another as shown in Fig.2.8.

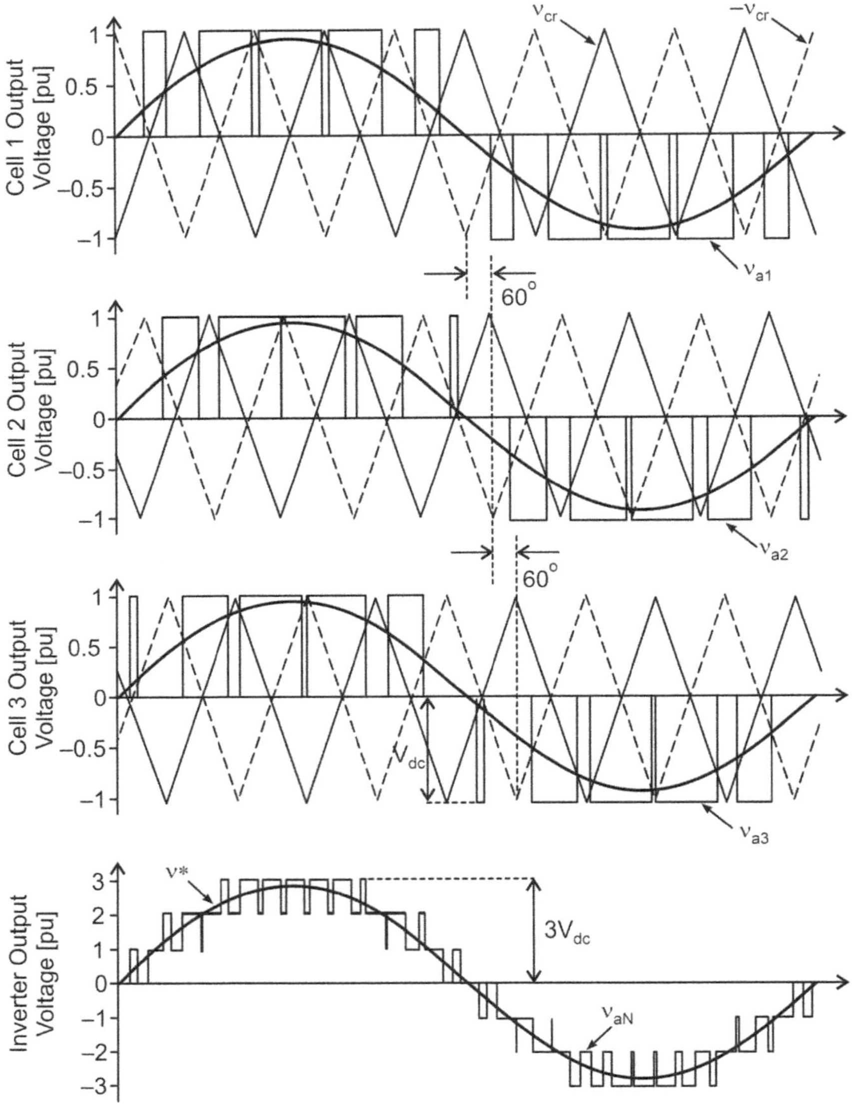

Fig. 2.7 Three H-bridge cell cascaded inverter PS-PWM waveform generation.

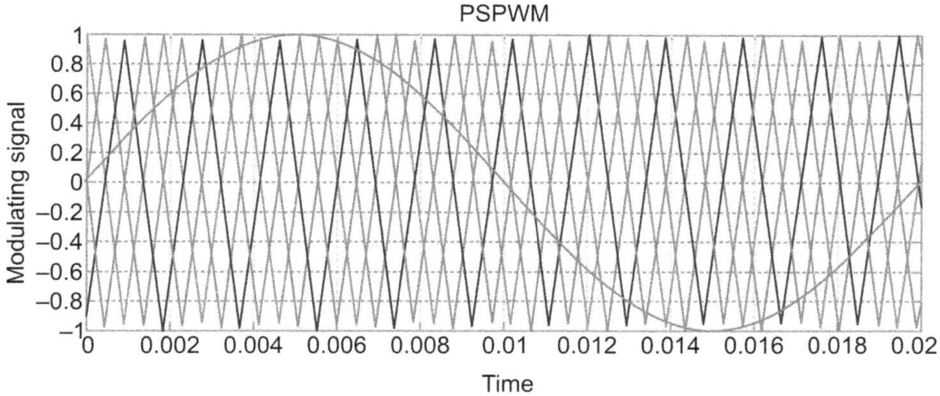

Fig. 2.8 Carrier waveform arrangement for PS-PWM strategy for two H-bridge cell-inverter.

2.3.2 Level-shifted Carrier Pulse Width Modulation Technique

Level shifted PWM is derived from the carrier disposition strategy. In this modulation method, the reference wave is compared with number of carrier waves to generate switching signals to the inverter. For 'm' levels in the output voltage of inverter, the sinusoidal reference wave is to be compared with 'm-1' triangular carrier waves with same amplitude and frequency so that they occupy contiguous band over $+V_{dc}$ to $-V_{dc}$ as shown in the figure. A single sinusoidal reference wave is then compared with each triangular carrier wave to determine output voltages for the level of inverter. There are many strategies to implement the carrier-based technology. They are classified as

Phase Opposition Disposition PWM, in which the carrier signals over the zero position are in phase but moved by 180° from the signals under zero reference which are in phase.

Alternative Phase Opposition Disposition PWM, in this each carrier signal is moved by 180° from the neighboring signal.

In Phase Disposition PWM, the carrier signals are in phase and have same magnitude and frequency.

As an example, implementation of Sinusoidal Carrier Based technology for five level Neutral point clamped inverter is explained. In order to realize five level output, four triangular signals are correlated with reference or modulated wave. The following are the rules for realization of various stages in the output waveform.

1. When the reference wave is bigger than both upper waves, the output is $+ V_{dc}/2$.
2. When the modulating wave is higher than the first upper carrier wave, the converter output is $+ V_{dc}/4$.

3. The converter output is '0' (zero) when sine wave is higher than the lower carrier and lower than upper wave.
4. When the reference wave is lower than first lower carrier wave, the output switches to –Vdc/4.
5. When the reference is lower than both lower carrier signals, the output is - $V_{dc}/2$. The same rule can be adopted for all the remaining PWM strategies i.e. PD, POD, and APOD.

2.3.2.1 In-Phase Disposition Carrier based PWM Technique (IPD or PD)

In this, the carrier signals have equal frequency, amplitude and are in phase as shown in Fig.2.9.

For Phase Disposition carrier-based pulse width modulation, the Fourier analysis of voltage is given by equation (2.20).

$$\begin{aligned} V_{PD\_AN} &= kMV_{dc}\cos(\omega_0 t) + \frac{8V_{dc}}{\pi^2}\sum_{m=0}^{\infty}\left\{\frac{A_{m\_}odd0}{2m+1}\cos((2m+1)\omega_c t)\right\} \\ &+ \frac{4V_{dc}}{\pi^2}\sum_{m=1}^{\infty}\sum_{\substack{n=-\infty \\ n=0}}^{\infty}\left\{\frac{A_{m\_}oddeven}{2m+1}\cos((2m+1)\omega_c t + 2n\omega_c t)\right\} \\ &+ \frac{V_{dc}}{\pi}\sum_{m=1}^{\infty}\sum_{\substack{n=-\infty \\ n=0}}^{\infty}\left\{\begin{matrix}\frac{1}{m}J_{2n-1}(2km\pi M)\cos((n-1)\pi) \\ \{\cos(2m\omega_c t + (2n-1\omega_0 t)\}\end{matrix}\right\} \end{aligned} \quad .....(2.20)$$

where M = Modulation index,
m, n = Harmonic index variables,
J = Bessel function,
V_{dc} = dc voltage

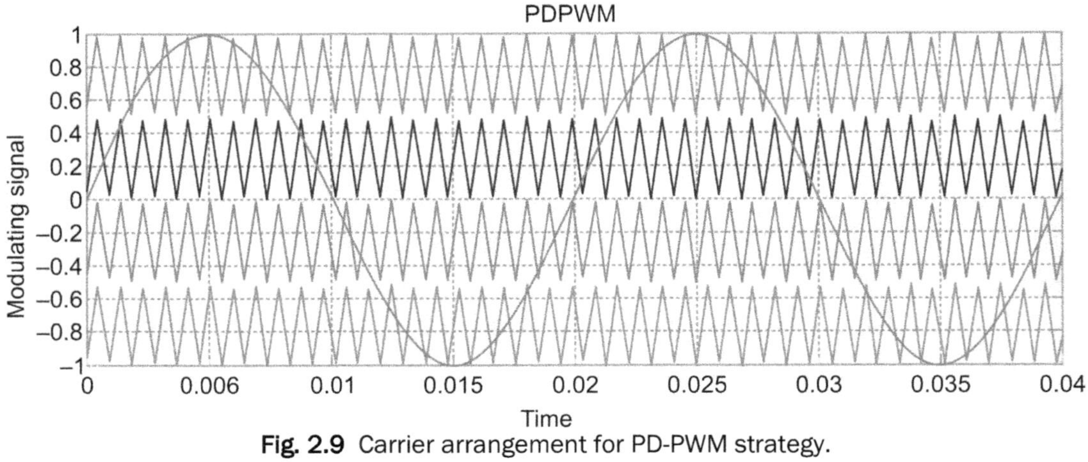

Fig. 2.9 Carrier arrangement for PD-PWM strategy.

2.3.2.2 Phase Opposite Disposition Carrier PWM Technique (POD)

In this technique, the upper signals are in phase with each other while lower signals are also in phase with each other. However, the upper group is phase shifted by 180° from lower group as shown in Fig. 2.10.

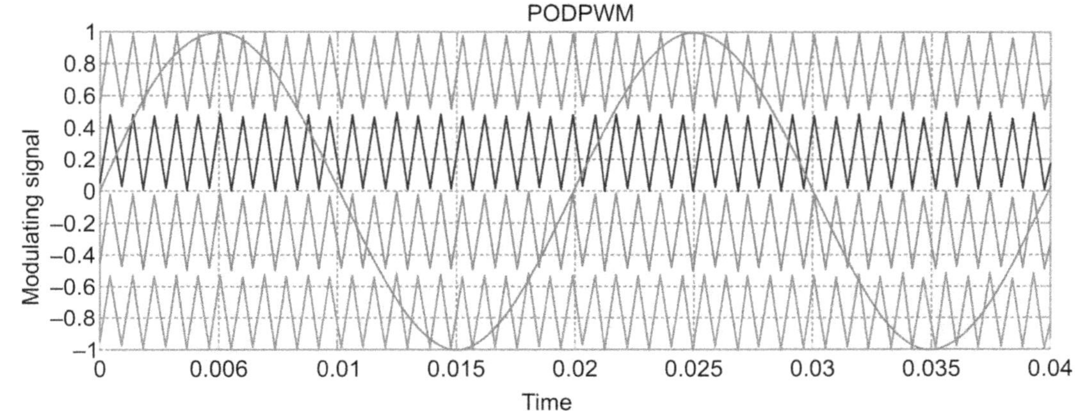

Fig. 2.10 Carrier arrangement for POD-PWM.

2.3.2.3 Alternate Phase Opposite Disposition Carrier PWM Technique (APOD):

In this scheme, all the carriers are placed in such a way that each wave is out of phase with its adjacent by 180° degrees as shown in Fig. 2.11.

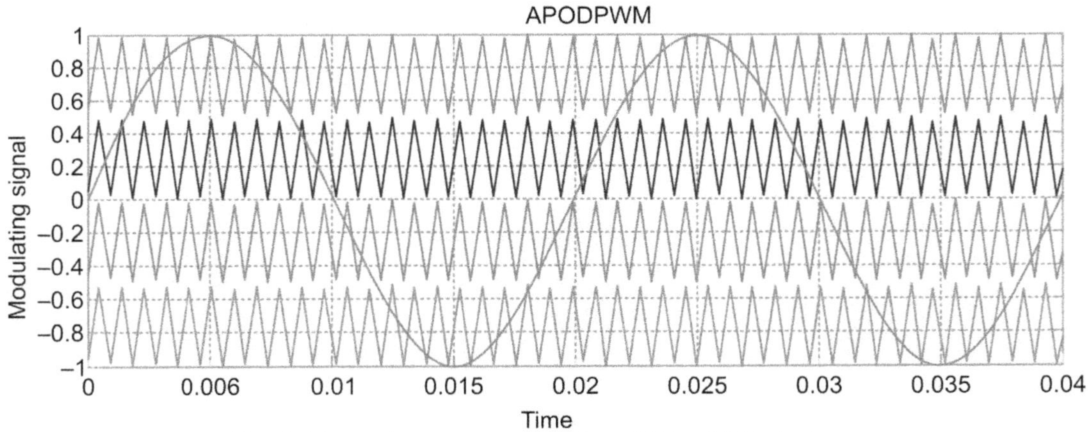

Fig. 2.11 Carrier arrangement for APOD - PWM strategy.

For Alternative Phase Disposition (APOD) carrier-based pulse width modulation, the Fourier analysis of voltage is given by equation (2.21).

$$V_{APOD\_AN}(t) = kMV_{dc}\cos(\omega_0 t) + \frac{2V_{dc}}{\pi}$$

$$\times\sum_{m=1}^{\infty}\sum_{\substack{n=-\infty \\ n=0}}^{\infty}\left\{\frac{\cos((k-1)m+n-1)\pi}{\pi}\left\{J_{2n-1}.(km\pi M)\cos(m\omega_c t+(2n-1)\omega_0 t)\right\}\right\} \qquad .....(2.21)$$

The detailed solution of equations (2.18) to (2.21) is given in [15]-[16]. Upon observing these equations, it is very clear that PD modulation places its harmonic energy into a carrier harmonic in each phase and cancellation of this harmonic in line to line voltage.

Example E.2.4

Consider a three phase five-level diode-clamped inverter with balanced wye-connected RL loads and employed all three methods of carrier based sinusoidal PWM techniques (i) Phase Disposition (PD) method, (ii) Phase Opposition Disposition (POD) Method, and (iii) Alternative Phase Opposition Disposition (APOD) method. The output voltage waveforms for one-phase of load is presented for all the three methods.

The three phase five level diode-clamped inverter in Simulink block diagram is given in Fig.E.2.6.

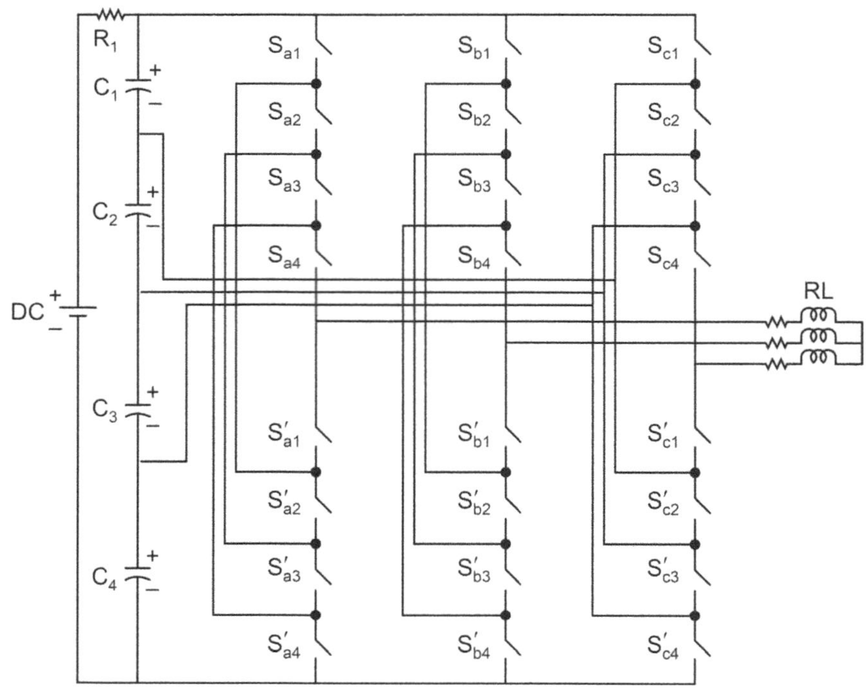

Fig. E.2.6 Simulink block diagram of Three phase five-level diode-clamped inverter.

The simulated output load voltages are presented in Fig.E.2.7 (a) for PD method, (b) POD method, and (c) APOD method along with their respective modulation schemes.

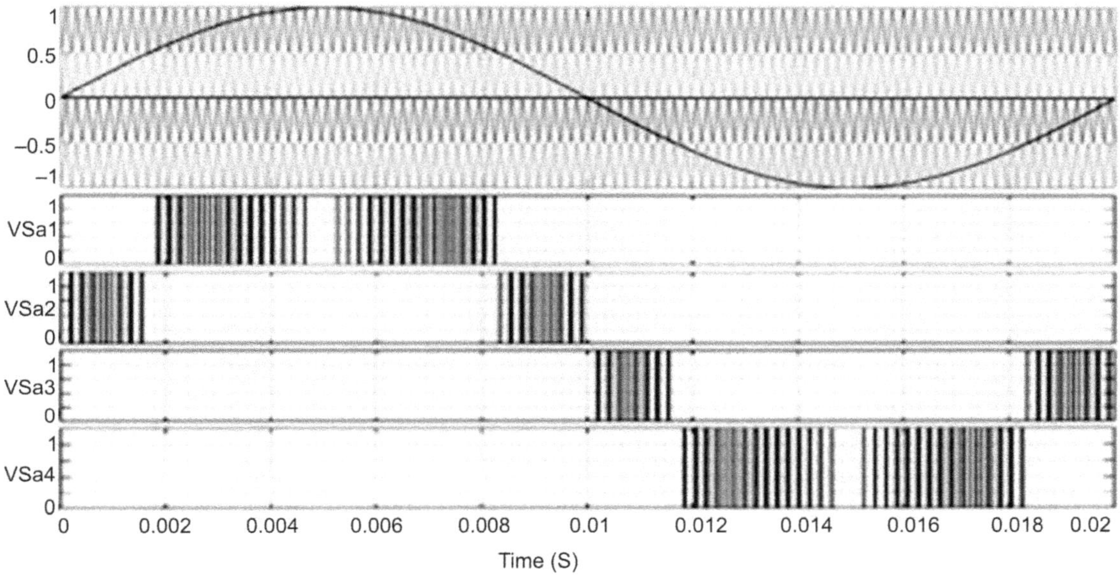

Fig. E.2.7(a) Modulation signal on PD Technique for 5-level Inverter and Output Load voltage.

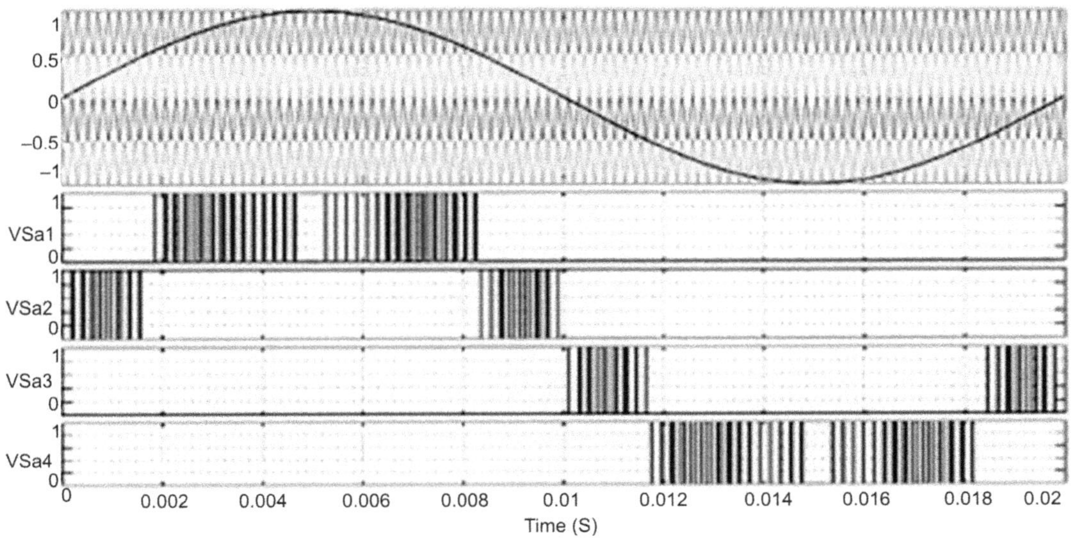

Fig. E.2.7(b) Modulation signal on POD technique for 5-level Inverter and Output Load voltage.

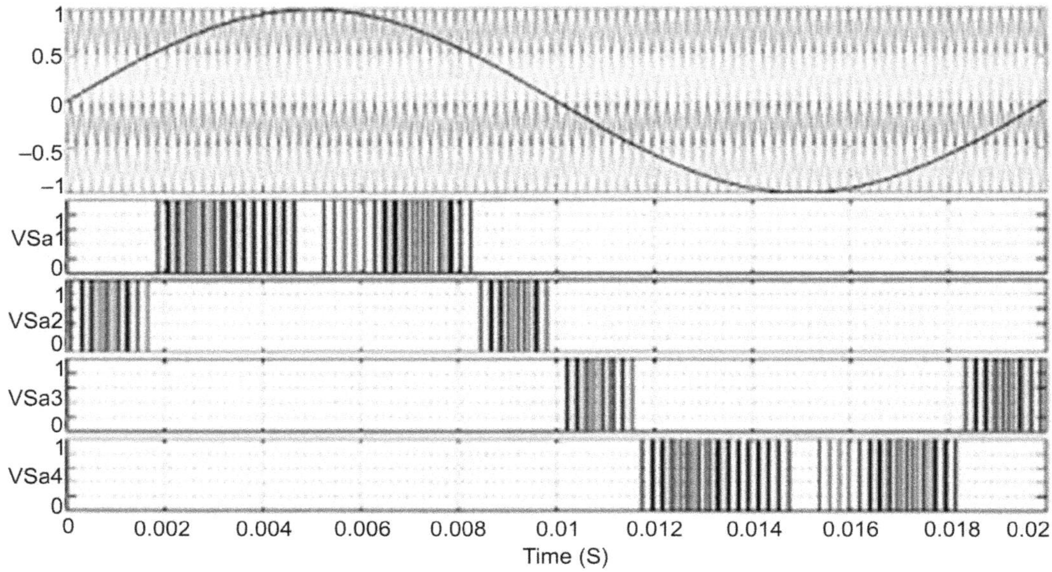

Fig. E.2.7 (c) Modulation signal on APOD technique for 5-level inverter and Output Load voltage.

Summary of LS-PWM: If all carriers are selected with the same phase, the method is known as PD method. A sinusoidal reference waveform is compared with vertically shifted carrier waveforms.

In POD method, the carriers above the zero line of reference voltage are out of phase and with the carriers below this line is by 180°. Compared to the PD method, this method has better results with reference to harmonic performances in lower modulation indices. In POD method, there is no harmonic at the carrier frequency.

The third group is known as APOD method. Each carrier of this method is phase shifted by 180° from its adjacent one. POD and APOD methods are exactly the same for a 3-level Inverter. The major difference in APOD is the presence of the larger amount of third order harmonics. But the third order harmonics is not considered because it gets cancelled in line voltages. This method results in a better THD for line voltages when compared with the POD method.

2.4 Space Vector Pulse width Modulation Technique

SVM-PWM for 2-level and 3-level derivations of reference voltage vectors, time duration, switching times, etc.

2.4.1 Two-dimensional Algorithm for SVPWM

2.4.1.1 SVM-PWM for three-phase 2-level voltage source inverter

The basic 2-level PWM Voltage Source Inverter (VSI) is shown in Figure 2.12. SVPWM control technique became a standard and well-established switching strategy for the power converters. In spite of, various PWM techniques like SINPWM and SHEPWM and some modified triangular carrier-based modulation schemes, SVM stood as the most popular one due to its relatively good performance and simplicity with inverters of any levels. In the other PWM techniques, the PWM patterns are generated for each phase independently. But the modulated wave generated in SVM is the combined effect of all the three phases of the output voltages of the inverter. Thus, the generation of such a modulated wave of SVM necessitates vectors' calculation and switching states selection.

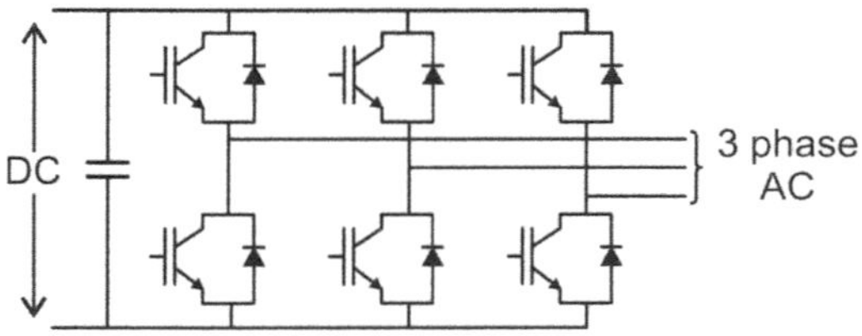

Fig. 2.12 Three-phase 2-level VSI.

- Determination of the reference voltage vector Vref from the equivalent 3 -2 conversion.
- Determination of different voltage vectors from the permissible switching combinations of the 2-level inverter and representation in the form of hexagon.
- Determination of the position of the reference space vector during its rotation through all the sectors.
- Determination of duration of the stay of each voltage vector in each sector.
- Determination of the total switching time of each switch in each sector.
- Calculation of the three phase reference voltages with the help of the total switching times through volt-time relationship.
- Generation of the PWM pulses by comparing the three reference voltages with a carrier wave.

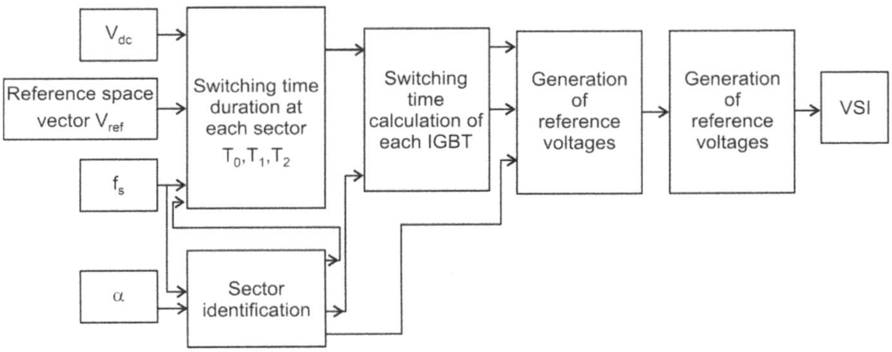

Fig. 2.13 Block Diagram of Space Vector PWM Scheme. The Space Vector PWM Scheme for a 2-level VSI can be divided into 7 detailed steps.

Space Vector Representation: The 2-Level SVPWM control scheme follows a decision procedure to obtain the angle and magnitude of the reference voltage vector. The three instantaneous line-neutral voltages V_{an}, V_{bn}, V_{cn} displaced by 120° are transformed into two mutually perpendicular vectors V_d and V_q and along the d and q axes respectively by the Parks Transformation defined in Equations (2.22) and (2.23). The three phase voltages in the parks transformation are collected instantaneously from the load or line terminal of the AC system.

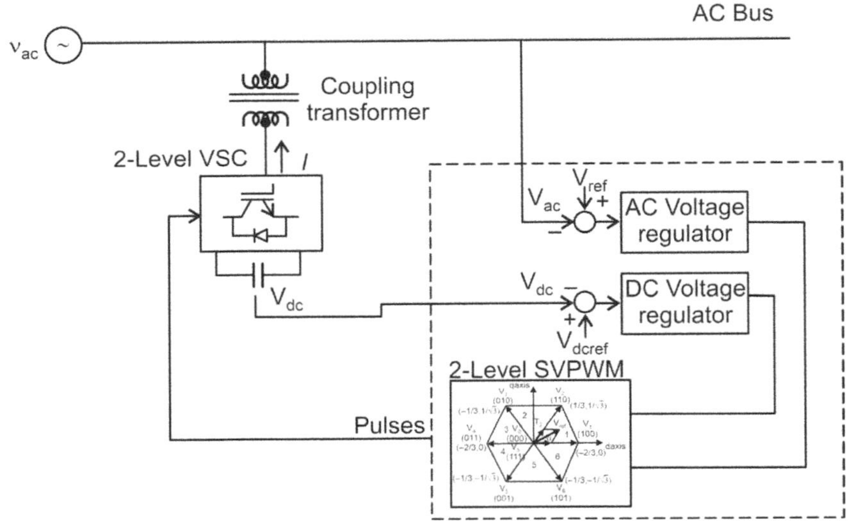

Fig. 2.14 2-Level SVPWM Controlled VSC.

The occurrence of any sudden contingency in the power system is fed to the controller in the form of line voltages and currents, so that it is compensated through the appropriate switching states of the VSC. The feedback loop control is followed as shown in the Fig. 2.14

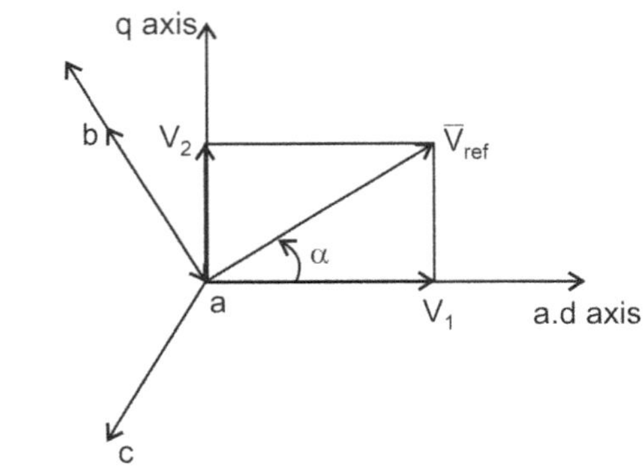

Fig. 2.15 Reference Voltage Space Vector.

$$V_d = (2/3)\, V_{an} - (1/3)\, V_{bn} - (1/3)\, V_{cn} \quad .....(2.22)$$

$$V_q = (1/\sqrt{3})\, V_{bn} - (1/\sqrt{3})\, V_{cn} \quad .....(2.23)$$

$$V_{ref} = V_d + j\,V_q \qquad \text{.....(2.24)}$$

$$\alpha = \tan^{-1}(V_q/V_d) \qquad \text{.....(2.25)}$$

$$\omega = 2\pi f \qquad \text{.....(2.26)}$$

where f = fundamental frequency.

Equation (2.24) denotes the reference voltage space vector obtained from the resultant of the V_d and V_q. Figure 2.15 shows the reference voltage space vector Vref that rotates with constant magnitude and uniform angular velocity ω. The angle obtained from equation (2.25) determines the position of the reference space vector at any particular instant, during its rotation.

The assumption of continuous rotation of the reference voltage space vector is limited to the ON-OFF states of the six –switches of the 2-Level Voltage Source Converter. Therefore, to determine the exact position of the reference space vector, one has to utilize different permissible combinations of ON-OFF states of the switches of a 2-Level Voltage Source Converter.

A. **Space Vector Distribution:** The ON-OFF states of six power switches of a 2- level VSI are predetermined to obtain the optimum PWM output voltage wave. The IGBT (Integrated Gate Bipolar Transistor) switch of a 2-Level Voltage Source Converter has two distinct switching states denoted as 1 or 0. 1 represents the ON state of the top switch in an inverter leg with the resultant voltage V_{dc} and 0 represents that the respective switch is OFF and so generates 0V. The output line-line voltage is expressed in terms of DC link voltage V_{dc}. S_1, S_3, S_5 denote the switching state of the three upper switches of the VSI as shown in the Table 2.2. The switching state of the respective lower switches is complementing to that of the upper switches. There are eight permissible combinations of ON-OFF patterns which include six active states and two null states. The voltage vectors V_0 and V_7 are considered as Zero or Null vectors, as they do not contribute any voltage. The other vectors V_1 to V_6 are considered as active states.

Table 2.2 Two Level Inverter Voltage Vector.

Voltage Vectors	Switching States			Line-Neutral Voltages		
	S_1	S_3	S_5			
V_0	0	0	0	0	0	0
V_1	1	0	0	$+V_{dc}$	0	$-V_{dc}$
V_2	1	1	0	0	$+V_{dc}$	$-V_{dc}$
V_3	0	1	0	$-V_{dc}$	$+V_{dc}$	0
V_4	0	1	1	$-V_{dc}$	0	$+V_{dc}$
V_5	0	0	1	0	$-V_{dc}$	$+V_{dc}$
V_6	1	0	1	$+V_{dc}$	$-V_{dc}$	0
V_7	1	1	1	0	0	0

The six space vectors and two null vectors describe a space vector plan with hexagon boundary, centered at the origin of the dq-plane as shown in the Fig. 2.16. The reference sinusoidal line modulating space vector is approximated from the eight switching vectors and can be distributed anywhere within the hexagon during its cycle of rotation. The voltage vectors V1, V2, V3, V4, V5 and V6 correspond to each vertex of the hexagon.

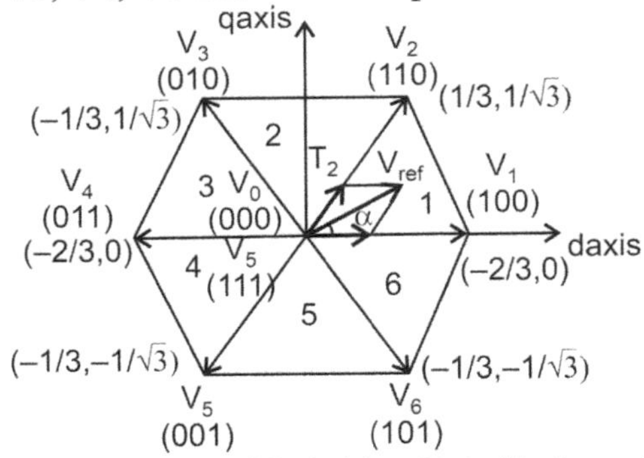

Fig. 2.16 Hexagon Representation of Switching State Vectors of 2 -Level VSI.

B. Sector Identification: The six triangles noted in Fig. 2.16, as 1, 2, 3, 4, 5 or 6 are termed as sectors. The six sectors, each covering 60º, frame the hexagon. The angle α by which the voltage space vector Vref, is displaced from d-axis and is given in equation (2.21). The MATLAB/ SIMULINK figure of sector identification is shown in Fig. 2.17.

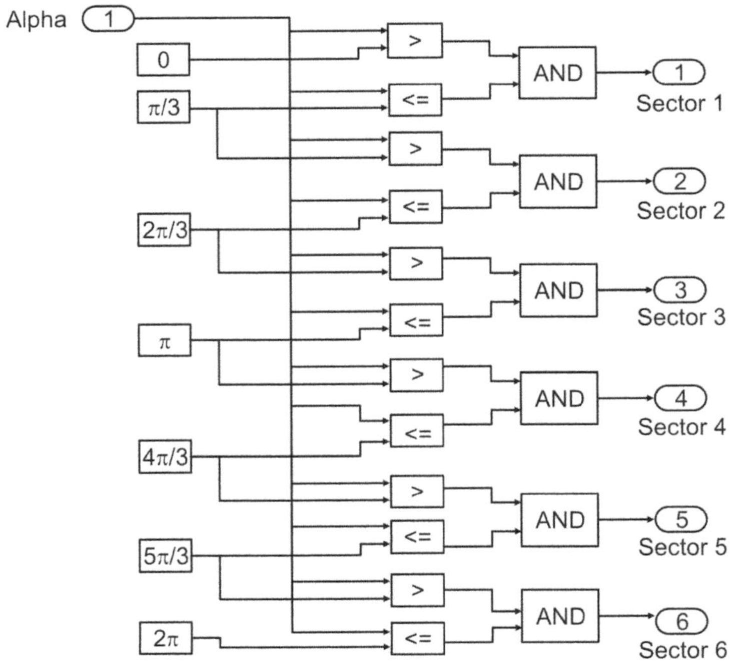

Fig. 2.17 Sector Identification.

If the angle α is between $[0 \le \alpha \le \pi/3]$, the space vector is in sector 1.

If the angle α is between $[\pi/3 \le \alpha \le 2\pi/3]$, the space vector is in sector 2.

If the angle α is between $[2\pi/3 \le \alpha \le \pi]$, the space vector is in sector 3.

If the angle α is between $[\pi \le \alpha \le 4\pi/3]$, the space vector is in sector 4.

If the angle α is between $[4\pi/3 \le \alpha \le 5\pi/3]$, the space vector is in sector 5.

If the angle α is between $[5\pi/3 \le \alpha \le 2\pi]$, the space vector is in sector 6.

In order to determine the switching time of each switch, the switching time of the reference voltage space vector is to be determined by utilizing the nearest 3 voltage vectors.

C. **ON Duration Time of Switching Vector:** The continuously rotating reference vector V_{ref} is sampled at a sampling frequency f_s. The total sampling time T_S is divided into T_0, T_1, T_2. The duration of the reference voltage space vector Vref in a particular sector is calculated with the help of three nearest space vectors and respective extended periods.

The reference voltage vector is considered as a stationary vector in sector1 with constant magnitude during the sampling time T_S. By the Volt –Time balancing principle, the sum of the voltage vectors with their respective stay periods can be equated to the reference voltage vector with its sampling period as in equation (2.27).

$$V_1 T_1 + V_2 T_2 + V_3 T_3 = V_{ref} T_s \quad .....(2.27)$$

where, $$T_0 + T_1 + T_2 = T_s \quad .....(2.28)$$

The line-neutral voltages are calculated from Table 2.2 and represented in equations (2.29), (2.30), and (2.31)

$$V_1 = (2/3)\, V_{dc} \quad .....(2.29)$$

$$V_2 = (2/3)\, V_{dc}\, e^{j\pi/2} = (2/3)\, V_{dc}\, [\cos \pi/2 + j \sin \pi/2] \quad .....(2.30)$$

$$V_0 = 0 \quad .....(2.31)$$

$$V_{ref} = V_{ref.}\, e^{j\alpha} = V_{ref}\, [\cos \alpha + \sin \alpha]$$

Equating real and imaginary parts on both sides,

$$(2/3)\, T_1 + (1/3)\, T_2 = (V_{ref}/V_{dc}) \cos \alpha\, T_s \quad .....(2.32)$$

$$(1/\sqrt{3}\,)\, T_2 = (V_{ref}/V_{dc}) \sin \alpha\, T_s \quad .....(2.33)$$

Solving for T_1 and T_2

$$T_1 = (\sqrt{3}\ \ T_s\, V_{ref}/V_{dc}) \sin (\pi/3 - \alpha) \quad .....(2.34)$$

$$T_2 = (\sqrt{3}\ \ T_s\, V_{ref}/V_{dc}) \sin (\alpha) \quad .....(2.35)$$

$$T_0 = T_s - T_1 - T_2 \quad .....(2.36)$$

Similarly, the ON time durations of the space vectors with the reference space vector in the other sectors is determined. The generalized equation for calculating time duration in any of the sectors is given by equations (2.27) and (2.28).

$$T_1 = (\sqrt{3}\ T_s \mid V_{ref} \mid / V_{dc})\ [\sin (\pi\ \alpha/3 + (n\text{-}1)\ \pi/3)] \quad .....(2.37)$$

$$T_2 = (\sqrt{3}\ T_s \mid V_{ref} \mid / V_{dc})\ [\sin (\alpha - (n\text{-}1)\ \pi/3)] \quad .....(2.38)$$

$$T_0 = T_s - T_1 - T_2 \quad .....(2.39)$$

N refers to sector $1 \rightarrow 6$

and $0 \leq \alpha \leq 60^o$

The ON time durations T_0, T_1, T_2 of the 3 nearest voltage vectors with respect to the position of reference vector in each sector are utilized in the calculation of total switching time of each switch. Also, to generate the reference space vector with minimum switching transition losses, a particular switching sequence is demanded.

D. Switching Sequences: A switching sequence is decided that reduces the switching losses and total harmonic distortion. To achieve this only one phase is switched during a transition from one state to the other as shown in Figure 2.18. A symmetric switching sequence is followed in order to reduce even harmonics through quarter wave symmetry. Suppose the reference space vector is considered in sector 1, then the forward switching sequence applied in the present sub cycle is V0 →V1→V2→V7 and the reverse sequence in the next sub cycle is V7→V2→V1→V0. Both forward and reverse switching sequences pertaining to all six sectors are tabulated in Table 2.3. The total time, for which each switch is ON in each sector, is determined by the space vector PWM switching pattern. Examining the switching sequence in voltage vectors of sector 1, in the counter clock wise the voltage vectors are selected from $V_0(000)$ to $V_7(111)$ and then reversed from $V_7(111)$ to $V_0(000)$ in clock wise direction. By representing this sequence in the form of a switching pattern as shown in the Figure 2.19, the total switching time for each switch in each sector can be easily determined. The total switching or sampling time is symmetrically distributed into the respective switching periods of the voltage vectors in each sector.

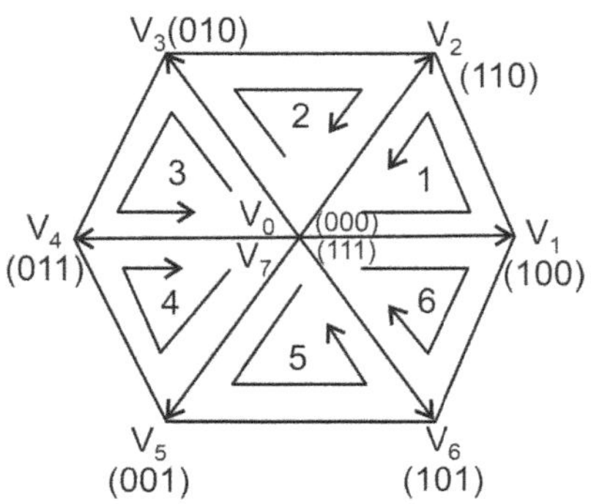

Fig. 2.18 Switching Transition with Directions.

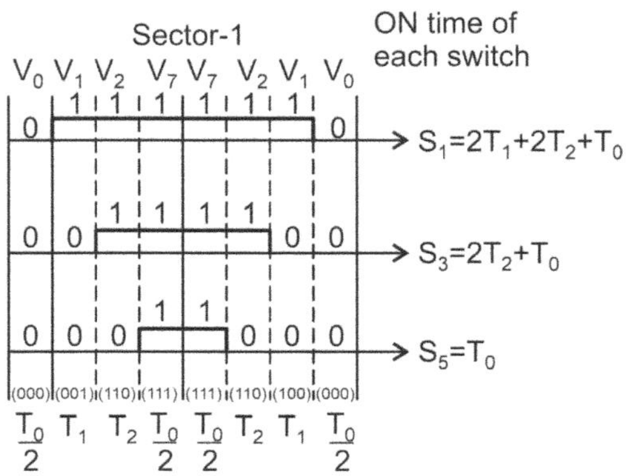

Fig. 2.19 Switching Pattern in Sector 1.

Table 2.3 Switching Sequence of Voltage Vectors.

Sector	Forward Direction	Reverse Direction
1	$V_0 \rightarrow V_1 \rightarrow V_2 \rightarrow V_7$	$V_7 \rightarrow V_2 \rightarrow V_1 \rightarrow V_0$
2	$V_0 \rightarrow V_3 \rightarrow V_2 \rightarrow V_7$	$V_7 \rightarrow V_2 \rightarrow V_3 \rightarrow V_0$
3	$V_0 \rightarrow V_3 \rightarrow V_4 \rightarrow V_7$	$V_7 \rightarrow V_4 \rightarrow V_3 \rightarrow V_0$
4	$V_0 \rightarrow V_5 \rightarrow V_4 \rightarrow V_7$	$V_7 \rightarrow V_4 \rightarrow V_5 \rightarrow V_0$
5	$V_0 \rightarrow V_5 \rightarrow V_6 \rightarrow V_7$	$V_7 \rightarrow V_6 \rightarrow V_5 \rightarrow V_0$
6	$V_0 \rightarrow V_1 \rightarrow V_6 \rightarrow V_7$	$V_7 \rightarrow V_6 \rightarrow V_1 \rightarrow V_0$

The pattern covers both forward and reverse switching sequence including zero vectors. This helps in switching one switch at a time, thus minimizing the switching losses. The total ON time calculation of each upper switch in all sectors for each phase in a total sampling period of $2T_S$, are shown in the Table 2.4.

Table 2.4 Switching Time of Upper Switches of 2-Level VSI.

Sector	S1	S3	S5
1	$T_0 + 2T_1 + 2T_2$	$T_0 + 2T_2$	T_0
2	$T_0 + 2T_2$	$T_0 + 2T_1 + 2T_2$	T_0
3	T_0	$T_0 + 2T_1 + 2T_2$	$T_0 + 2T_2$
4	T_0	$T_0 + 2T_1$	$T_0 + 2T_1 + 2T_2$
5	$T_0 + 2T_2$	T_0	$T_0 + 2T_1 + 2T_2$
6	$T_0 + 2T_1 + 2T_2$	T_0	$T_0 + 2T_1$

E. Generation of Reference Voltages: The sum of the respective switching times of S1, S3 and S5 in all the sectors gives the total operation times of the three phases, respectively as shown here under. The total ON time of each phase is utilized in the generation of PWM pulses, given as

$$T_a = (\text{sector} ==1) * [T_0 + 2T_1 + 2T_2] + (\text{sector} ==2) * [T_0 + 2T_2] + (\text{sector} ==3) * [T_0] + (\text{sector} ==4) * [T_0] + (\text{sector} ==5) * [T_0 + 2T_2] + (\text{sector} ==6) * [T_0 + 2T_1 + 2T_2] \quad .....(2.40)$$

$$T_b = (\text{sector} ==1) * [T_0 + 2T_2] + (\text{sector} ==2) * [T_0 + 2T_1 + 2T_2] + (\text{sector} ==3) * [T_0 + 2T_1 + 2T_2] + (\text{sector} ==4) * [T_0 + 2T_1] + (\text{sector} ==5) * [T_0] + (\text{sector} ==6) * [T_0] \quad .....(2.41)$$

$$T_c = (\text{sector} ==1) * [T_0] + (\text{sector} ==2) * [T_0] + (\text{sector} ==3) * [T_0 + 2T_2] + (\text{sector} ==4) * [T_0 + 2T_1 + 2T_2] + (\text{sector} ==5) * [T_0 + 2T_1 + 2T_2] + (\text{sector} ==6) * [T_0 + 2T_1] \quad .....(2.42)$$

F. Generation of PWM Pulses: The total operating times of the three phases T_a, T_b, T_c are compared with a triangular carrier wave to generate space vector pulse width modulated pulses PWM1, PWM3, PWM5 respectively. The MATLAB/Simulink in Fig. 2.20 illustrates pulse generation model. Thus, generated space vector modulated pulses are fed to VSI to provide an optimum PWM output.

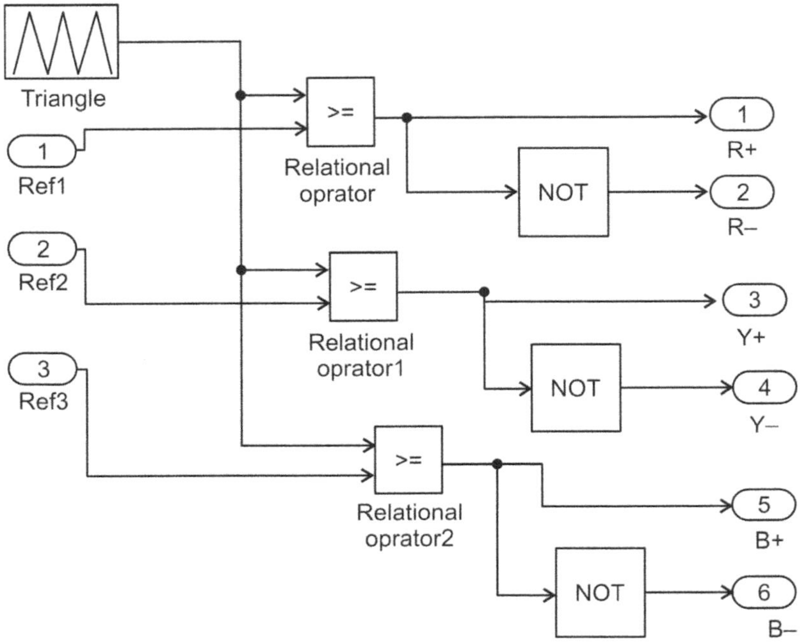

Fig. 2.20 MATLAB Simulink Model for Generation of SVPWM Pulses.

2.4.1.2 Three-phase 3-Level Space Vector PWM Controlled Voltage Source Inverter

A. Concept of SVPWM for Three-Level DCMC: SVPWM control technique became a standard and well-established switching strategy for the power converters. In spite of various PWM techniques like Sinusoidal PWM (SPWM), Selective Harmonic Elimination PWM (SHEPWM) and some modified triangular carrier-based modulation schemes, SVPWM stood as the most popular one due to its relatively good performance and simplicity with inverters of any levels. In the other PWM techniques, the PWM patterns are generated for each phase independently. But the modulated wave generated in SVPWM is the combined effect of all the three phases of the output voltages of the inverter. Thus, the generation of such a modulated wave of SVPWM necessitates vectors' calculation and switching states selection. This section presents a SVPWM based switching strategy for a three-level DCMC.

The SVPWM exhibits the following salient features in the application of three-level DCMC.

- Higher voltage magnitude on the AC side of the VSC compared to that generated by conventional SPWM technique.

- Flexibility in the selection of switching pattern without additional controllers that helps in the balance of DC capacitor voltage and minimize harmonics. A comprehensive methodology using fractal structure is presented that acts as a tool in designing the SVPWM controller for a three-level DCMC.

B. Methodology of SVPWM: The flow in the methodology continues by considering the concern of some functional inputs as, voltage across the capacitor, switching frequency used to calculate switching time and the resultant reference space vector calculated from the three phase output voltages at the point of connection. The methodology of three-level SVPWM technique is explained as follows and its block diagram representation is shown in fig. 2.21.

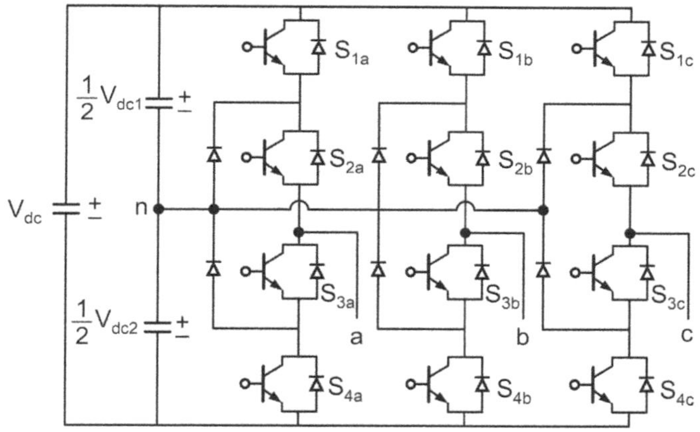

Fig. 2.21 Block diagram of three-level SVPWM technique.

Detailed steps:

1. Computation of the reference voltage space vector Vref from the equivalent 3 - 2 conversion
2. Synthesis of voltage space vectors from the permissible switching combinations of the three-level converter
3. Hexagon representation of switching state vectors
4. Identification of sectors in which the reference space vector is positioned
5. Determination of region in selected sector where the reference vector is located
6. 'ON' time calculation of switching vectors in each region of sector using optimal switching sequence
7. Synthesis of reference space vector and calculation of total ON time of each switch using optimal switching pattern

C. Computation of Reference Voltage Space Vector: The three-level SVPWM control scheme follows a decision procedure to obtain the angle and magnitude of the reference voltage space vector. The motivation for the reference space vector generation comes from the AC bus three phase balanced sinusoidal signals displaced by 120° represented in equations 2.43, 2.44 and 2.45,

$$v_a(t) = V_s \sin \omega t \qquad .....(2.43)$$

$$v_b(t) = V_s \sin(\omega t - 2\pi/3) \qquad .....(2.44)$$

$$v_c(t) = V_s \sin(\omega t - 4\pi/3) \qquad .....(2.45)$$

Where, $\omega = 2\pi f_s$

The reference space vector is the resultant of the three phase signals defined as in equation 2.46,

$$V(t) = 2/3\,[v_a(t) + v_b(t)\alpha + v_c(t)\alpha_2] \qquad .....(2.46)$$

Where, $\alpha = \exp(j\,2\pi/3)$

By substituting 2.43, 2.44, 2.45 and α in equation 2.46, the resultant voltage is as shown in equation 2.47,

$$V(t) = V_s \exp(j\omega t - (j\,\pi/2)) \qquad .....(2.47)$$

That is, V(t) is the reference space vector in the complex reference frame with magnitude V_s and rotating in a counter clock wise direction with a frequency ω rad/sec.

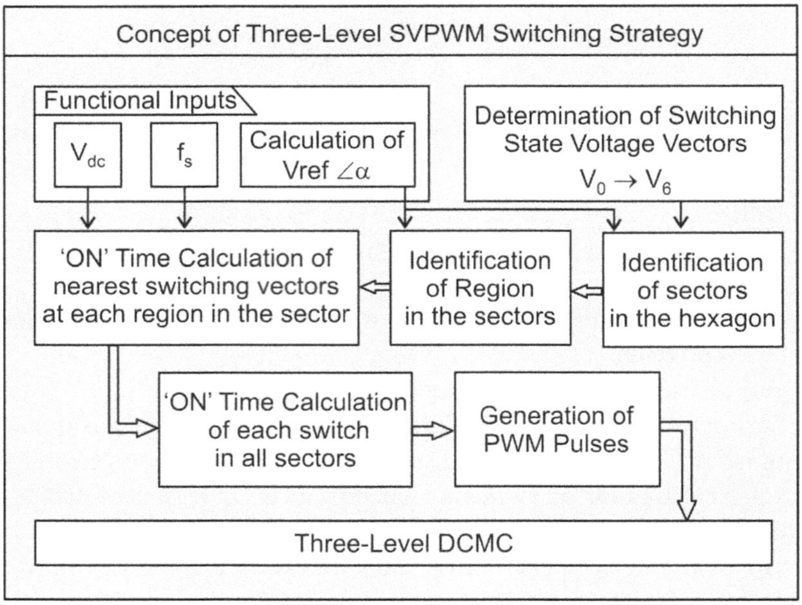

Fig. 2.22 Reference voltage space vector block diagram of three-level SVPWM technique.

The reference space vector V_{ref} as shown in figure 2.22 is generated by the transformation of three phase signals into two mutually perpendicular vectors V_d and V_q along the real and imaginary axes respectively, in the d-q complex reference frame known as Parks transformation.

The three phase voltages in the parks transformation are collected instantaneously from the load or line terminal of the AC system. The occurrence of any sudden contingency in the power system is fed to the controller in the form of line voltages and currents, so that it is compensated through the appropriate switching states of the MLI.

This transformation from $v_a(t)$, $v_b(t)$, $v_c(t)$ into V_d and V_q clearly shown in equations 2.48 and 2.49.

$$V_d = 2/3\ v_a(t) - 1/3\ v_b(t) - 1/3\ v_c(t) \quad .....(2.48)$$

$$V_q = 1/\sqrt{3}\ v_b(t) - 1/\sqrt{3}\ v_c(t) \quad .....(2.49)$$

$$V_{ref} = V_d + jV_q \quad .....(2.50)$$

$$\alpha = \tan^{-1}(V_q/V_d) \quad .....(2.51)$$

Equation 2.50 denotes the reference voltage space vector obtained from the resultant of the V_d and V_q. The reference voltage space vector V_{ref} rotates with constant magnitude and uniform angular velocity ω. The angle α obtained from equation 2.51 determines the position of the reference space vector at any particular instant, during its rotation.

The vectors $v_a(t)$, $v_b(t)$, *and* $v_c(t)$ are recovered from equation 2.50,

$$\begin{bmatrix} v_a(t) \\ v_b(t) \\ v_c(t) \end{bmatrix} = \begin{bmatrix} 1 & 0 \\ -1/2 & \sqrt{3}/2 \\ -1/2 & -\sqrt{3}/2 \end{bmatrix} \begin{bmatrix} V_d(t) \\ V_q(t) \end{bmatrix} \quad .....(2.52)$$

The assumption of continuous rotation of the reference voltage space vector is limited to the ON-OFF states of the twelve–switches of the three-level DCMC. Therefore, to determine the exact position of the reference space vector, one has to utilize different permissible combinations of ON-OFF states of the switches of a three-level DCMC.

D. Synthesis of Voltage Space Vectors: Under the synthesis of voltage space vectors, initially, the permissible switching combinations are determined, that represent the operating status of a three-level DCMC. Then, they are grouped into different types of switching state space vectors. Finally, the reference vector obtained from the AC side voltages of the diode clamped power converter, is synthesized by the three nearest switching vectors.

E. Switching Combinations in Three-Level Converter: The ON-OFF switching combinations represent the operating status of the switches in the three-level DCMC. Different switching states gives three levels of output as 1 (positive), 0 (zero), -1 (negative). When two upper switches are turned ON in any leg, the switching state is '1', which means the output phase to neutral voltage is + $V_{dc}/2$. If two lower switches are ON, then the switching state is '-1', with output voltage as - $V_{dc}/2$. And if the inner two switches are turned ON, the switching state is '0', which means the output voltage becomes zero depending on the direction of load current through the clamping diodes. The different switching states of the switches in leg 'a' and their respective terminal voltages are tabulated in table 2.5.

Table 2.5 Different switching states of leg 'a' in three-level DCMC.

Switching State	Status of switches in leg 'a'				Terminal Voltage V_{an}
	S_{1a}	S_{2a}	S_{3a}	S_{4a}	
1	ON	ON	OFF	OFF	$+V_{dc}/2$
0	OFF	ON	ON	OFF	0
-1	OFF	OFF	ON	ON	$-V_{dc}/2$

F. Switching States of Space V ectors: The magnitudes of different voltage space vectors along with their respective angular positions are computed and tabulated in table 2.6.

Table 2.6 Magnitudes of vectors at different switching states.

Sl.no.	Angle	Space Vector	Vector Type	Switching State			Magnitude	No. of Switching Combinations
1.	0	V_0	Zero	1	1	1	0	3
				0	0	0		
				-1	-1	-1		
2.	0	V_1	Small	1	0	0	$V_{dc}/3$	2
				0	-1	-1		
3.	π/3	V_2	Small	1	1	0	$V_{dc}/3$	2
				0	0	-1		
4.	2π/3	V_3	Small	0	1	0	$V_{dc}/3$	2
				-1	0	-1		
5.	π	V_4	Small	0	1	1	$V_{dc}/3$	2
				-1	0	0		
6.	4π/3	V_5	Small	0	0	1	$V_{dc}/3$	2
				-1	-1	0		
7.	5π/3	V_6	Small	1	0	1	$V_{dc}/3$	2
				0	-1	0		
8.	0	V_7	Large	1	-1	-1	$2V_{dc}/3$	1
9.	π/3	V_9	Large	1	1	-1	$2V_{dc}/3$	1
10.	2π/3	V_{11}	Large	-1	1	-1	$2V_{dc}/3$	1

Contd...

Sl.no.	Angle	Space Vector	Vector Type	Switching State			Magnitude	No. of Switching Combinations
11.	π	V_{13}	Large	-1	1	1	$2V_{dc}/3$	1
12.	4π/3	V_{15}	Large	-1	-1	1	$2V_{dc}/3$	1
13.	5π/3	V_{17}	Large	1	-1	1	$2V_{dc}/3$	1
14.	π/6	V_8	Medium	1	0	-1	$V_{dc}/\sqrt{3}$	1
15.	π/2	V_{10}	Medium	0	1	-1	$V_{dc}/\sqrt{3}$	1
16.	5π/6	V_{12}	Medium	-1	1	0	$V_{dc}/\sqrt{3}$	1
17.	-5π/6	V_{14}	Medium	-1	0	1	$V_{dc}/\sqrt{3}$	1
18.	-π/2	V_{16}	Medium	0	-1	1	$V_{dc}/\sqrt{3}$	1
19.	-π/6	V_{18}	Medium	1	-1	0	$V_{dc}/\sqrt{3}$	1

The total switching combination states of 12 switches in a three-level converter has 33 = 27 voltage vectors [22]. The different possible combinations of 27 switching vectors are classified into four groups as zero vectors, small vectors, medium vectors and large vectors. The possibility of three phase switching states represented by 1, 0, -1 are grouped into four types of voltage vectors. The corresponding switching states voltage vectors are described as follows:

Zero Vector (V_0): The corresponding switching states are [111], [000], [-1-1-1] which gives voltage magnitude as zero.

Small Vector (V_1 to V_6): Each small vector corresponds to two switching states, and all having a magnitude of ($V_{dc}/3$). In total there are 12 small vectors.

Medium Vectors: There are 6 medium vectors each of magnitude ($V_{dc}/\sqrt{3}$).

Large Vectors: There are 6 large vectors each of magnitude ($2V_{dc}/3$)

G. **Hexagon Representation of Switching State Vectors:** The 27 switching vectors are structured into a two-layer regular hexagon design centered at the origin of d-q plane. The six bigger triangles of the outer hexagon noted as I, II, III, IV, V and VI denote individually a sector. Each sector is divided into four regions numbered as 1, 2, 3 and 4 as shown in figure 2.23. There are, in total, 24 regions in all the sectors. The trajectory of the reference voltage space vector V_{ref} can expand from the origin of inner hexagon to that of the vertices of the larger or outer hexagon covering all the 24 regions. The Large Vector (LV) points to the vertex of the outer hexagon, the Small Vector (SV) points to the vertex of the inner hexagon and the Medium Vector (MV) points in between outer hexagon & inner hexagon. To obtain the duty cycle of the V_{ref}, it requires identifying its position at every sampling instant.

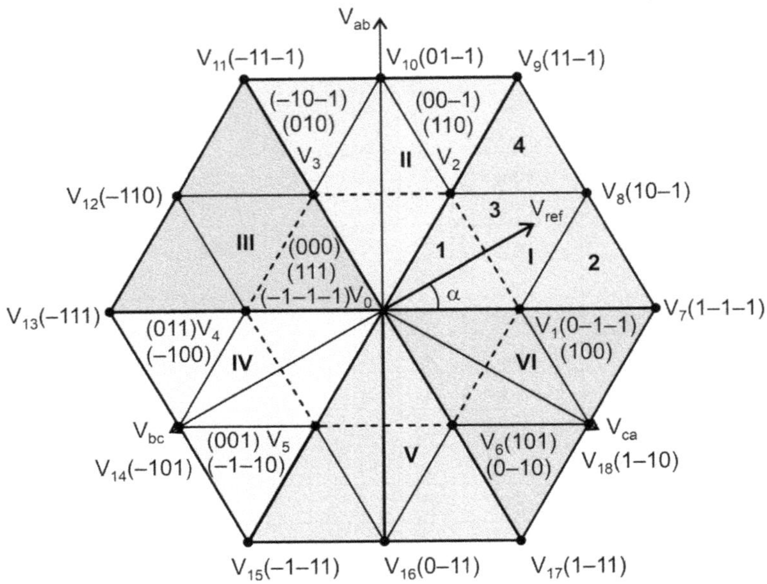

Fig. 2.23 Hexagon representation of three-level VSC.

H. **Identification of Sectors and Regions in a Hexagon:** The larger hexagon plane is divided into 6 sectors (I to VI), each sector corresponding to 60^0. The sector is further split into four regions (1 to 4). The angle of the voltage space vector V_{ref}, is displaced from d-axis as given in equation 2.47. The determination of the sector with the calculation, in which the reference space vector is located, is followed as,

If the angle is between $[0 \le \alpha < \pi/3]$, the space vector is in sector-I.

If the angle α is between $[\pi/3 \le \alpha < 2\pi/3]$, the space vector is in sector-II.

If the angle α is between $[2\pi/3 \le \alpha < \pi]$, the space vector is in sector-III.

If the angle α is between $[\pi \le \alpha < 4\pi/3]$, the space vector is in sector-IV.

If the angle α is between $[4\pi/3 \le \alpha < 5\pi/3]$, the space vector is in sector-V.

If the angle α is between $[5\pi/3 \le \alpha < 2\pi]$, the space vector is in sector-VI.

I. **Identification of Regions in a Sector of a Hexagon:** For deciding the region in which the reference vector falls, it is required to determine the switching times or the duty cycles of the three nearest vectors in the region. Keeping in view the symmetry of all the sectors, it is sufficient to reflect the reference vector into sector-I to identify its projection in any of the four regions.

It is assumed in this case that the amplitude of the reference vector fits within the hexagon in which the sides of the triangular regions have lengths equal to unity as shown in figure 2.24. The length of the reference vector projecting out of the larger hexagon leads to over modulation. The reference vector m_n is decomposed into two vectors m_l and m_2 along the zero- and sixty-degrees axes as shown in figure 2.25. The sides, p, q and r are the extended projections made to ease the calculation of the values m_l and m_2.

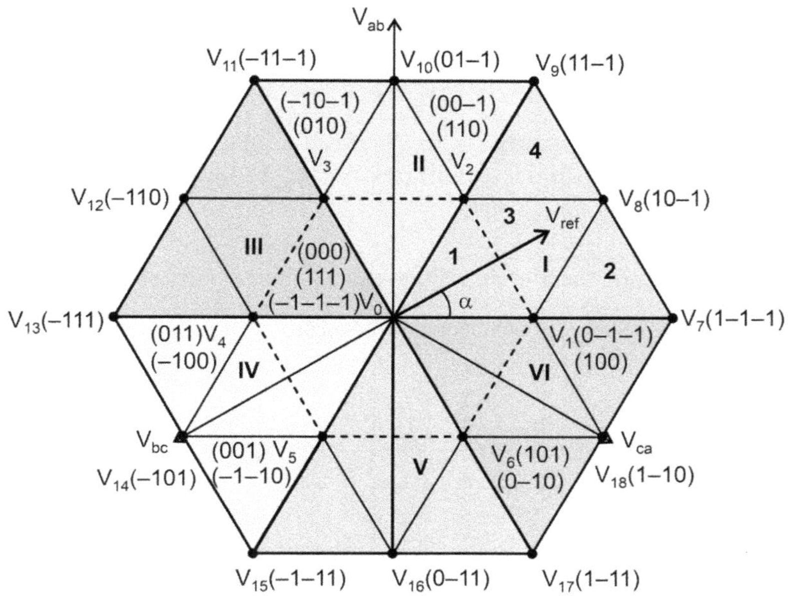

Fig. 2.24 Projection of reference vector in sector-I.

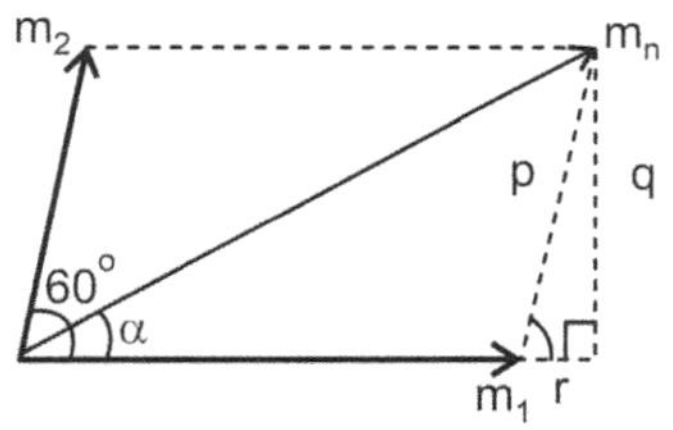

Fig. 2.25 Resolution of reference vector.

In order to obtain the values of m_1 and m_2 in figure 2.25, Sine rule is applied as shown in equations 2.54 and 2.55 and calculated as follows:

$$p/(\sin \pi/2) = q/(\sin \pi/3) \qquad \text{.....(2.53)}$$

$$p = m_2 = q\,(2/\sqrt{3}\,) = 2/\sqrt{3}\ m_n \sin \alpha$$

$$\therefore \quad m_2 = (2/\sqrt{3}\,)\, m_n \sin \alpha \qquad \text{.....(2.54)}$$

$$m_1 = m_n \cos \alpha - r$$

$$m_1 = m_n \cos \alpha - p \cos \pi/3$$

$$m_1 = m_n \cos \alpha - [(2/\sqrt{3}\ m_n \sin \alpha) \cos \pi/3]$$

$$\therefore \quad m_1 = m_n (\cos \alpha - \sin \alpha/\sqrt{3}\,) \qquad \text{.....(2.55)}$$

The values of m_1 and m_2 are derived and shown in equation 2.54 and 2.55 respectively are taken as the duty cycles of the three nearest vectors to decide the region where the reference vector lies.

Table 2.7 summarizes the conditions to identify the placement of the reference vector in any of the four regions. Similar conditions can be followed for different regions in the entire sectors.

Table 2.7 Conditions to determine the location of reference vector.

Sl.no.	Condition	Region
1.	$m_1 \leq 1$ $m_2 \leq 1$ $m_1 + m_2 \leq 1$	1
2.	$m_1 > 1$	2
3.	$m_1 \leq 1$ $m_2 \leq 1$ $m_1 + m_2 > 1$	3
4.	$m_2 > 1$	4

J. 'ON' Time Calculation of Switching Vectors: The rotating reference vector V_{ref} is sampled at a sampling frequency f_s. The total sampling time T_S is divided into T_a, T_b, and T_c. The duration of the reference voltage vector V_{ref} in a particular region of a sector is calculated with the help of three nearest vectors V_a, V_b, & V_c and their respective extended periods T_a, T_b, and T_c.

The time duration of the reference vector in each region of a sector is calculated with the assistance of volt-time balance relationship shown in equation 2.56.

$$V_a T_a + V_b T_b + V_c T_c = V_{ref} T_S \quad .....(2.56)$$

Suppose the V_{ref} falls in region 3 of sector-I, the three nearest voltage vectors of region 3 are, small vectors V_1, V_2 and medium vector V_8.

$$V_1 = (1/3)\, V_{dc} \quad .....(2.57)$$

$$V_2 = (1/3)\, V_{dc}\, e^{j\pi/3} \quad .....(2.58)$$

$$V_8 = (1/3)\, V_{dc}\, e^{j\pi/6} \quad .....(2.59)$$

Substituting equations 2.57, 2.58 and 2.59 into volt-time balance relationship equation 2.56,

$$(1/3)\, V_{dc}\, T_a + (1/\sqrt{3})\, V_{dc}\, (\cos \pi/6 + j\sin \pi/6)\, T_b + (1/3)\, V_{dc}\, (\cos \pi/3 + j\sin \pi/3)\, T_c = V_{ref}\, (\cos\theta + j\sin\theta)\, T_s \quad .....(2.60)$$

Dividing the equation 2.60 into real and imaginary parts eases the calculations for the duty cycle.

Equating the real parts,

$$1/3\, V_{dc}\, T_a + 1/2\, V_{dc}\, T_b + 1/6\, V_{dc}\, T_c = V_{ref} \cos\theta\, T_s \quad .....(2.61)$$

Multiplying above equation with $3\, V_{dc}$

$$T_a + 3/2\, T_b + 1/2\, T_c = 3(V_{ref}/V_{dc}) \cos\theta\, T_s \quad .....(2.62)$$

Equating the imaginary parts,

$$(1/\sqrt{3})(1/2)\, V_{dc}\, T_b + (1/3)(\sqrt{3}/2)\, V_{dc}\, T_c = V_{ref}\, sin\theta\, T_s \quad .....(2.63)$$

Multiplying above equation with $3/V_{dc}$

$$(\sqrt{3}/2)\, T_b + (\sqrt{3}/2)\, T_c = 3\, (V_{ref}/\, V_{dc})\, sin\theta\, T_s \quad .....(2.64)$$

The uniform distribution of sampling time is expressed as

$$T_S = T_a + T_b + T_c \quad .....(2.65)$$

By solving equations 2.63, 2.64 and 2.65,

$$T_a = T_S\, [1 - 2Msin\theta] \quad .....(2.66)$$

$$T_b = T_S\, [2Msin\, (\pi/3 + \theta) - 1] \quad .....(2.67)$$

$$T_c = T_S\, [1 - 2Msin\, (\pi/3 + \theta)] \quad .....(2.68)$$

Where M is the Modulation index and expressed as,

$$M = \sqrt{3}\, (V_{ref}/V_{dc}) \quad .....(2.69)$$

Similarly, the time duration of reference vector can be calculated in remaining regions of all the sectors. The ON time duration of switching vectors in each region T_a, T_b, T_c can be represented with respect to the switching pattern to generate the phase voltage waveforms.

K. Synthesis of Reference Space Vector: The reference space vector trajectory is uniform in all the sectors. So, it is sufficient to consider sector-I of $0 \leq \theta \leq 60^{o}$ for the synthesis of the reference space vector. The switching frequency is maintained that only one switch is operated while transiting from one switching state to the next as shown in figure 2.26.

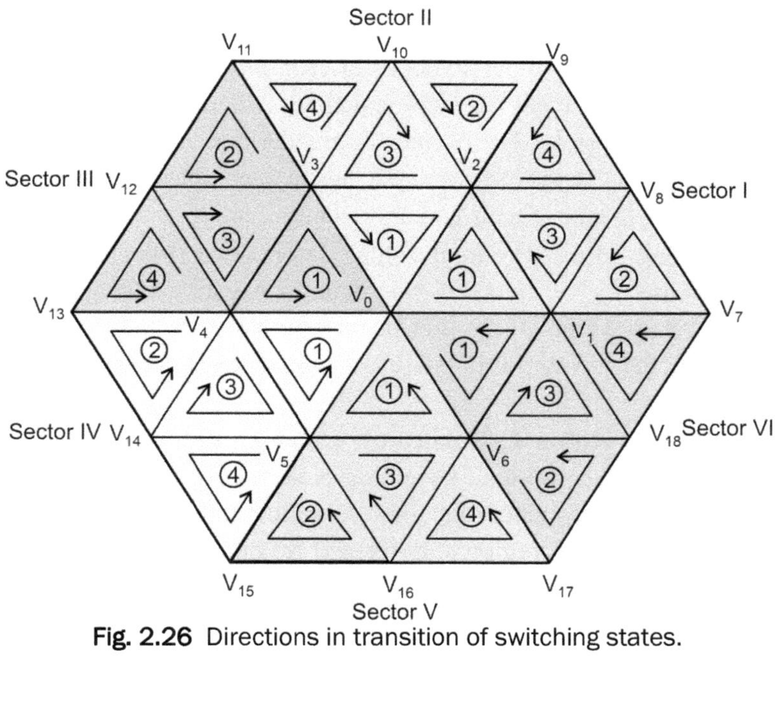

Fig. 2.26 Directions in transition of switching states.

Suppose the reference voltage vector is located in region 3, it is synthesized by the three nearest switching vectors V_1, V_2 and V_8. The switching pattern in region 3 which leads to respective phase voltages and switching status is shown in figure 2.27.

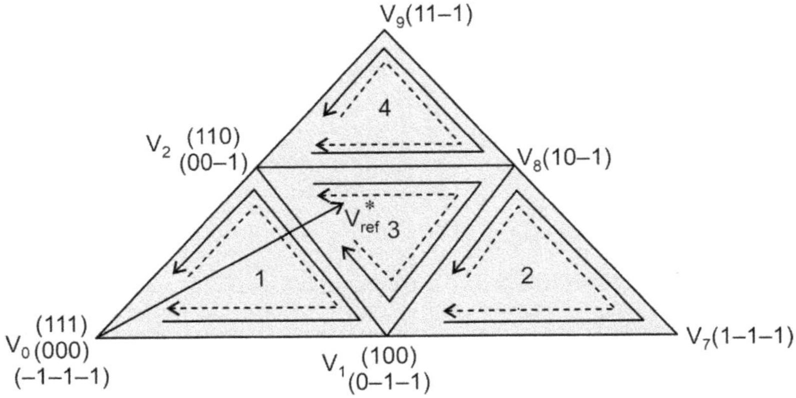

Fig. 2.27 Sequence of selecting voltage vectors in each region.

L. Switching pattern of each region's switching sequence in sector-I: Since each voltage vector is available for different switching states, the uniform distribution of sampling time into individual switching vector time periods is not alike for all the regions in each sector. In all the sectors, region 1 is divided into 14 switching segments, region 2 and 4 has eight switching segments and region 3 has ten switching segments.

The demanded states of the upper switches of each phase leg S_{1a} and S_{2a}; S_{1b} and S_{2b}; S_{1c} and S_{2c} and their respective ON times calculation depending on the time periods of nearest switching vectors as T_a, T_b and T_c are derived for all the regions in sector-I. However, the switching states of lower switches in each phase that is S_{3a}, S_{4a}; S_{3b}, S_{4b}; S_{3c}, S_{4c}; are complement to that of their upper switches.

In a given switching status '0' output voltage is obtained if the middle two switches S_{2a} and S_{3a} in the phase leg 'a' are OFF. '-1' denotes negative voltage i.e., the bottom two switches S_{3a} and S_{4a} are ON. '1' denotes positive voltage i.e., the top two switches S_{1a} and S_{2a} are ON. Thus, the ON and OFF status of the switches determine the output voltages V_a, V_b and V_c.

Switching pattern in region 1: The nearest switching vectors of V_{ref} located in region 1 of sector-I are arranged in optimal switching pattern as shown in figure 2.28. To follow symmetrical switching sequence the total sampling time is divided into 14 switching segments. The boundary of region 1 is formed by one null vector with three switching states $V_0[V_0(000); V_{0P}(111); V_{0N}(-1-1-1)]$, and two short vectors each with two switching states as $V_1[V_{1P}(100); V_{1N}(0-1-1)]$, V_2 $[V_{2P}(110); V_{2N}(00-1)]$.

The figure shows the output voltages V_a, V_b, V_c and the ON time of the corresponding leg's upper switches. However, the ON time of the lower switches is complementing to that of upper switches. In upper switches of all legs, S_{2a} is ON for longest period extending from switching segments 2 to 13 for a time $T_a/3 + T_b/3 + T_c/3$. And S_{1a} is ON from segments 5 to 10 and giving raise to the total ON time of $2T_a/3 + T_b + T_c$. In phase leg 'b' the ON state of S_{2b} extend from switching segments 3 to 12 for period equal to $2T_a/3 + T_b + T_c$ and Switch S_{1b} ON time is $T_a/3 + T_c/2$. In phase leg 'c', S_{2c} extends from switching segments 4 to 11 for a period $2T_a/3 + T_b/2 + T_c/2$ and S_{1c} is ON for a shortest period out of all leg's upper switches for a time $T_a/3$.

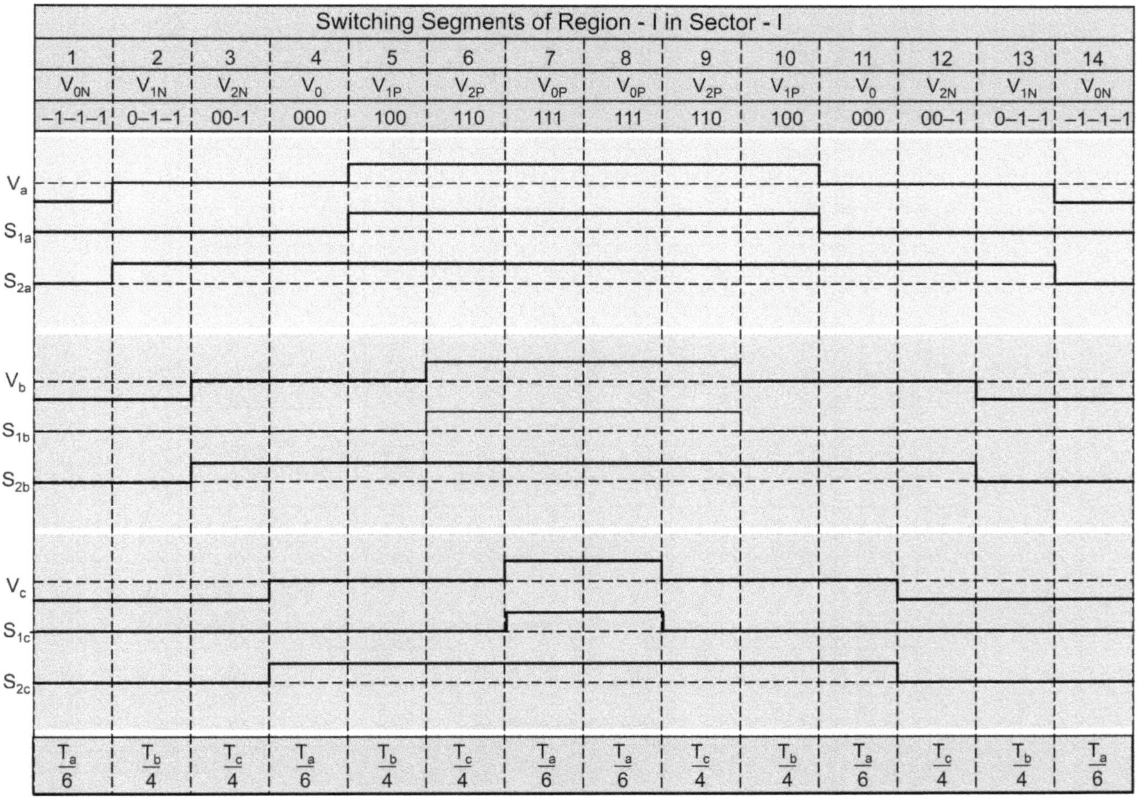

Fig. 2.28 Switching pattern followed in region1 of sector-I.

Switching pattern in region 2: Figure 2.29 shows the switching pattern of the optimal symmetrical switching sequence followed in region 2 of sector-I. The total sampling time is divided into 8 switching segments the boundaries of region 2 are one large vector V_7(1-1-1), one medium vector V_8 (10-1) and one short vector V_1 having two switching states as [V_{1N} (0-1-1), V_{1P} (-100)]. The output voltages V_a, V_b, V_c shown in the figure are obtained from the sequence switching pattern. From the magnitudes of V_a, V_b and V_c the corresponding ON times of the upper switches is calculated. Out of all the upper switches in all the legs, S_{2a} is ON for maximum time through all the switching segments 1 to 8 and time equal to $T_a + T_b + T_c$. And as vice-versa, S_{1b} and S_{1c} are OFF through out the sampling period i.e., its ON time is equal to zero. And similarly, the ON time calculation of S_{1a}, S_{2b} and S_{2c} are $T_a/2 + T_b + T_c$, $T_a/2 + T_c$ and $T_a/2$ respectively.

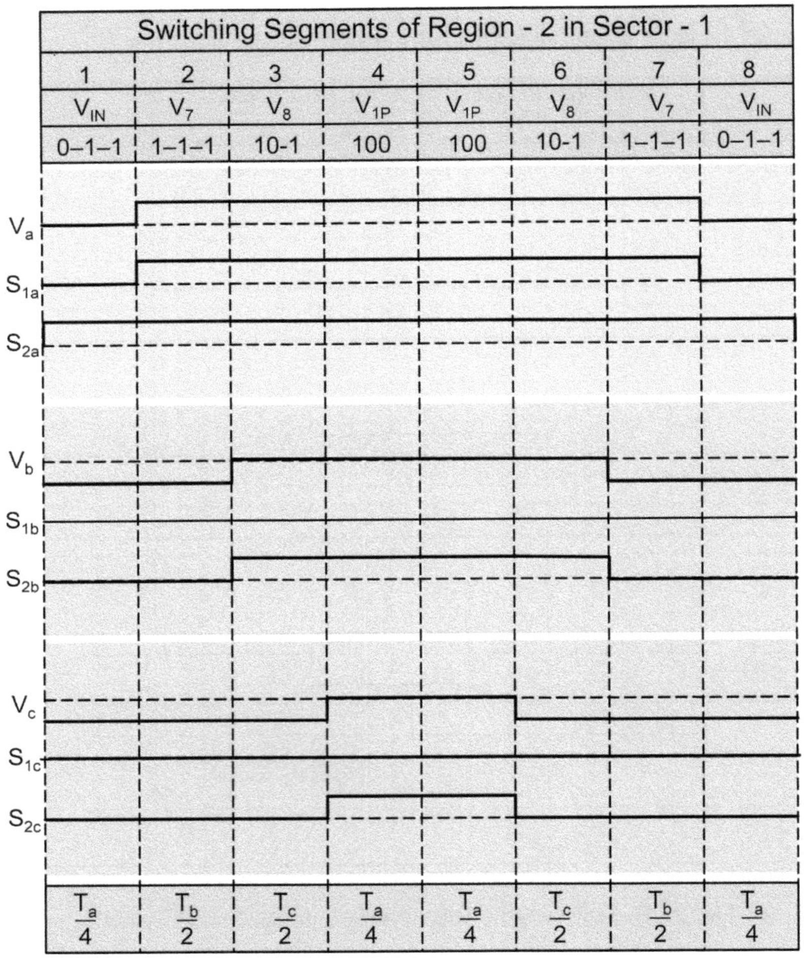

Fig. 2.29 Switching pattern followed in region 2 of sector-I.

Switching pattern in region 3: The switching pattern of the sequence followed with the boundaries of the region is formed into 10 switching segments as shown in the figure 2.30. The output phase voltage V_a is +ve and forms from segments 3 to 8 and suits the ON time of S_{1a} equal to $T_a/2 + T_b + T_c/2$. To obtain this voltage, S_{2a} is ON throughout the sampling period i.e., from segment 1 to 10 with ON time equal to $T_a + T_b + T_c$. Similarly, the output voltage V_b and V_c are obtained from the corresponding switching states which give rise to the following ON times of the corresponding upper switches as, S_{1b} is ON for a shorter period equal to $T_a/2$. S_{2b} extends from segments 2 to 9 for a period equal to $T_a + T_b + T_c/2$. And S_{2c} of leg 'c' is ON from segments 4 to 7 equal to time $T_a/2 + T_c/2$. The only upper switch which is OFF completely in this region is S_{1c}. Two short vectors and one medium vector of the two switching states are $V_1[V_{1N}$ (0-1-1); V_{1P} (100)], $V_2[V_{2N}$ (00-1); V_{2p} (110)] and V_8 (10-1).

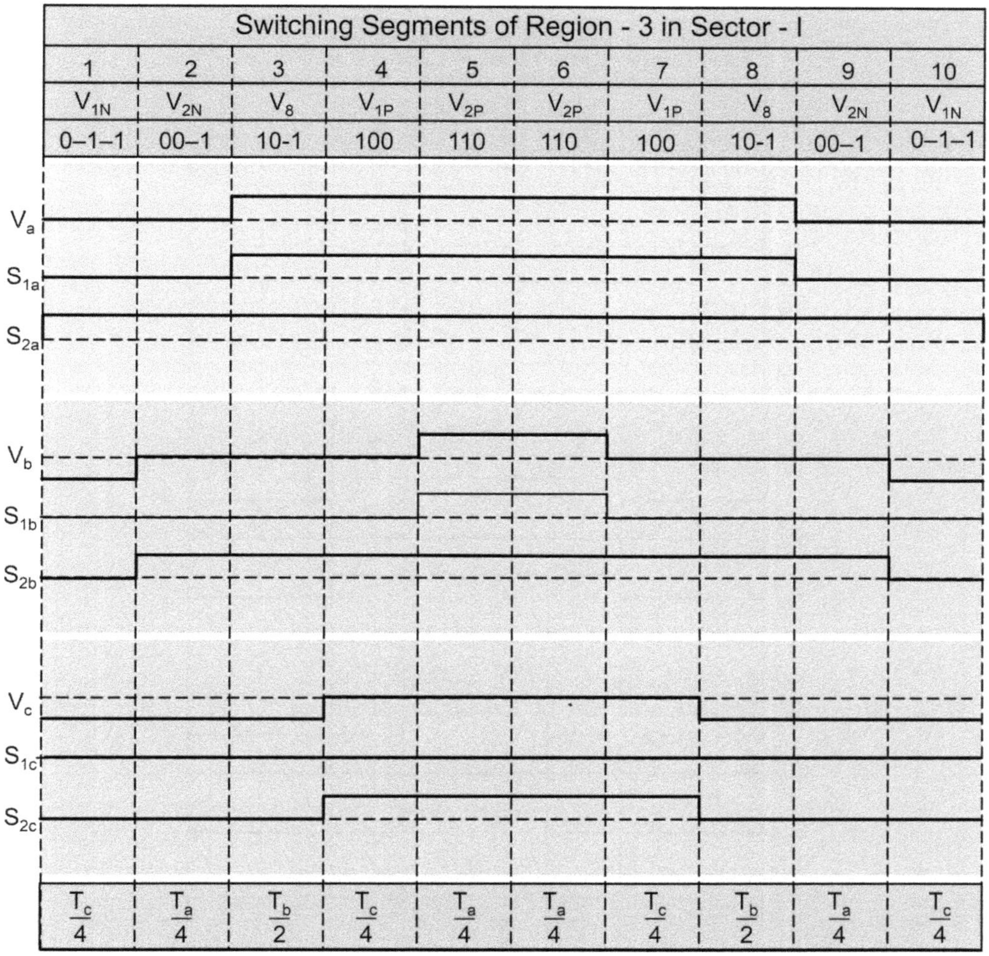

Fig. 2.30 Switching pattern followed in region3 of sector-I.

Switching pattern in region 4: The boundaries of this region are one short vector $V_2[(V_{2P}$ (110), V_{2N} (00-1)] one medium vector V_8 (10-1) and one large vector V_9 (11-1). The switching vector following these switching vectors in a symmetrical sequence is shown in figure 2.31.

The pattern is divided into 8 switching segments from which the positive, 0 or negative magnitudes of V_a, V_b and V_c are obtained. Now, analyzing the corresponding ON times of upper switches, S_{2a} is observed to be ON completely from switching segment 1 to 8 with time $T_a + T_b + T_c$ and S_{1a} is ON for a time $T_a/2 + T_b + T_c$, whereas S_{1b}, S_{1c} are seen to be completely OFF. Whereas S_{2b} ON time follows the +ve voltage period of V_b as $T_a/2 + T_c$ and S_{2c} follows the zero-voltage period of V_c which is the shortest period equal to $T_a/2$.

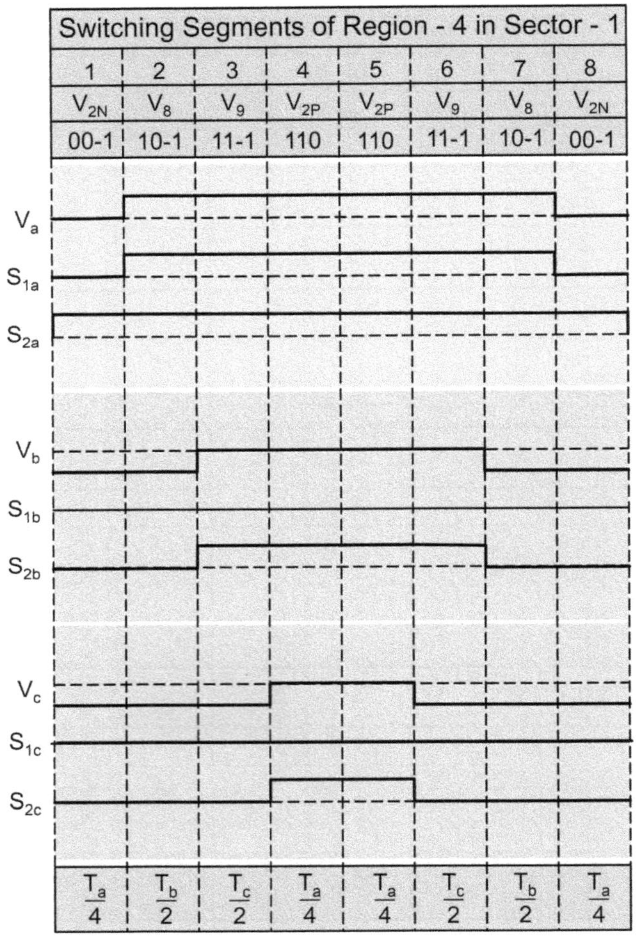

Fig. 2.31 Switching pattern followed in region 4 of sector-I.

The total ON time of each upper switch obtained from the switching pattern of all the regions of sector-I are listed in Table 2.8.

Table 2.8 Switching times of upper switches in sector-I of three-level diode-clamped inverter.

Sector-I	Region 1	Region 2	Region 3	Region 4
S_{1a}	$T_a/3 + T_b/2 + T_c/2$	$T_a/2 + T_b + T_c$	$T_a/2 + T_b + T_c/2$	$T_a/2 + T_b + T_c$
S_{2a}	$T_a/3 + T_b/2 + T_c/2$	$T_a + T_b + T_c$	$T_a + T_b + T_c$	$T_a + T_b + T_c$
S_{1b}	$T_a/3 + T_c/2$	0	$T_a/2$	0
S_{2b}	$T_a/3 + T_b + T_c$	$T_a/2 + T_c$	$T_a + T_b + T_c/2$	$T_a/2 + T_c$
S_{1c}	$T_a/3$	0	0	0
S_{2c}	$2T_a/3 + T_b/2 + T_c/2$	$T_a/2$	$T_a/2 + T_c/2$	$T_a/2$

Similar procedure is applied to decide the total switching time of each switch in the remaining regions of all the sectors [24]. The total ON time of each switch with respect to the tip of V_{ref} in any region of sector over a sampling time T_S is used to generate respective high pulses as SVPWM pulses.

2.4.2 Three-dimensional (3-D) Algorithm for SVPWM

Three-dimensional Space Vector Modulation for Four-Leg Voltage Source Inverters

2.4.2.1 Four-Leg Voltage Source Inverter

The schematic diagram of a four-leg inverter is shown in Fig. 2.32. This topology is known to produce balanced output voltages even under unbalanced load conditions [10]. Due to the additional leg, a four-leg inverter can assume sixteen topologies which is twice the number of topologies a conventional three-leg inverter can assume. These topologies are shown in Fig. 2.33. The topologies are similar to the three-leg inverter with the fourth leg being connected either to the negative rail (Fig. 2.33(a)) or to the positive rail (Fig. 2.33(b)). Each of these topologies is referred to as a switching state. Thus, a four-leg inverter can produce sixteen switching states.

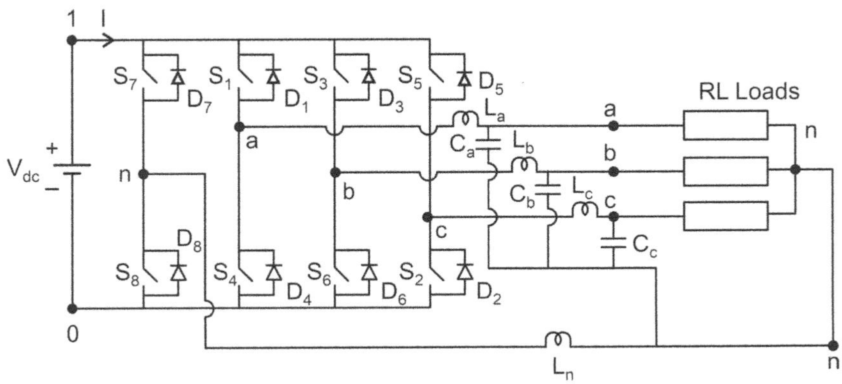

Fig. 2.32 Topologies of a four-leg voltage source inverter.

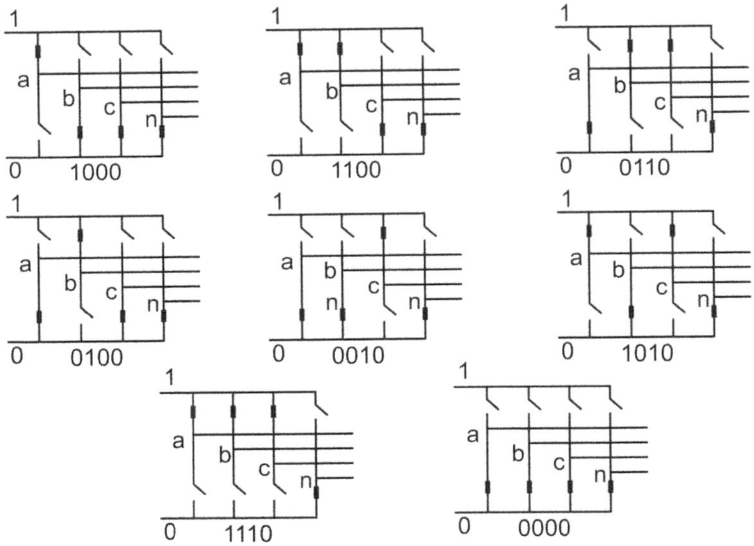

Fig. 2.33 (a) Topologies of a four-leg voltage source inverter. (Fourth leg to negative rail)

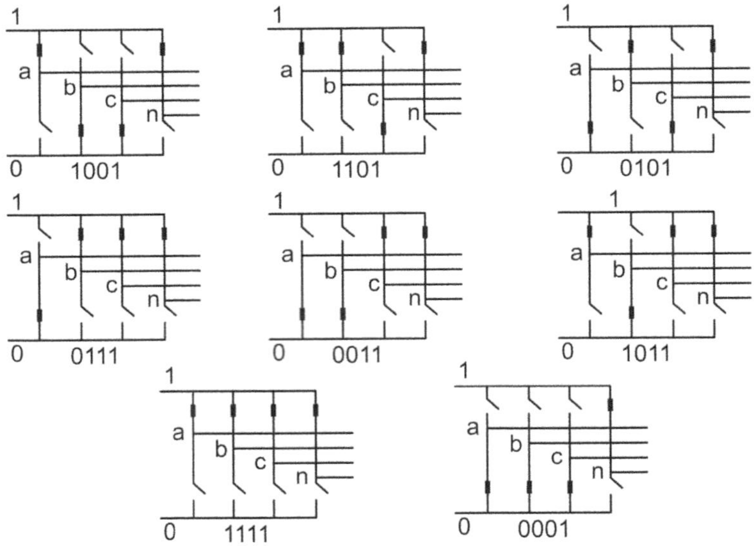

Fig. 2.33 (b) Topologies of a four-leg voltage source inverter. (Fourth leg to positive rail)

2.4.2.2 Voltage Space Vectors

Space vector modulation for a four-leg voltage source inverter is based on the representation of the phase voltage space vectors in the α, β, γ plane. Considering for example the first topology of Fig. 2.33 (a) which is repeated in Fig. 2.34 for the sake of convenience, we see that the phase voltages V_{an}, V_{bn}, V_{cn} are given by

$$V_{an} = V_{dc}$$

$$V_{bn} = 0$$

$$V_{cn} = 0 \qquad \text{.....(2.70)}$$

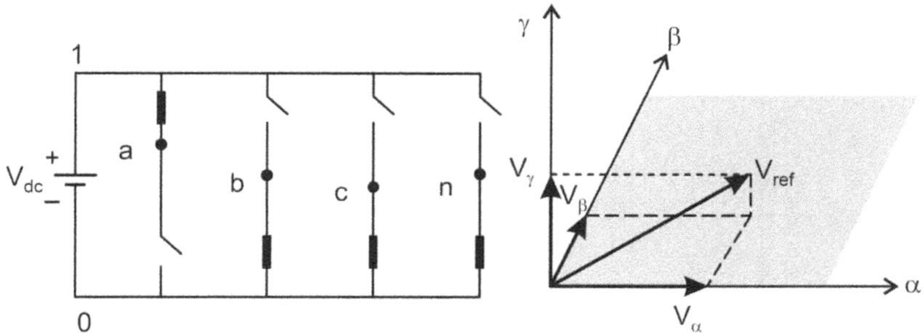

Fig. 2.34 (a) Topology 1 (1000) of a four-leg voltage source inverter. (b) Representation of topology 1 in α, β, ϒ space.

The transformed voltages in the α, β, ϒ space can be obtained using the transformation.

$$\begin{bmatrix} V_\alpha \\ V_\beta \\ V_\gamma \end{bmatrix} = [T]\begin{bmatrix} V_{an} \\ V_{bn} \\ V_{cn} \end{bmatrix}; \quad T = \frac{2}{3}\begin{bmatrix} 1 & -\frac{1}{2} & -\frac{1}{2} \\ 0 & \frac{\sqrt{3}}{2} & -\frac{\sqrt{3}}{2} \\ \frac{1}{2} & \frac{1}{2} & \frac{1}{2} \end{bmatrix} \qquad \text{.....(2.71)}$$

Thus, topology of Fig. 2.34 could be represented by the vector '1000' as shown in Fig. 2.34(b) and in Fig 2.35 (a). A similar procedure could be adopted for all the topologies of Fig. 2.33. This would generate sixteen vectors. These are known as switching state vectors (SSVs). These SSV's are shown in Fig. 2.35, in the space of transformed phase voltages $\{V_{an}, V_{bn}, V_{cn}\}$, where α, β-plane is the plane in which $V_{an} + V_{bn} + V_{cn} = 0$, and γ is the axis of the zero-sequence component. The tips of these vectors when projected onto the α, β-plane would form a regular hexagon as shown in Fig. 2.35(b).

2.4.2.3 Three-dimensional Space Vector Modulation

The required phase voltages at the output of the inverter could be represented by an equivalent vector, rotating in a 3-D vector space (unlike the three-leg VSI where the required output line voltages are represented by a vector rotating in the 2-D space). Space vector modulation for four-leg inverter, as in a three-leg inverter involves identification of adjacent switching state vectors, calculation of their duty-cycles and sequencing of these vectors. The first two steps are described in detail in [10,13] are reviewed here for the sake of completeness.

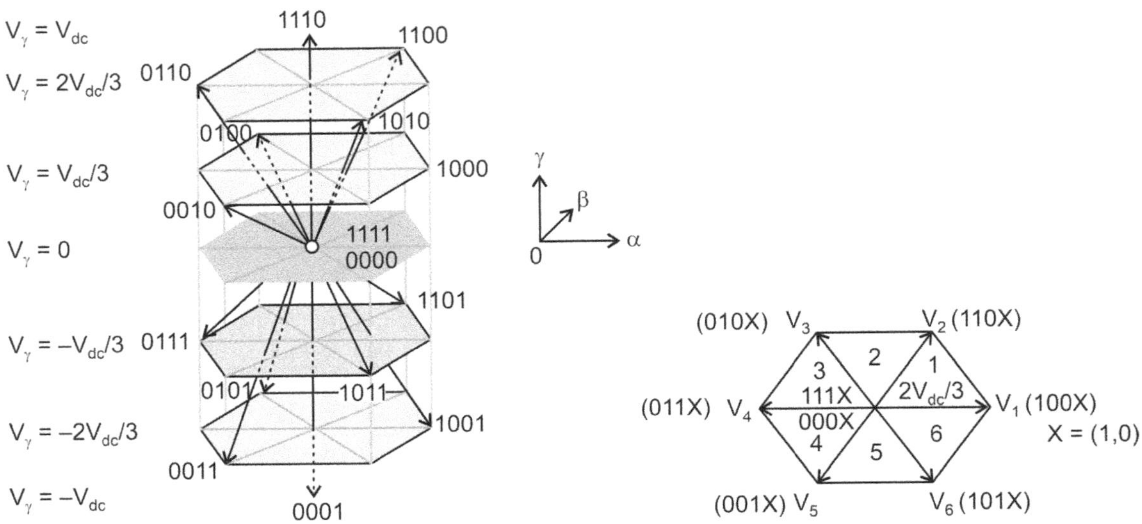

Fig. 2.35 (a) Switching state vectors of a four-leg inverter and (b) Projection of the sixteen vectors into α, β plane.

Identification of Adjacent Vectors

Identification of adjacent vectors is a two-step process. At first, the vector space of Fig. 2.35(a) is divided into six triangular prisms (P1-P6) (Fig. 2.39). At any given instant the required output voltage vector is present in one of these six prisms as can be seen in Fig. 2.36, where the reference vector happens to be in prism1. The flowchart used to determine the prism information at any given instant is shown in Fig. 2.37, where $V_{\alpha\ \beta,\ \gamma}$ are obtained from (Fig. 2.36), and is essentially the same as sector identification for the three-leg inverter.

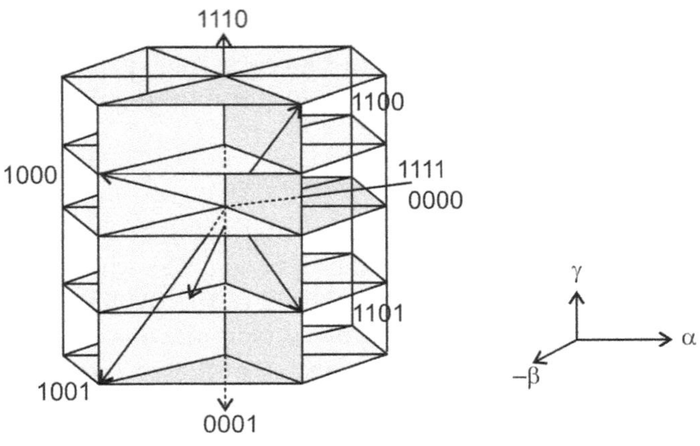

Fig. 2.36 Reference vector in prism1.

Once the prism information is determined the next step is to identify the tetrahedron in which the reference vector is present. To do this each prism is further subdivided into four tetrahedrons, thus leading to a total of 24 tetrahedrons. Each tetrahedron is formed by three non-zero switching state vectors [11,13] as shown in Fig. 2.38 for prism1(P1). Table 2.9 shows how these tetrahedrons could be identified based on the sign of required phase voltages at the output of the inverter. Thus, the three adjacent active vectors are determined.

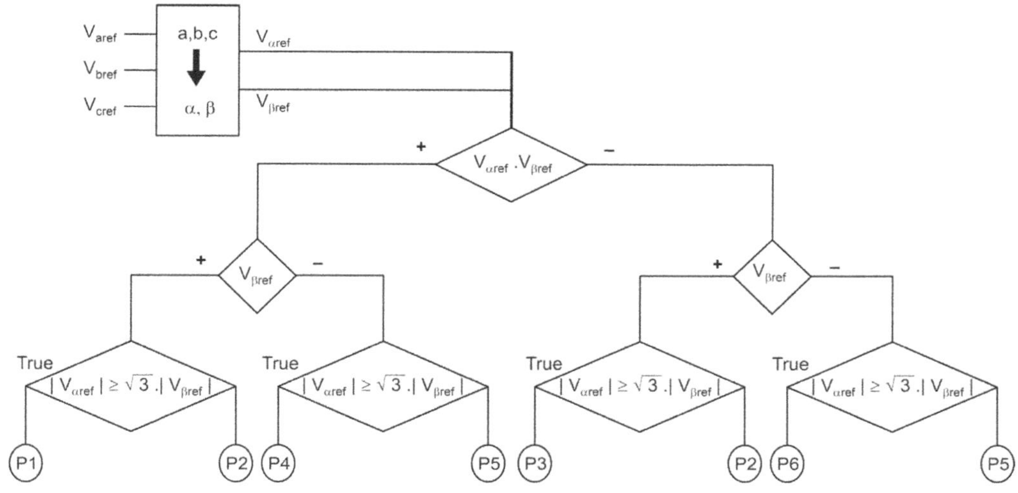

Fig. 2.37 Flowchart used to determine prism information.

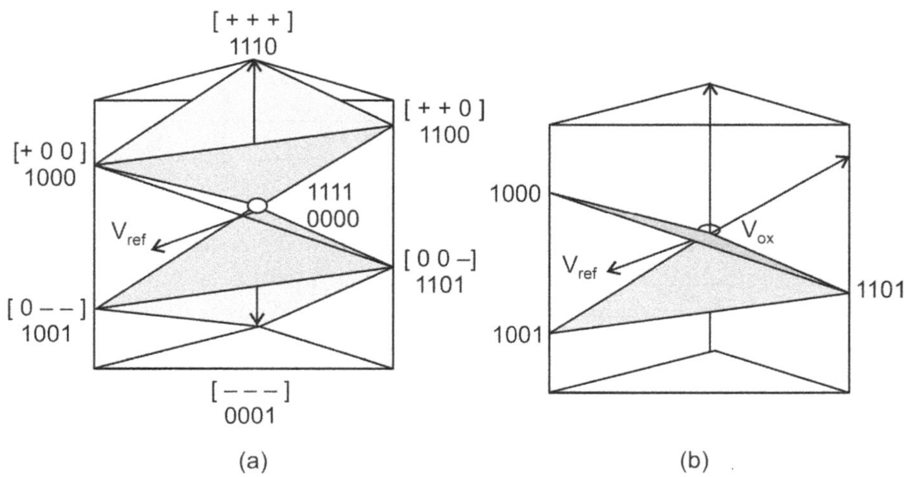

Fig. 2.38 (a) Prism (PI) consisting of four tetrahedrons (T1, T2, T13, T14) (b) Reference vector in tetrahedron-1 (T1).

Table 2.9 Identification of Tetrahedrons.

Prism	Tetrahedron	Adjacent Active Vectors			Conditions		
					V_a	V_b	V_c
P1	T-1	1101	1001	1000	$V_a > 0$	$V_b < 0$	$V_c < 0$
	T-2	1000	1100	1101	$V_a > 0$	$V_b > 0$	$V_c < 0$
	T-13	0001	1001	1101	$V_a < 0$	$V_b < 0$	$V_c < 0$
	T-14	1110	1100	1000	$V_a > 0$	$V_b > 0$	$V_c > 0$
P2	T-3	0100	1100	1101	$V_a > 0$	$V_b > 0$	$V_c < 0$
	T-4	1101	0101	0100	$V_a < 0$	$V_b > 0$	$V_c < 0$
	T-15	0001	0101	1101	$V_a < 0$	$V_b < 0$	$V_c < 0$
	T-16	1110	1100	0100	$V_a > 0$	$V_b > 0$	$Vc > 0$
P3	T-5	0111	0101	0100	$V_a < 0$	$V_b > 0$	$Vc < 0$
	T-6	0100	0110	0111	$V_a < 0$	$V_b > 0$	$Vc > 0$
	T-17	0001	0101	0111	$V_a < 0$	$V_b < 0$	$Vc < 0$
	T-18	1110	0100	0110	$V_a > 0$	$V_b > 0$	$V_c > 0$
P4	T-7	0010	0110	0111	$V_a < 0$	$V_b > 0$	$V_c > 0$
	T-8	0111	0011	0010	$V_a < 0$	$V_b < 0$	$V_c > 0$
	T-19	0001	0011	0111	$V_a < 0$	$V_b < 0$	$V_c < 0$
	T-20	1110	0110	0010	$V_a > 0$	$V_b > 0$	$V_c > 0$
P5	T-9	1011	0011	0010	$V_a < 0$	$V_b < 0$	$V_c > 0$
	T-10	0010	1010	1011	$V_a > 0$	$V_b < 0$	$V_c > 0$
	T-21	0001	1011	0011	$V_a < 0$	$V_b < 0$	$V_c < 0$
	T-22	1110	0010	1010	$V_a > 0$	$V_b > 0$	$V_c > 0$
P6	T-11	1000	1010	1011	$V_a > 0$	$V_b < 0$	$V_c > 0$
	T-12	1010	1010	1000	$V_a > 0$	$V_b < 0$	$V_c < 0$
	T-23	0001	1001	1011	$V_a < 0$	$V_b < 0$	$V_c < 0$
	T-24	1101	1000	1010	$V_a > 0$	$V_b > 0$	$V_c > 0$

Prism Identification: There are six triangular prisms obtained (triangular prism 1 to triangular prism 6) as shown in Fig. 2.39, the angle of each triangular prism section is equal to 60 degrees, each triangular prism contains eight switching vectors, two of them is zero switching vectors and other six non-zero switching vectors are distributed according to the value of $V_{\Upsilon\text{-ref}}$.

Tetrahedron Identification: After prism identification, it is then divided into four regions with four faces called tetrahedron. Each tetrahedron contains three NZSV and two ZSV as shown in Fig. 2.40. The selection of adjacent switching vectors in tetrahedron occurs according to the minimum circulating energy and non-conflicting line to neutral voltage. Each tetrahedron has a particular matrix, which means there are 24 matrices which cover all switching states for different operation condition. The four Tetrahedrons for prism 1 is presented in Fig. 2.40 for quick reference.

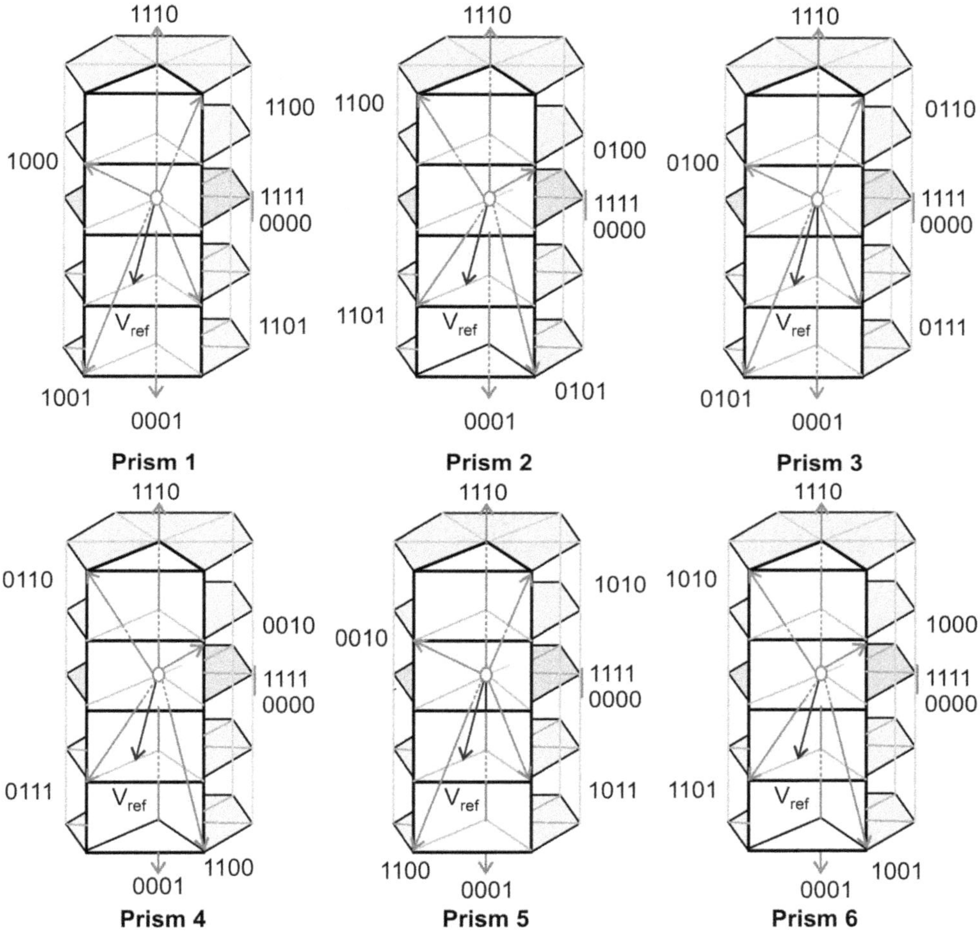

Fig. 2.39 Representation of six prism triangles with reference Vector.

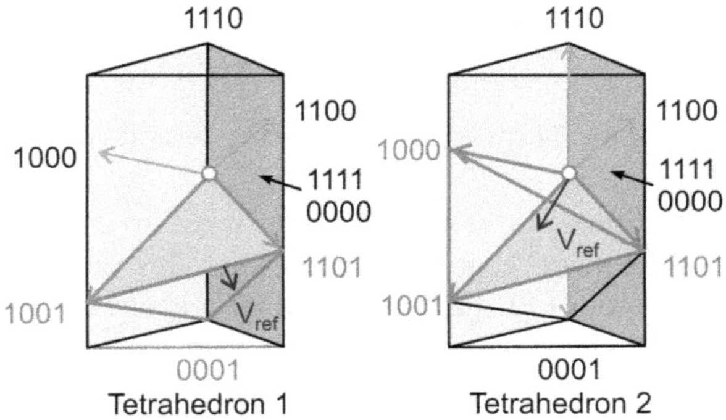

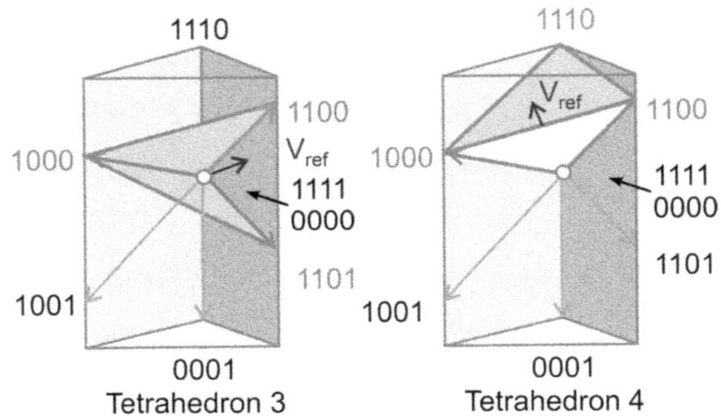

Fig. 2.40 Tetrahedron representation in prism 1 (T1, T2, T13, and T14).

Duty Cycle Calculation: Once the active vectors are determined, their duty-cycles are calculated based on projections. This is illustrated for the case when the reference vector is in tetrahedron-1 (T1) in Fig. 2.41, where the duty-cycles d1, d2, d3 for the active vectors '1101', '1001', '1000' are obtained based on projections and the duty-cycle for the zero vector is obtained from

$$d0 = 1 - d1 - d2 - d3 \quad .....(2.72)$$

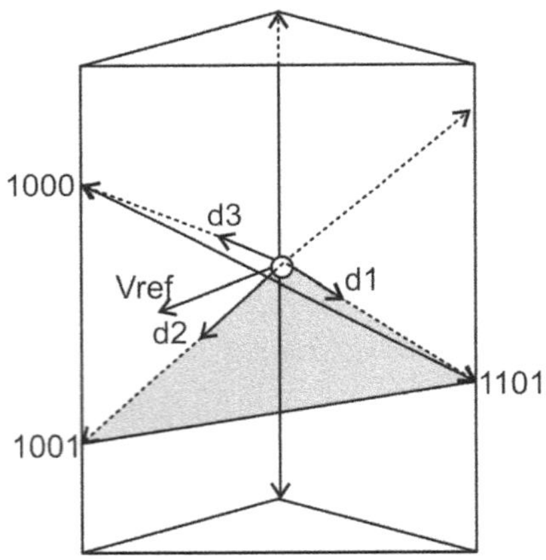

Fig. 2.41 Duty cycles for the active vectors in T1.

Under balanced load conditions the reference vector takes up a circular trajectory in the α, β plane with the Υ component being equal to zero. Under this condition the reference vector traverses only twelve tetrahedrons (T1-T12) and the time it spends in each tetrahedron is equal. This is indicated by the stair-case waveform in Fig. 2.42 (a) where the length of all the stairs is equal, indicating that the time spent by the reference vector in each tetrahedron is the same. Fig. 2.42 (b) shows the time spent by the reference vector in each prism. It can be noted that since the reference vector traverses through two tetrahedrons of a given prism consecutively; the sum of two consecutive stairs in Fig. 2.42 (a) would give rise to one stair in Fig. 2.42 (b).

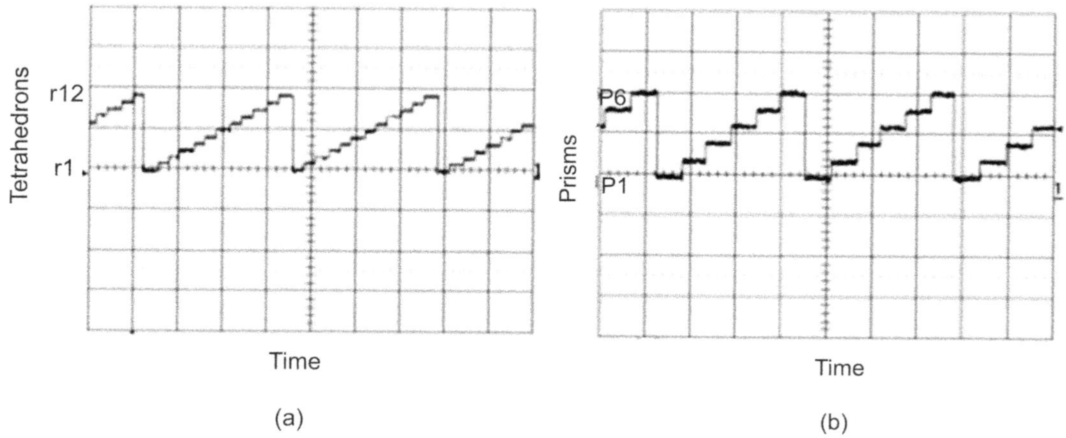

Fig. 2.42 Traversal of (a) tetrahedrons and (b) prisms for a balanced load.

Under small unbalance, the trajectory of the reference vector becomes elliptic, with a varying γ component. Under this condition the time spent by the reference vector in each of these twelve tetrahedrons is not a constant as can be seen from Fig. 2.43 (a). Fig. 2.43 (b) shows the traversal of prisms under this condition.

Under balanced or small unbalanced operating conditions, the reference vector remains in only twelve tetrahedrons(T1-T12). However, during transients or during extreme unbalance the reference vector could traverse tetrahedrons beyond twelve (T13- T24).

Sequencing of Vectors

Once the active vectors are determined and the duty cycles calculated, the degrees of freedom we have in the choice of a given modulation algorithm are

1. The choice of the zero vector - whether we would like to use 0000 or 1111 or both,
2. Sequencing of the vectors,
3. Splitting of the duty cycles of the vectors without introducing additional commutations.

Three such SVM algorithms are considered, namely:

1. The symmetric sequence - Scheme1
2. The highest current not switched sequence - Scheme2
3. The alternating zero vector sequence - Scheme3

The schemes analyzed here are similar to the ones found in literature for a three-leg inverter operating under balanced load conditions as described in the previous section. Scheme1 corresponding to SVM2, scheme2 corresponding to symmetric SVM4, and scheme3 corresponding to SVM3. The scheme corresponding to SVM1 is not analyzed here because it did not have any advantage under hard-switched conditions when compared to the other schemes.

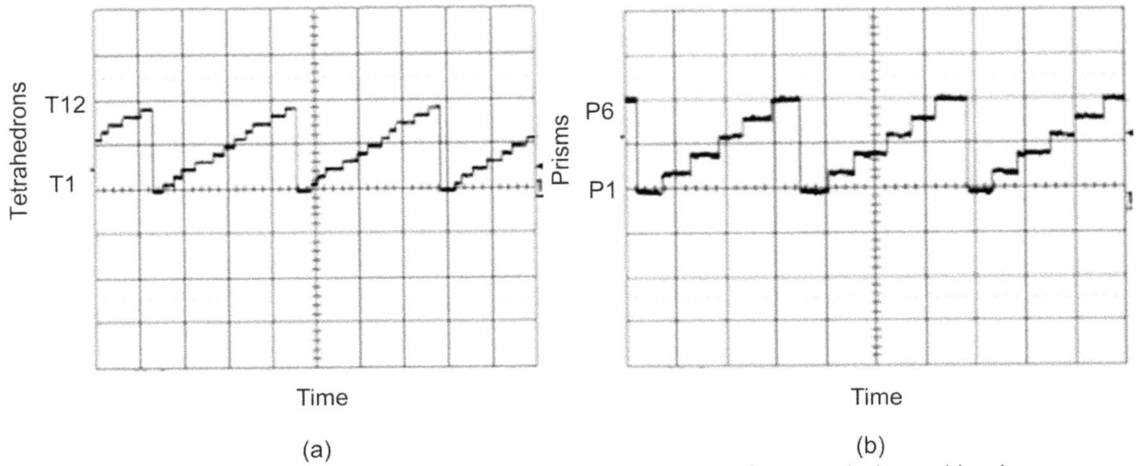

Fig. 2.43 Traversal of (a) tetrahedrons and (b) prisms for an unbalanced load.

2.4.2.4 Modulation Schemes

A. Symmetric sequence: Scheme 1: The sequence of vectors applied in this scheme is shown in Fig. 2.44 for the case when the reference vector is in tetrahedron 1. This scheme is expected to have low THD, as in three-leg voltage source inverter, because of the symmetry in the waveforms. The number of commutations in one sampling period is eight.

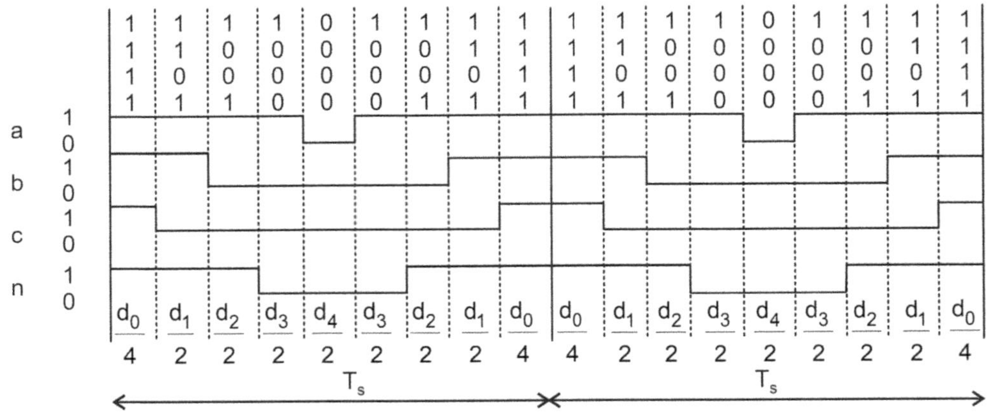

Fig. 2.44 Phase gating signals in scheme 1.

B. Highest Current Not-Switched Sequence: Scheme 2: This scheme is based on the fact that the switching losses are approximately proportional to the magnitude of the current being switched and hence it would be advantageous to avoid switching the inverter leg with the highest instantaneous current. This is possible in most cases, because all adjacent SSV's in a given tetrahedron differ in the state of switches in only one leg. Hence, by using only one zero vector, 0000 or 1111 within a given tetrahedron, one of the legs does not have to be switched at all, as shown in Fig. 2.45.

However, since the choice of the non-zero SSVs is made based on the desired output voltage vector and the phase and magnitude of the current is determined by the load, it is not always possible to avoid switching the phase carrying the highest current. In such a case the phase carrying the second highest current is not switched but the switching loss savings are reduced. The splitting of active vectors about the zero vector as shown below results in a significant reduction in the THD. The number of commutations in one cycle is six.

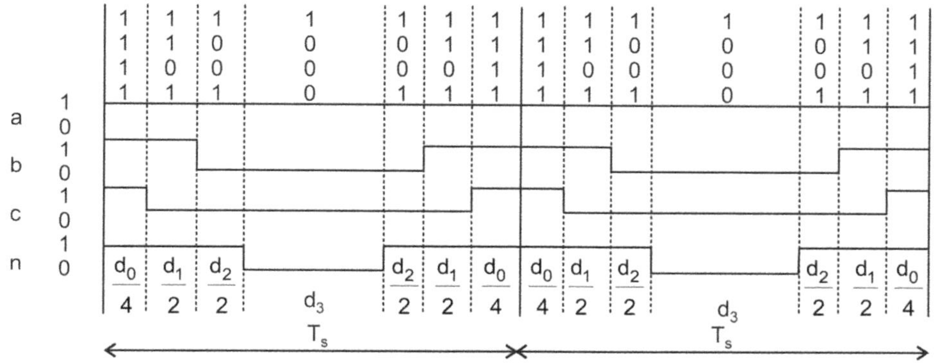

Fig. 2.45 Phase gating signals in scheme 2.

C. Alternating Zero Vector Sequence: Scheme 3: In this scheme the zero vectors (1111, 0000) are used alternatively in adjacent switching cycles so that the effective switching frequency is halved, as shown in Fig. 2.46. However, the sampling period is still Ts, same as in the other schemes. The number of commutations in one Ts is four and hence the switching losses are expected to reduce. Current THD is significantly higher due to the existence of the harmonics at half the sampling frequency.

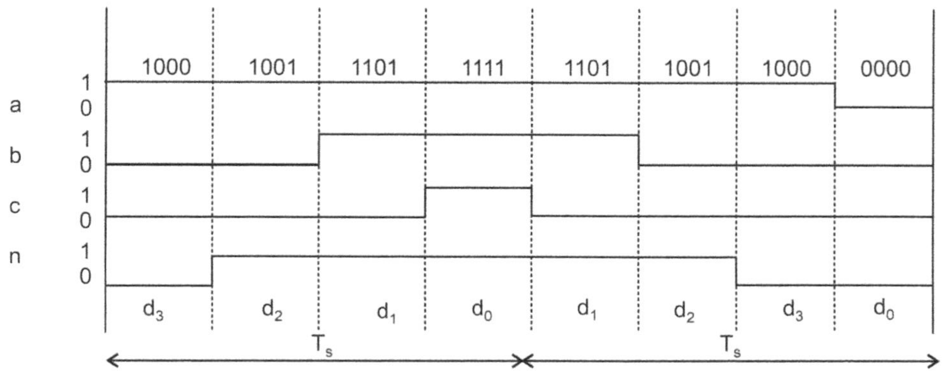

Fig. 2.46 Phase gating signals in scheme 3.

Table 2.10 Relative performance of various schemes (four-leg).

Modulation Schemes	Scheme-1 Symmetric	Scheme-2. Highest Current not Switched	Scheme-3. Alternating Zero Vector
No. of commutations in Ts	8	6	4
No. of switching states in Ts	9	7	4
Dominant Harmonic	fs	fs	fs/2

Table 2.11 Comparison of 2-D SVM and 3-D SVM Techniques.

Parameters	2-D (α-β) SVM	3-D (α-β-Υ) SVM
Circuit configuration:	Three -leg six switches	Four-leg eight switches
Type of Loads:	Balanced	Balanced, Unbalanced, Transient Conditions
Switching States: Switching Vectors:	$2^3 = 8$ NZSV = 6; ZSV = 2	$2^4 = 16$ NZSV = 14 ; ZSV =2
Transformation required:	*a,b,c* - to – α,β	*a,b,c* - to – α,β,Υ
Sectors/Prisms:	Six-Sectors	Six-Prisms
Angle of each sector/Prism	60 degrees (each Sector)	60 degrees (each Prism section)
Sequence of Steps:	**Seven-Steps:** • Computation of the reference voltage space vector V_{ref} from the equivalent 3 phase – 2 phase conversion. • Synthesis of voltage space vectors from the permissible switching combinations of the three-level converter. • Hexagon representation of switching state vectors. • Identification of sectors in which the reference space vector is positioned. • Determination of region in selected sector where the reference vector is located. • 'ON' time calculation of switching vectors in each region of sector using optimal switching sequence. • Synthesis of reference space vector and calculation of total ON time of each switch using optimal switching pattern	**Four-Steps:** • Selection of adequate switching vector sequence pattern. • The transformation of the reference vector (V_{ref}) from V_{abc} coordinate to $V_{\alpha\beta\Upsilon}$ coordinate. • Determination of prism and tetrahedron. • Determine duty ratios of all four legs.

Advantages and applications

The four-leg VSI demands a smaller DC link capacitance and the use of the 3-D SVPWM technique allows a higher utilization of the DC link voltage; these features make them an interesting alternative for high power applications and devices. Another advantage of 3-Dimensional SVPWM is that it solves the harmonic contents problem in spite of solving the unbalance in 3-phase load. This technique is compatible with hybrid power networks and it will be the most dominant in the future.

2.5 Carrier-based Space Vector Pulse Width Modulation

The conventional SVPWM technique gives good performance but however increases the complexity in calculating angle and sector. The complexity involved in SVPWM can be reduced by using the concept of effective time which is known as Unified voltage modulation or carrier-based space vector pulse width modulation (CBSVPWM). The effective times directly synthesizes the inverter output voltage, simplifying the modulation task. To achieve SVPWM like performance a fixed common mode voltage is added to the reference voltage taking modulation index into account.

From the Fig 2.47 the effective voltage is applied to the load side, when each phase switching states changes from 0 to 1 at different time during one sampling interval. Let T_s denotes the total sampling time and Teff is the effective time which is the difference between two switching time T_{as}, and T_{bs}. When Tbs is greater than T_{as}, then T_{eff} has negative value and a negative voltage is applied to load side. For one leg operation the relation between voltage and times is given as

$$V_{as} : V_{dc} = T_{as} : T_s \qquad .....(2.73)$$

From the above relation

$$T_{as} = \frac{T_s}{V_{dc}} * V_{as} \qquad .....(2.74)$$

where Vas is the equivalent phase-A voltage for the load side. The effective time can be obtained from the different values of each imaginary switching time.

$$T_{eff} = T_{as} - T_{bs} = \frac{T_s}{V_{dc}}(V_{as} - V_{bs}) \qquad .....(2.75)$$

The average values of the three-phase leg voltages over an interval $\Delta T/2$, in terms of the midpoint of the DC link is given as,

$$V_{as} = V_{dc}\left(\frac{T_{sv1}}{\Delta T / 2} + \frac{T_{sv2}}{\Delta T / 2}\right) \qquad .....(2.76)$$

$$V_{bs} = V_{dc}\left(-\frac{T_{sv1}}{\Delta T / 2} + \frac{T_{sv2}}{\Delta T / 2}\right) \qquad .....(2.77)$$

$$V_{cs} = V_{dc}\left(-\frac{T_{sv1}}{\Delta T / 2} + \frac{T_{sv2}}{\Delta T / 2}\right) \qquad .....(2.78)$$

The maximum value of V_0 is expressed as $V_0 = MV_{dc}$ where $0 \le M \le 2/\sqrt{3}$. Therefore, the line-to-DC midpoint voltages can be written as

$$V_{as} = \frac{\sqrt{3}}{2} MV_{dc}\left[\cos\left(\theta_0 - \frac{\pi}{6}\right)\right] \qquad .....(2.79)$$

$$V_{bs} = \frac{\sqrt{3}}{2} MV_{dc} \left[\cos\left(\theta_0 - \frac{2\pi}{3} \right) \right] \quad .....(2.80)$$

$$V_{cs} = \frac{\sqrt{3}}{2} MV_{dc} \left[\cos\left(\theta_0 - \frac{5\pi}{6} \right) \right] \quad .....(2.81)$$

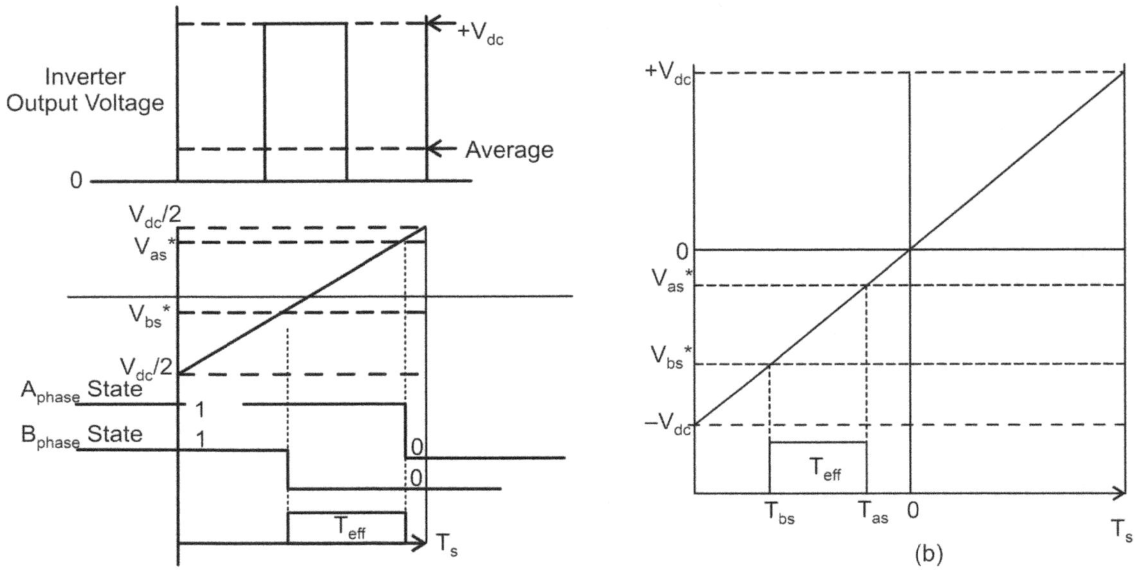

Fig. 2.47 (a) Relationship between the effective time and the output voltage, (b) Extended concept for effective time.

For the period $0 \leq \theta \leq \pi/3$

The common offset voltage is given as

$$V_{offset} = \frac{\max\left(V_{as}^*, V_{bs}^*, V_{cs}^*\right) + \min\left(V_{as}^*, V_{bs}^*, V_{cs}^*\right)}{2} \quad .....(2.82)$$

An additional common mode voltage which correctly positions the first and last switching transitions in each switching period can be given as

$$V'_{offset} = \frac{V_{dc}}{N-1} \frac{\max\left(V_{as}^*, V_{bs}^*, V_{cs}^*\right) + \min\left(V_{as}^*, V_{bs}^*, V_{cs}^*\right)}{2} \quad .....(2.83)$$

Now, the effective time can be defined as

$$T_{eff} = T_{max} - T_{min} \quad .....(2.84)$$

Where $T_{min} = \min(T_{as}, T_{bs}, T_{cs})$ .....(2.85)

$T_{max} = \max(T_{as}, T_{bs}, T_{cs})$ .....(2.86)

The actual gating times (T_{ga}, T_{gb}, T_{gc}) for each inverter arm can be obtained by adding the imaginary times as follows:

$$T_{ga} = T_{as} + T_{offset} \qquad .....(2.87)$$

$$T_{gb} = T_{bs} + T_{offset} \qquad .....(2.88)$$

$$T_{gc} = T_{cs} + T_{offset} \qquad .....(2.89)$$

where T_{offset} is the 'offset time'. The "OFF sequence" and the "ON sequence" are shown in Fig 2.48. The actual switching time will be replaced by the subtraction value, with sampling time as follows:

$$T_{ga} = T_s - T_{ga} \qquad .....(2.90)$$

$$T_{gb} = T_s - T_{gb} \qquad .....(2.91)$$

$$T_{gc} = T_s - T_{gc} \qquad .....(2.92)$$

Therefore, from this effective time relocation algorithm, full active voltage can be applied to the load like the space voltage vector PWM. If V* is the reference voltage vector of SVPWM, then this can be represented by the two nearest active vectors as

$$V^* = \frac{T_1, V_n + T_2, V_{n+1}}{T_s} \qquad .....(2.93)$$

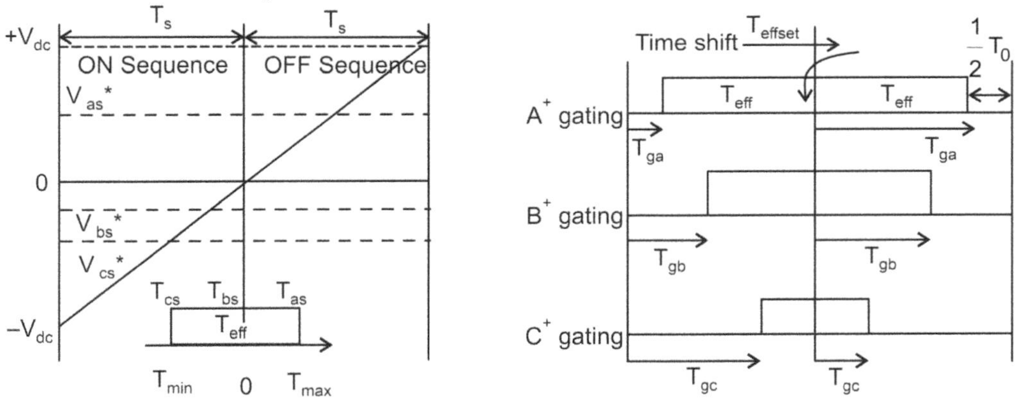

Fig. 2.48 (a) Effective Time Calculation (b) Gating Signals.

The values of T_1, T_2 are obtained by using reference voltage V_{ds}^*, V_{qs}^* of the stationary reference frame d-q axis as

$$T_1 = \frac{|V^*|.T_s}{V_{dc},\frac{2}{s}} \frac{\sin\left(\frac{\pi}{s} - \alpha\right)}{\sin\left(\frac{\pi}{s}\right)} = \frac{T_s.\sqrt{3}}{V_{dc}}\left[\sin\left(\frac{\pi}{3}m\right).V_{ds}^* - \cos\left(\frac{\pi}{3}.m\right).V_{qs}^*\right] \qquad .....(2.94)$$

$$T_2 = \frac{|V^*|.T_s}{V_{dc},\frac{2}{s}} \frac{\sin\left(\frac{\pi}{s} - \alpha\right)}{\sin\left(\frac{\pi}{s}\right)}$$

$$= \frac{T_s.\sqrt{3}}{V_{dc}}\left[-\sin\left(\frac{\pi}{3}(m-1)\right).V_{ds}^* - \cos\left(\frac{\pi}{3}.(m-1)\right).V_{qs}^*\right] \quad .....(2.95)$$

$$T_0 = T_s - T_1 - T_2 \quad .....(2.96)$$

Where $$V_{ds}^* = |V^*|.\cos(\theta),\ V_{qs}^* = |V^*|.\sin(\theta) \quad .....(2.97)$$

$$\theta = \alpha + \frac{\pi}{3}(m-1) \quad .....(2.98)$$

m = sector number (1, 2, 3........6)

$$T_1 = \sqrt{3}.T_s / V_{dc}\left[\frac{\sqrt{3}}{2}.V_{ds}^* - 1/2.V_{qs}^*\right]$$

$$= \frac{T_s}{V_{dc}}.V_{as}^* - \frac{T_s}{V_{dc}}.V_{bs}^* = T_{as} - T_{bs} \quad .....(2.99)$$

$$T_2 = \sqrt{3}.T_s / V_{dc}\left[0.V_{ds}^* + 1.V_{qs}^*\right]$$

$$= \frac{T_s}{V_{dc}}.V_{bs}^* - \frac{T_s}{V_{dc}}.V_{cs}^* = T_{bs} - T_{cs} \quad .....(2.100)$$

$$V_{as}^* - V_{ds}^*, V_{bs}^*$$

$$= -\frac{1}{2}V_{ds}^* + \frac{\sqrt{3}}{2}V_{qs}^*, \quad .....(2.101)$$

$$V_{cs}^* = -\frac{1}{2}V_{ds}^* - \frac{\sqrt{3}}{2}V_{qs}^*, \quad .....(2.102)$$

To relocate the effective times at the centre of the sampling interval, the time shifting value T_{offset} is

$$T_{offset} = \frac{1}{2}T_0 - T_{min} \quad .....(2.103)$$

$$T_{eff} = T_{max} - T_{min} \quad .....(2.104)$$

$$T_0 = T_s - T_{eff} \quad .....(2.105)$$

Finally, the generation of reference waveform in MATLAB/SIMULINK using CBSVPWM technique is shown in the Fig 2.49. It is a very simple method without having a need of calculating angle and sector as in case of conventional SVPWM.

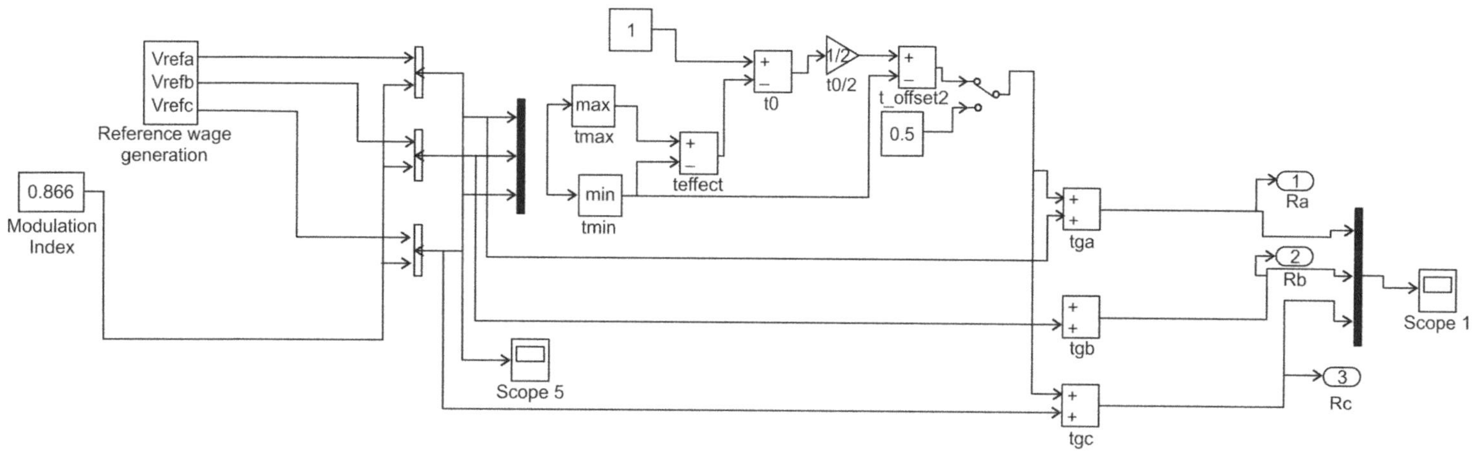

Fig. 2.49 Reference wave generation using CBSVPWM.

The resultant reference wave is compared with the carrier waves to generate the pulses. The reference wave comparing with the carrier waves along with the pulse generation is shown in the Fig 2.50.

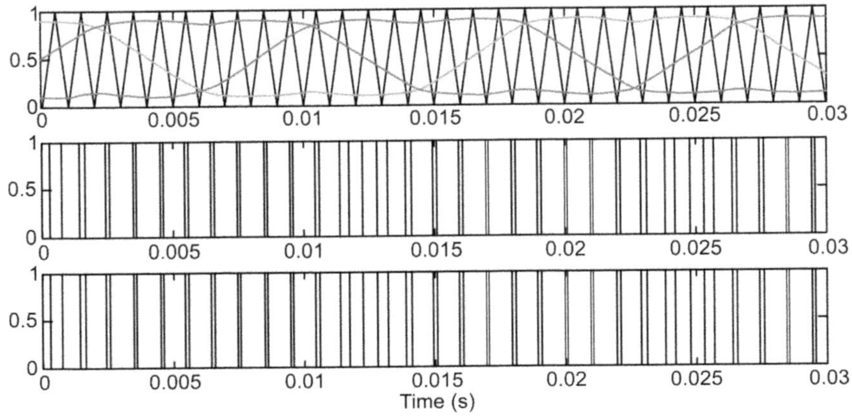

Fig. 2.50 Reference wave comparing with the carrier wave to generate pulses using CBSVPWM technique.

Example 2.5

A Three phase three-level diode-clamped inverter topology is considered to control the output voltages to the three phase resistive-inductive (RL) loads when different modulation techniques are applied, they are (i) Sinusoidal PWM (SPWM), (ii) Space-Vector PWM (SVPWM), and (iii) Carrier-based Space-Vector PWM (CBSVPWM), techniques. The simulated output phase voltage is presented here for the above three modulation schemes.

Solution: Switching sequences for three level diode-clamped inverter

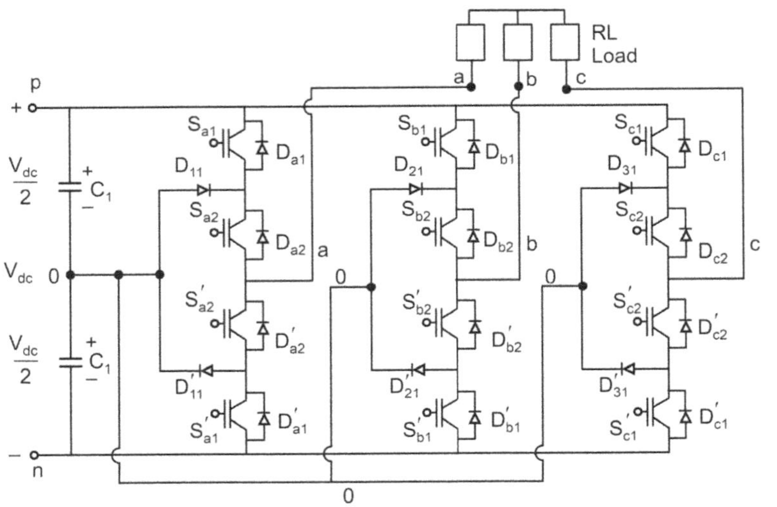

Fig. E2.8 Three phase Three-level diode clamped inverter fed RL loads.

Output Voltage	Switching Sequence			
	S_{a1}	S_{a2}	S_{a1}'	S_{a2}'
0	0	1	1	0
$V_{dc}/2$	1	1	0	0
$-V_{dc}/2$	0	0	1	1

The simulation has been carried out to the above circuit topology with SPWM, SVPWM, and CBSVPWM schemes using Simulink/MATLAB platform and obtained the reference signal waveforms, triangular carrier signal waveforms with respective reference waveforms, and one-phase output voltage waveforms and phase currents of the loads are presented in the following figures E2.9., E2.10, E2.11 and E2.12 respectively.

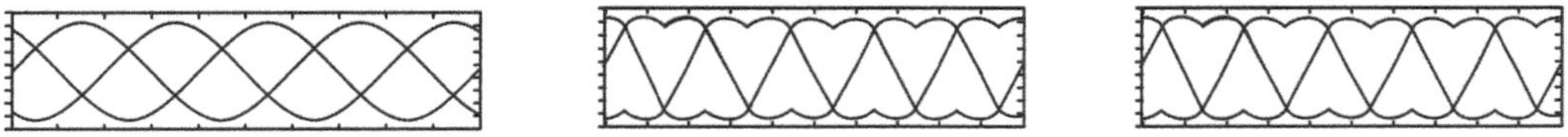

(a) (b) (c)

Fig. E2.9 Reference signal waveforms of (a) SPWM, (b) SVPWM, (c) CBSVPWM

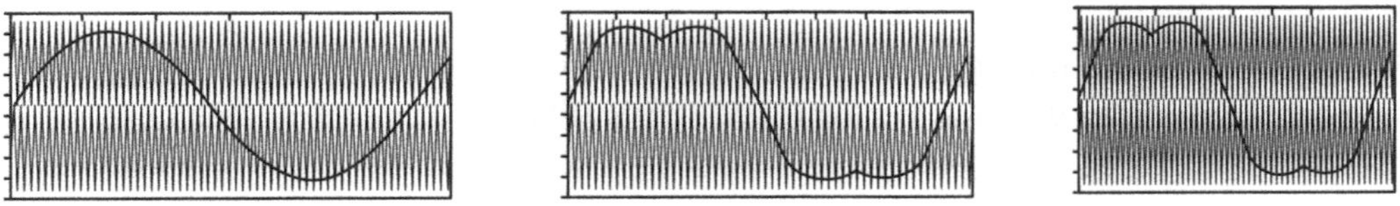

(a) (b) (c)

Fig. E2.10 Reference and triangular Carrier signal waveforms of (a) SPWM, (b) SVPWM, (c) CBSVPWM.

(a) (b) (c)

Fig. E2.11 Output phase voltage waveform of Three-level diode clamped inverter using (a) SPWM, (b) SVPWM, (c) CBSVPWM.

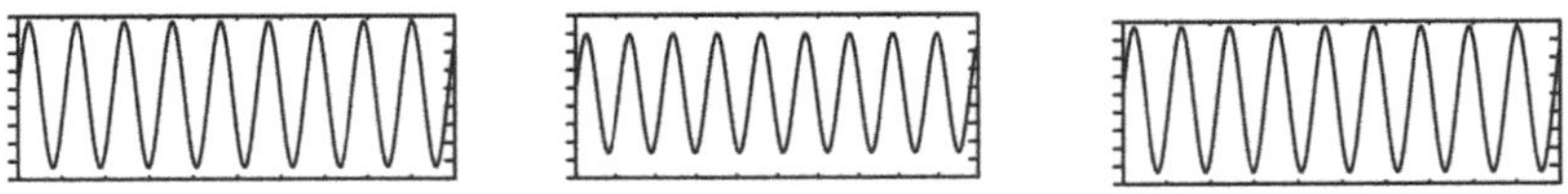

(a) (b) (c)

Fig. E2.12 Output phase current waveform of Three-level diode clamped inverter for RL load using (a) SPWM, (b) SVPWM, (c) CBSVPWM.

2.6 Summary

In this chapter various modulation techniques suitable for multilevel inverters are classified and explained with examples. The reader can easily understand and apply appropriate modulation technique for various multilevel inverters.

They are classified into three broad categories, namely, PWM, CBPWM and SVPWM techniques. PWM is further classified as regular Sinusoidal PWM (SPWM), Third-Harmonic Injected PWM (THIPWM), and Selected Harmonic Elimination PWM (SHEPWM) schemes and explained in detail with examples for applications to multilevel inverters.

The Multicarrier based PWM technique is classified into two categories, one is Phase-shifted carried based PWM (PS-CBPWM), another is Level-shifted carrier based PWM (LS-CBPWM). LS-CBPWM is further classified into three varieties, (a) Phase Disposition (PD) CBPWM, (b) Phase Opposition Disposition (POD) CBPWM, and (c) Alternative Phase Opposition Disposition (APOD) CBPWM schemes are explained in detail with examples by applying to multilevel inverters.

Third variety is Space Vector PWM (SVPWM) is classified into two categories, one is two-dimensional algorithm for space vector PWM and another one is three-dimensional algorithm for space vector PWM schemes are discussed with detailed procedures and illustrated an example at the end.

Finally, a combination of CBPWM and SVPWM with two-dimensional algorithm is also discussed as Carrier Based Space Vector PWM (CBSVPWM) technique and an example is also presented.

This chapter of various PWM schemes are very much essential and useful knowledge at one place and can be applied often to various topologies and categories of multilevel inverters which are dealt with in subsequent chapters.

Learning Outcomes

- To be able to acquire knowledge about various pulse width modulation techniques. (CO1) :: (PO1)
- To acquire ability to analyze various schemes of pulse width modulation for generations of control signals to switching devices of multilevel inverters. (CO4) :: (PO2)
- To be able to apply suitable pulse width modulation techniques for multilevel inverters. (CO3):: (PO1)
- To be able to develop the simulation model using software tools for generating pulses for multilevel inverters. (CO4) :: (PO5)
- To be able to evaluate the performance of different pulse width modulation techniques for multilevel inverters by simulation. (PWM, THI-PWM, SHE-PWM, CB-SPWM, PD, POD, APOD, SVM, CBSVPWM). (CO5) :: (PO3)
- To be able to extend the various PWM schemes concept to three-phase multilevel inverter and investigate the performance of Pulse width modulation and its suitability for three-phase multilevel inverters. (CO6) :: (PO4)

Questions

2.1 What is the importance of Pulse width modulation with respect to dc-ac power inverters?

2.2 Classify various pulse width modulation techniques suitable for multilevel inverter?

2.3 Explain the generation of control signals with sinusoidal pulse width modulation scheme?

2.4 What is the importance of THI-PWM scheme? Discuss the implementation of THI-PWM for single phase two-level inverter?

2.5 Describe the concept of selective harmonic elimination PWM technique, its advantages, for three-level inverter?

2.6 Name various methods and algorithms suitable for solving non-linear equations with respect to Selective harmonic elimination PWM technique?

2.7 Give merits, demerits and applications of Selective harmonic elimination PWM technique.

2.8 Discuss the phase-shifted carrier-based PWM technique suitable for multilevel inverters?

2.9 Classify level-shifted carrier-based PWM technique? Explain each of them elaborately for generation of control signals for three-level inverters?

2.10 Compare the three methods of level-shifted carrier-based PWM technique and bring out their features in case of five-level inverter?

2.11 What is the concept of Space vector modulation? How it is different from sinusoidal PWM technique?

2.12 Discuss the step-by-step procedure for implementation of space vector modulation for two-level inverter.

2.13 Describe the step-by-step procedure for implementation of space vector modulation for three-level inverter.

2.14 What is the importance of two-dimensional (2-D) space vector and three-Dimensional (3-D) space vector modulation techniques? And give their applications?

2.15 Explain the importance of carrier-based space-vector pulse width modulation (CBSVPWM) scheme employed for multilevel inverter?

Multiple Choice Questions

2.1 Why are the pulse width modulated outputs required in most of the applications?

(a) to control average value of input variables

(b) to control average value of output variables

(c) both (a) & (b)

(d) none of the above

2.2 What type of carrier waveform is used in pulse width modulation schemes?

(a) Square wave form (b) Sine wave form

(c) Triangular wave form (d) Rectangular wave form

2.3 Pulse width modulation controls the output voltage by producing pulses of
   (a) constant magnitude and variable duty cycle
   (b) variable magnitude and variable duty cycle
   (c) constant magnitude and constant duty cycle
   (d) variable magnitude and constant duty cycle

2.4 For multilevel inverter, the reference sine wave is compared with number of level-shifted triangular waves for pulse generation and for m-level inverter, the reference is compared with
   (a) m-carrier signals (b) (m-2) carrier signals
   (c) (m-1) carrier signals (d) 2(m–1) carrier signals

2.5 Maximum output voltage of inverter with THIPWM is
   (a) decreased (b) increased
   (c) constant (d) none of the above

2.6 In Selective harmonic elimination PWM technique, which of the output wave is "chopped" a number of times?
   (a) triangular wave (b) sinusoidal wave
   (c) saw tooth wave (d) square wave

2.7 The carrier-based technology can be easily extended to higher inverter levels and can be applied to
   (a) sinusoidal PWM modulation scheme
   (b) space vector modulation scheme
   (c) sinusoidal and space vector modulation schemes
   (d) none of the above

2.8 In phase-shifted carrier based PWM technique, the carrier waves have
   (a) same equal amplitude, frequency and are phase shifted by 90° from one another
   (b) unequal amplitude, frequency and are phase shifted by 90° from one another
   (c) same equal amplitude, frequency and are phase shifted by 60° from one another
   (d) unequal amplitude, frequency and are phase shifted by 60° from one another

2.9 Level shifted PWM technique is derived from the carrier disposition strategies, namely,
   (a) Phase Opposition Disposition
   (b) Alternative Phase Opposition Disposition
   (c) In Phase Disposition
   (d) All of the above

2.10 The space vector modulation scheme for 2-D algorithm is represented by 1____________ and for 3-D algorithm is represented by 2__________
   (a) 1. Six vectors and 2. Six sectors
   (b) 1. Six sectors and 2. Six prisms
   (c) 1. Six prisms and 2. Six sectors
   (d) 1. Six prisms and 2. Six sectors

2.11 CBSVPWM is combination of _______ and ________ in two-dimensional algorithm

(a) Sinusoidal PWM and Carrier based PWM

(b) Carrier based PWM and SHE-PWM

(c) Carrier based PWM and Space vector PWM

(d) THI-PWM and SHE-PWM

Exercise Problems

Ex.2.1: Simulate the single-phase two-level inverter with RL loads (R= 10 ohms and L= 10 mH) by considering Sinusoidal PWM technique and obtain the triangular carrier signal, sinusoidal reference signal waveforms, firing signals for switching devices and output load voltage waveform. Consider the circuit topology shown in Fig. Ex. 2.1 for two-level inverter.

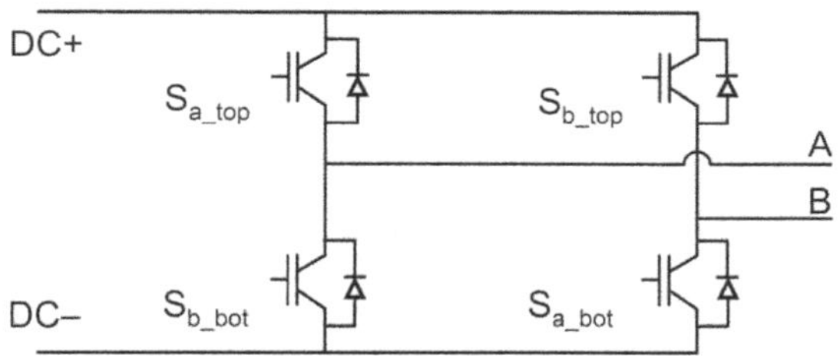

Fig. Ex.2.1 Single-phase two-level voltage source inverter.

Ex.2.2: Evaluate the single-phase two-level inverter with RL loads (R= 10 ohms and L= 10 mH) by using Third Harmonic Injected PWM and obtain the triangular carrier signal, third harmonic injected resultant reference signal waveforms, firing signals for switching devices, and output load voltage by simulation. Consider the circuit topology shown in Fig.Ex.2.1 for two-level inverter.

Ex.2.3: Simulate the single-phase two-level inverter with RL loads (R= 10 ohms and L= 10 mH) by considering SHE-PWM and obtain the triangular carrier signal, sinusoidal reference signal waveforms, firing signals for switching devices, and output load voltage waveform.

Ex.2.4: Evaluate the single-phase three-level inverter with RL loads (R= 10 ohms and L= 10 mH) by considering Phase Shifted-PWM (PS-PWM) and obtain the triangular carrier signals, sinusoidal reference signal waveforms, firing signals for switching devices, and output load voltage by simulation. Consider the circuit topology shown in Fig.Ex.2.2 for three-level inverter.

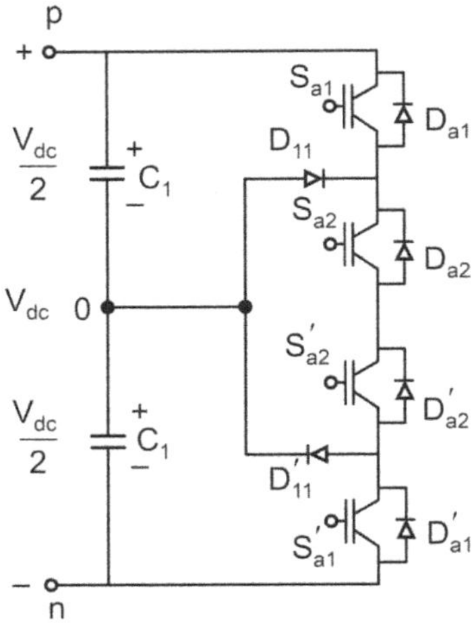

Fig. Ex.2.2 Single-phase three - level Voltage source inverter.

Ex.2.5: Simulate the single-phase three-level inverter with RL loads (R = 10 ohms and L = 10 mH) by considering Phase Disposition scheme of Level Shifted-PWM (LS-PWM) and obtain the triangular carrier signals, sinusoidal reference signal waveforms, firing signals for switching devices, output load voltage. Consider the circuit topology shown in Fig.Ex.2.2 for three-level inverter.

Ex.2.6: Evaluate the single-phase three-level inverter with RL loads (R = 10 ohms and L = 10 mH) by considering Phase Opposition Disposition scheme of Level Shifted-PWM (LS-PWM) and obtain the triangular carrier signals, sinusoidal reference signal waveforms, firing signals for switching devices, output load voltage. Consider the circuit topology shown in Fig.Ex.2.2 for three-level inverter by simulation.

Ex.2.7: Simulate the single-phase three-level inverter with RL loads (R = 10 ohms and L= 10 mH) by considering Alternative Phase Opposition Disposition scheme of Level Shifted-PWM (LS-PWM) and obtain the triangular carrier signals, sinusoidal reference signal waveforms, firing signals for switching devices, output load voltage. Consider the circuit topology shown in Fig.Ex.2.2 for three-level inverter.

Ex.2.8: Evaluate the single-phase two-level inverter with RL loads (R= 10 ohms and L= 10 mH) by using 2-Dimentional Space Vector PWM (SV-PWM) and obtain the firing signals for switching devices, and output load voltage. Consider the circuit topology shown in Fig.Ex.2.1 for two-level inverter by simulation.

Ex.2.9: Simulate the single-phase three-level inverter with RL loads (R= 10 ohms and L= 10 mH) by Carrier-based Space-Vector PWM (CBSV-PWM) and obtain the triangular carrier signal, sinusoidal reference signal waveforms, firing signals for switching devices, output load voltage. Consider the circuit topology shown in Fig.Ex.2.2 for three-level inverter.

Answers to Multiple Choice Questions

2.1	(b)	2.2	(c)	2.3	(a)	2.4	(c)	2.5	(b)
2.6	(d)	2.7	(c)	2.8	(a)	2.9	(d)	2.10	(b)
2.11	(c)								

CHAPTER

3 Diode-Clamped Multilevel Inverters

LEARNING OBJECTIVES

- ✓ *To understand and acquire knowledge about construction of diode-clamped multilevel inverters.*
- ✓ *To prepare learners to analyze the operation of three-level, five-level, seven-level, and m-level diode-clamped inverters.*
- ✓ *To prepare learners to practice through case studies on diode-clamped multilevel inverters.*
- ✓ *To prepare learners to simulate the performance of three-level, five-level, seven-level, and m-level diode-clamped inverters with suitable pulse width modulation techniques.*
- ✓ *To understand the applications of various levels of diode-clamped inverters.*

3.1 Construction of Single-phase Three-level Diode-clamped Inverter

The multilevel converters can immediately replace the existing systems that use traditional multi-pulse converters without the need for transformers. For a 3-phase system, the relationship between the number of levels, m, and the number of pulses, p, can be formulated by $p = (m - 1) \times 6$. The multilevel inverters are being used in the medium to high voltage range of industrial applications.

A single-phase three-level voltage source inverter equivalent circuit is drawn with single-pole triple throw switch for RL load for single-leg arrangement with one switch shown in Fig.3.1(a) and with single-phase full-bridge circuit with two switches shown in Fig.3.1(b). The operation of the circuits is explained as follows:

(a) **For single-phase with one single-pole triple throw switch for RL load:** If switch S_a is connected to terminal 1 then $V_{an}= +V_{dc}$, and when it is connected to mid-point terminal 0, then $V_{an}= +V_{dc}/2$, in both cases the load current is positive flowing from load terminal 'a' to terminal 'n'. If switch S_a is connected to terminal 2, then $V_{an}=0$. Thus three-level positive voltage is produced as V_{an} and shown in the Fig.3.1(a).

(b) **For single-phase with two single-pole triple throw switches for RL Load:** If switch S_a is connected to terminal 1 and switch S_b is connected to terminal 4 then $V_{ab} = +V_{dc}$ and current will be positive flowing from load terminal 'a' to terminal 'b' of RL load. When switch S_a kept connected to terminal 1 and Switch S_b is connected to terminal 3 then $V_{ab} = 0$. If switch S_a remain connected to terminal 1 and switch S_b is connected to terminal 0, then $V_{ab} = +V_{dc}/2$.

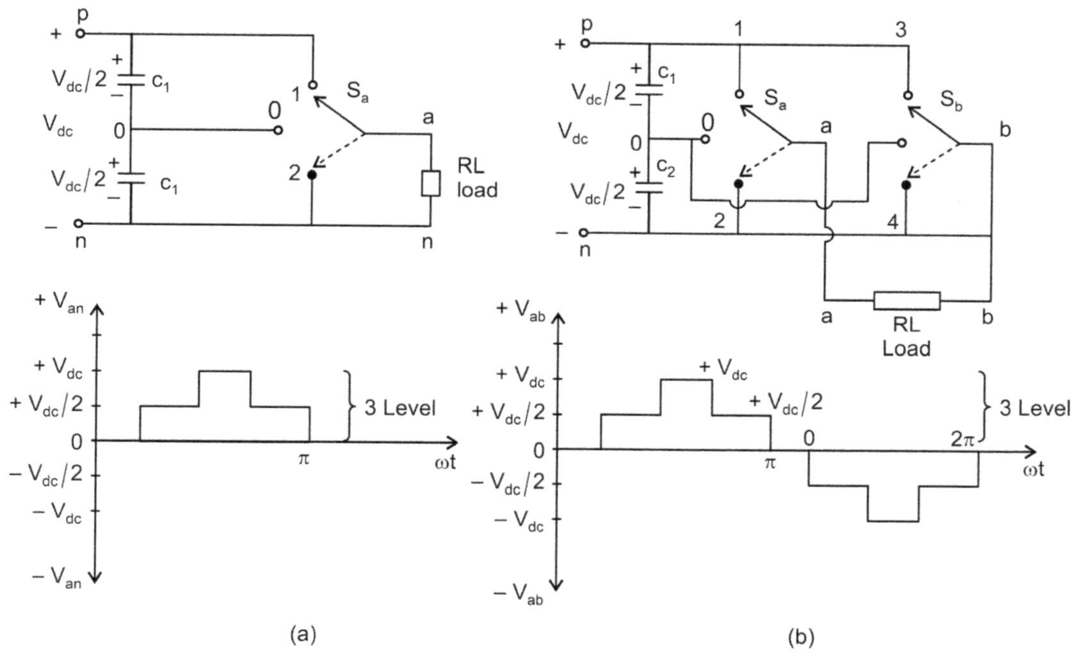

Fig. 3.1 Single phase three-level voltage source inverter equivalent circuit (a) Single-leg circuit and three-level phase voltage waveform, V_{an}. (b) Single-phase full bridge circuit and three-level line voltage waveform, V_{ab}.

When switch S_a is connected to terminal 2 and S_b is connected to terminal 3, then $V_{ab} = -V_{dc}$. And switch S_a is kept connected to terminal 2 and S_b is connected to terminal 4, then $V_{ab} = 0$. If switch S_a remain connected to terminal 2 and S_b is connected to terminal 0, then $V_{ab} = -V_{dc}/2$. The working of the circuit is summarized in the Table 3.1.

Table 3.1(a) Switching pattern for single phase one single-pole triple throw switch circuit.

Mode	Switch S_a Connected to	Output Voltage, V_{an}	Load Current, Io
M1	Terminal 1	+ V_{dc}	Positive (*a* to *n*)
M2	Terminal 0	+ $V_{dc}/2$	Positive (*a* to *n*)
M3	Terminal 2	0	Zero

Table 3.1(b) Switching pattern for single phase two single-pole triple throw switches in bridge-configuration.

Mode	Switch S_a	Switch S_b	Output Voltage, V_{ab}	Load current, Io	V across off-switches	V across off-switches
M1	Terminal 1	Terminal 4	+ V_{dc}	Positive (*a* to *b*)	$V_{a0} = +V_{dc}/2$	$V_{a2} = +V_{dc}$
M2	Terminal 1	Terminal 3	0	Zero	----	----
M3	Terminal 1	Terminal 0	+ $V_{dc}/2$	Positive (*a* to *b*)	----	----
M4	Terminal 2	Terminal 3	$-V_{dc}$	Negative (*b* to *a*)	$V_{a0} = -V_{dc}/2$	$V_{a1} = +V_{dc}$
M5	Terminal 2	Terminal 4	0	Zero	----	----
M6	Terminal 2	Terminal 0	$-V_{dc}/2$	Negative (*b* to *a*)	----	----
M7	Terminal 0	Terminal 0	0	Zero	----	----

The expected output voltage across RL load is a stepped voltage giving a three-level voltage (0, $\pm V_{dc}/2$ and $\pm V_{dc}$) as given in Fig. 3.1 (a) and (b). For single phase full-bridge configuration, when switch S_a is connected to terminal T_1 (top), blocking voltage across off-switches are V_{a0} (terminal '*a*' of the load and '*0*' terminal) is $+V_{dc}/2$ and V_{a2} (terminal '*a*' of the load and terminal T_2 (bottom)) is $+V_{dc}$. Similarly, when switch S_b is connected to terminal 4 (bottom) voltage across other two off-switches V_{b0} (terminal '*b*' of load and terminal '*0*') is $-V_{dc}/2$ and V_{b3} (terminal '*b*' of load and terminal T_3 (top)) is $-V_{dc}$.

The three-level operation of switch S_a (for terminal 1 and 2 operations) and S_b (for terminal 3 and 4 operations) can be realized using a bidirectional electronic device which can be capable to conduct in both directions for inductive loads and to block voltages across off-switches to $\pm V_{dc}$. An IGBT device with anti-parallel diode arrangement can be used here to explain the realization by considering only single-leg for simplicity. And for operations of S_a and S_b when connected to terminals 0, can be realized by using one-directional conducting IGBT with series diode arrangement as shown in the Fig. 3.2. For medium voltage applications where voltage ratings of ac motors are ranging from 2kV to 5kV for which one device may not be sufficient to match the voltage requirements, hence more than one device is used in series, as shown in the Fig.3.2.

Due to two IGBTs connected in series, namely, S_{a1} and S_{a2} in the upper limb and S'_{a3} and S'_{a4} in the lower limb to realize switches S_a and S_b respectively, the operation of one-directional IGBTs S_2' and S_3' have become redundancy, when diode D_{11} is directly connected to the mid-point of S_{a1} and S_{a2} and diode D_{11}' to mid-point of S_{a3} and S_{a4} to exhibit the operation of terminals '*a*' and '*b*' of load with terminal 0, hence they are removed from the circuit. The final circuit in standard configuration for diode-clamped three-level inverter is shown in the Fig.3.3 for single-leg arrangement.

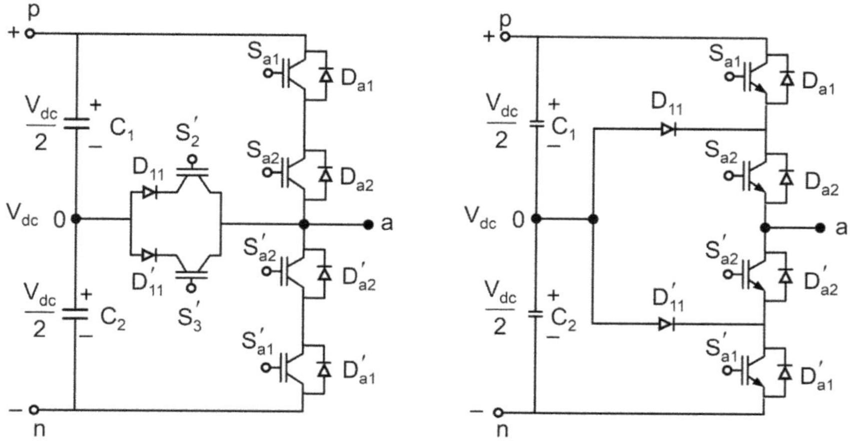

(a) (b)

Fig. 3.2 Three-level Inverter with IGBTs for single-leg arrangement (a) for first-step realization circuit and (b) for final-step circuit realization.

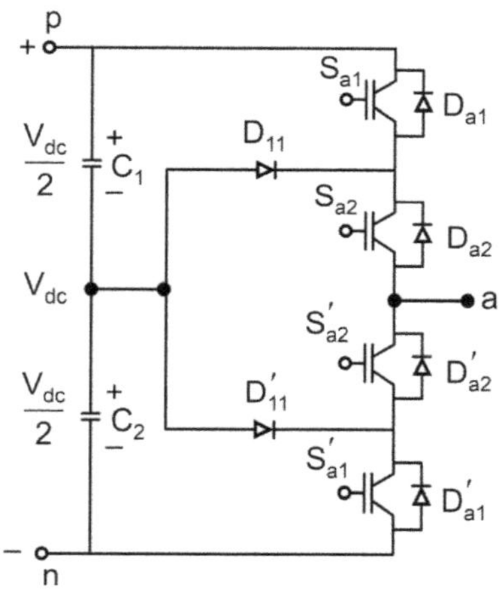

Fig. 3.3 Single-phase three-level diode-clamped inverter with single-leg arrangement.

Similarly, a single-phase three-level full-bridge diode-clamped inverter and load voltage waveform is shown in Fig.3.4 and Fig.3.5 respectively.

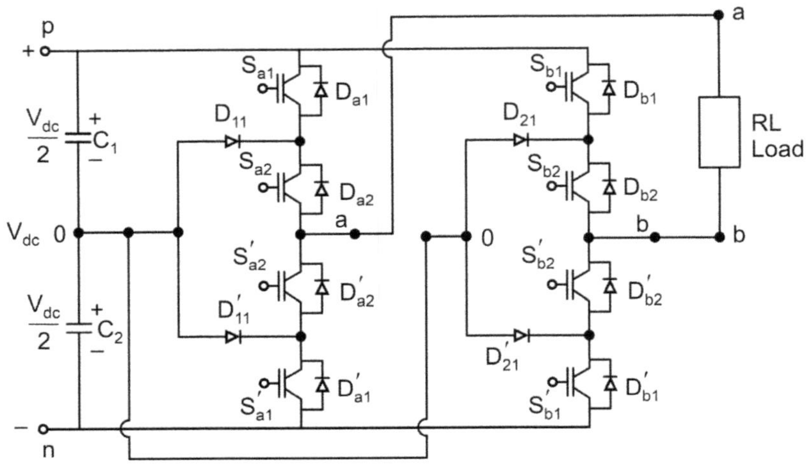

Fig. 3.4 Single-phase three-level diode-clamped inverter full-bridge circuit.

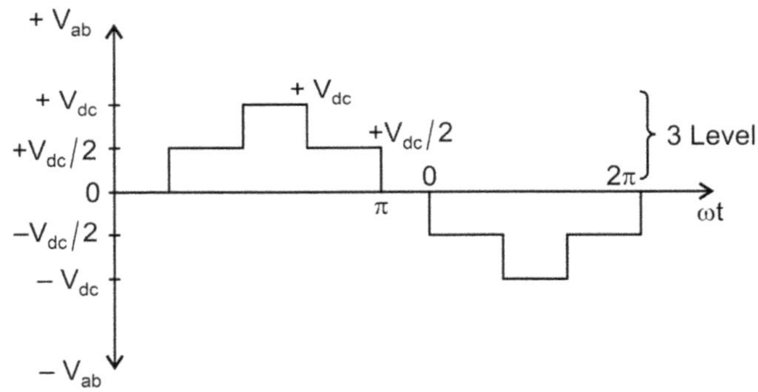

Fig. 3.5 Three-level single-phase diode-clamped bridge inverter load voltage waveform.

3.2 Three Level Diode-Clamped Inverter

The basic configuration of a three phase three level diode-clamped (or NPC) inverter is shown in the Figure 3.6 (complete three-phase, three-level, inverter diagram). Each phase leg consists of four switching devices with four anti-parallel diodes, two for upper limb and two for lower limb, connected to positive and negative terminals of the source, respectively. Two neutral-clamped diodes are connected in series between upper and lower pairs of switching devices and the center of the pair is connected to the neutral point.

Total three phase three level inverter (Fig.3.6) consists of 12 switching devices, 12 anti-parallel diodes, 6 neutral-clamped diodes, 2 capacitors are connected at the source splitting the dc voltage into two halves. The switching devices are denoted as S_{a1}-S_{a12}, anti-parallel diodes as D_{a1}-D_{a12}, neutral-clamped diodes as D_{11} – D_{13} and D_{11}'-D_{13}' and capacitors as C_1-C_2 carry $+V_{dc}/2$ each. The switching devices S_{a1} and S_{a1}' function as main devices and S_{a2} and S_{a2}' function as auxiliary devices which actually facilitates the clamp the output potential to the neutral point through neutral clamped diodes D_{11} and D_{11}'.

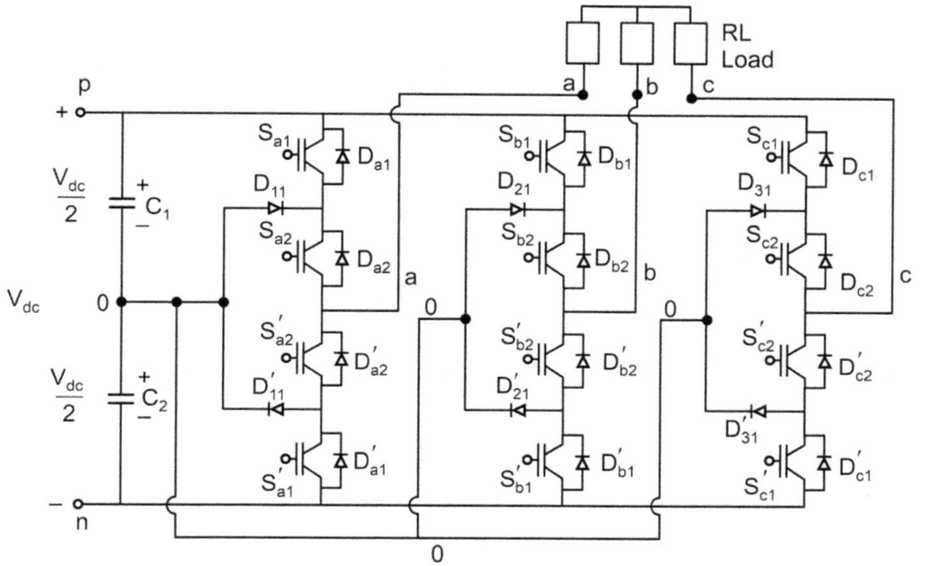

Fig. 3.6 Complete circuit diagram of three-level neutral point diode- clamped inverter.

The phase voltages V_{a0}, V_{b0}, and V_{c0} are obtained with phase difference of 120° and 240° with respect to V_{a0}. The levels of line voltage waveform V_{ab}, are $+V_{dc}$, $+V_{dc}/2$, 0, $-V_{dc}/2$, $-V_{dc}$, similarly, V_{bc} and V_{ca} waveforms can also be obtained as shown in Fig.3.7.

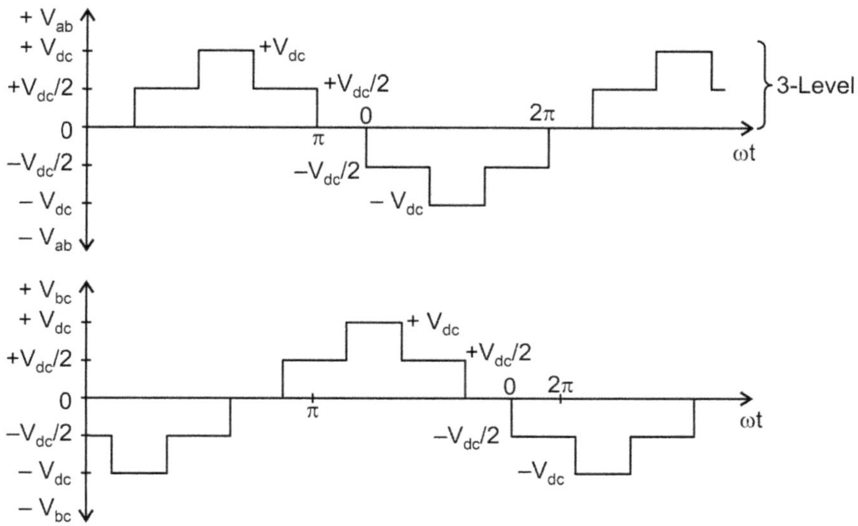

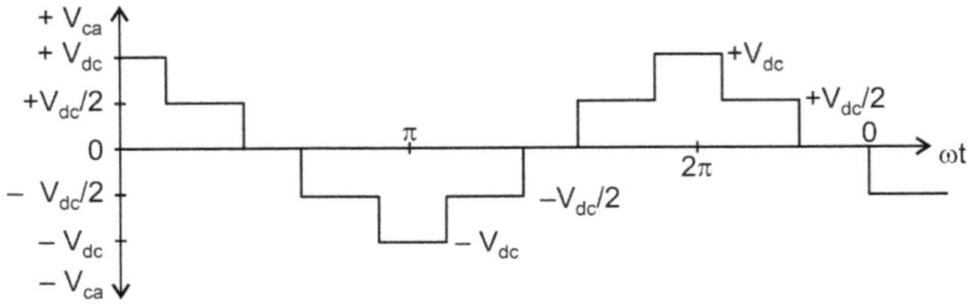

Fig. 3.7 Load line voltage waveforms for three-phase three-level diode-clamped full-bridge inverter fed wye-connected loads.

For explaining the operation of single-phase diode-clamped inverter, a single-leg arrangement is considered as shown in the Fig. 3.8 (a). The inverter operation is explained in detail with Fig.3.8 (a) – (d) three modes with the influence of dc source voltage for producing three-level voltages across the load V_{a0} ($+V_{dc}/2$, 0, $-V_{dc}/2$) and Fig.3.8 (e) – (f) explains the operation of inverter with the influence of induced emf in the load inductance which is acting as the source to force a current I_a into the input terminals '*p*' and '*n*' to feed back the stored energy. Considering pulse width modulation technique for firing the switching devices to obtain the output a.c. voltage of the three-level inverter. The gating pulse signals are provided for S_{a1} and S_{a1}' as main devices and complementary gate pulse signals are given to S_{a2} and S_{a2}' to obtain a.c. phase voltage V_{a0}. The three phase load voltages are V_{a0}, V_{b0} and V_{c0} and phase currents are I_a, I_b and I_c, respectively.

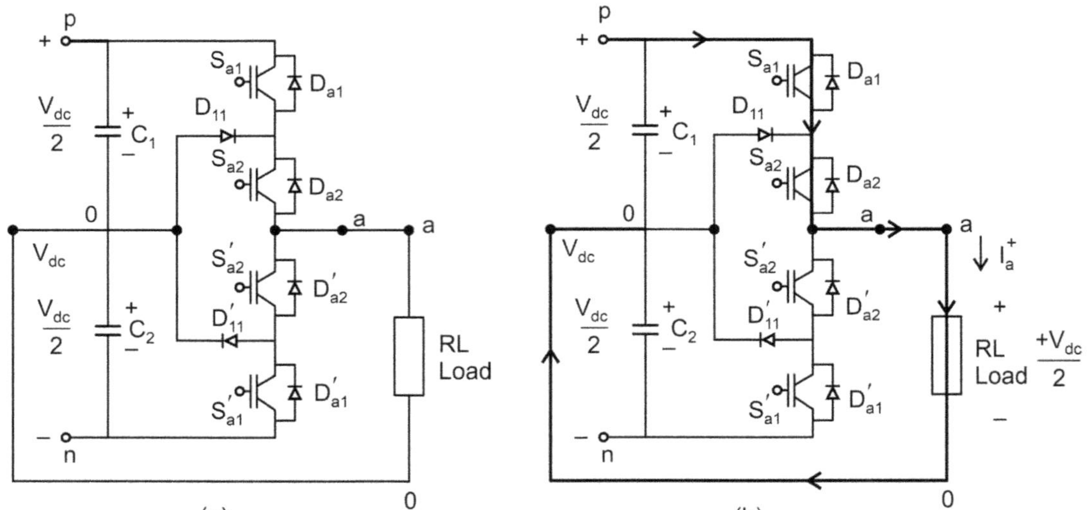

Fig. 3.8 (a) Main circuit configuration of single-phase single-leg three-level diode-clamped inverter, (b) Conduction path for positive V_{a0}, for positive current I_a.

The conduction is from dc source voltage: Mode1: $V_{a0} = +V_{dc}/2$, the devices S_{a1} and S_{a2} are fired to generate positive phase voltage V_{a0} across the load and load current is positive when I_a flows from load terminal '*a*' to '*0*' as shown in Fig. 3.8(b).

Mode 2: $V_{a0} = -V_{dc}/2$, the devices S_{a2}' and S_{a1}' are fired to produce negative load voltage V_{a0} across the load and load current is negative when I_a flows load terminal '*0*' to '*a*' as shown in Fig.3.8(c).

Mode 3: $V_{a0} = 0$, when S_{a1} is off, S_2 is turned on and D_{11} is made to conduct as shown in Fig. 3.8(d), then the conduction path initiates from source terminal '*0*', diode D_{11}, switch S_2, load terminal '*a*' to '*0*' resulting V_{a0} is zero voltage with zero current I_a.

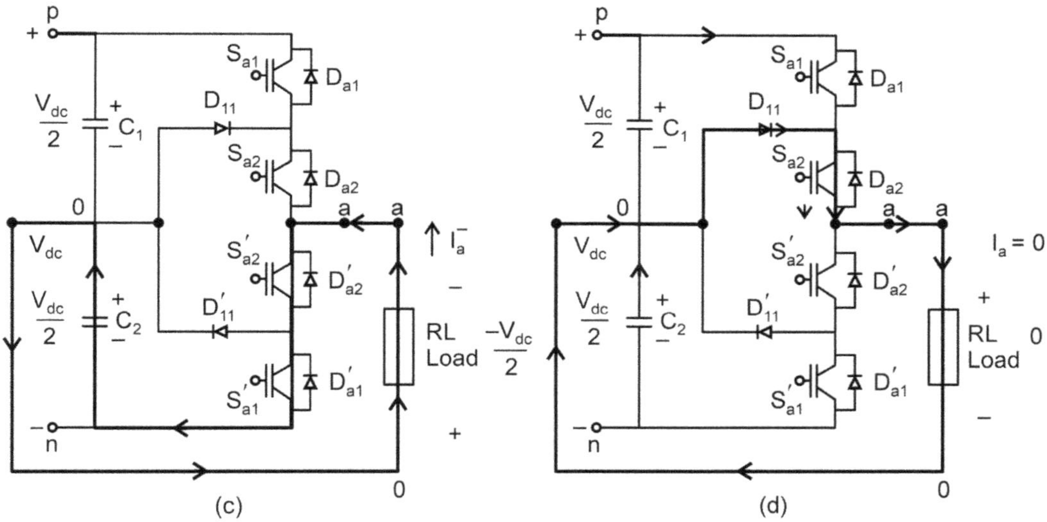

Fig. 3.8 (c) Conduction path for negative V_{a0} for negative load current I_a and (d) conduction path for zero V_{a0} for zero load current I_a, for Single-phase single-leg three-level diode-clamped inverter.

The conduction is from ac load induced emf (stored energy) when the load is resistive-inductive:

Mode 4: $V_{a0} = +V_L$, the induced emf in the load inductance V_L is acting as a source and forcing a reverse (negative) current from load through terminal '*a*', diodes D_{a2}, D_{a1} to the dc source terminals '*p*' to '*0*', through RL load, until the stored energy in the inductance is dissipated or exhausted (Fig. 3.8f).

Mode 5: $V_{a0} = -V_L$, the induced emf in the load inductance V_L is acting as a source and forcing a reverse (positive) current from load through terminal '*0*', source capacitor C_2, terminal '*n*', diodes D_{a1}', D_{a2}' to the load terminals '*a*' to '*0*', through RL load, until the stored energy in the inductance is dissipated or exhausted (Fig. 3.8e).

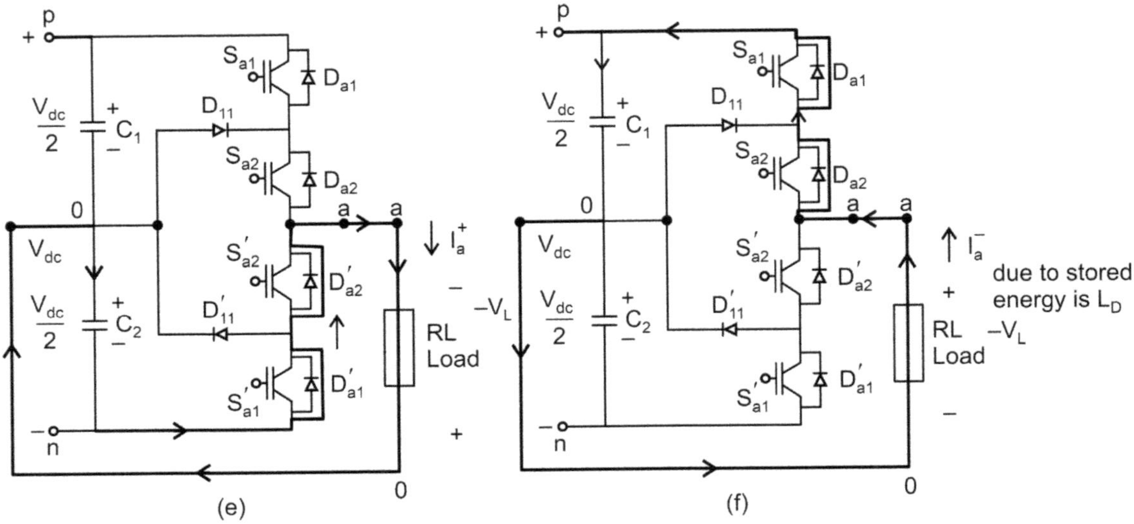

Fig. 3.8 (e) Conduction path for positive load current, I_a flow from load terminal 'a' to '0' forced by V_L, the stored energy (induced emf) in the load inductance, and (f) Conduction path for negative load current, I_a flow from load terminal '0' to 'a' forced by V_L, the stored energy (induced emf) in the load inductance for three - level NPC inverter.

The operation modes explained above generates three voltage levels $+V_{dc}/2$, 0, $-V_{dc}/2$ and the phase voltage, Va0 waveform is given in Fig.3.9. The switching states are presented in the Table 3.2.

Table 3.2 The switching states for one leg of inverter.

Symbol of States	States of the switches				Diodes	Output Phase Voltage, V_{a0}	Load current I_a
	S_{a1}	S_{a2}	S_{a2}'	S_{a1}'			
P	1	1	0	0	----	$+V_{dc}/2$	Positive
0	0	1	0	0	D_{11}	0	Zero
N	0	0	1	1	----	$-V_{dc}/2$	Negative

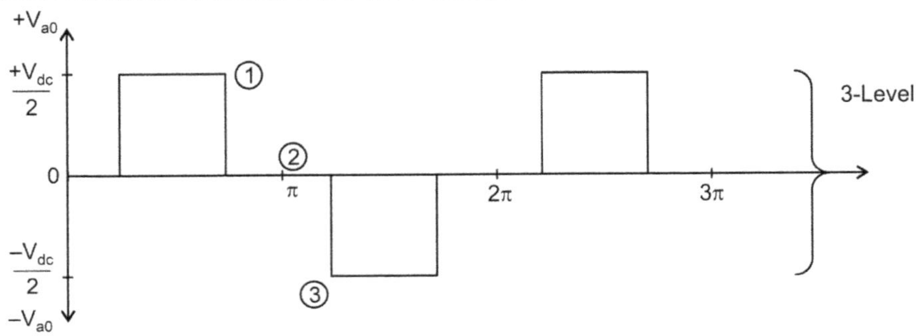

Fig. 3.9 Phase voltage (Va0) waveform of one-leg single-phase diode-clamped inverter.

The blanking time interval δ_B: The blanking time interval δ is deliberately introduced during transition of operation, for ensuring successful commutation of current I_a. To explain the commutation of current I_a, in the circuit is shown in Fig.3.10, two operations are considered, one for positive current I_a and another for negative current I_a, when transition between $V_{a0} = 0$ to $V_{a0} = +V_{dc}/2$, respectively. For positive load current I_a, consider operation of the circuit from $V_{a0} = 0$ to $V_{a0} = +V_{dc}/2$ with blanking time interval δ as shown in Fig.3.10 (a) and similarly for negative load current I_a, shown in Fig.3.10 (b).

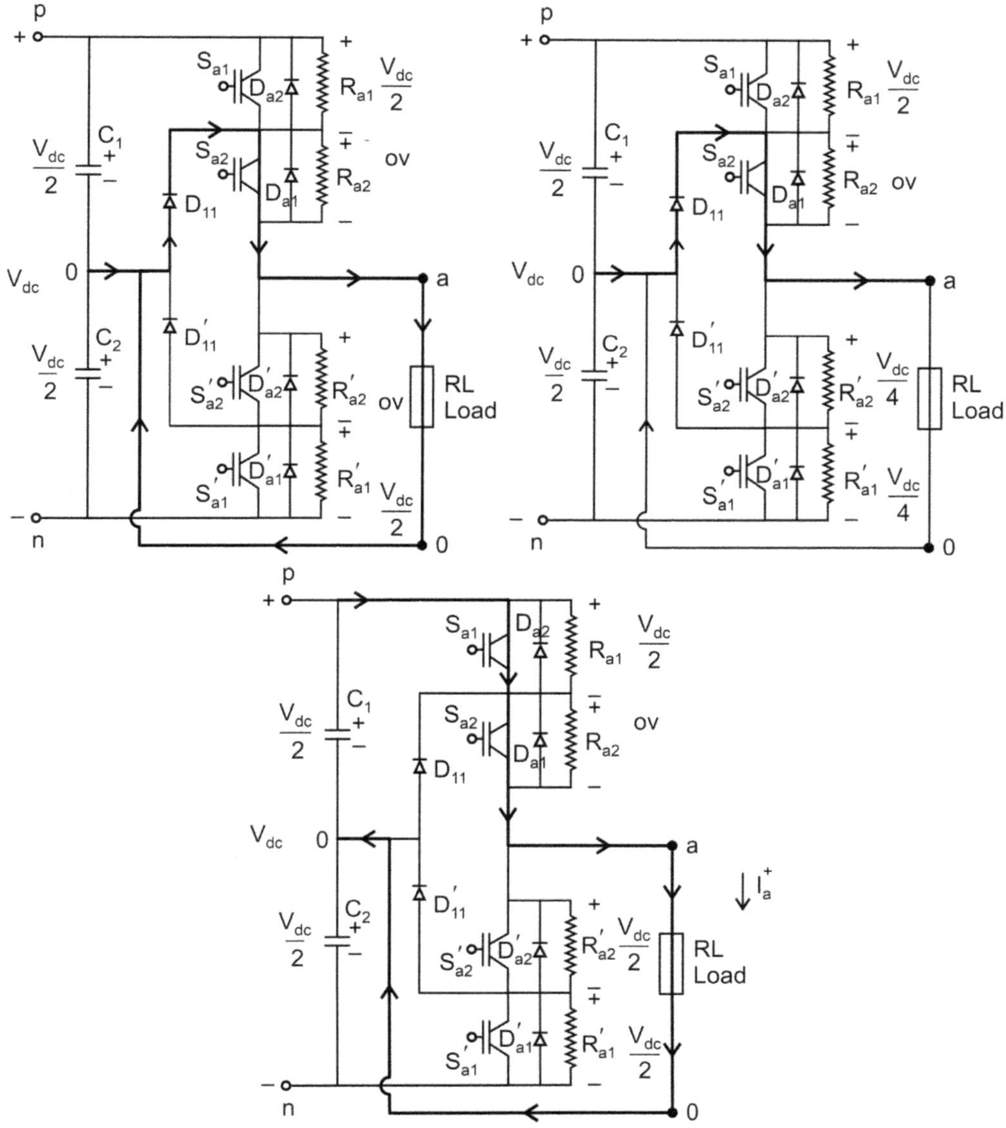

(a)

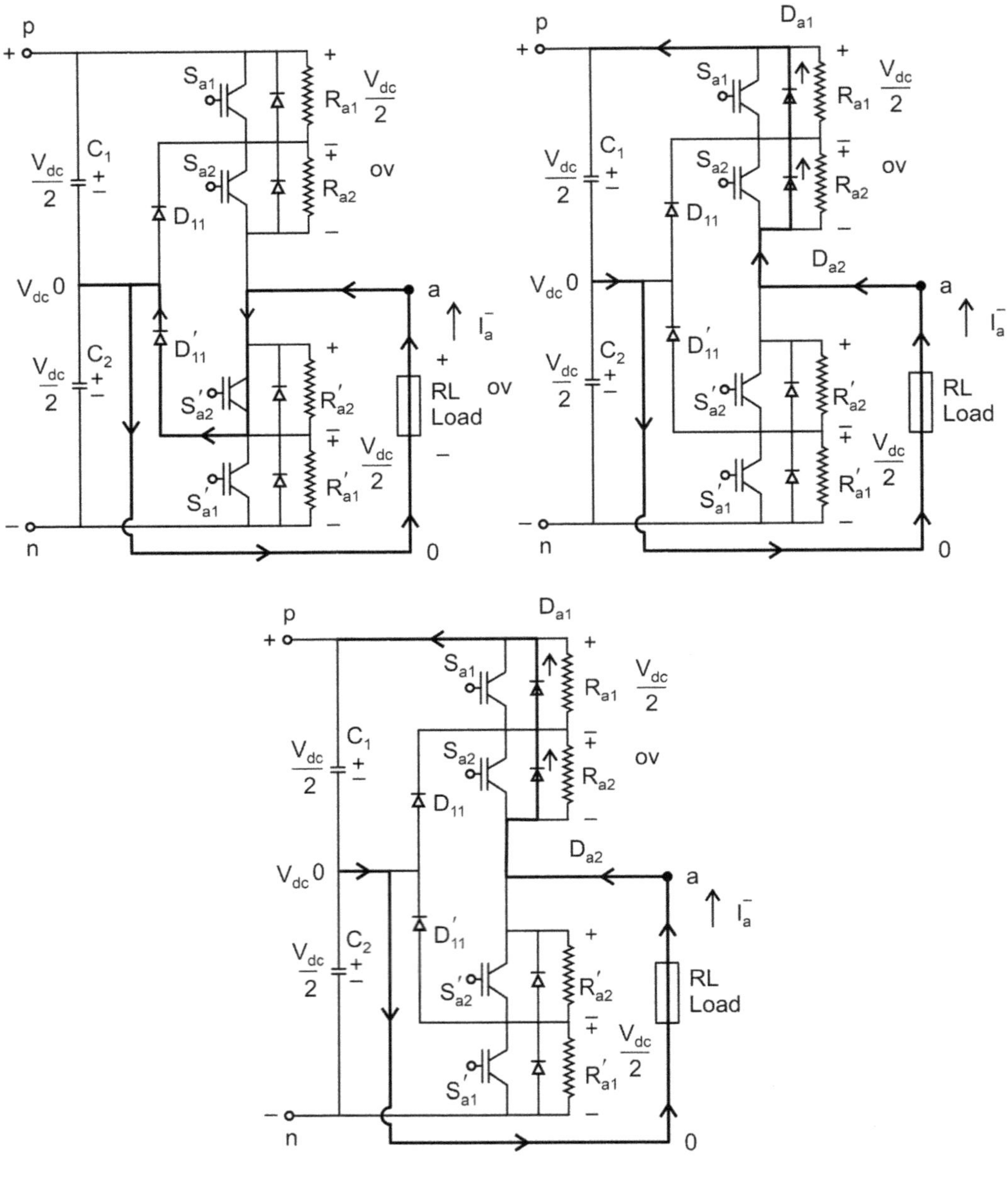

(b)

Fig. 3.10 Commutation during transition of operation from $V_{a0}=0$ to $V_{a0}=+V_{dc}/2$ for (a) positive load current I_a and (b) negative load current I_a, respectively.

As per Fig. 3.10 (a), when $V_{a0} = 0$, switching devices S_{a1} is off and S_{a2} is made ON, diode D_{11} is forward biased by the positive direction of the current, and clamped to the neutral, resulting output phase voltage V_{a0} is zero volts. The current I_a passes through Neutral, diode D_{11}, switch S_{a2} and phase '*a*' terminal of the load. Now it is explained how the current I_a gets commutated from diode D_{11} and transfer to switch S_{a1} for V_{a0} is $+V_{dc}/2$ by introducing the blanking time interval δ for ensuring complete turn-off of switch S_{a2} and turn-on of switch S_{a1} is on along with switch S_{a2} but no current passes through this device and now allow sufficient time δ to turn-off switch S_{a2}, as shown in intermediate circuit (Fig.3.10 (a)(ii)). Once switch S_{a2} is completely turns-off and switch S_{a1} is completely on, the current now transfers from diode D_{11} to switch S_{a1} and connects positive potential of the source to phase '*a*' of the load through switches S_{a1} and S_{a2} resulting $V_{a0} = +V_{dc}/2$ when load current I_a is positive. Diode D_{11} is reverse biased by conducting switch S_{a1}.

Similarly, for negative load current I_a, as per Fig. 3.10 (b), when $V_{a0}=0$, switching devices S_{a2}' and S_{a1}' are made ON, diode D_{11}' is forward biased and terminal phase '*a*' is clamped to the neutral and current now passes through load terminal '*a*', switch S_{a2}', diode D_{11}' and neutral point '0', resulting $V_{a0}=0$ volts. Now see how the current I_a gets commutated from D_{11}' to diodes D_{a2} and D_{a1} and with blanking time interval δ as shown in intermediate circuit (Fig.3.10(b) (ii)) for S_{a2}' completely turns off and S_{a1} is on. The current I_a, after blanking time δ interval passes through load terminal '*a*', D_{a2}, D_{a1} and positive potential of the source 'p', connecting $V_{a0}=+V_{dc}/2$. Though S_{a1} and S_{a2} are on, current cannot reverse its direction momentarily for inductive load, the current passes through their anti-parallel diodes D_{a2} and D_{a1}. The current commutations can be carried out similarly between remaining operations of the three-level neutral point inverter for other phases as well.

Advantages of three level diode-clamped inverters over two level inverters:

- Inverter can have high power handling capability.
- Several voltage levels are used, dv/dt of output voltage is smaller.
- Less percentage Total harmonic distortion in output voltage and current waveforms.
- Inverter can work with very low switching frequency and switching losses are reduced.

Applications of three level inverter are

- Variable speed and torque motor drives
- Static Var Compensation
- Traction applications

3.3 Five level Diode-Clamped Inverter

Now consider single phase five level diode clamped inverter for resistive-inductive load. The Single phase five-level diode clamped inverter consists of 8 switching devices, 8 anti-parallel diodes, 12 clamped diodes per leg and 4 capacitors for splitting input dc voltage. The circuit is shown in Fig. 3.11 for Single phase with resistive-inductive load.

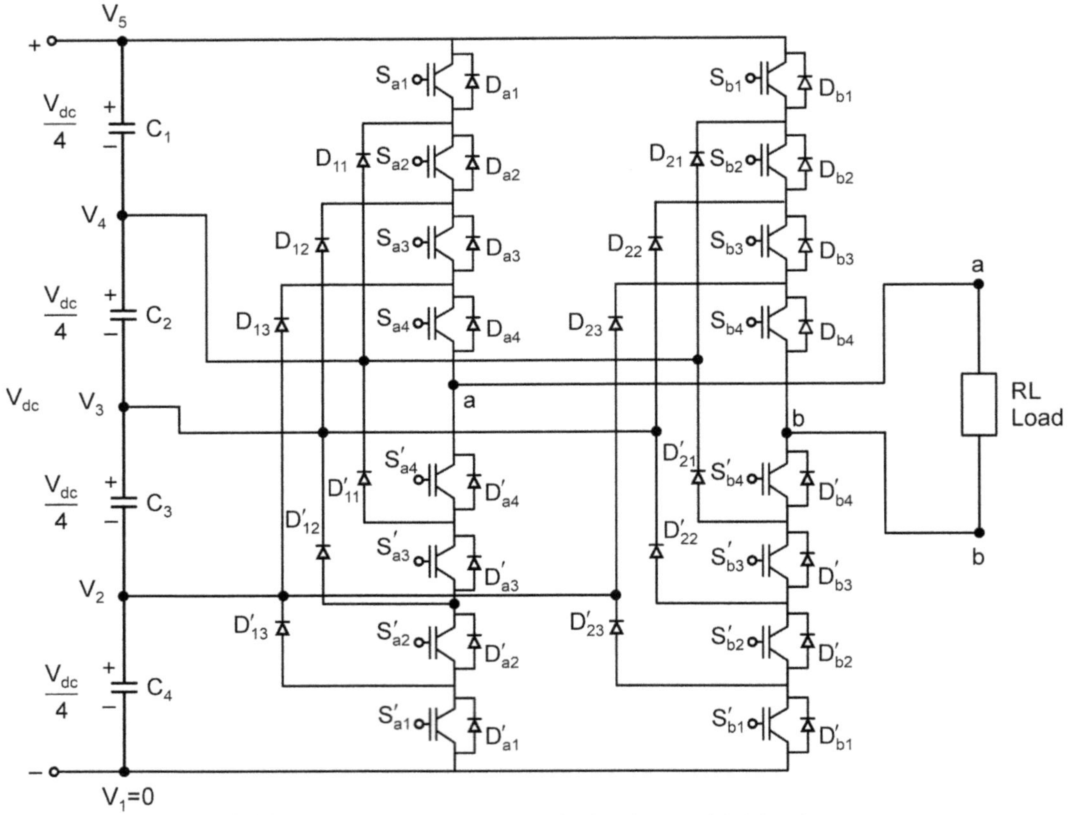

Fig. 3.11 Single phase five level diode-clamped bridge inverter.

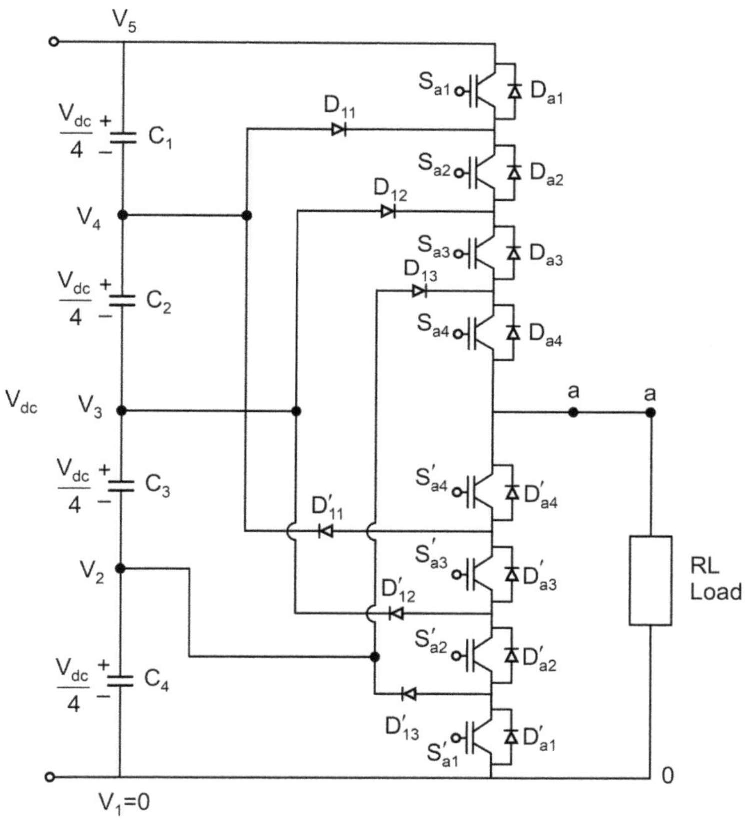

(a)

Fig. 3.12 Main circuit configuration of single phase five level diode-clamped one-leg circuit.

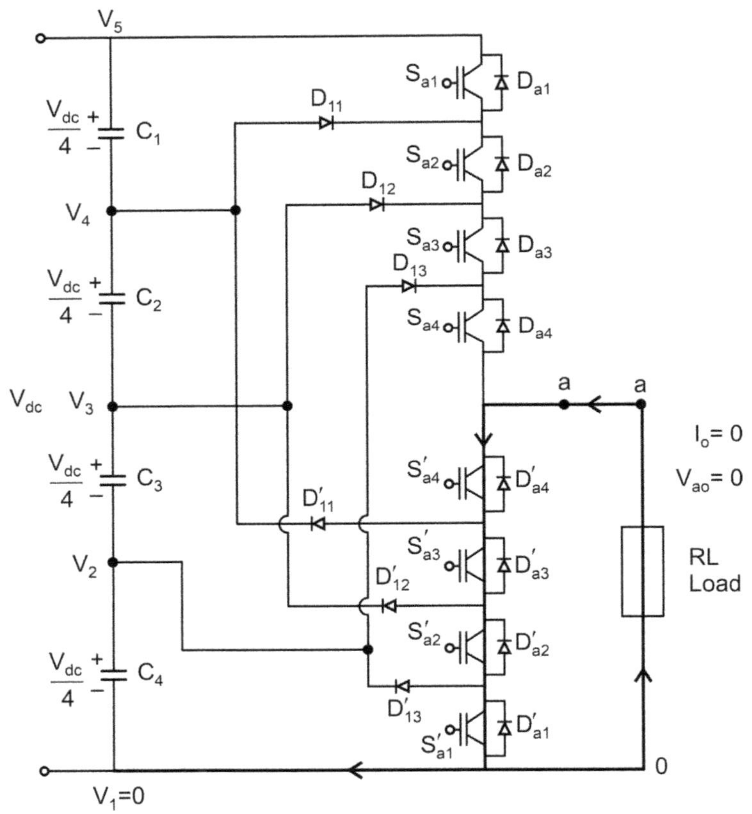

(b)

Fig. 3.12 (a) Main circuit configuration, (b) (i) For V_1=0, when D_{a1}', D_{a2}', D_{a3}', D_{a4}' are conducting for forward path. (ii) For V_1=0, when S_{a1}', S_{a2}', S_{a3}' and S_{a4}' are ON, for reverse path.

The working of the circuit is explained in details with the help of Fig. 3.12 for various levels of input voltages, namely, $V_0 = V_1, V_2, V_3, V_4, V_5$ ($V_1=0$, $V_2=V_{dc}/4$, $V_3=V_{dc}/2$, $V_4=3V_{dc}/4$, and $V_5=V_{dc}$) by considering one-leg of the circuit for ease of understanding. Fig. 3.12 (a) main circuit configuration and (b) shows for $V_0 = V_1 = 0$ volts, through forward conduction path formed by D_{a1}', D_{a2}', D_{a3}', D_{a4}', and the reverse conduction path formed by firing switching devices S_{a1}', S_{a2}', S_{a3}' and S_{a4}' shown by thick-line.

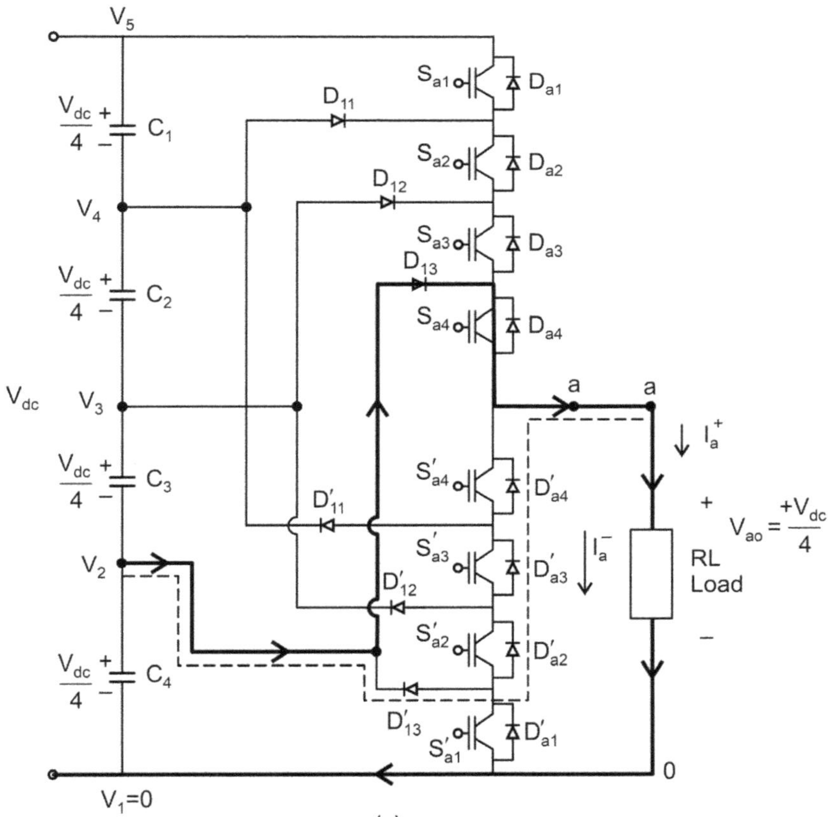
V5
Sa1
Da1
D11
Vdc/4
C1
Sa2
Da2
D12
V4
Sa3
Da3
D13
Vdc/4
C2
Sa4
Da4
a
a
Vdc
V3
Ia+
S'a4
D'a4
Vdc/4
C3
D'11
Vao = +Vdc/4
S'a3
Ia−
D'a3
RL
Load
V2
D'12
S'a2
D'a2
Vdc/4
C4
D'13
D'a1
S'a1
0
V1=0

(c)

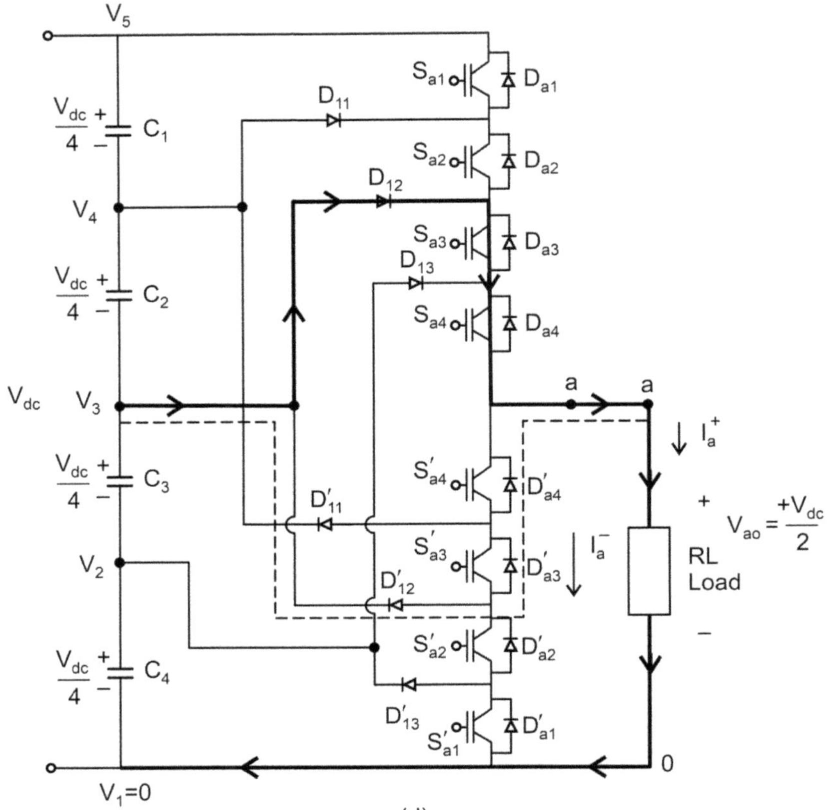

(d)

Fig. 3.12 (c) (i) For V_2= $+V_{dc}/4$, when S_{a4}, D_{13} are ON for positive load current I_a. (ii) For V_2 = $+V_{dc}/4$, when S_{a4}', S_{a3}', S_{a2}' and D'_{13} are ON for negative I_a. (d) (i) For V_3= $+V_{dc}/2$, when S_{a3}, S_{a4}, and D_{12} are ON for positive current I_a. (ii) For V_3= $+V_{dc}/2$, when S_{a4}', S_{a3}', D_{12}' are ON for negative current I_a.

For $V_o = V_2 = +V_{dc}/4$ is obtained by firing switching devices S_{a4} and conduction of clamping diode D_{13} as shown in the Fig. 3.12 (c)(i) current path is marked with dark line. The input dc voltage terminal V_2 is connected to the load through D_{13} and S_{a4}, since load current I_a is positive. (For negative load current, load terminal '*a*' is connected to V_2 terminal through S_{a4}', S_{a3}', S_{a2}' and D_{13}', the negative current path is shown by broken line).

For V_o=V_3= $+V_{dc}/2$, where V_3 point is clamped by conduction of D_{12} along with switching devices S_{a3}, S_{a4}, as shown in the Fig. 3.12 (d)(i). The input dc terminal V_3 is connected to the load through D_{12}, S_{a3} and S_{a4} for positive load current I_a. (For negative load current I_a, load terminal V_a is connected to V_3 terminal through S_{a4}', S_{a3}' and D_{12}', the negative current path is shown by broken line).

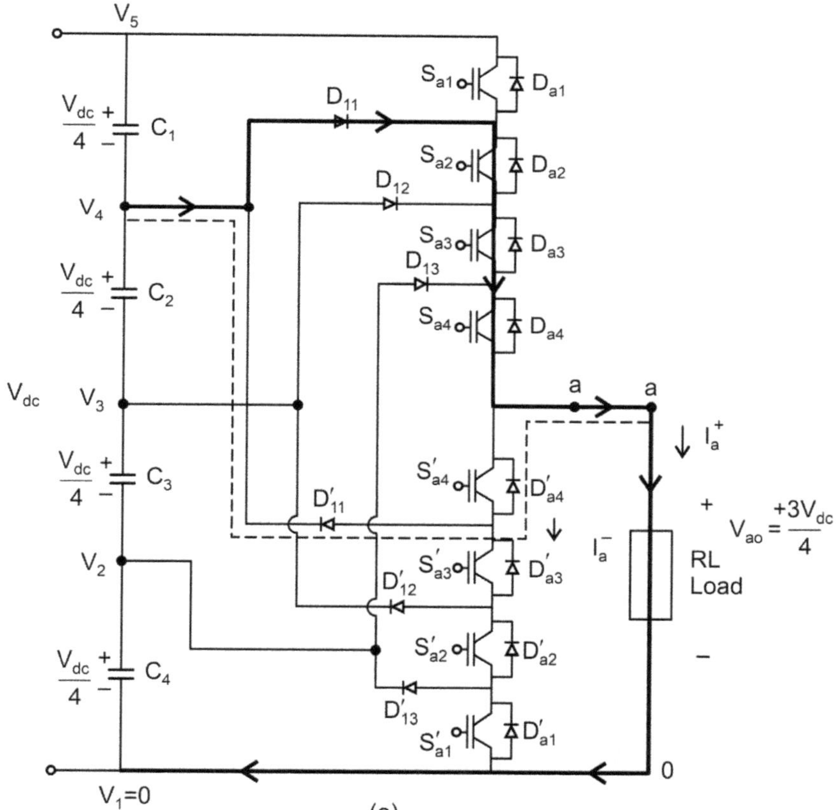
V5
Vdc/4
C1
D11
Sa1
Da1
Sa2
Da2
D12
V4
Sa3
Da3
D13
C2
Sa4
Da4
Vdc
V3
a
a
I+a
C3
S'a4
D'a4
D'11
+
Vao = +3Vdc/4
I−a
V2
S'a3
D'a3
RL
Load
D'12
C4
S'a2
D'a2
−
D'13
S'a1
D'a1
V1=0
0

(e)

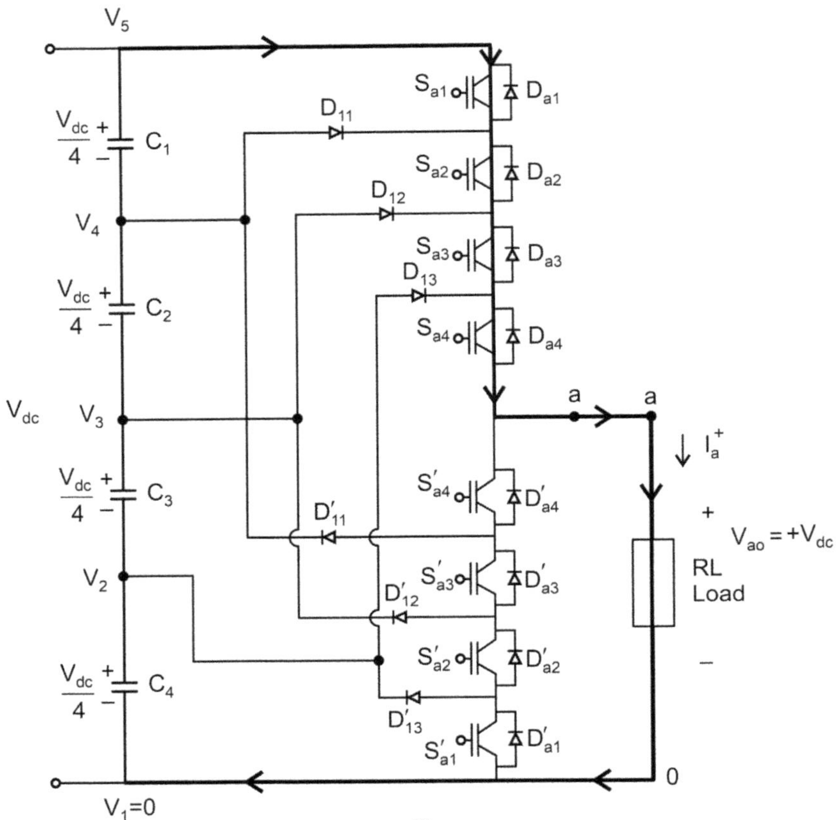

(f)

Fig. 3.12 (e) (i) For V_4= $+3V_{dc}/4$, when S_{a2}, S_{a3}, S_{a4}, and D_{11} are ON for positive I_a. (ii) For V_4= $+3V_{dc}/4$, when S_{a4}' and D'_{11}, are ON for negative I_a. (f) (i) For V_5= $+V_{dc}$, when S_{a1}, S_{a2}, S_{a3}, S_{a4} are ON for positive load current I_a. (ii) For V_5= $+V_{dc}$, when D_{a4}, D_{a3}, D_{a2}, D_{a1} are conducting negative I_a.

Fig. 3.12 (e) depicts the $V_o = V_4 = +3V_{dc}/4$ voltage level, by conducting the switching devices S_{a2}, S_{a3}, S_{a4}, along with clamping diode D_{11}. The input dc terminal V_4 is connected to the load through D_{11}, S_{a2}, S_{a3}, and S_{a4} for positive load current I_a is marked by thick-line. (For negative load current I_a, load terminal V_a is connected to V_4 terminal through S_{a4}' and D_{11}' is shown by the broken-line).

For $V_o = V_5 = +V_{dc}$ voltage level, by conducting switching devices S_{a1}, S_{a2}, S_{a3}, and S_{a4} directly and there is no need of clamping diodes as shown in Fig. 3.12 (f). The input dc terminal V_5 is connected to the load through S_{a1}, S_{a2}, S_{a3}, and S_{a4} for positive load current I_a marked by thick-line. (For negative load current I_a, load terminal V_a is connected to V_5 terminal through D_{a4}, D_{a3}, D_{a2}, D_{a1} is shown by the broken-line).

Fig. 3.13 Shows the stepped-waveform depicting 5-levels of voltages 0, $V_{dc}/4$, $V_{dc}/2$, $3V_{dc}/4$ and V_{dc} in both positive and negative half-waves for a cycle including fundamental waveform.

The switching states of five-level diode clamped inverter for single leg arrangement are presented in Table 3.3.

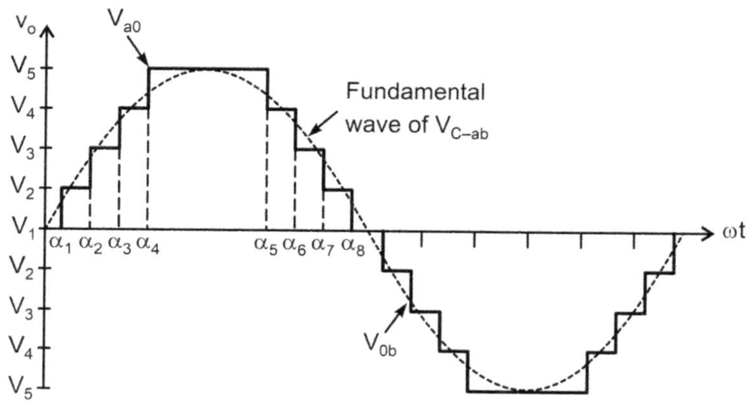

Fig. 3.13 Phase Voltage and fundamental voltage waveforms for five level diode clamped inverter.

Table 3.3 The switching states of five-level diode clamped inverter for single leg arrangement.

V_{ao}	S_{a1}	S_{a2}	S_{a3}	S_{a4}	S_{a4}'	S_{a3}'	S_{a2}'	S_{a1}'	Clamping diodes	Anti-parallel diodes	Load current I_a
V_5=+V_{dc}	1	1	1	1	0	0	0	0	-----	-----	Positive
V_4= +3V_{dc}/4	0	1	1	1	0	0	0	0	D_{11}	-----	Positive
V_3= +V_{dc}/2	0	0	1	1	0	0	0	0	D_{12}	-----	Positive
V_2= +V_{dc}/4	0	0	0	1	0	0	0	0	D_{13}	-----	Positive
V_1= 0	0	0	0	0	1	1	1	1	-----	-----	Zero
V_5= -V_{dc}	0	0	0	0	0	0	0	0	-----	D_{a4}, D_{a3}, D_{a2}, D_{a1}	Negative
V_4= -3V_{dc}/4	0	0	0	0	1	0	0	0	D_{11}'	-----	Negative
V_3= -V_{dc}/2	0	0	0	0	1	1	0	0	D_{12}'	-----	Negative
V_2= -V_{dc}/4	0	0	0	0	1	1	1	0	D_{13}'	-----	Negative
V1= 0	0	0	0	0	0	0	0	0	-----	D_{a4}', D_{a3}', D_{a2}', D_{a1}'	Zero

Single phase five-level diode-clamped inverter with bridge circuit configuration: The working of the circuit is explained in details with the help of Fig. 3.14 for various levels of input voltages, namely, $V_0 = V_1$, V_2, V_3, V_4, V_5 ($V_1 = 0$, $V_2 = V_{dc}/4$, $V_3 = V_{dc}/2$, $V_4 = 3V_{dc}/4$, and $V_5 = V_{dc}$) by considering full-bridge circuit. Fig. 3.14 (a) main circuit configuration for single phase five-level diode-clamped full-bridge inverter. The circuit operation is explained in detail by considering various modes for resistive-inductive load.

The conduction of inverter with dc source voltage:

Mode 1: For $V_{ab} = V_5 = +V_{dc}$, the switching devices S_{a1}, S_{a2}, S_{a3}, S_{a4} and S_{b4}', S_{b3}', S_{b2}', S_{b1}' are turned-on and conduction takes place from the path formed by S_{a1}, S_{a2}, S_{a3}, S_{a4} and load terminal '*a*' to '*b*' and S_{b4}', S_{b3}', S_{b2}', S_{b1}' marked with thick-line as shown in Fig. 3.14 (b), and load current I_a is positive when it flows from load terminal '*a*' to '*b*'.

Mode 2: For $V_{ab} = V_4 = +3V_{dc}/4$, the switching devices S_{a2}, S_{a3}, S_{a4} and S_{b4}', S_{b3}', S_{b2}', S_{b1}' are turned-on and conduction takes place from the path formed by V_4, clamping diode D_{11}, switches S_{a2}, S_{a3}, S_{a4} and load terminal '*a*' to '*b*' and S_{b4}', S_{b3}', S_{b2}', S_{b1}' marked with thick-line as shown in Fig.3.14 (c), and load current I_a is positive when it flows from load terminal '*a*' to '*b*'.

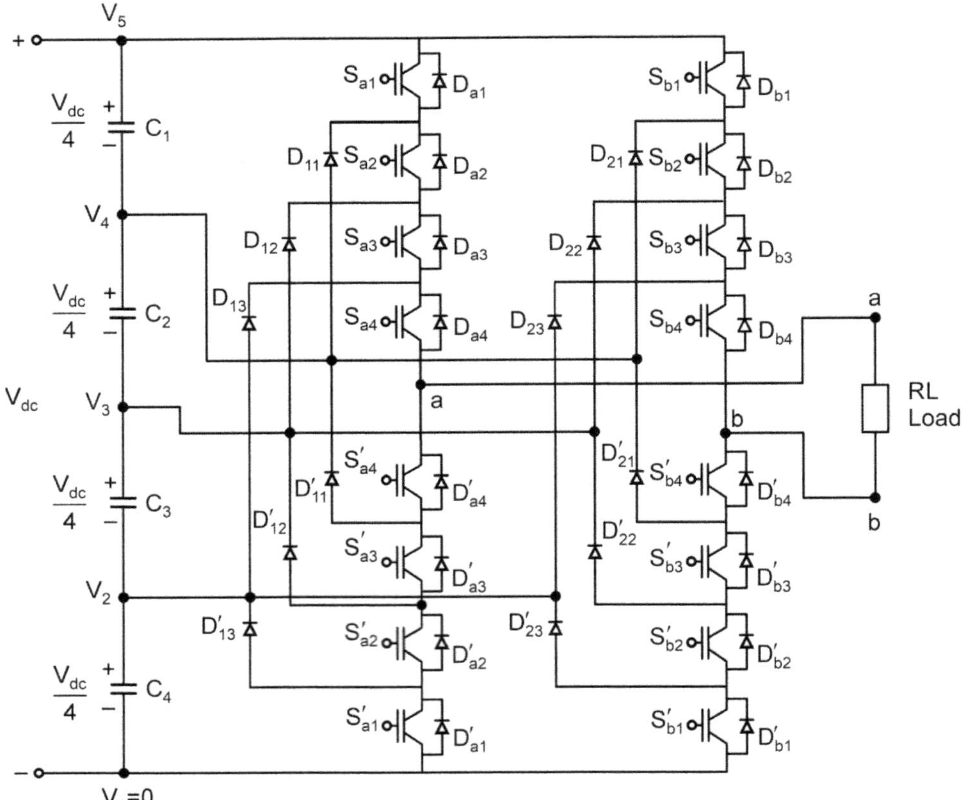

(a)

Fig. 3.14 (a) Main circuit configuration.

Mode 3: For $V_{ab} = V_3 = +V_{dc}/2$, the switching devices S_{a3}, S_{a4} and S_{b4}', S_{b3}', S_{b2}', S_{b1}' are turned-on and conduction takes place from the path formed by V_3, clamping diode D_{12}, switches S_{a3}, S_{a4} and load terminal '*a*' to '*b*' and S_{b4}', S_{b3}', S_{b2}', S_{b1}' marked with thick-line as shown in Fig.3.14 (d), and load current I_a is positive when it flows from load terminal '*a*' to '*b*'.

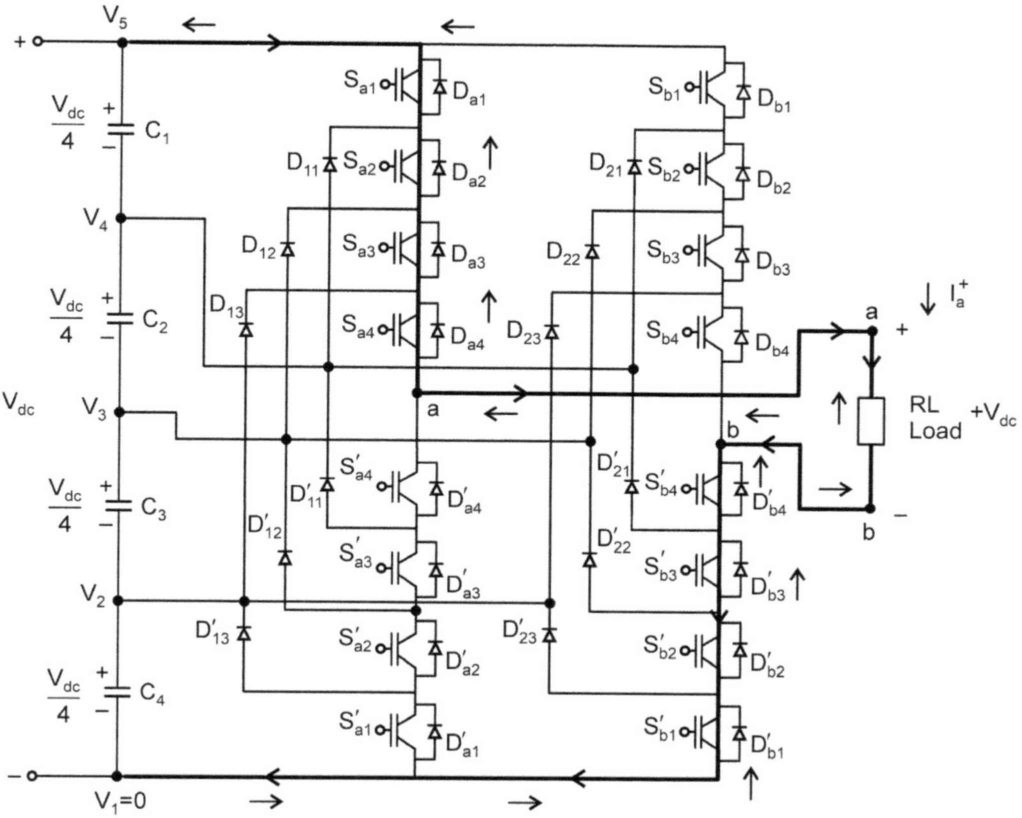

(b)

Fig. 3.14 (b) Mode1: $V_{ab} = +V_{dc}$ operation.

Mode 4: For $V_{ab} = V_2 = +V_{dc}/4$, the switching devices S_{a4} and S_{b4}', S_{b3}', S_{b2}', S_{b1}' are turned-on and conduction takes place from the path formed by V_2, clamping diode D_{13}, switches S_{a4} and load terminal '*a*' to '*b*' and S_{b4}', S_{b3}', S_{b2}', S_{b1}' marked with thick-line as shown in Fig.3.14 (e), and load current I_a is positive when it flows from load terminal '*a*' to '*b*'.

Mode 5: For $V_{ab}=V_1= 0$ volts, the switching devices S_{b4}', S_{b3}', S_{b2}', S_{b1}' are turned-on and conduction takes place from terminal V_1, through anti-parallel diodes D_{a1}', D_{a2}' D_{a3}', D_{a4}', and load terminal '*a*' to '*b*' and S_{b4}', S_{b3}', S_{b2}', S_{b1}' marked with thick-line as shown in Fig.3.14 (f), and load current I_a is zero.

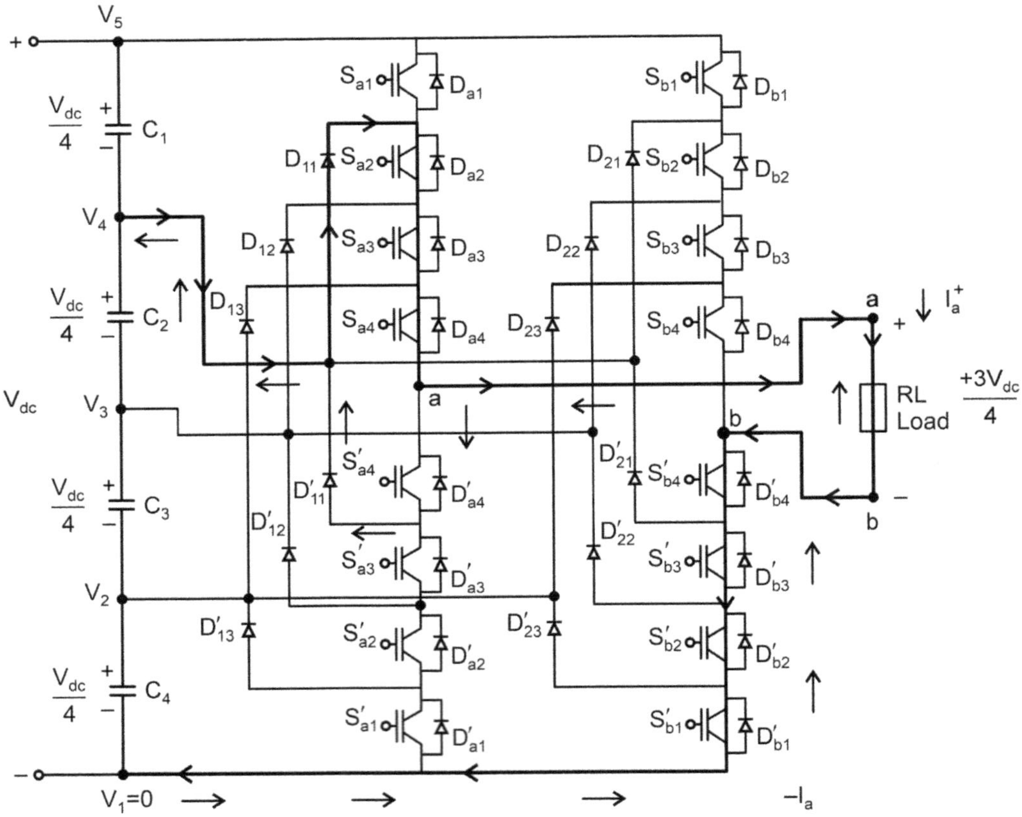

(c)

Fig. 3.14 (c) Mode2: V_{ab}=+3V_{dc}/4 operation.

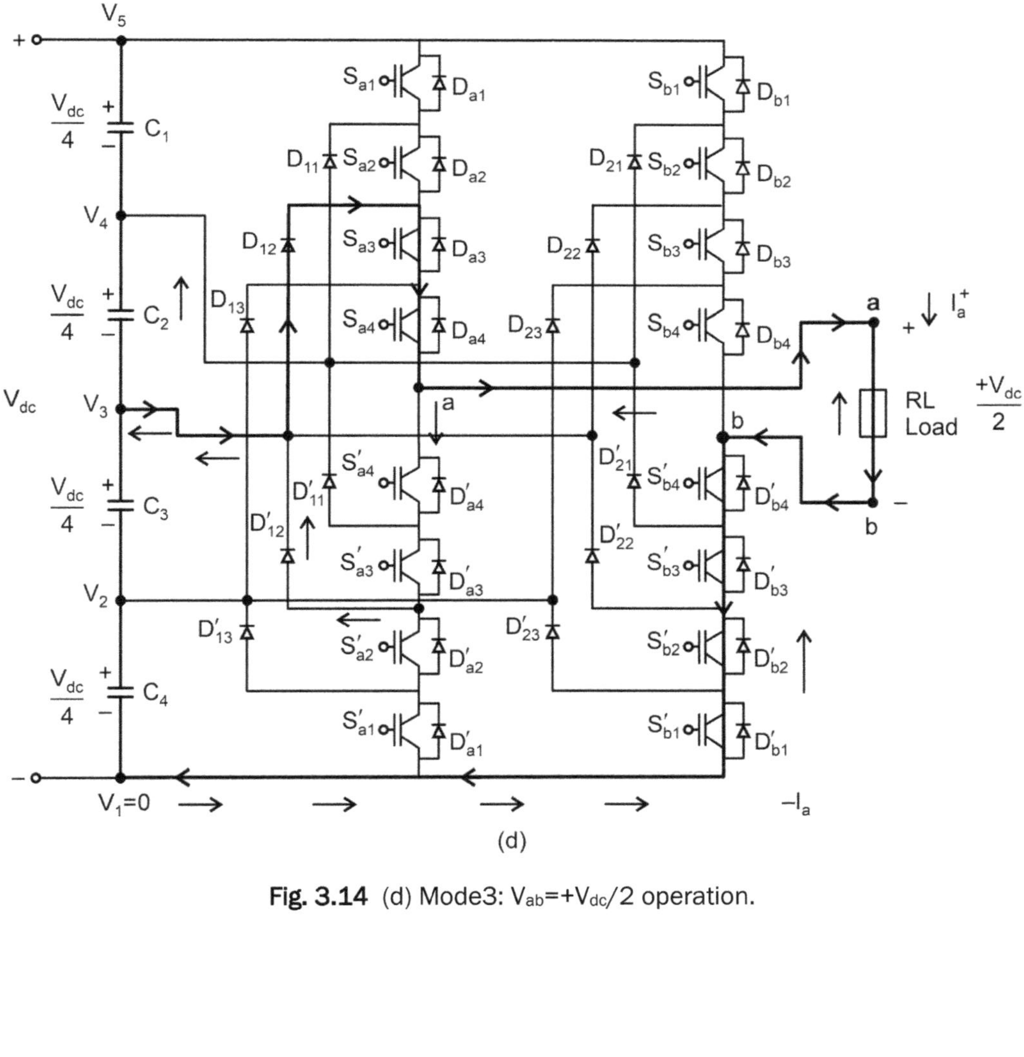

(d)

Fig. 3.14 (d) Mode3: V_{ab}=+V_{dc}/2 operation.

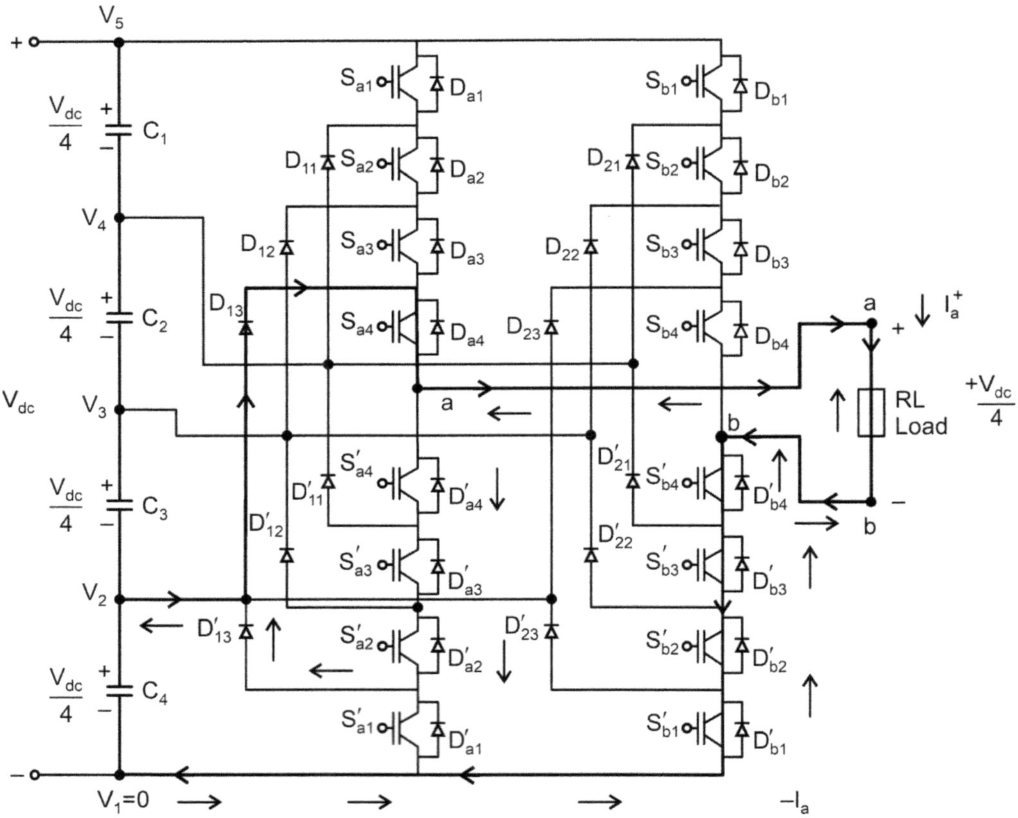

(e)

Fig. 3.14 (e) Mode4: V_{ab}=+V_{dc}/4 operation.

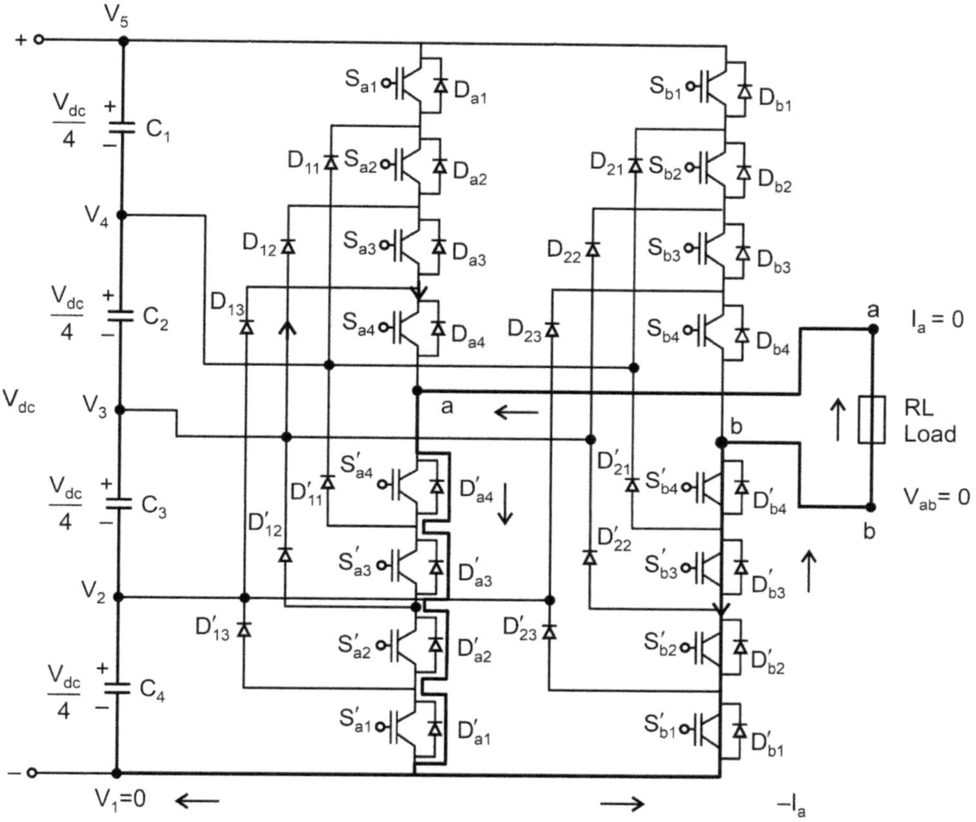

(f)

Fig. 3.14 (f) Mode5: V_{ab}=0 operation.

Mode 6: For $V_{ab} = V_2 = -V_{dc}/4$, the switching devices S_{b4} and S_{a4}', S_{a3}', S_{a2}', S_{a1}' are turned-on and conduction takes place from the path formed by V_2, clamping diode D_{23}, switches S_{b4} and load terminal '*b*' to '*a*' and S_{a4}', S_{a3}', S_{a2}', S_{a1}' marked with thick-line as shown in Fig.3.14 (g), and load current I_a is negative when it flows from load terminal '*b*' to '*a*'.

Mode 7: For $V_{ab} = V_3 = -V_{dc}/2$, the switching devices S_{b3}, S_{b4} and S_{a4}', S_{a3}', S_{a2}', S_{a1}' are turned-on and conduction takes place from the path formed by V_3, clamping diode D_{22}, switches S_{b3}, S_{b4} and load terminal '*b*' to '*a*' and S_{a4}', S_{a3}', S_{a2}', S_{a1}' marked with thick-line as shown in Fig.3.14 (h), and load current I_a is negative when it flows from load terminal '*b*' to '*a*'.

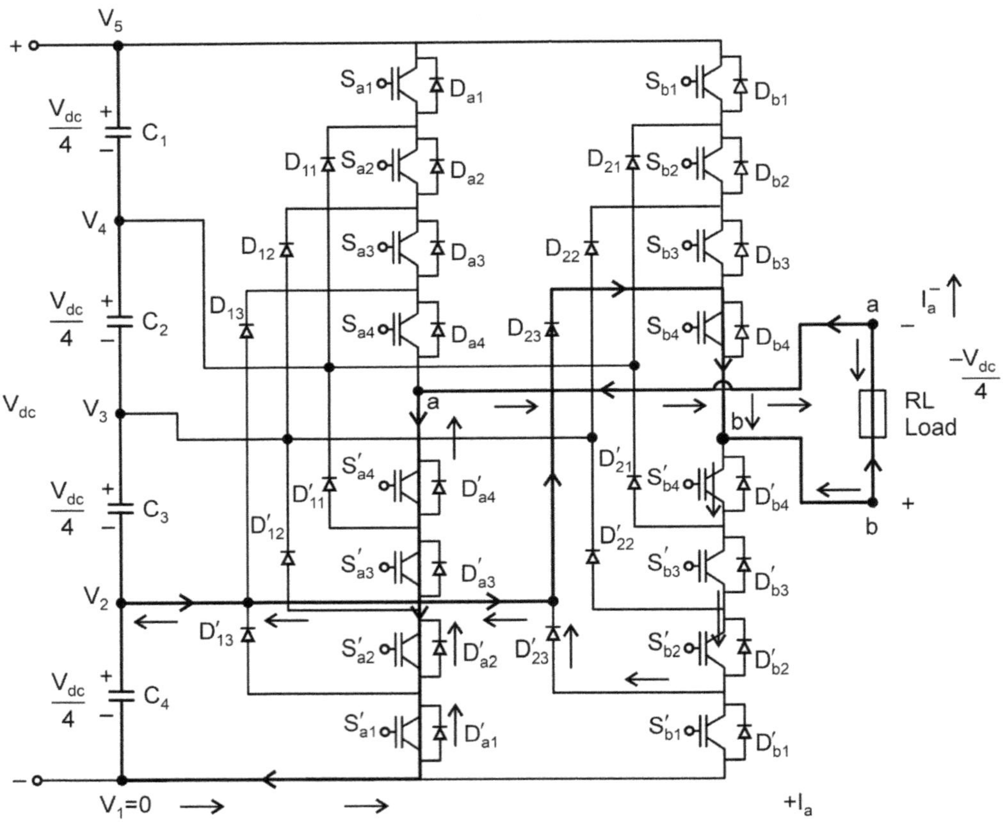

(g)

Fig. 3.14 (g) Mode6: $V_{ab}=-V_{dc}/4$ operation.

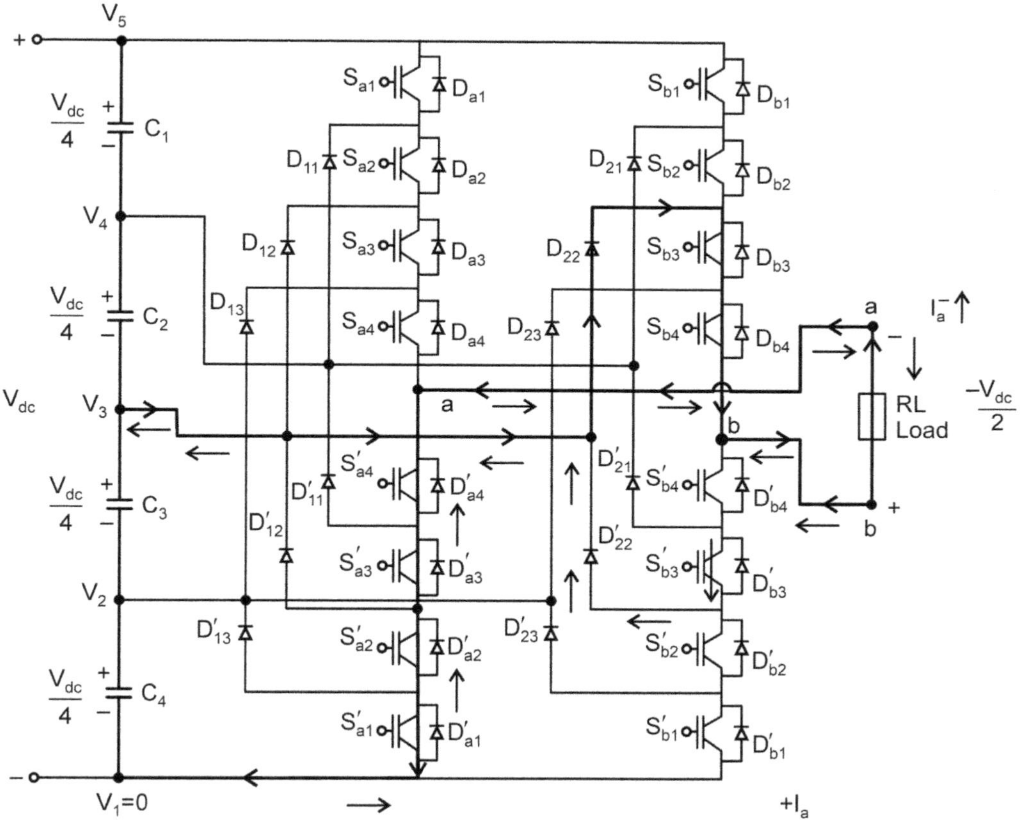

(h)

Fig. 3.14 (h) Mode7: V_{ab}=-V_{dc}/2 operation.

Mode 8: For $V_{ab} = V_4 = -3V_{dc}/4$, the switching devices S_{b2}, S_{b3}, S_{b4} and S_{a4}', S_{a3}', S_{a2}', S_{a1}' are turned-on and conduction takes place from the path formed by V_4, clamping diode D_{21}, switches S_{b2}, S_{b3}, S_{b4} and load terminal '*b*' to '*a*' and S_{a4}', S_{a3}', S_{a2}', S_{a1}' marked with thick-line as shown in Fig.3.14 (i), and load current I_a is negative when it flows from load terminal '*b*' to '*a*'.

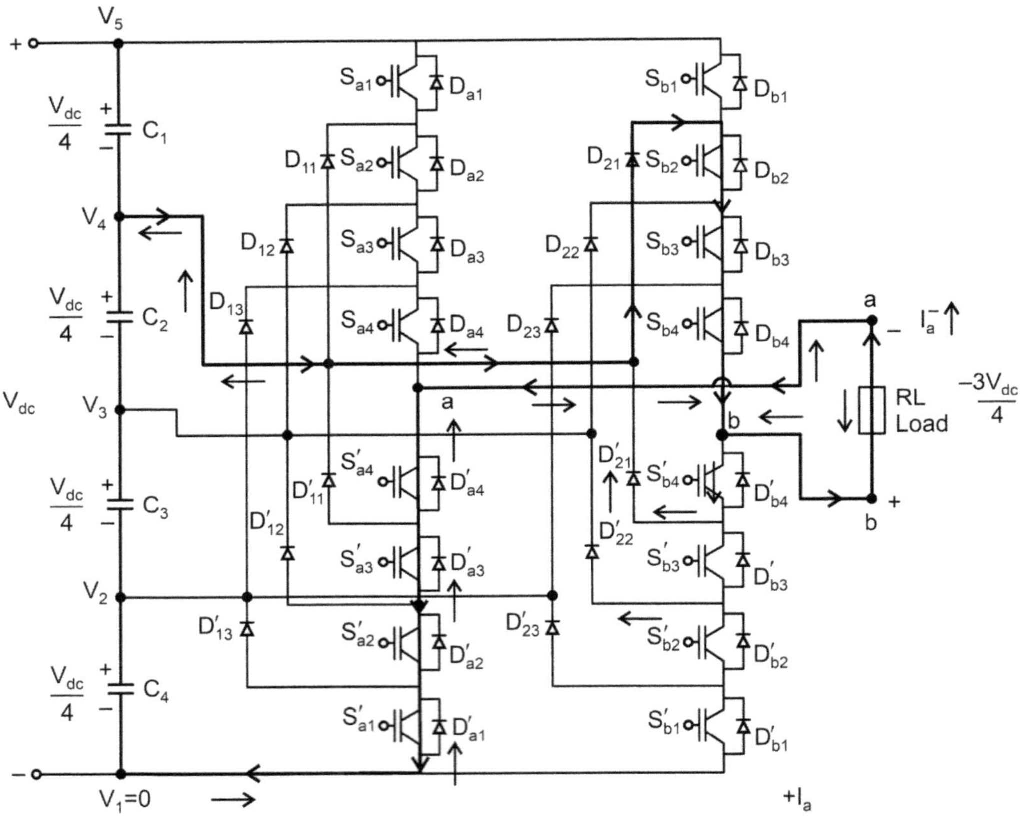

(i)

Fig. 3.14 (i) Mode8: V_{ab}=-3V_{dc}/4 operation.

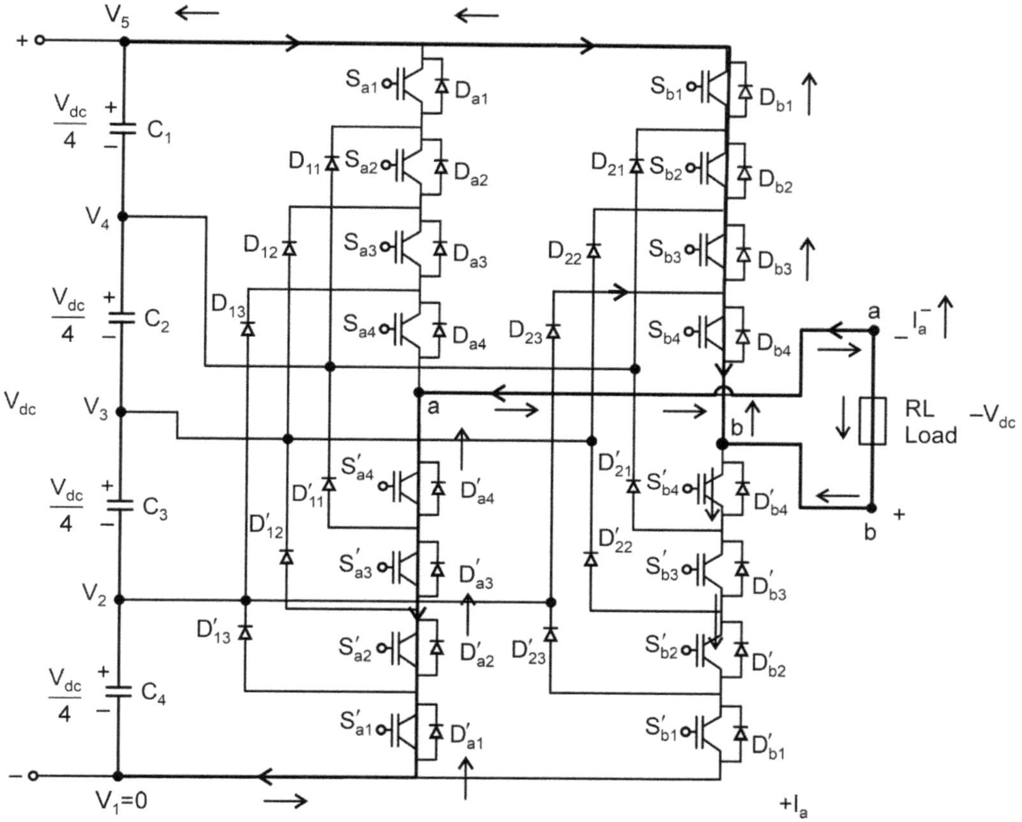

(j)

Fig. 3.14 (j)Mode9: V_{ab}=-V_{dc} operation.

Mode 9: For $V_{ab} = V_5 = -V_{dc}$, the switching devices S_{b1}, S_{b2}, S_{b3}, S_{b4} and S_{a4}', S_{a3}', S_{a2}', S_{a1}' are turned-on and conduction takes place from the path formed by S_{b1}, S_{b2}, S_{b3}, S_{b4} and load terminal '*b*' to '*a*' and S_{a4}', S_{a3}', S_{a2}', S_{a1}' marked with thick-line as shown in Fig.3.14 (j), and load current I_a is negative when it flows from load terminal '*b*' to '*a*'.

One more leg is added to form a three-phase full-bridge inverter. The circuit diagram for three phase five-level diode-clamped full-bridge inverter is shown in Fig.3.15, and three phase five-level load line voltage waveform given in Fig.3.16.

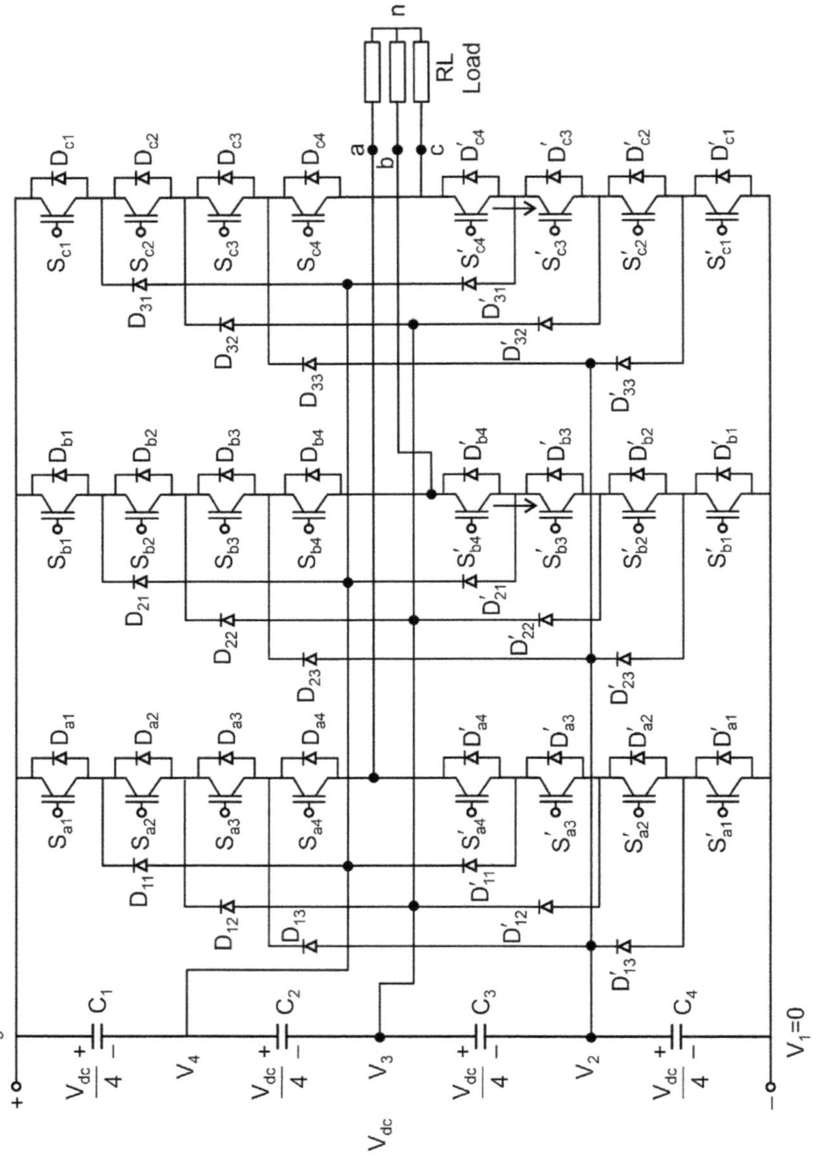

Fig. 3.15 Three phase five-level full bridge diode-clamped inverter for star-connected loads.

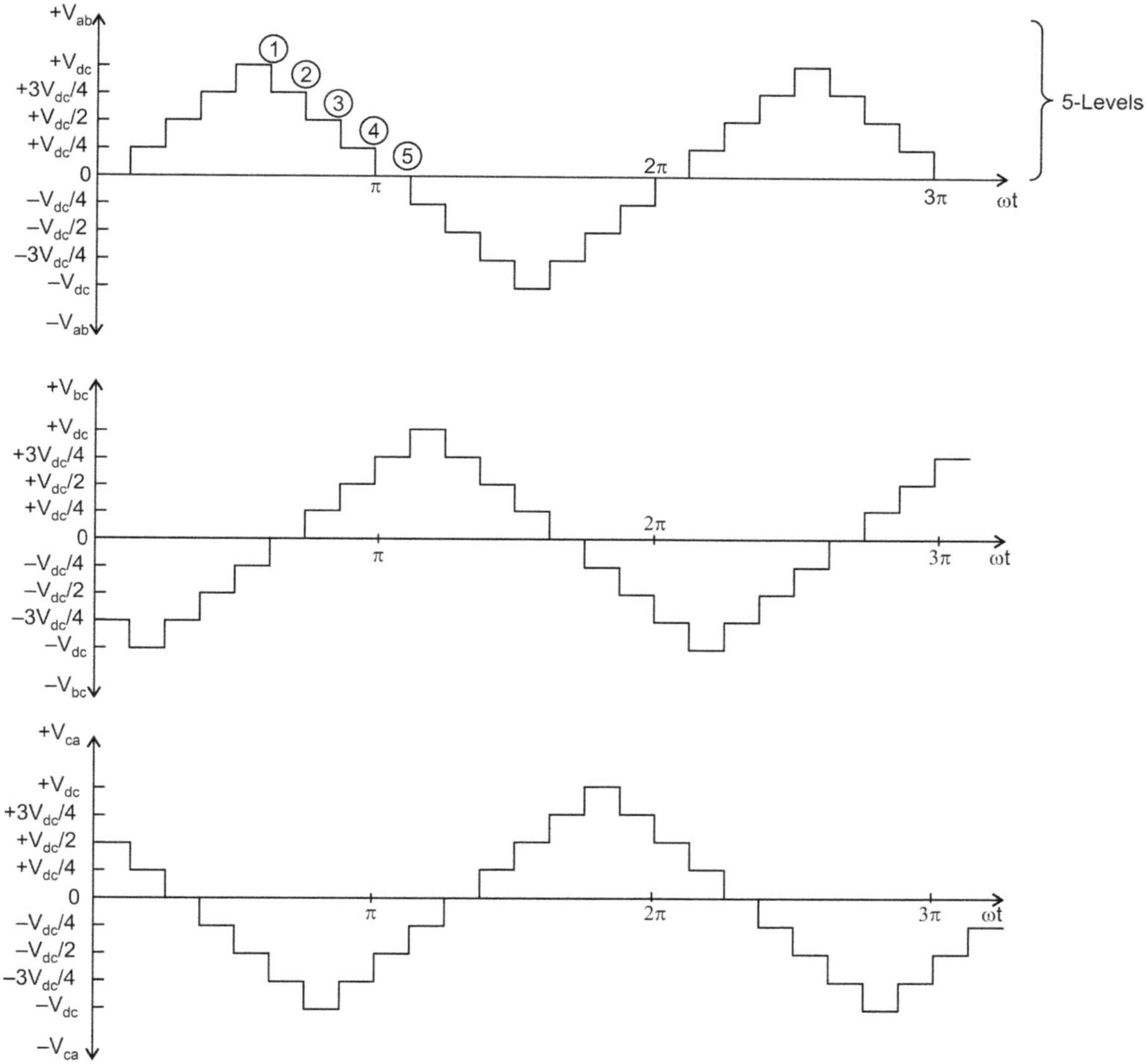

Fig. 3.16 Load line voltage waveforms for three phase five-level full-bridge diode-clamped inverter with star-connected loads.

The conduction of inverter with the influence of induced emf in load inductance (stored energy): During the operation of the inverter for each mode, the flow of current into the resistive-inductive load results in an induced emf of appropriate magnitude depends both on the strength of the inductance and the magnitude of the dc source voltage, is now the load is acting as a source to force a current in the reverse direction to find a conduction path to feedback the energy stored in the inductor in the form of emf (V_L). The nine-modes are considered for explaining the stored energy feedback operations as follows.

Mode 1: During $V_{ab}= +V_{dc}$ with positive load current I_a, the load inductor induced appropriate magnitude of emf and after this period the load induced emf acts as a source and forces a current in the opposite direction into the input terminals for feedback of stored energy until the energy is completely dissipated or exhausted. The reverse current (I_a is negative) path is formed through input terminal V_1, diodes D_{b1}', D_{b2}', D_{b3}', D_{b4}', load terminals '*b*' to '*a*', diodes D_{a4}, D_{a3}, D_{a2}, D_{a1}, to input terminal V_5. This path is shown with broken-lines in Fig.3.14 (b).

Mode 2: During $V_{ab}= +3V_{dc}/4$ with positive load current I_a, the load inductor induced appropriate magnitude of emf and after this period the load induced emf acts as a source and forces a current in the opposite direction into the input terminals to feedback the stored energy until the energy is completely dissipated or exhausted. The reverse current (I_a is negative) path is formed through input terminal V_1, diodes D_{b1}', D_{b2}', D_{b3}', D_{b4}', load terminals '*b*' to '*a*', switch S_{a4}', diode D_{11}' to input terminal V_4. This path is shown with broken-lines in Fig.3.14 (c).

Mode 3:During $V_{ab}= +V_{dc}/2$ with positive load current I_a, the load inductor induced appropriate magnitude of emf and after this period the load induced emf acts as a source and forces a current in the opposite direction into the input terminals to feedback the stored energy until the energy is completely dissipated or exhausted. The reverse current (I_a is negative) path is formed through input terminal V_1, diodes D_{b1}', D_{b2}', D_{b3}', D_{b4}', load terminals '*b*' to '*a*', switches S_{a4}', S_{a3}', diode D_{12}' to input terminal V_3. This path is shown with broken-lines in Fig.3.14 (d).

Mode 4: During $V_{ab}= +V_{dc}/4$ with positive load current I_a, the load inductor induces appropriate magnitude of emf and after this period the load induced emf acts as a source and forces a current in the opposite direction into the input terminals to feedback the stored energy until the energy is completely dissipated or exhausted. The reverse current (I_a is negative) path is formed through input terminal V_1, diodes D_{b1}', D_{b2}', D_{b3}', D_{b4}', load terminals '*b*' to '*a*', switches S_{a4}', S_{a3}', S_{a2}', diode D_{13}' to input terminal V_2. This path is shown with broken-lines in Fig.3.14 (e).

Mode 5: During $V_{ab}= 0$ with zero load current, there is no induced emf in the load inductor, hence there is no stored energy to feedback to the source as in Fig. 3.14(f).

Mode 6: During $V_{ab}= -V_{dc}/4$ with negative load current I_a, the load inductor induces appropriate magnitude of emf in the opposite polarity and after this period the load induced emf acts as a source and forces a current in the opposite direction (feedback current I_a is positive). The conduction path is initiated from input terminal V_1, through D_{a1}', D_{a2}', D_{a3}', D_{a4}', load terminals '*a*' to '*b*', switches S_{b4}', S_{b3}', S_{b2}', diode D_{23}', to input terminal V_2. This path is shown with broken-lines in Fig.3.14 (g).

Mode 7: During $V_{ab}= -V_{dc}/2$ with negative load current I_a, the load inductor induces appropriate magnitude of emf in the opposite polarity and after this period the load induced emf acts as a source and forces a current in the opposite direction (feedback current Ia is positive). The conduction path is initiated from input terminal V_1, through D_{a1}', D_{a2}', D_{a3}', D_{a4}', load terminals 'a' to 'b', switches S_{b4}', S_{b3}', diode D_{22}', to input terminal V_3. This path is shown with broken-lines in Fig.3.14 (h).

Mode 8: During V_{ab}= $-3V_{dc}/4$ with negative load current I_a, the load inductor induces appropriate magnitude of emf in the opposite polarity and after this period the load induced emf acts as a source and forces a current in the opposite direction (feedback current I_a is positive). The conduction path is initiated from input terminal V_1, through D_{a1}', D_{a2}', D_{a3}', D_{a4}', load terminals 'a' to 'b', switches S_{b4}', diode D_{21}', to input terminal V_4. This path is shown with broken-lines in Fig.3.14 (i).

Mode 9: During V_{ab}= $-V_{dc}$ with negative load current I_a, the load inductor induces appropriate magnitude of emf in the opposite polarity and after this period the load induced emf acts as a source and forces a current in the opposite direction (feedback current I_a is positive). The conduction path is initiated from input terminal V_1, through D_{a1}', D_{a2}', D_{a3}', D_{a4}', load terminals 'a' to 'b', diodes D_{b4}, D_{b3}, D_{b2}, D_{b1}, to input terminal V_5. This path is shown with broken-lines in Fig.3.14 (j).

Single phase seven-level diode-clamped inverter with single-leg arrangement: The working of the single phase seven-level diode-clamped inverter circuit is explained in details with the help of Fig. 3.17 for various levels of input voltages, namely, V_0= V_1, V_2, V_3, V_4, V_5, V_6, V_7 (V_1=0, V_2=$V_{dc}/6$, V_3=$V_{dc}/3$, V_4=$V_{dc}/2$, V_5=$2V_{dc}/3$, V_6=$5V_{dc}/6$ and V_7=V_{dc}) by considering one-leg of the circuit for ease of understanding with terminal V_4 as mid-point connection. Fig. 3.17 (a) main circuit configuration and (b) shows for V_{a0} = +Vdc/2, through forward conduction path formed by firing switching devices S_{a1}, S_{a2}, S_{a3}, S_{a4}, S_{a5} and S_{a6} as shown by thick-line. The detailed operations are discussed with the help of seven- modes as follows.

The conduction of seven-level inverter with dc source voltage:

Mode 1: For V_{a0} = $+V_{dc}/2$, the switching devices S_{a1}, S_{a2}, S_{a3}, S_{a4}, S_{a5} and S_{a6} are to be fired so that the conduction takes place from dc source terminal V_7, switches S_{a1}, S_{a2}, S_{a3}, S_{a4}, S_{a5}, S_{a6}, load terminal 'a' to '0' to terminal V_4 with load current I_a is positive as given in Fig.3.17 (b), shown with thick lines.

Mode 2: For V_{a0} = $+V_{dc}/3$, the switching devices S_{a2}, S_{a3}, S_{a4}, S_{a5} and S_{a6} are to be fired so that the conduction takes place from dc source terminal V_6, clamping diode D_{11}, switches S_{a2}, S_{a3}, S_{a4}, S_{a5}, S_{a6}, load terminal 'a' to '0' to terminal V_4 with load current Ia is positive as given in Fig.3.17 (c), shown in thick lines.

Mode 3: For V_{a0} = $+V_{dc}/6$, the switching devices S_{a3}, S_{a4}, S_{a5} and S_{a6} are to be fired so that the conduction takes place from dc source terminal V_5, clamping diodes D_{22}, D_{21}, switches S_{a3}, S_{a4},

S_{a5}, S_{a6}, load terminal '*a*' to '*0*' to terminal V_4 with load current Ia is positive as given in Fig.3.17 (d), shown with thick lines.

Mode 4: For $V_{a0} = 0$, the switching devices S_{a4}, S_{a5} and S_{a6} are to be fired so that the conduction takes place from dc source terminal V_4, clamping diodes D_{33}, D_{32}, D_{31}, switches S_{a4}, S_{a5}, S_{a6}, load terminal '*a*' to '*0*' to terminal V_4 with load current I_a is zero as given in Fig.3.17 (e), shown with thick lines.

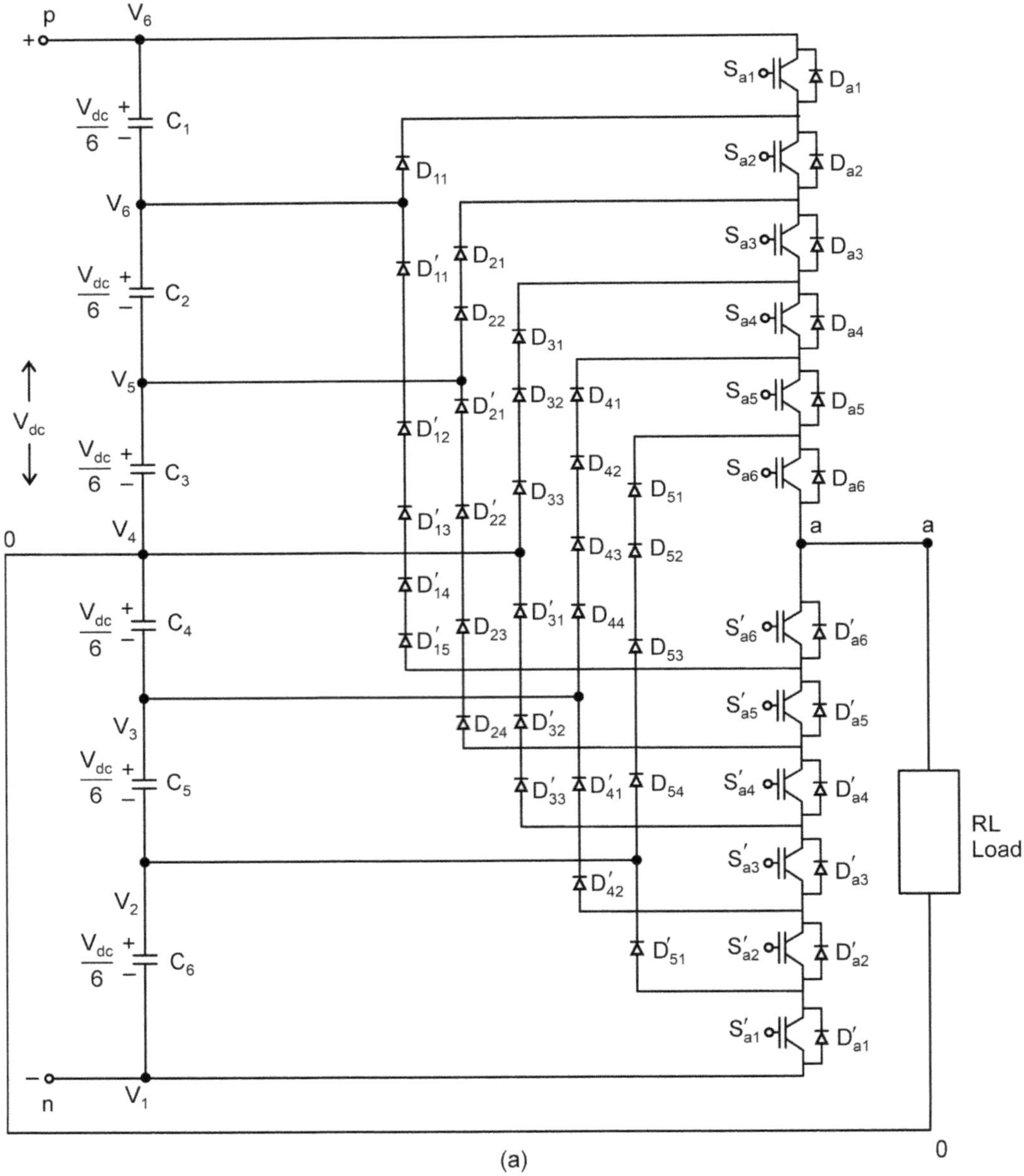

(a)

Fig. 3.17 (a) Main circuit configuration.

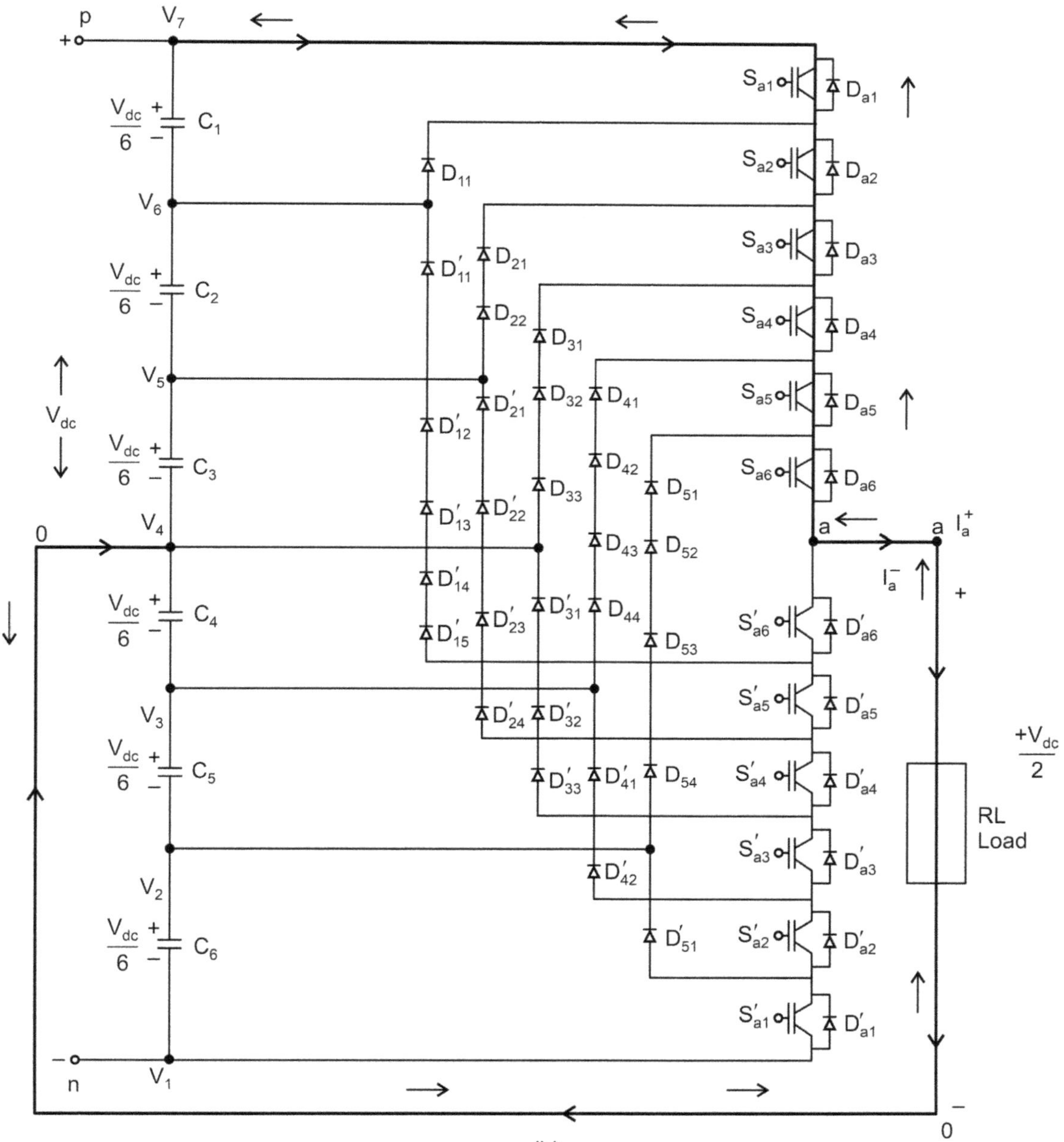

(b)

Fig. 3.17 (b) Mode1: $V_{a0} = +V_{dc}/2$ with I_a positive.

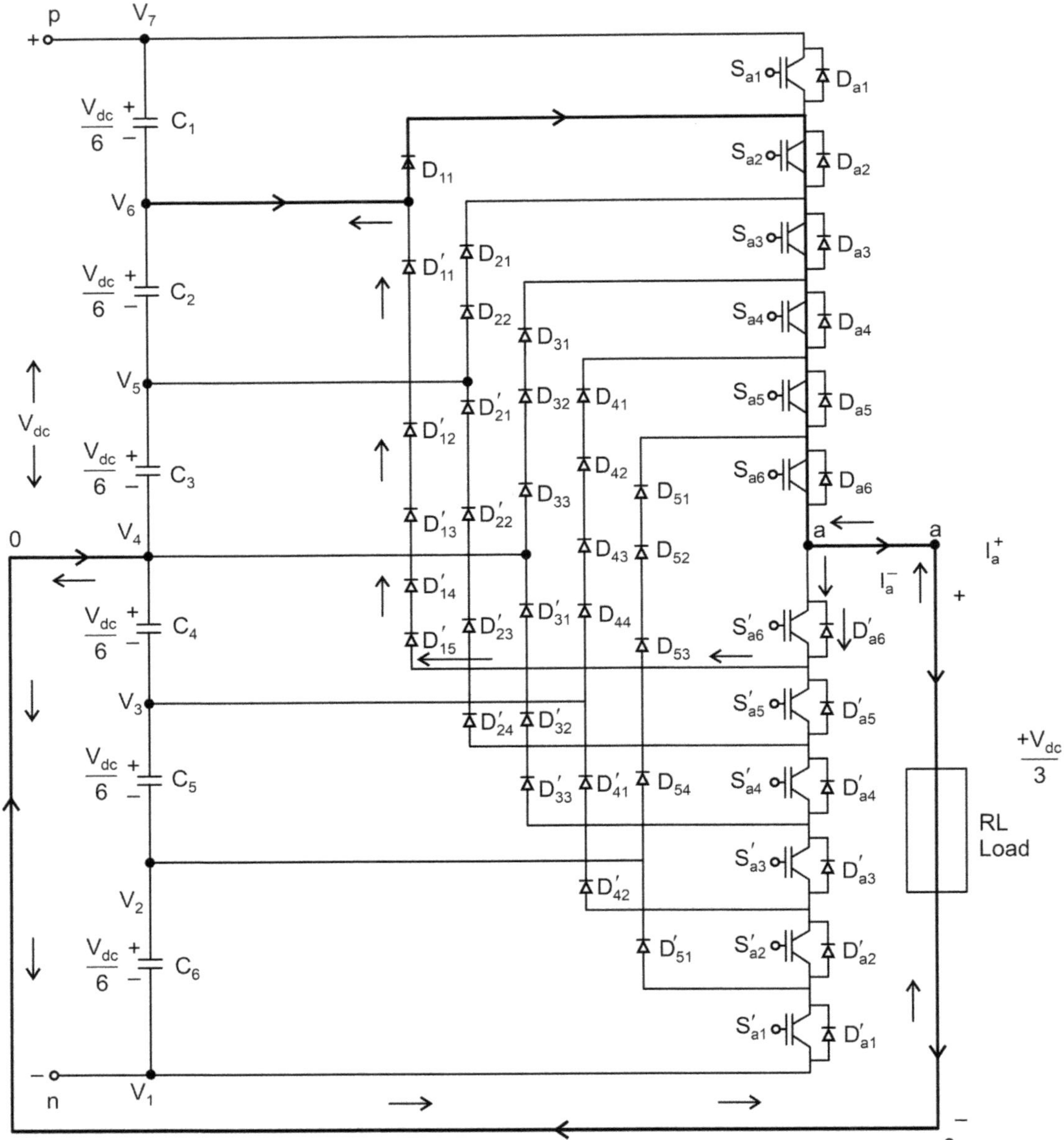

(c)

Fig. 3.17 (c) Mode2: V_{a0}=+V_{dc}/3 with I_a positive.

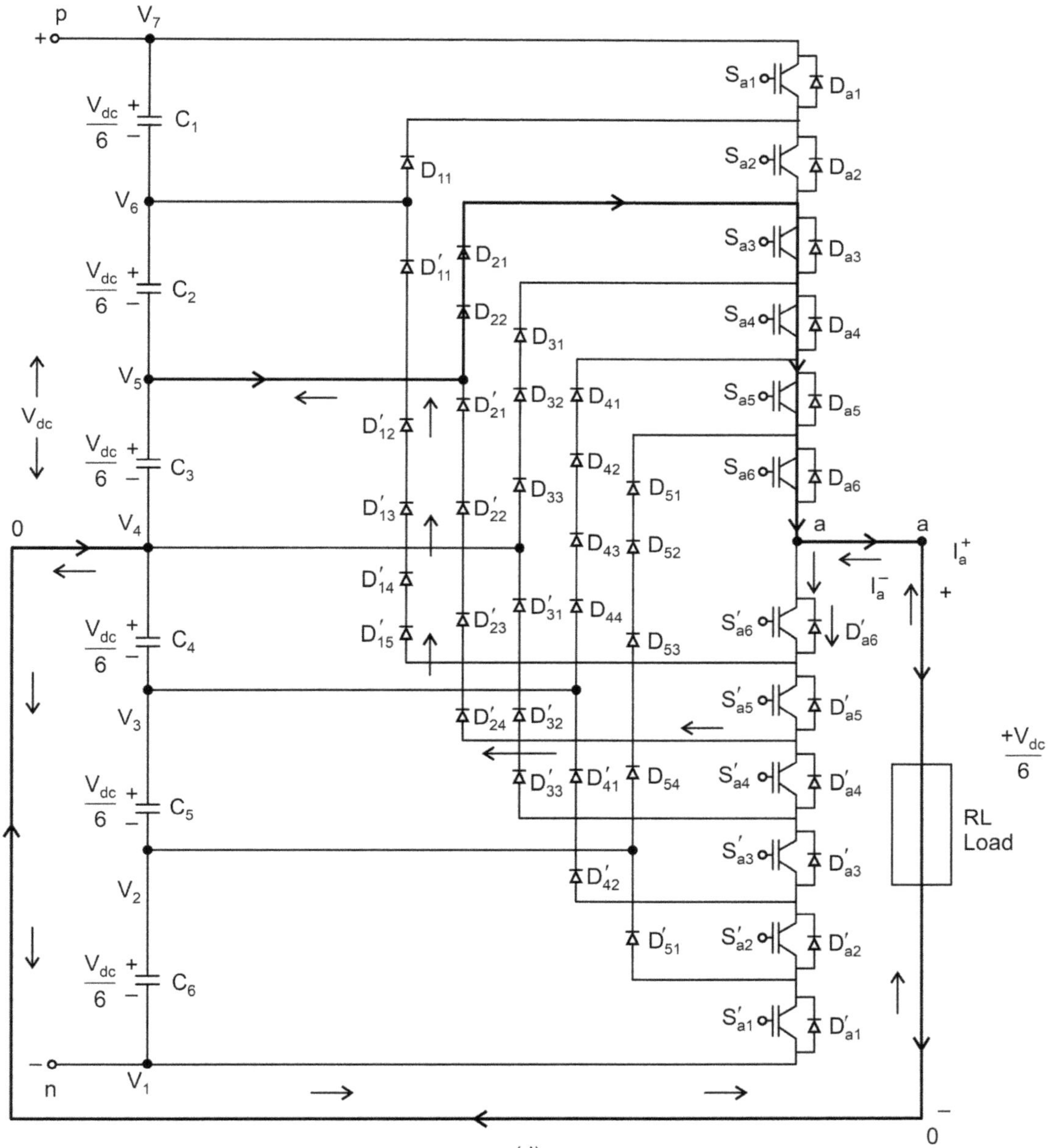

(d)

Fig. 3.17 (d) Mode3: $V_{a0} = +V_{dc}/6$ with I_a positive.

Mode 5: For V_{a0}= $-V_{dc}/6$, the switching devices S_{a6}',S_{a5}',S_{a4}' and S_{a3}' are to be fired so that the conduction takes place from dc source terminal V_4, load terminals '*0*' to '*a*', switches S_{a6}',S_{a5}',S_{a4}',S_{a3}', clamping diodes D_{42}', D_{41}' source terminal V_3 with load current I_a is negative as given in Fig.3.17 (f), shown with thick lines.

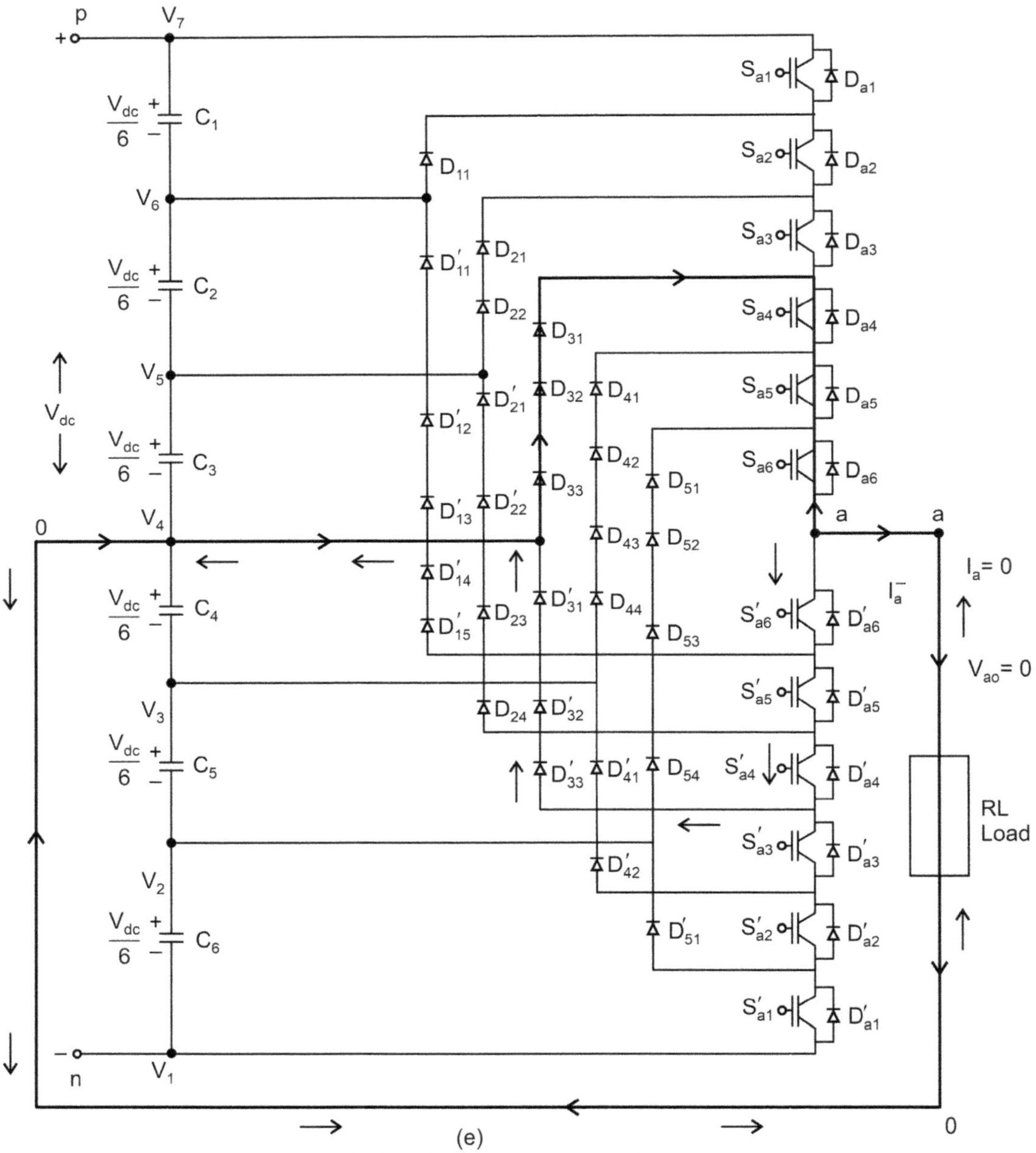

(e)

Fig. 3.17 (e) Mode4: V_{a0}= 0 with I_a zero.

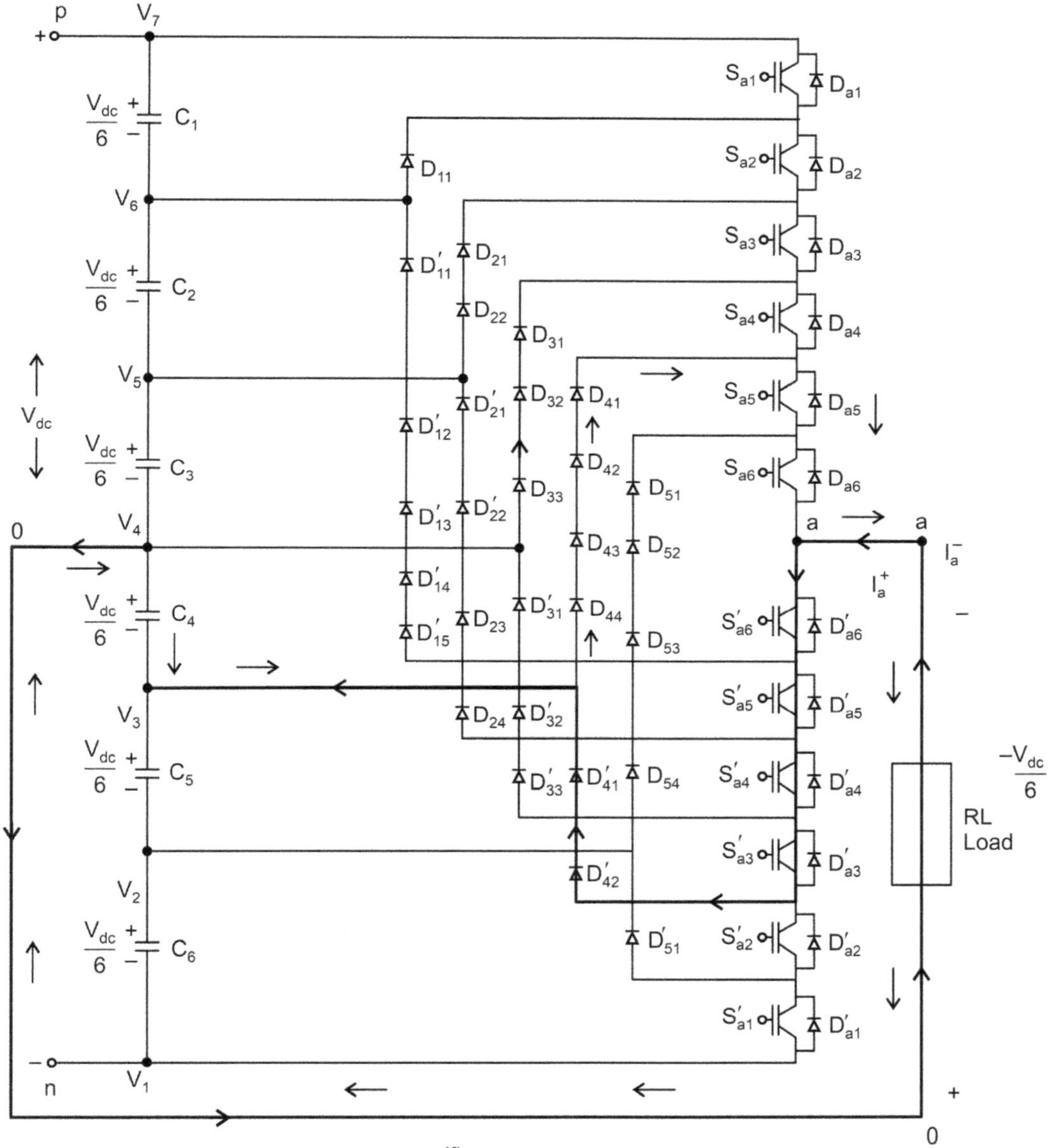

(f)

Fig. 3.17 (f) Mode 5: $V_{a0} = -V_{dc}/6$ with I_a negative.

Mode 6: For $V_{a0}= -V_{dc}/3$, the switching devices S_{a6}',S_{a5}',S_{a4}',S_{a3}' and S_{a2}' are to be fired so that the conduction takes place from dc source terminal V_4, load terminals '*0*' to '*a*', switches S_{a6}',S_{a5}',S_{a4}',S_{a3}',S_{a2}', clamping diode D_{51}' to source terminal V_2 with load current I_a is negative as given in Fig.3.17 (g), shown with thick lines.

Mode 7: For V_{a0}= -V_{dc}/2, the switching devices S_{a6}',S_{a5}',S_{a4}',S_{a3}',S_{a2}' and S_{a1}' are to be fired so that the conduction takes place from dc source terminal V_4, load terminals '*0*' to '*a*', switches S_{a6}',S_{a5}',S_{a4}',S_{a3}',S_{a2}',S_{a1}', source terminal V_1 with load current I_a is negative as given in Fig.3.17 (h), shown with thick lines.

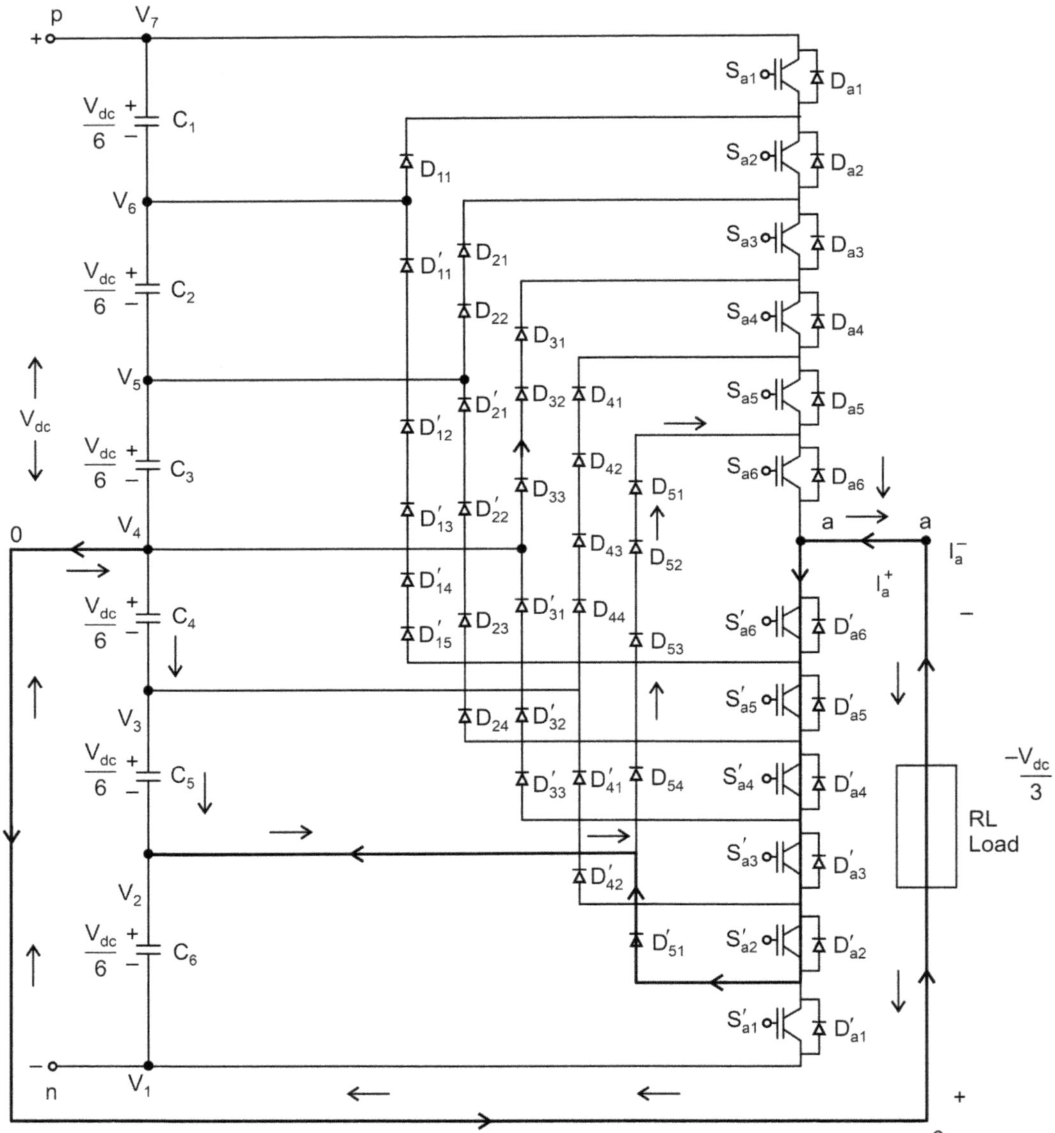

(g)

Fig. 3.17 (g) Mode 6: V_{a0}= $-V_{dc}/3$ with I_a negative.

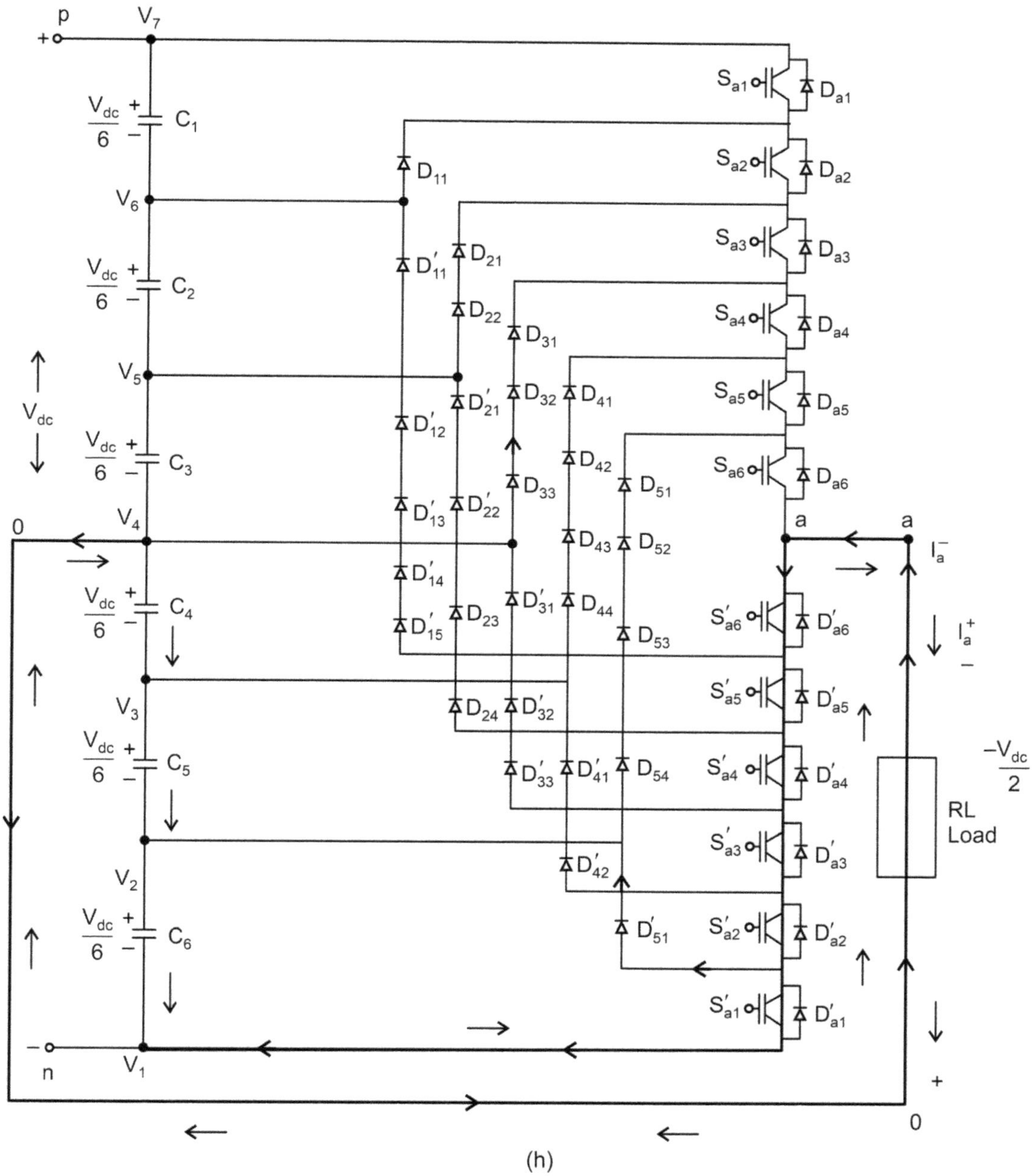

Fig. 3.17 (h) Mode 7: $V_{a0} = -V_{dc}/2$ with I_a negative.

The conduction of inverter with the influence of induced emf in load inductance (stored energy): During the operation of the inverter for each mode, the flow of current into the resistive-inductive load results in an induced emf of appropriate magnitude depends both on the strength of the inductance and the magnitude of the dc source voltage, is now the load is acting as a source to force a current in the reverse direction to find a conduction path to feedback the energy stored in the inductor in the form of emf (V_L). The seven-modes are considered for explaining the stored energy feedback operations as follows.

Mode 1: During $V_{a0}= +V_{dc}/2$ with positive load current I_a, the load inductor induced appropriate magnitude of emf and after this period the load induced emf acts as a source and forces a current in the opposite direction into the input terminals for feedback of stored energy until the energy is completely dissipated or exhausted. The reverse current (I_a is negative) path is formed through input terminal V_4, load terminals '*0*' to '*a*', diodes D_{a6}, D_{a5}, D_{a4}, D_{a3}, D_{a2}, D_{a1}, to input terminal V_7. This path is shown with broken-lines in Fig.3.17 (b).

Mode 2: During $V_{a0}= +V_{dc}/3$ with positive load current I_a, the load inductor induced appropriate magnitude of emf and after this period the load induced emf acts as a source and forces a current in the opposite direction into the input terminals for feedback of stored energy until the energy is completely dissipated or exhausted. The reverse current (I_a is negative) path is formed through input terminal V_4, load terminals '*0*' to '*a*', switch S_{a6}', clamping diodes D_{15}', D_{14}', D_{13}', D_{12}', D_{11}', to input terminal V_6. This path is shown with broken-lines in Fig.3.17 (c).

Mode 3: During $V_{a0}= +V_{dc}/6$ with positive load current I_a, the load inductor induced appropriate magnitude of emf and after this period the load induced emf acts as a source and forces a current in the opposite direction into the input terminals for feedback of stored energy until the energy is completely dissipated or exhausted. The reverse current (I_a is negative) path is formed through input terminal V_4, load terminals '*0*' to '*a*', switches S_{a6}', S_{a5}', clamping diodes D_{24}', D_{23}', D_{22}', D_{21}', to input terminal V_5. This path is shown with broken-lines in Fig.3.17 (d).

Mode 4: During $V_{a0} = 0$ with zero load current, emf in the load inductor, hence here is no stored energy to feed back to the source, as shown in Fig. 3.17(e), the path is shown with broken lines.

Mode 5: During $V_{a0}= -V_{dc}/6$ with negative load current I_a, the load inductor induced appropriate magnitude of emf and after this period the load induced emf acts as a source and forces a current in the opposite direction into the input terminals for feedback of stored energy until the energy is completely dissipated or exhausted. The reverse current (I_a is positive) path is formed through input terminal V_3, clamping diodes D_{44}, D_{43}, D_{42}, D_{41}, S_{a5}, S_{a6} load 'a' to '0' to V_4. This path is shown with broken-lines in Fig.3.17 (f).

Mode 6: During $V_{a0}= -V_{dc}/3$ with negative load current I_a, the load inductor induced appropriate magnitude of emf and after this period the load induced emf acts as a source and forces a current in the opposite direction into the input terminals for feedback of stored energy until the energy is completely dissipated or exhausted. The reverse current (I_a is positive) path is formed through input terminal V_2, clamping diodes, D_{54}, D_{53}, D_{52}, D_{51}, S_{a6}, 'a' to '0' to V_4. This path is shown with broken-lines in Fig.3.17 (g).

Mode 7: During $V_{a0} = -V_{dc}/2$ with negative load current I_a, the load inductor induced appropriate magnitude of emf and after this period the load induced emf acts as a source and forces a current in the opposite direction into the input terminals for feedback of stored energy until the energy is completely dissipated or exhausted. The reverse current (I_a is positive) path is formed through input terminal V_1, Diodes D'_{a1}, D'_{a2}, D'_{a3}, D'_{a4}, D'_{a5}, D'_{a6}, load terminal 'a' to '0' to V_4. This path is shown with broken-lines in Fig.3.17 (h).

The switching pattern of switching devices and clamping diodes for single phase single-leg seven-level diode-clamped inverter with various modes are given in Table 3.4 and load phase voltage waveform of seven-level is presented in Fig. 3.18.

Table 3.4 Switching pattern of switching devices and clamping diodes for single phase single-leg seven-level diode-clamped inverter.

	V_{a0}	S_{a1}	S_{a2}	S_{a3}	S_{a4}	S_{a5}	S_{a6}	S_{a6}'	S_{a5}'	S_{a4}'	S_{a3}'	S_{a2}'	S_{a1}'	Clamping diode	Load current I_a
M_1	$+V_{dc}/2$	1	1	1	1	1	1	0	0	0	0	0	0	-----	Positive
M_2	$+V_{dc}/3$	0	1	1	1	1	1	0	0	0	0	0	0	D_{11}	Positive
M_3	$+V_{dc}/6$	0	0	1	1	1	1	0	0	0	0	0	0	D_{22}, D_{21}	Positive
M_4	0	0	0	0	1	1	1	0	0	0	0	0	0	D_{33}, D_{32}, D_{31}	Zero
M_5	$-V_{dc}/6$	0	0	0	0	0	0	1	1	1	1	0	0	D_{42}', D_{41}'	Negative
M_6	$-V_{dc}/3$	0	0	0	0	0	0	1	1	1	1	1	0	D_{51}'	Negative
M_7	$-V_{dc}/2$	0	0	0	0	0	0	1	1	1	1	1	1	-----	Negative

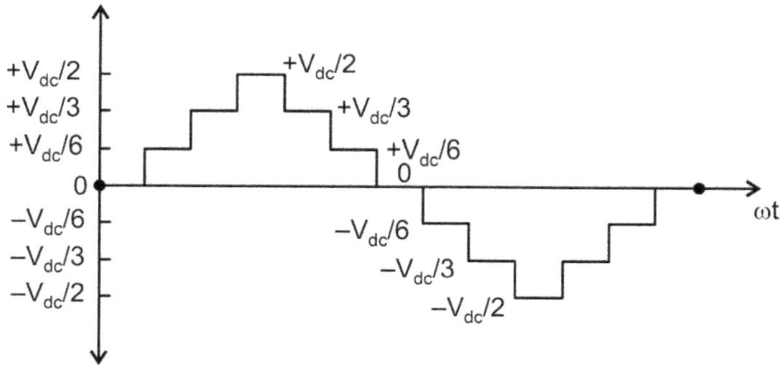

Fig. 3.18 Load phase Voltage (Va0) waveform for single phase single-leg seven-level diode-clamped inverter.

Single phase seven-level diode-clamped full-bridge inverter: The single phase seven-level diode-clamped full-bridge inverter standard configuration is shown in Fig. 3.19 (a) for resistive-inductive load. The detailed circuit operation is explained by considering 10 - modes for producing seven-levels of output voltage, V_{ab} as $V_7 = +V_{dc}$, $V_6 = +5V_{dc}/6$, $V_5 = +2V_{dc}/3$, $V_4 = +V_{dc}/2$, $V_3 = +V_{dc}/3$, $V_2 = +V_{dc}/6$, and $V_1 = 0$ volts.

Mode 1: For $V_{ab} = +V_{dc}$ with load current I_a positive, the switching devices S_{a1}, S_{a2}, S_{a3}, S_{a4}, S_{a5}, S_{a6}, and S_{b6}', S_{b5}', S_{b4}', S_{b3}', S_{b2}', S_{b1}' are to be fired so that the conduction can takes place from the source terminal V_7, through switches S_{a1}, S_{a2}, S_{a3}, S_{a4}, S_{a5}, S_{a6}, load terminals '*a*' to 'b', switches S_{b6}', S_{b5}', S_{b4}', S_{b3}', S_{b2}', S_{b1}', to the source terminal V_1. The load current I_a is positive when it flows from load terminals '*a*' to '*b*' as shown in the Fig.3.19(b).

Mode 2: For $V_{ab}= +5V_{dc}/6$ with load current I_a positive, the switching devices S_{a2}, S_{a3}, S_{a4}, S_{a5}, S_{a6}, and S_{b6}', S_{b5}', S_{b4}', S_{b3}', S_{b2}', S_{b1}' are to be fired so that the conduction can takes place from the source terminal V_6, through clamping diode D_{11}, switches S_{a2}, S_{a3}, S_{a4}, S_{a5}, S_{a6}, load terminals '*a*' to '*b*', switches S_{b6}', S_{b5}', S_{b4}', S_{b3}', S_{b2}', S_{b1}', to the source terminal V_1. The load current I_a is positive when it flows from load terminals '*a*' to '*b*' as shown in the Fig.3.19(c).

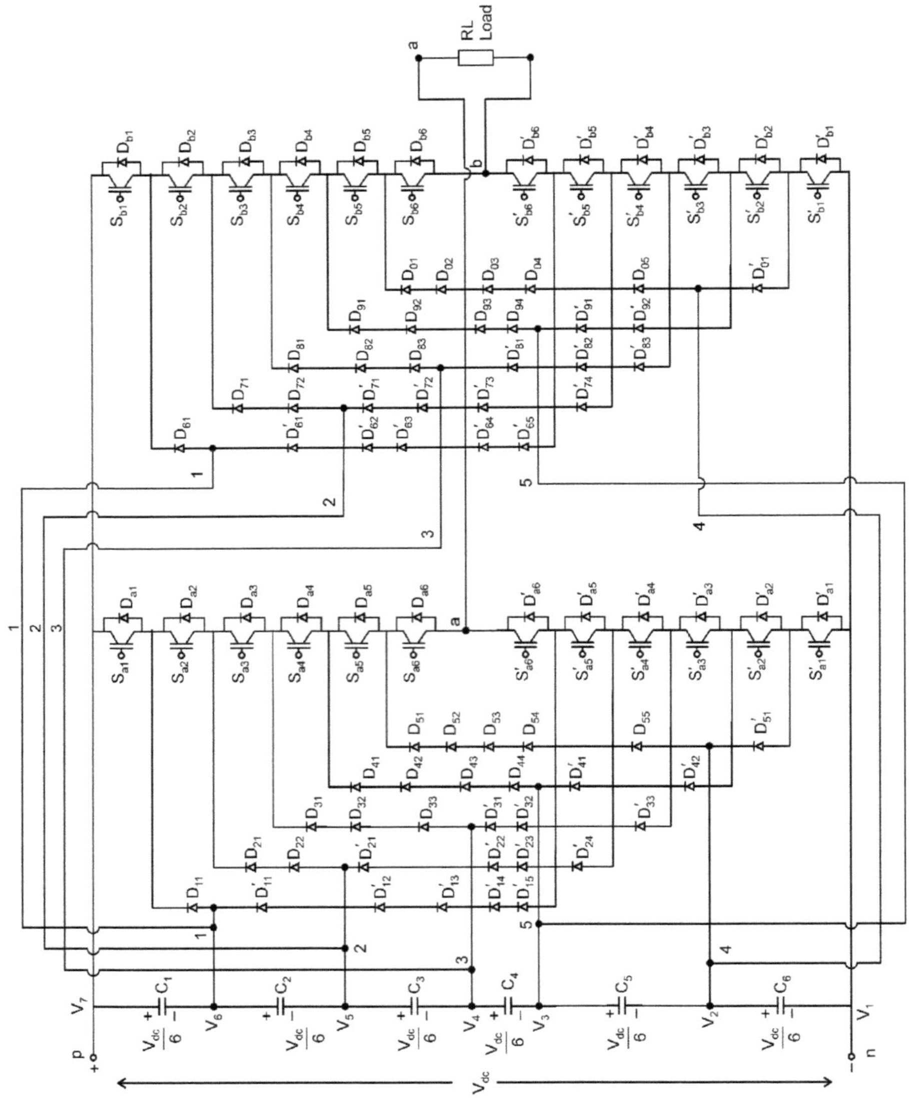

(a)

Fig. 3.19 (a) Main circuit configuration for single phase seven-level diode-clamped full-bridge inverter.

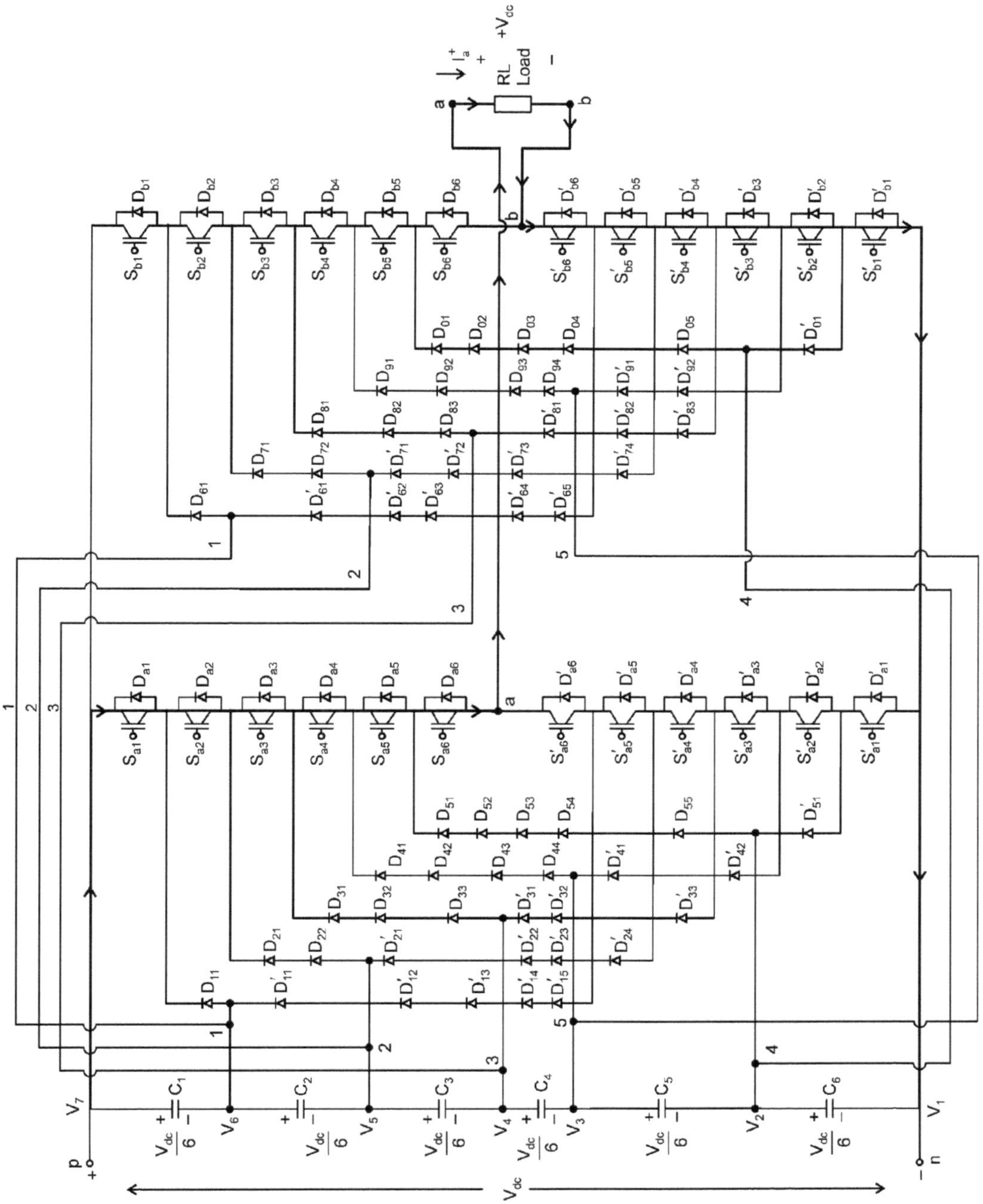

Fig. 3.19 (b) Mode 1: $V_{ab} = +V_{dc}$ (I_a +Ve).

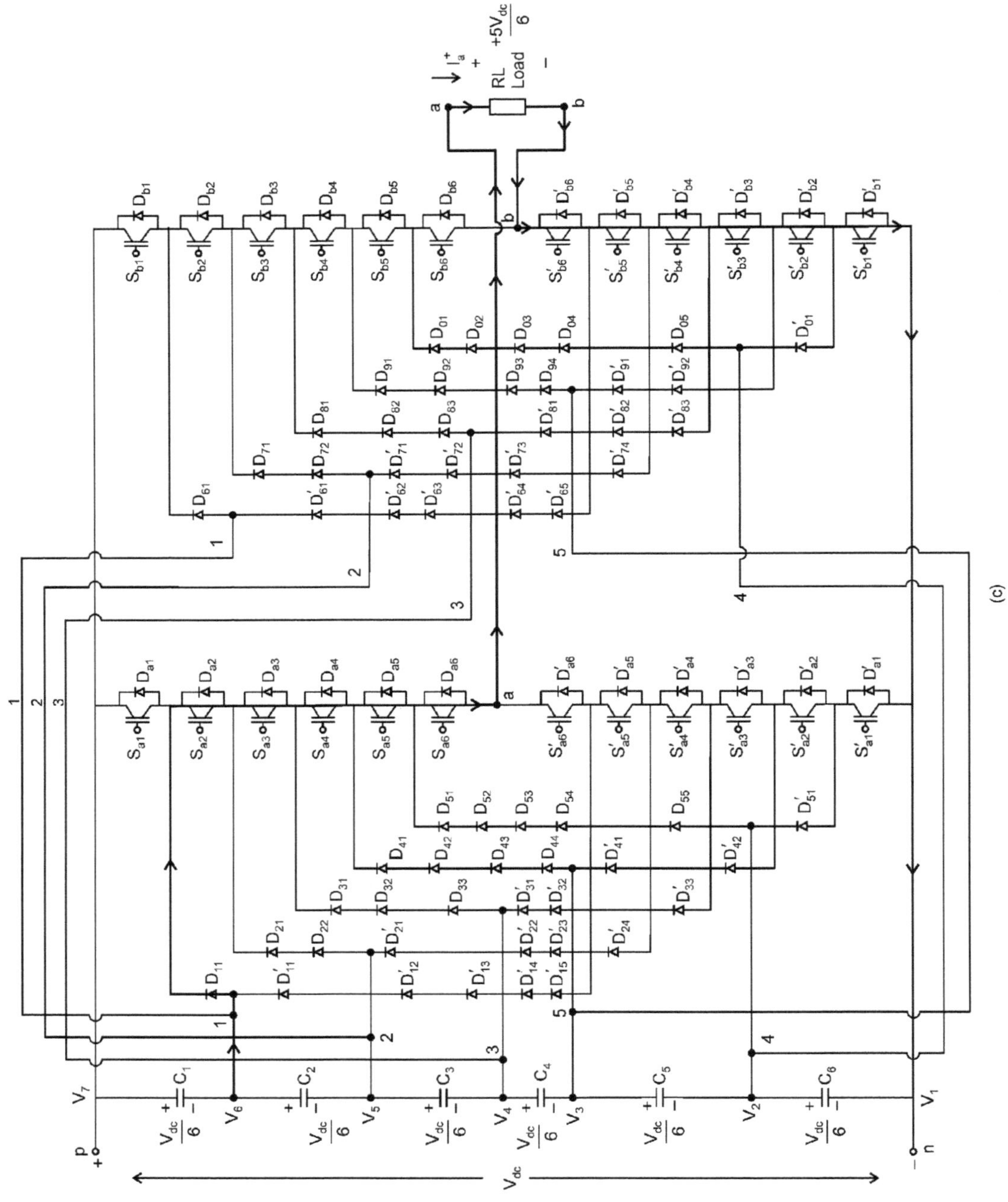

Fig. 3.19 (c) Mode2: V_{ab}= $+5V_{dc}/6$ with I_a positive.

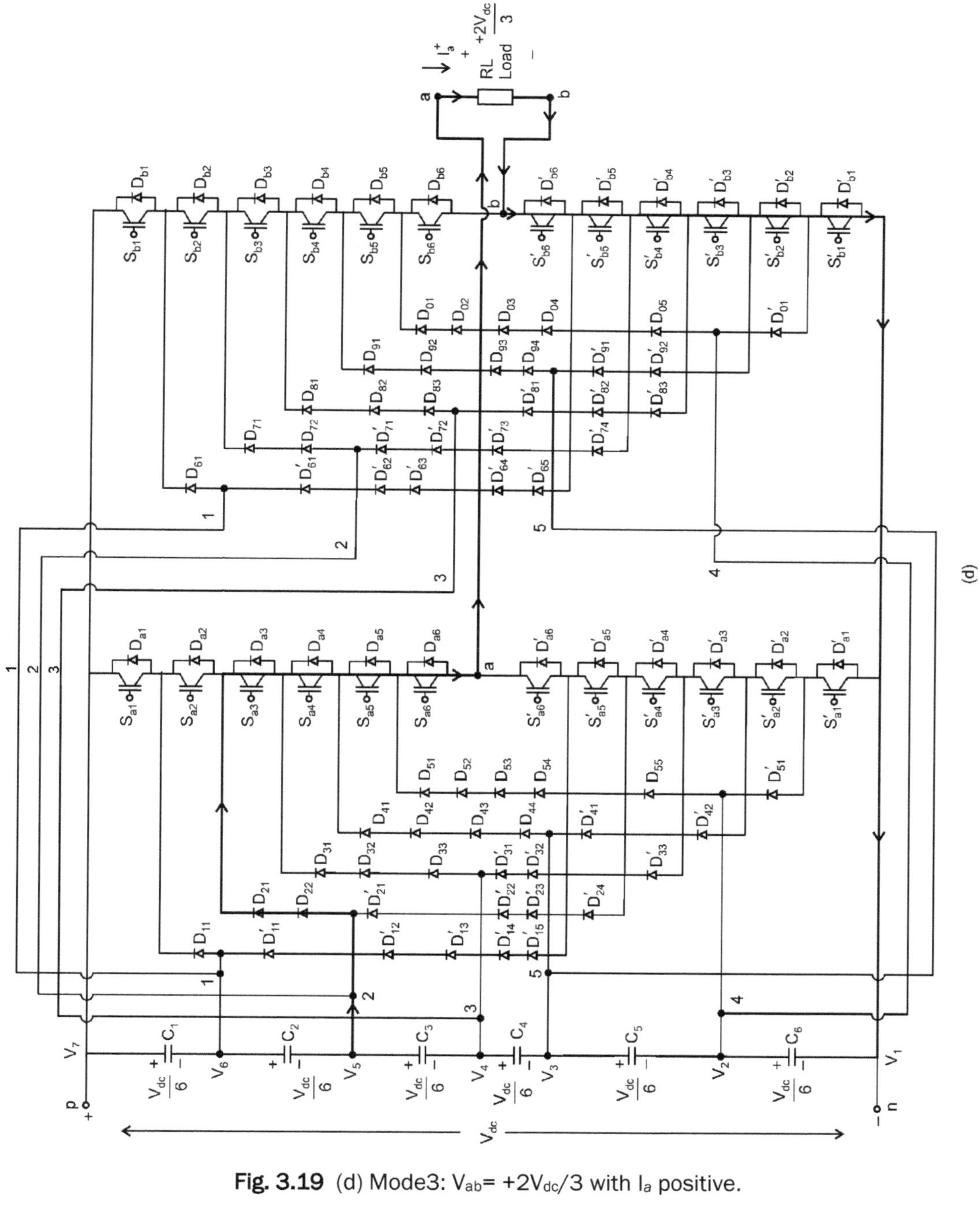

Fig. 3.19 (d) Mode3: $V_{ab} = +2V_{dc}/3$ with I_a positive.

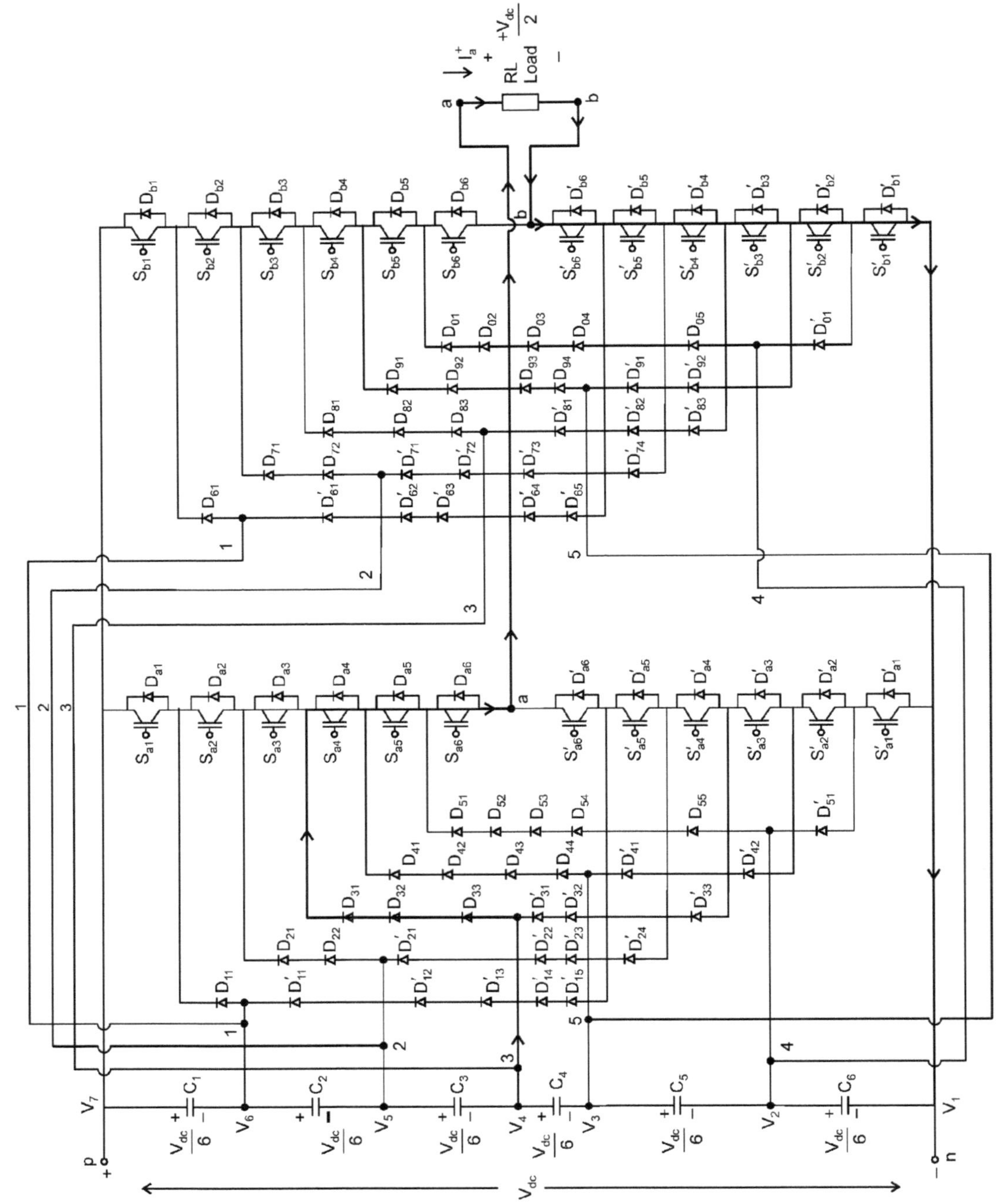

Fig. 3.19 (e) Mode4: $V_{ab}= +V_{dc}/2$ with I_a positive.

Mode 3: For $V_{ab} = +2V_{dc}/3$ with load current I_a positive, the switching devices S_{a3}, S_{a4}, S_{a5}, S_{a6}, and S_{b6}', S_{b5}', S_{b4}', S_{b3}', S_{b2}', S_{b1}' are to be fired so that the conduction can takes place from the source terminal V_5, through clamping diodes D_{22}, D_{21}, switches S_{a3}, S_{a4}, S_{a5}, S_{a6}, load terminals '*a*' to '*b*', switches S_{b6}', S_{b5}', S_{b4}', S_{b3}', S_{b2}', S_{b1}', to the source terminal V_1. The load current I_a is positive when it flows from load terminals '*a*' to '*b*' as shown in the Fig.3.19(d).

Mode 4: For $V_{ab} = +V_{dc}/2$ with load current I_a positive, the switching devices S_{a4}, S_{a5}, S_{a6}, and S_{b6}', S_{b5}', S_{b4}', S_{b3}', S_{b2}', S_{b1}' are to be fired so that the conduction can takes place from the source terminal V_4, through clamping diodes D_{33}, D_{32}, D_{31}, switches S_{a4}, S_{a5}, S_{a6}, load terminals '*a*' to '*b*', switches S_{b6}', S_{b5}', S_{b4}', S_{b3}', S_{b2}', S_{b1}', to the source terminal V_1. The load current I_a is positive when it flows from load terminals '*a*' to '*b*' as shown in the Fig.3.19(e).

Mode 5: For $V_{ab}= +V_{dc}/3$ with load current I_a positive, the switching devices S_{a5}, S_{a6}, and S_{b6}', S_{b5}', S_{b4}', S_{b3}', S_{b2}', S_{b1}' are to be fired so that the conduction can takes place from the source terminal V_3, through clamping diodes D_{44}, D_{43}, D_{42}, D_{41}, switches S_{a5}, S_{a6}, load terminals '*a*' to '*b*', switches S_{b6}', S_{b5}', S_{b4}', S_{b3}', S_{b2}', S_{b1}', to the source terminal V_1. The load current I_a is positive when it flows from load terminals '*a*' to '*b*' as shown in the Fig.3.19(f).

Mode 6: For $V_{ab}= +V_{dc}/6$ with load current I_a positive, the switching devices S_{a6}, and S_{b6}', S_{b5}', S_{b4}', S_{b3}', S_{b2}', S_{b1}' are to be fired so that the conduction can takes place from the source terminal V_2, through clamping diodes D_{55}, D_{54}, D_{53}, D_{52}, D_{51}, switch S_{a6}, load terminals '*a*' to '*b*', switches S_{b6}', S_{b5}', S_{b4}', S_{b3}', S_{b2}', S_{b1}', to the source terminal V_1. The load current I_a is positive when it flows from load terminals '*a*' to '*b*' as shown in the Fig.3.19(g).

Mode 7: For $V_{ab}= 0$ with load current I_a zero, the switching devices S_{b6}', S_{b5}', S_{b4}', S_{b3}', S_{b2}', S_{b1}' are to be fired so that the conduction can takes place from the source terminal V_1, through anti-parallel diodes D_{a1}', D_{a2}', D_{a3}', D_{a4}', D_{a5}', D_{a6}', load terminals '*a*' to '*b*', switches S_{b6}', S_{b5}', S_{b4}', S_{b3}', S_{b2}', S_{b1}', to the source terminal V_1. The load current I_a is zero as shown in the Fig.3.19(h).

Mode 8: For $V_{ab}= -V_{dc}/6$ with load current I_a negative, the switching devices S_{b6}, and S_{a6}', S_{a5}', S_{a4}', S_{a3}', S_{a2}', S_{a1}' are to be fired so that the conduction can takes place from the source terminal V_2, through clamping diodes D_{05}, D_{04}, D_{03}, D_{02}, D_{01}, switch S_{b6}, load terminals '*b*' to '*a*', switches S_{a6}', S_{a5}', S_{a4}', S_{a3}', S_{a2}', S_{a1}', to the source terminal V_1. The load current I_a is negative when it flows from load terminals '*b*' to '*a*' as shown in the Fig.3.19(i).

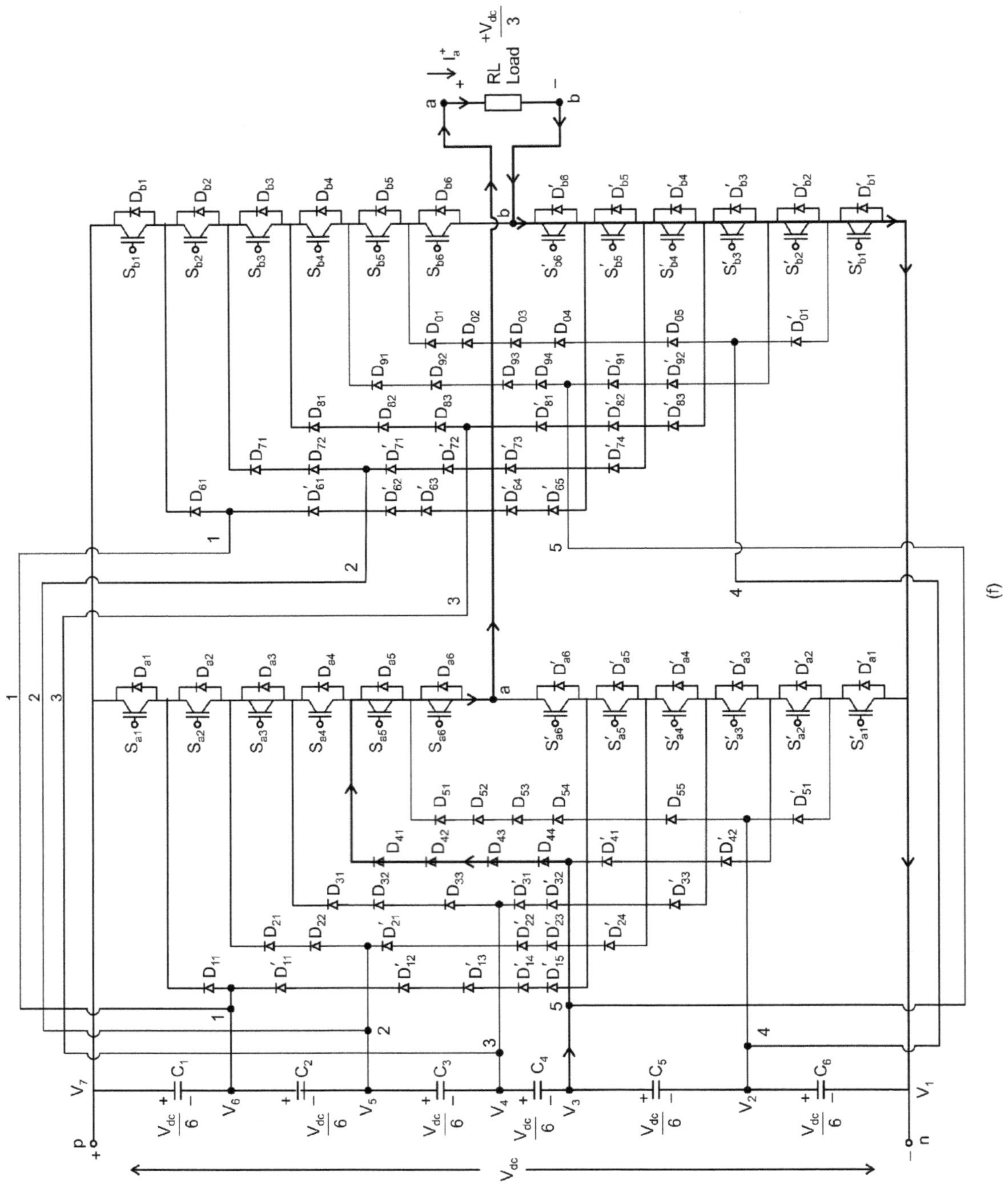

Fig. 3.19 (f) Mode 5: $V_{ab} = +V_{dc}/3$ with I_a positive.

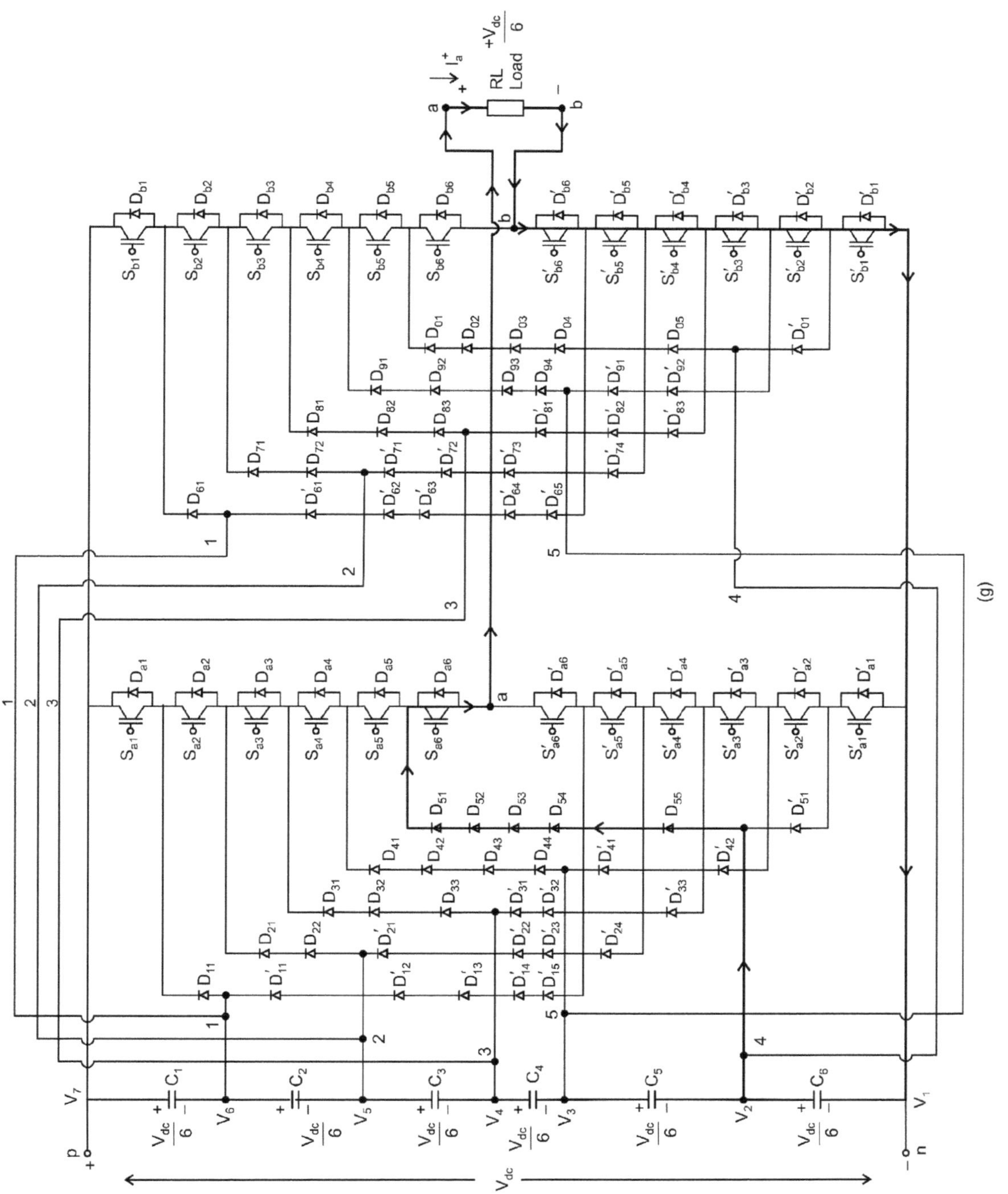

Fig. 3.19 (g) Mode6: $V_{ab}= +V_{dc}/6$ with I_a positive.

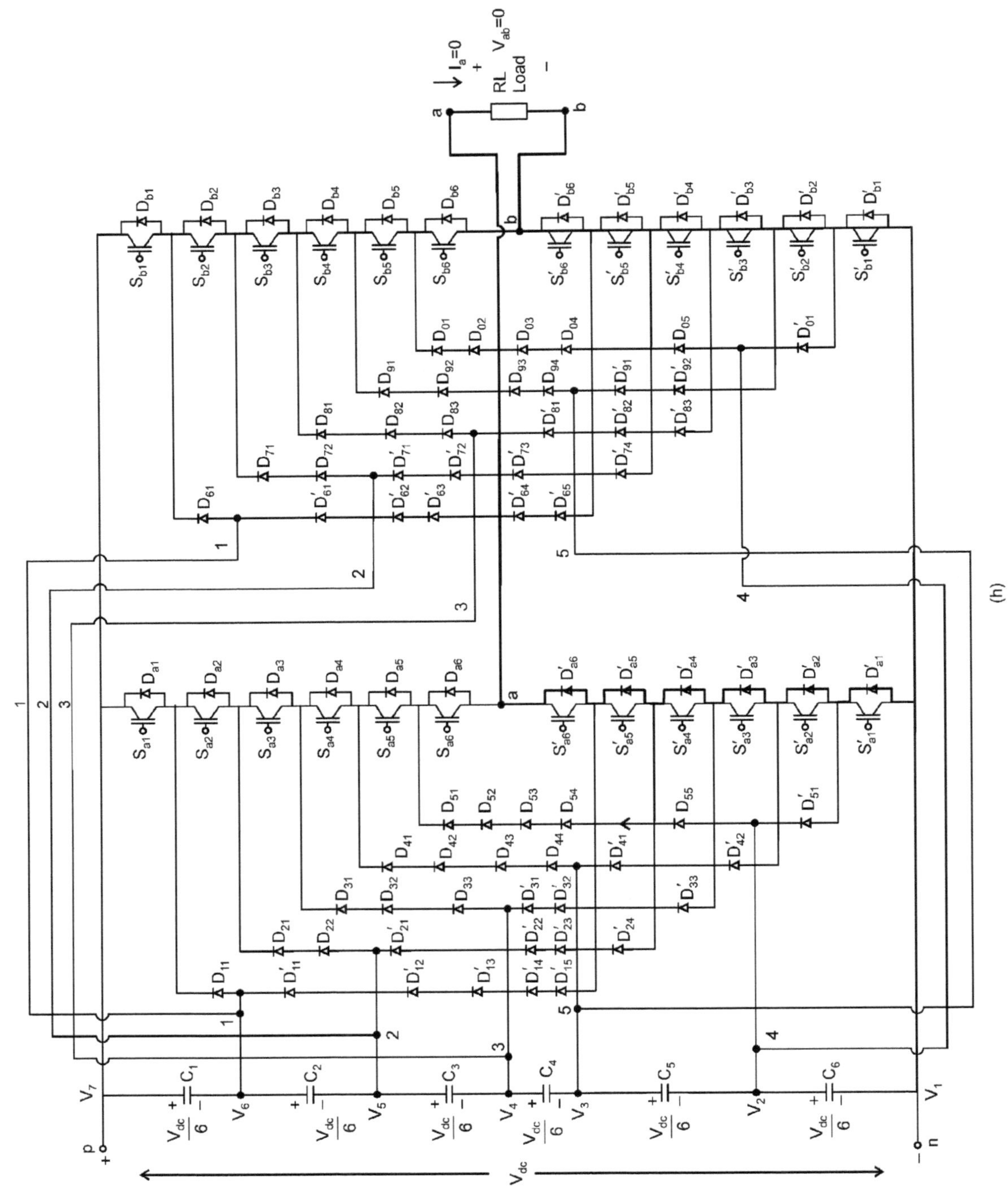

Fig. 3.19 (h) Mode7: V_{ab}= 0 with I_a zero.

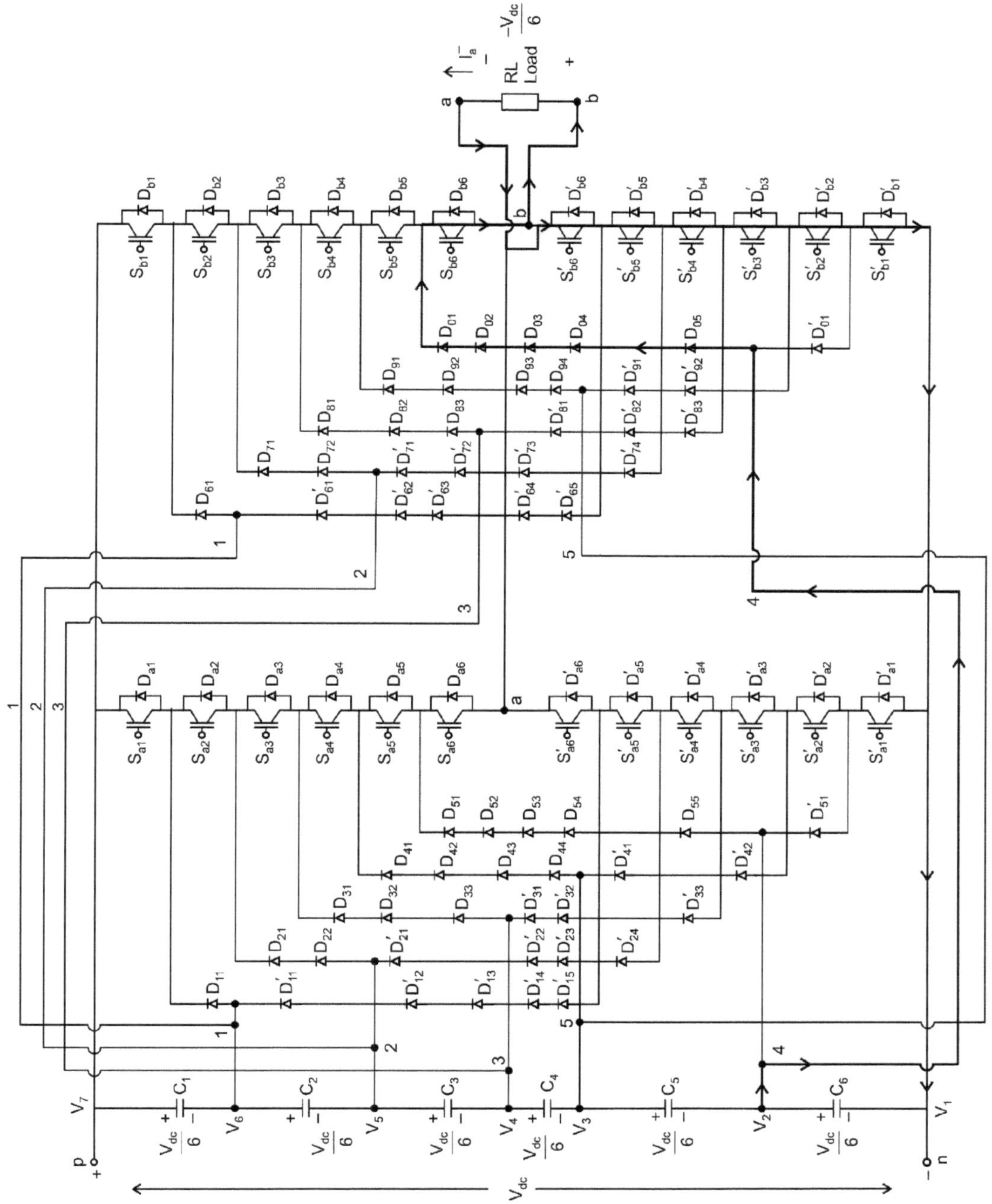

Fig. 3.19 (i) Mode 8: $V_{ab} = -V_{dc}/6$ with I_a negative.

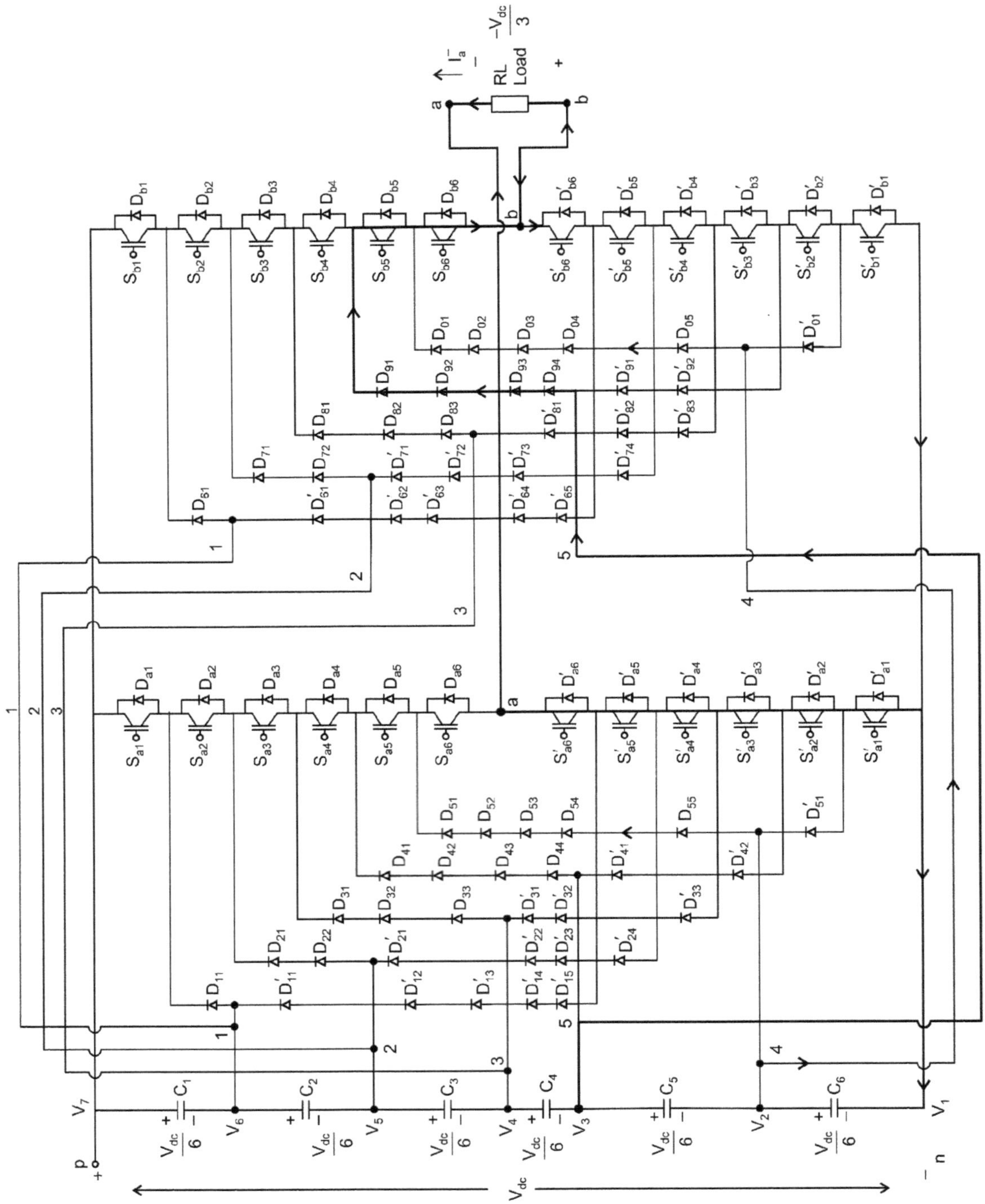

Fig. 3.19 (j) Mode 9: $V_{ab} = -V_{dc}/3$ with I_a negative.

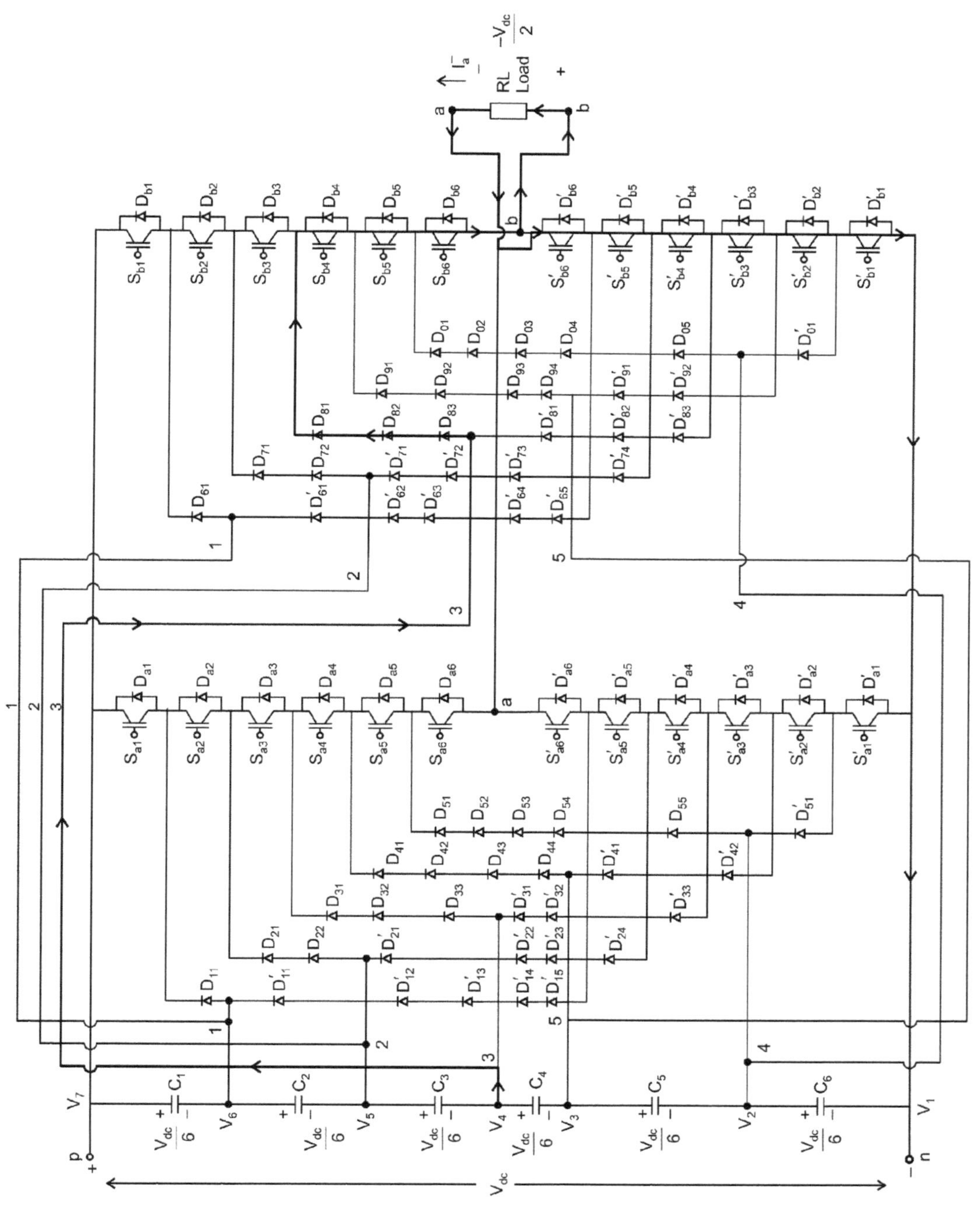

(k)

Fig. 3.19 (k) Mode 10: $V_{ab} = -V_{dc}/2$ with I_a negative.

Mode 9: For V_{ab}= $-V_{dc}/3$ with load current I_a negative, the switching devices S_{b5}, S_{b6}, and S_{a6}', S_{a5}', S_{a4}', S_{a3}', S_{a2}', S_{a1}' are to be fired so that the conduction can takes place from the source terminal V_3, through clamping diodes D_{94}, D_{93}, D_{92}, D_{91}, switches S_{b5}, S_{b6}, load terminals '*b*' to '*a*', switches S_{a6}', S_{a5}', S_{a4}', S_{a3}', S_{a2}', S_{a1}', to the source terminal V_1. The load current I_a is negative when it flows from load terminals '*b*' to '*a*' as shown in the Fig.3.19(j).

Mode 10: For V_{ab}= $-V_{dc}/2$ with load current I_a negative, the switching devices S_{b4}, S_{b5}, S_{b6}, and S_{a6}', S_{a5}', S_{a4}', S_{a3}', S_{a2}', S_{a1}' are to be fired so that the conduction can takes place from the source terminal V_4, through clamping diodes D_{83}, D_{82}, D_{81}, switches S_{b4}, S_{b5}, S_{b6}, load terminals '*b*' to '*a*', switches S_{a6}', S_{a5}', S_{a4}', S_{a3}', S_{a2}', S_{a1}', to the source terminal V_1. The load current I_a is negative when it flows from load terminals '*b*' to '*a*' as shown in the Fig.3.19(k).

Mode 11: For V_{ab}= $-2V_{dc}/3$ with load current I_a negative, the switching devices S_{b3}, S_{b4}, S_{b5}, S_{b6}, and S_{a6}', S_{a5}', S_{a4}', S_{a3}', S_{a2}', S_{a1}' are to be fired so that the conduction can takes place from the source terminal V_5, through clamping diodes D_{72}, D_{71}, switches S_{b3}, S_{b4}, S_{b5}, S_{b6}, load terminals '*b*' to '*a*', switches S_{a6}', S_{a5}', S_{a4}', S_{a3}', S_{a2}', S_{a1}', to the source terminal V_1. The load current I*a* is negative when it flows from load terminals '*b*' to '*a*' as shown in the Fig.3.19(l).

Mode 12: For V_{ab}= $-5V_{dc}/6$ with load current I*a* negative, the switching devices S_{b2}, S_{b3}, S_{b4}, S_{b5}, S_{b6}, and S_{a6}', S_{a5}', S_{a4}', S_{a3}', S_{a2}', S_{a1}' are to be fired so that the conduction can takes place from the source terminal V_6, through clamping diode D_{61}, switches S_{b2}, S_{b3}, S_{b4}, S_{b5}, S_{b6}, load terminals '*b*' to '*a*', switches S_{a6}', S_{a5}', S_{a4}', S_{a3}', S_{a2}', S_{a1}', to the source terminal V_1. The load current I*a* is negative when it flows from load terminals '*b*' to '*a*' as shown in the Fig.3.19(m).

Mode 13: For V_{ab}= $-V_{dc}$ with load current I_a negative, the switching devices S_{b1}, S_{b2}, S_{b3}, S_{b4}, S_{b5}, S_{b6}, and S_{a6}', S_{a5}', S_{a4}', S_{a3}', S_{a2}', S_{a1}' are to be fired so that the conduction can takes place from the source terminal V_7, through switches S_{b1}, S_{b2}, S_{b3}, S_{b4}, S_{b5}, S_{b6}, load terminals '*b*' to '*a*', switches S_{a6}', S_{a5}', S_{a4}', S_{a3}', S_{a2}', S_{a1}', to the source terminal V_1. The load current I_a is negative when it flows from load terminals '*b*' to 'a' as shown in the Fig.3.19(n).

The switching pattern for switching devices, clamping diodes and anti-parallel diodes for single phase seven-level diode-clamped full-bridge inverter for various modes is given in Table 3.5

The load line voltage waveform of seven-level is presented in Fig. 3.20 obtained from single phase seven-level diode-clamped full-bridge inverter.

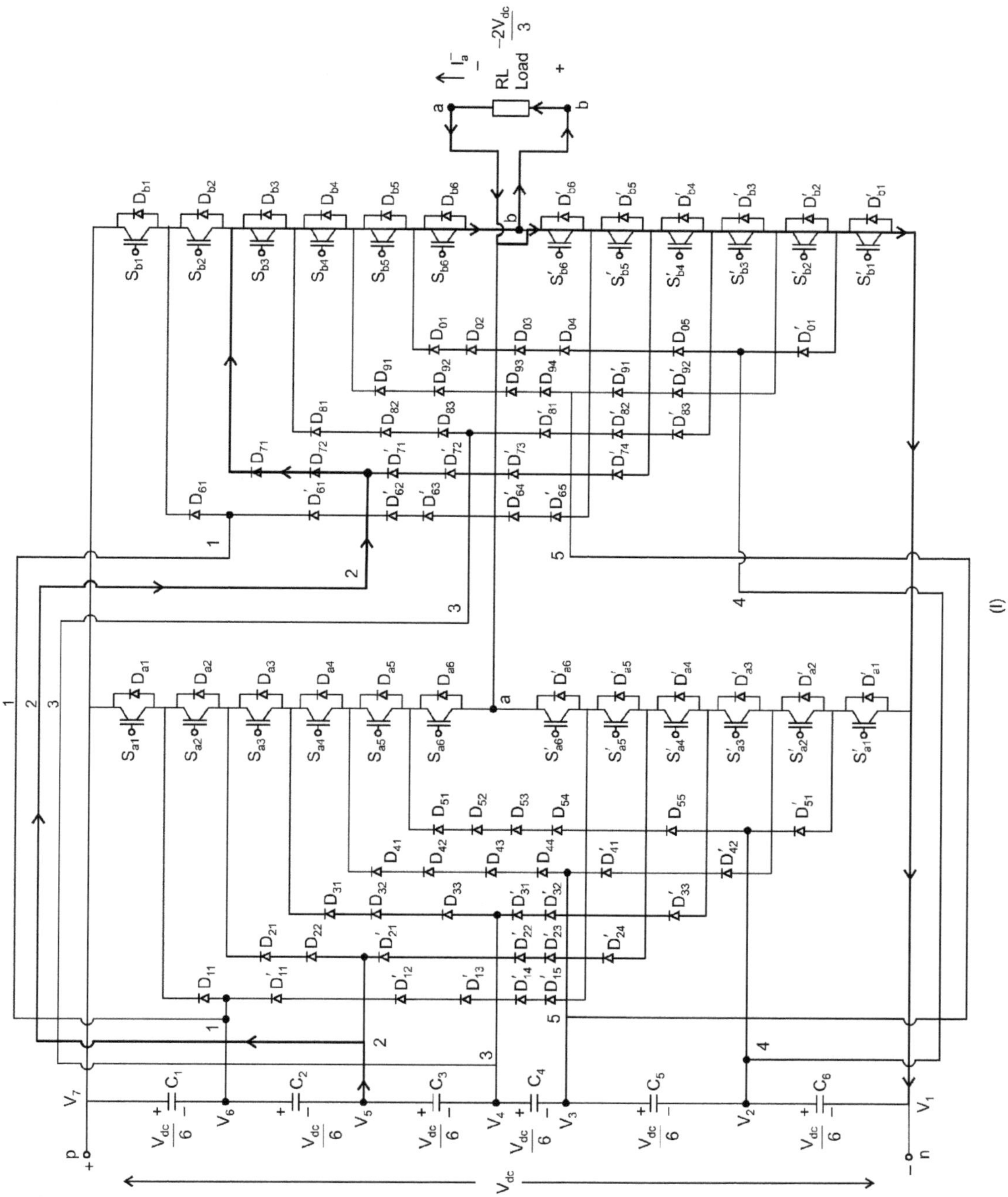

Fig. 3.19 (l) Mode 11: $V_{ab} = -2V_{dc}/3$ with I_a negative.

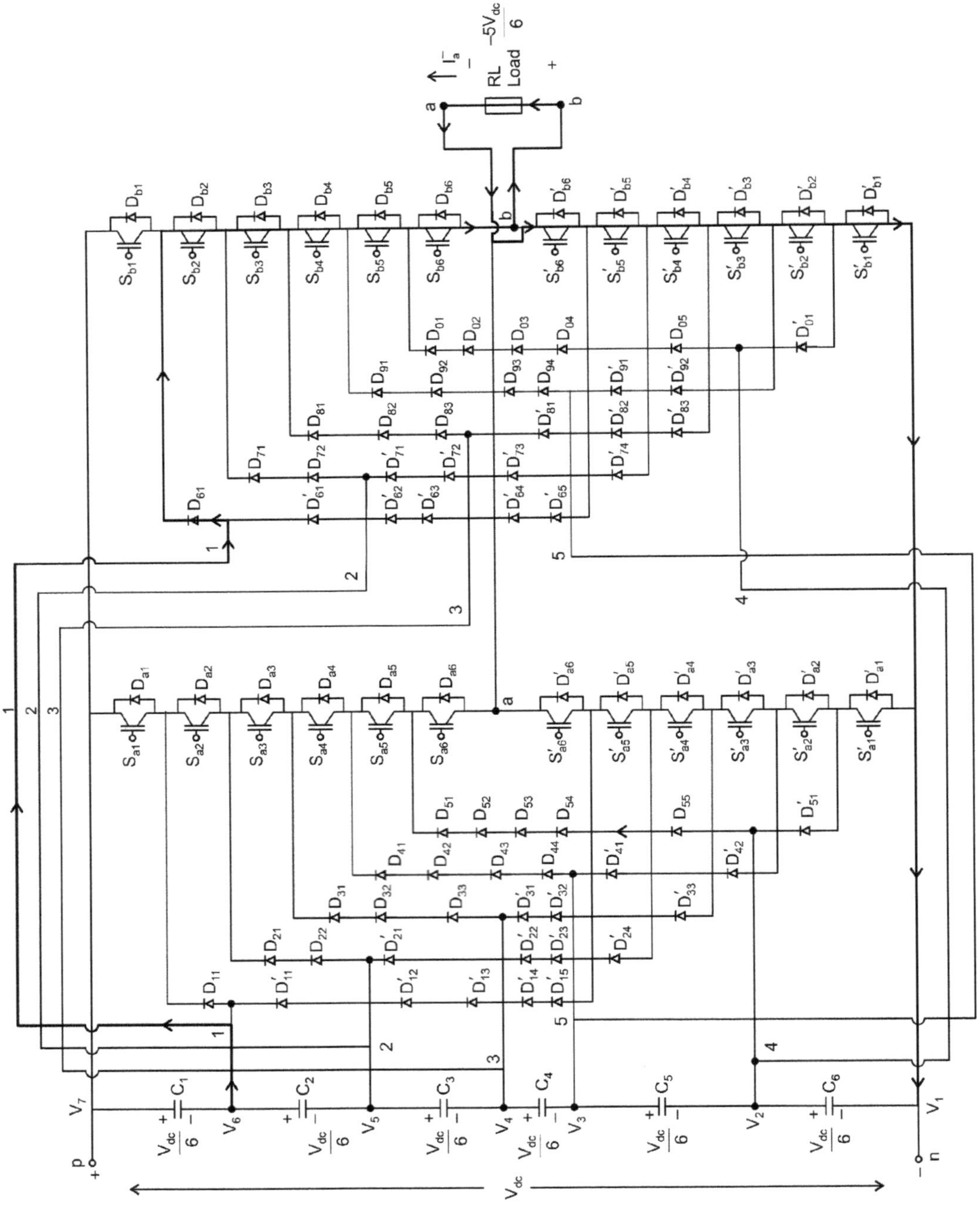

Fig. 3.19 (m) Mode 12: $V_{ab} = -5V_{dc}/6$ with I_a negative.

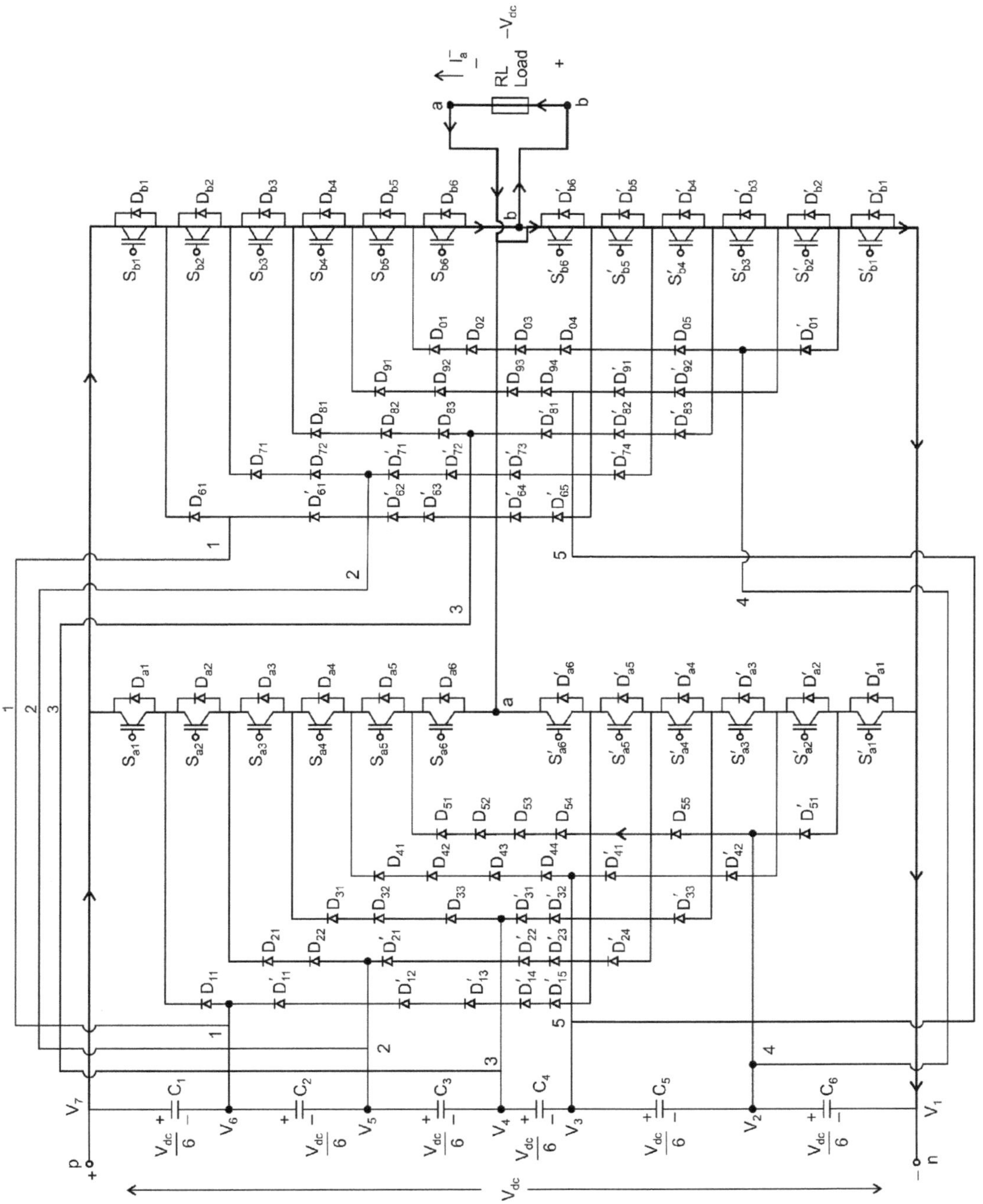

Fig. 3.19 (n) Mode13: $V_{ab}= -V_{dc}$ with I_a negative.

Table 3.5 Switching pattern for single phase seven-level diode-clamped full-bridge inverter.

V_{ab} =	$+V_{dc}$	$+5V_{dc}/6$	$+2V_{dc}/3$	$+V_{dc}/2$	$+V_{dc}/3$	$+V_{dc}/6$	0	$-V_{dc}/6$	$-V_{dc}/3$	$-V_{dc}/2$	$-2V_{dc}/3$	$-5V_{dc}/6$	$-V_{dc}$
Switches:	M_1	M_2	M_3	M_4	M_5	M_6	M_7	M_8	M_9	M_{10}	M_{11}	M_{12}	M_{13}
S_{a1}/S_{a1}'	1/0	0/0	0/0	0/0	0/0	0/0	0/0	0/1	0/1	0/1	0/1	0/1	0/1
S_{a2}/S_{a2}'	1/0	1/0	0/0	0/0	0/0	0/0	0/0	0/1	0/1	0/1	0/1	0/1	0/1
S_{a3}/S_{a3}'	1/0	1/0	1/0	0/0	0/0	0/0	0/0	0/1	0/1	0/1	0/1	0/1	0/1
S_{a4}/S_{a4}'	1/0	1/0	1/0	1/0	0/0	0/0	0/0	0/1	0/1	0/1	0/1	0/1	0/1
S_{a5}/S_{a5}'	1/0	1/0	1/0	1/0	1/0	0/0	0/0	0/1	0/1	0/1	0/1	0/1	0/1
S_{a6}/S_{a6}'	1/0	1/0	1/0	1/0	1/0	1/0	0/0	0/1	0/1	0/1	0/1	0/1	0/1
S_{b6}/S_{b6}'	0/1	0/1	0/1	0/1	0/1	0/1	0/1	1/0	1/0	1/0	1/0	1/0	1/0
S_{b5}/S_{b5}'	0/1	0/1	0/1	0/1	0/1	0/1	0/1	0/0	1/0	1/0	1/0	1/0	1/0
S_{b4}/S_{b4}'	0/1	0/1	0/1	0/1	0/1	0/1	0/1	0/0	0/0	1/0	1/0	1/0	1/0
S_{b3}/S_{b3}'	0/1	0/1	0/1	0/1	0/1	0/1	0/1	0/0	0/0	0/0	1/0	1/0	1/0
$S_{b2}/S_{b}2'$	0/1	0/1	0/1	0/1	0/1	0/1	0/1	0/0	0/0	0/0	0/0	1/0	1/0
S_{b1}/S_{b1}'	0/1	0/1	0/1	0/1	0/1	0/1	0/1	0/0	0/0	0/0	0/0	0/0	1/0
Clamping diodes	—	D_{11}	D_{22}, D_{21}	D_{33}, D_{32}, D_{31}	D_{44}, D_{43}, D_{42}, D_{41}	D_{55}, D_{54}, D_{53}, D_{52}, D_{51}	—	D_{05}, D_{04}, D_{03}, D_{02}, D_{01}	D_{94}, D_{93}, D_{92}, D_{91}	D_{83}, D_{82}, D_{81}	D_{72}, D_{71}	D_{61}	—
Anti-parallel diodes	—	—	—	—	—	—	D_{a1}' to D_{a6}'						—
Current I_a	+ve	+ve	+ve	+ve	+ve	+ve	0	-ve	-ve	-ve	-ve	-ve	-ve

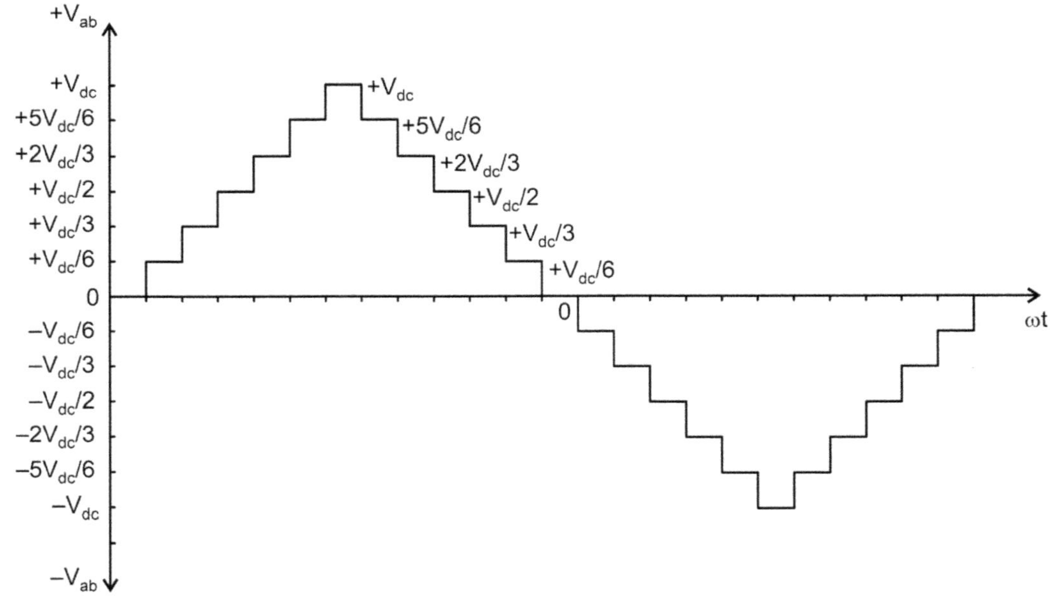

Fig. 3.20 Load line Voltage waveform for single phase seven-level diode-clamped full bridge inverter.

Although each active switching device is only required to block a voltage level of $V_{dc}/(m-1)$, the clamping diodes have different voltage ratings for reverse voltage blocking. Assuming that each blocking diode voltage rating is the same as the active device voltage rating, the number of diodes required for each phase will be (m-1) (m-2). This number represents a quadratic increase in m. When m is sufficiently high, the number of diodes required will make the system impractical to implement. If the inverter runs under PWM, the diode reverse recovery of these clamping diodes becomes the major design challenge in high-voltage high-power applications.

One more leg is added to the above circuit to get a three phase seven-level diode-clamped inverter for star-connected load, the circuit configuration is shown in Fig.3.21(a), and its load line voltage waveforms are given in Fig.3.21 (b).

Modified single-phase five-level diode-clamped inverter with distributed clamping diodes: A Single phase five-level Modified diode clamped inverter circuit with distributed clamping diodes, a single-leg arrangement is shown in Fig. 3.22.

This modified diode-clamped inverter with distributed clamping diodes require a total of 8 controlled switches and 12 clamping diodes of equal voltage sharing. It is almost similar to diode clamping Inverter with series connected diodes.

A single phase five-level diode-clamped inverter leg requires (m-1) dc source capacitors, 2 x (m-1) controlled switches, and (m-1) x (m-2) clamping diodes. As an example, the five-level inverter is demonstrated here, which requires 4 dc source capacitors, 8 controlled switches per leg, and 12 clamping-diodes per leg and they are distributed in a pyramid architecture as shown in the Fig.3.22.

The problem of multi-blocking voltages of the clamping-diodes can be added by connecting an appropriate number of diodes in series as shown in Figs. 3.3 - 3.19. Due to mismatching of diode characteristics used for clamping operation, the voltage sharing is not equal, thus, an improved version of diode-clamped inverter is proposed with distribution of clamping diodes in a pyramid architecture, which is called a modified diode-clamped inverter. This modified architecture can be used to any level of diode-clamped inverter.

The circuit operation is explained for single-leg topology as follows, by considering the source terminal V_3 as mid-point and designated as '*0*'.

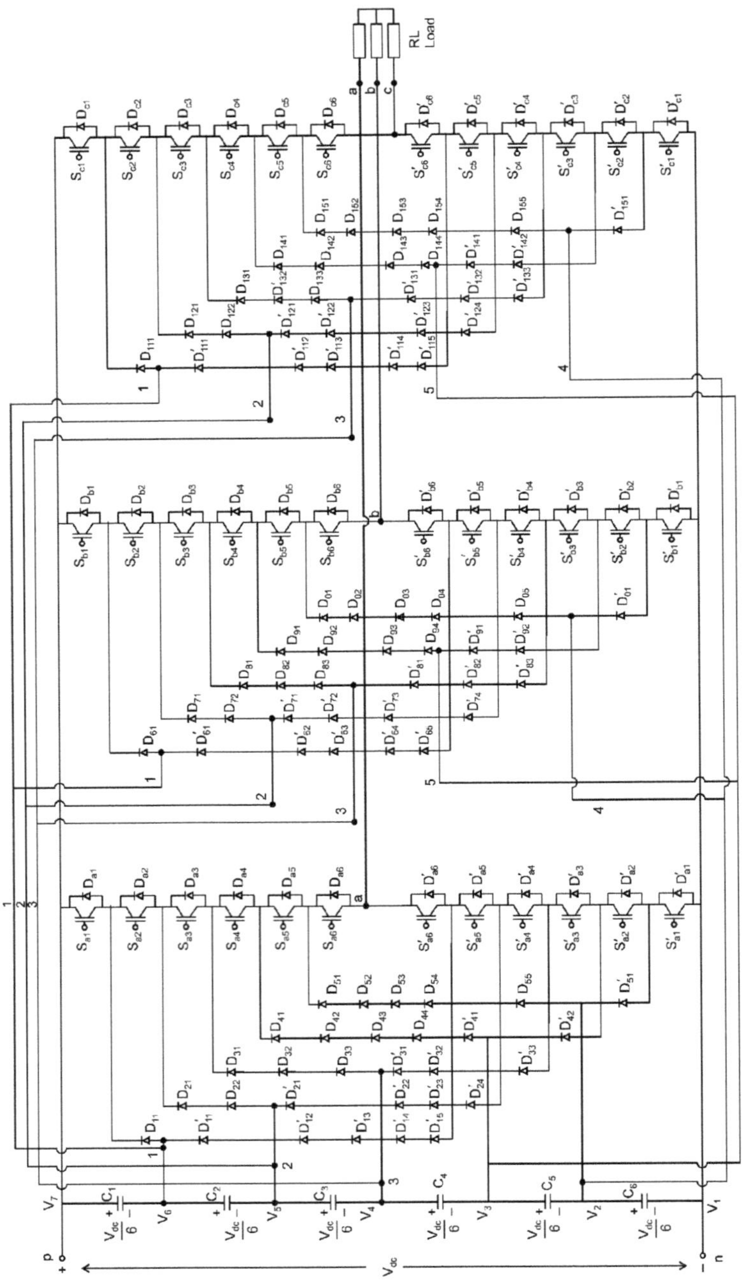

Fig. 3.21 (a) Three phase seven-level diode-clamped inverter for star-connected loads.

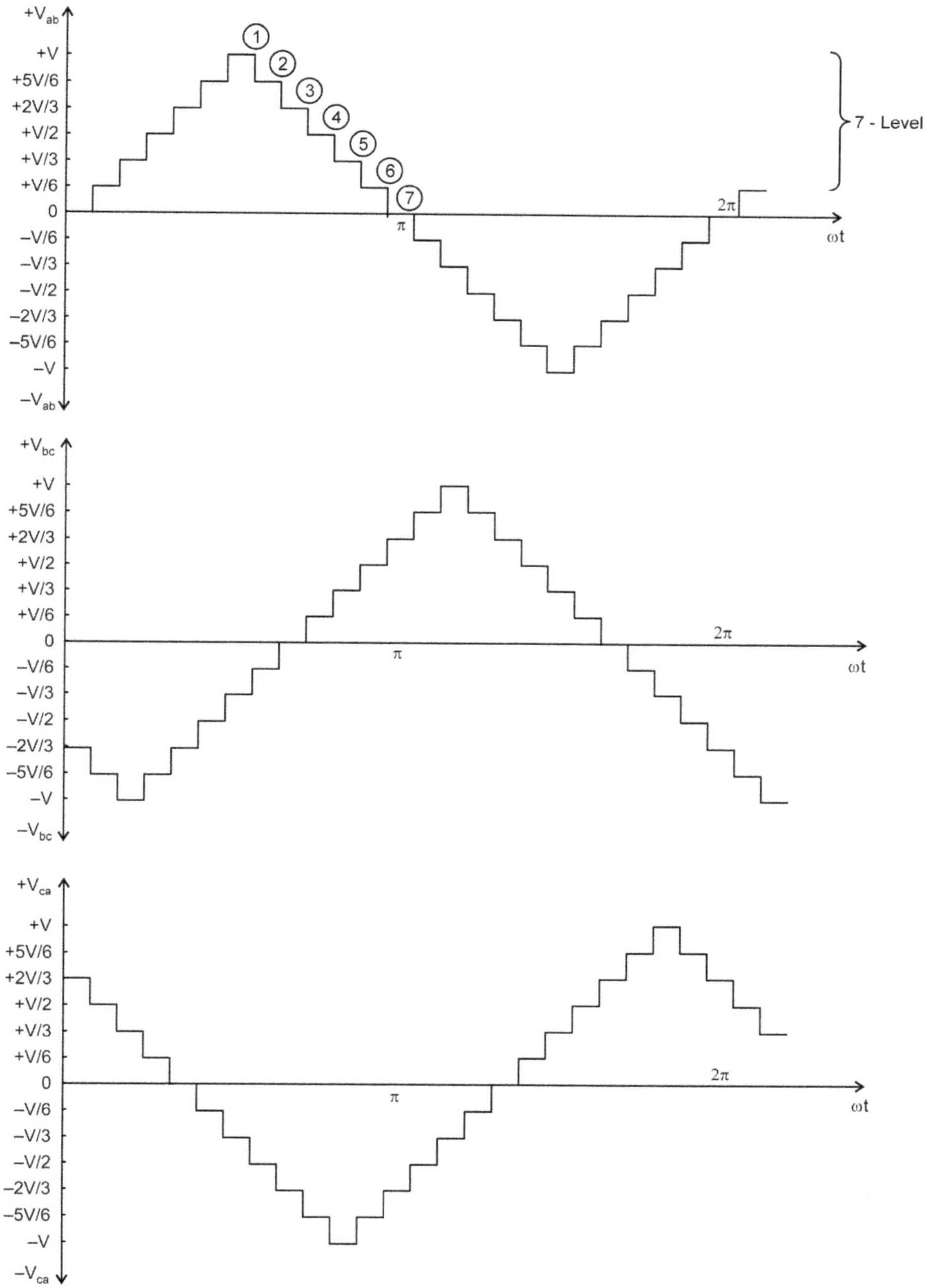

Fig. 3.21 (b) Load line voltage waveforms for three phase seven-level diode-clamped inverter with star-connected loads.

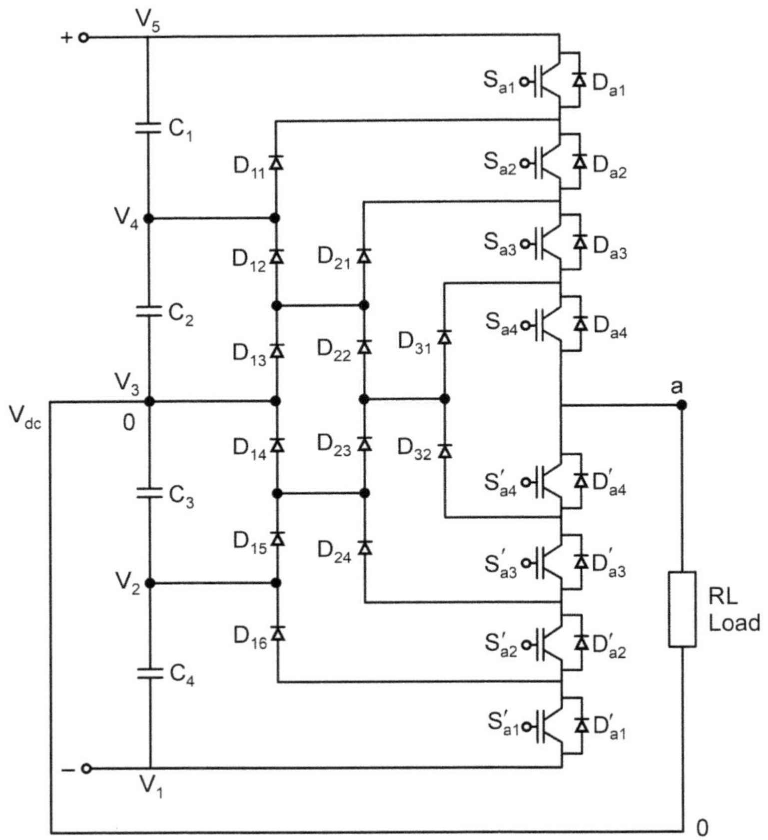

Fig. 3.22 Single phase five-level modified diode clamped inverter with distributed clamping diodes.

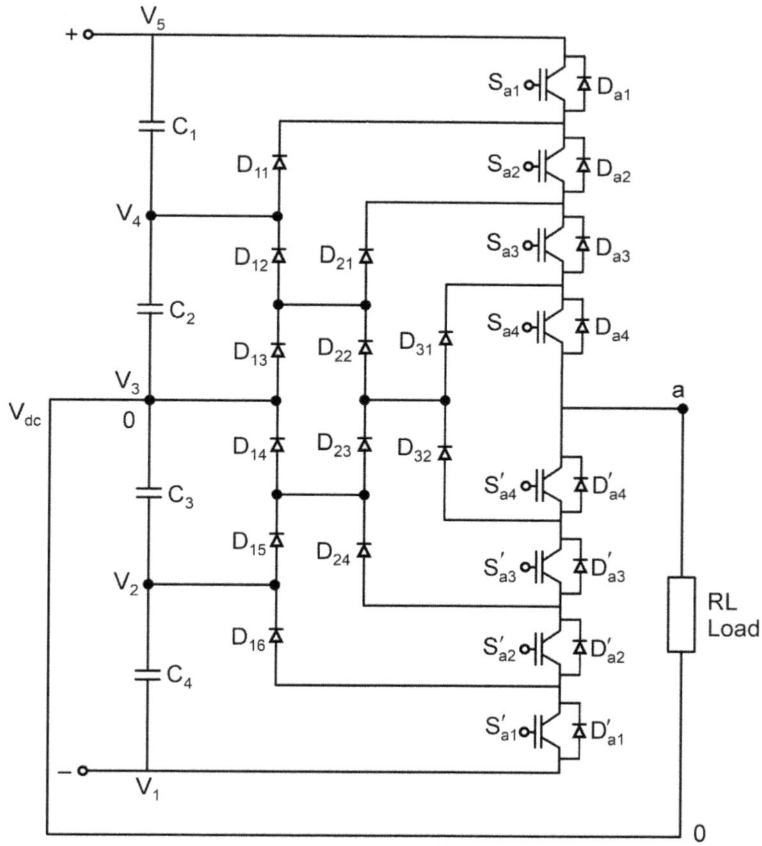

(a)

Fig. 3.22 (a) Main circuit topology.

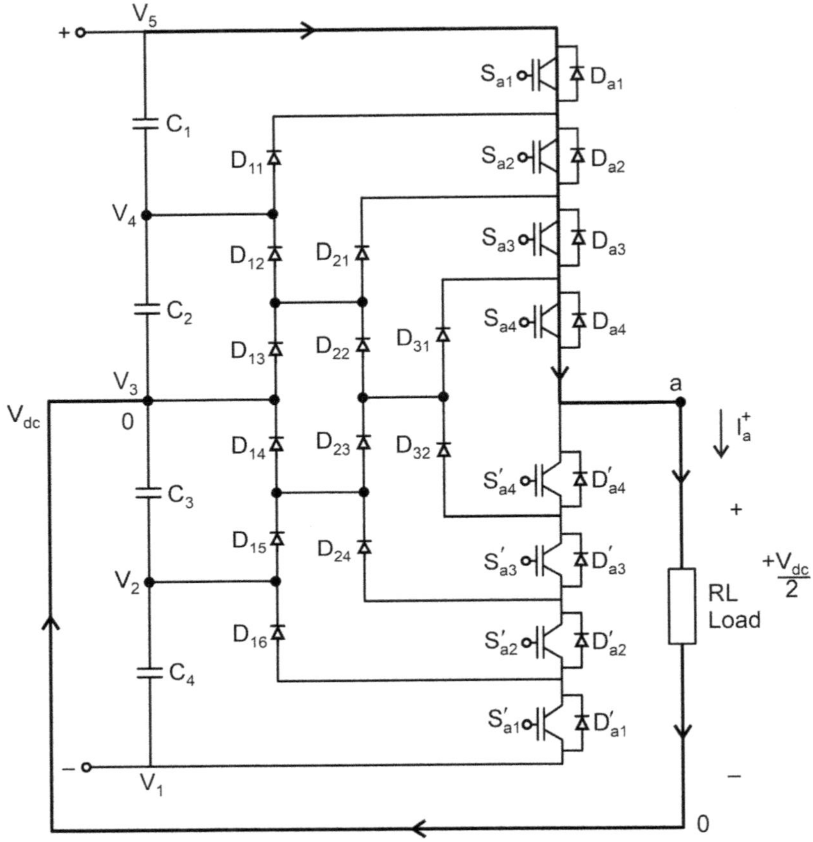

(b)

Fig. 3.22 (b) Mode1: V_{a0}=+V_{dc}/2 with positive I_a.

Mode 1: $V_{a0} = +V_{dc}/2$ with positive load current I_a, the switches S_{a1}, S_{a2}, S_{a3}, S_{a4} are to be turned-on to make the conduction from source terminal V_5, switches S_{a1}, S_{a2}, S_{a3}, S_{a4}, load terminal '*a*' to '*0*' and terminate at V_3 and I_a is positive when it flows from load terminals '*a*' to '*0*' as shown in Fig.3.22(b).

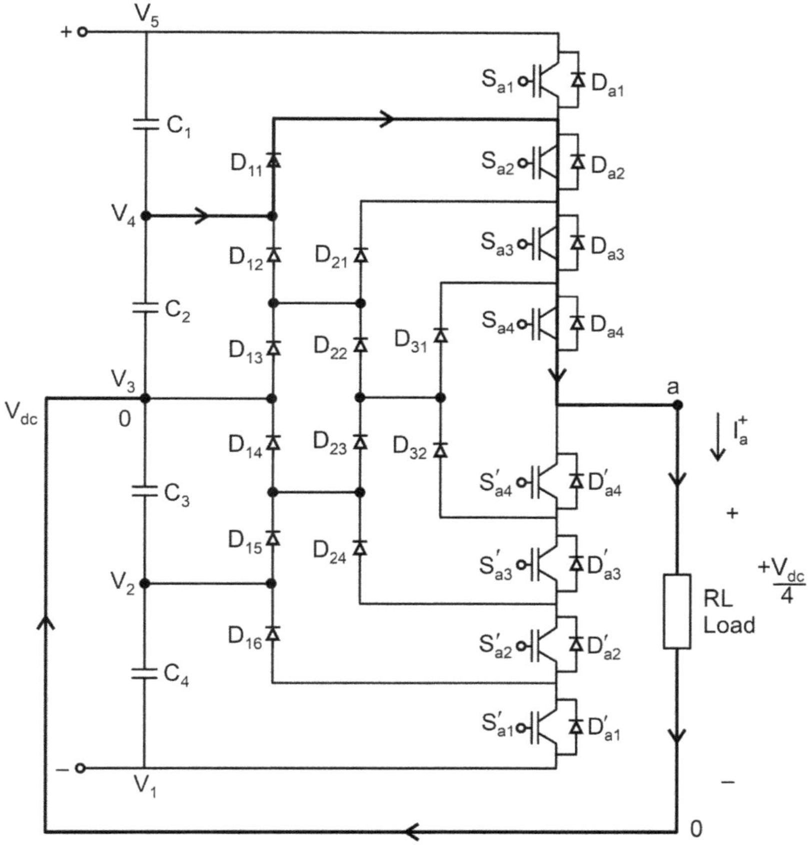

(c)

Fig. 3.22 (c) Mode2: $V_{a0}=+V_{dc}/4$ with positive I_a.

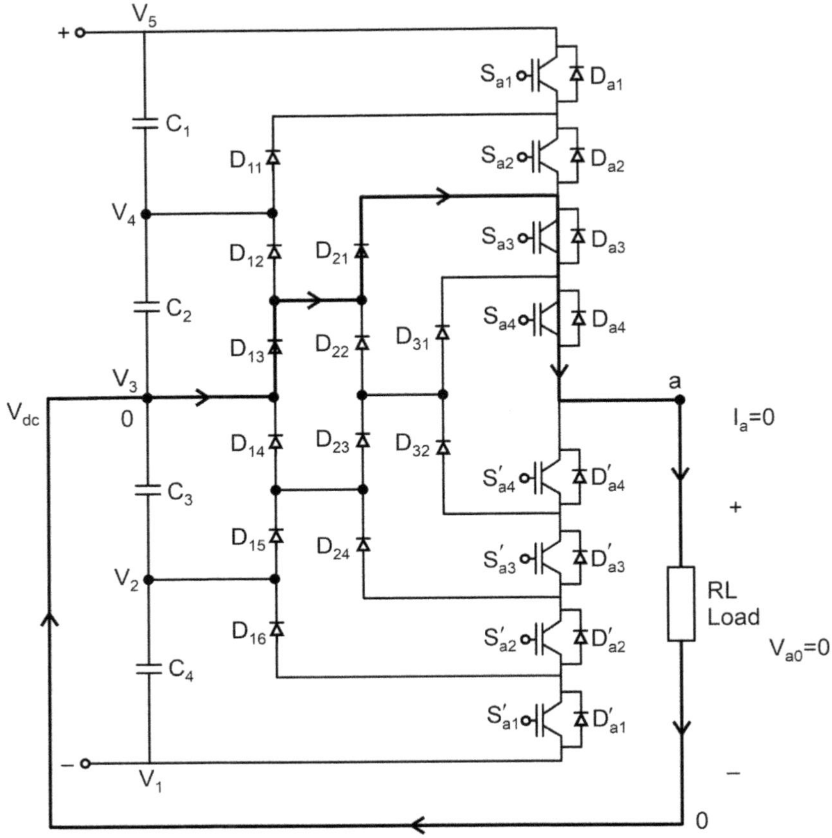

(d)

Fig. 3.22 (d) Mode3: V_{ao}=0 with zero I_a.

Mode 2: $V_{a0} = +V_{dc}/4$ with positive load current I_a, the switches S_{a2}, S_{a3}, S_{a4} are to be turned-on to make the conduction from source terminal V_4, through clamping diode D_{11}, switches S_{a2}, S_{a3}, S_{a4}, load terminal '*a*' to '*0*' and terminate at V_3, and I_a is positive when it flows from load terminals '*a*' to '*0*' as shown in Fig.3.22(c).

Mode 3: $V_{a0} = 0$, with zero load current I_a, the switches S_{a3}, S_{a4} are to be turned-on to make the conduction from source terminal V_3, through clamping diodes D_{13}, D_{21}, switches S_{a3}, S_{a4}, load terminal '*a*' to '*0*' and terminal at V_3 and I_a is zero as shown in Fig.3.22(d).

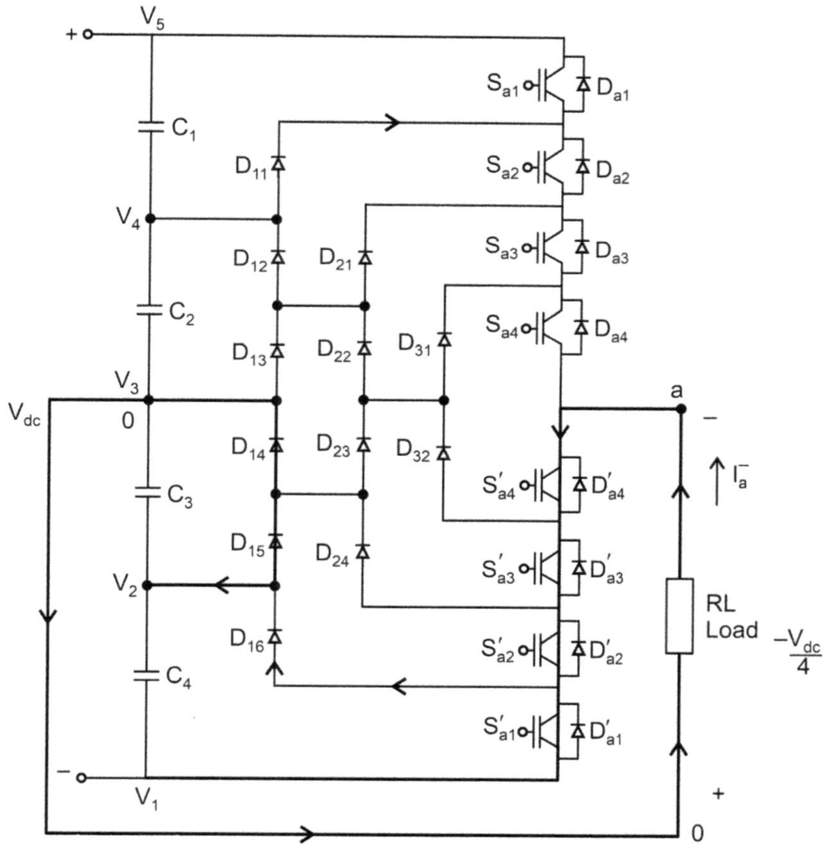

(e)

Fig. 3.22 (e) Mode 4: V_{a0}=-V_{dc}/4 with negative I_a.

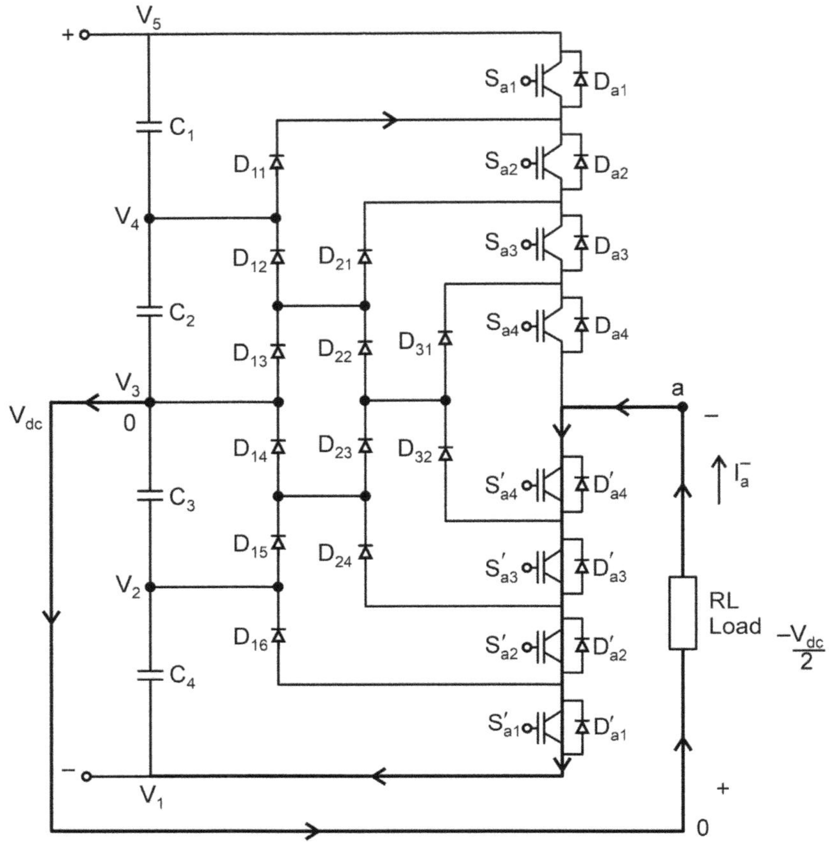

(f)

Fig. 3.22 (f) Mode5: V_{a0}=-V_{dc}/2 with negative I_a.

Mode 4: $V_{a0} = -V_{dc}/4$ with negative load current I_a, the switches S_{a4}', S_{a3}', S_{a2}' are to be turned-on to make the conduction from source terminal V_3, through load terminals '*0*' to '*a*', and switches S_{a4}', S_{a3}, S_{a2}', D_{16} to source terminal V_2, and I_a is negative when it flows from load terminals '*0*' to '*a*' as shown in Fig.3.22(e).

Mode 5: $V_{a0} = -V_{dc}/2$ with negative load current I_a, the switches S_{a4}', S_{a3}', S_{a2}', S_{a1}' are to be turned-on to make the conduction from source terminal V_3, through load terminals '*0*' to '*a*', switches S_{a4}', S_{a3}', S_{a2}', S_{a1}', to source terminal V_1, and I_a is negative when it flows from load terminals '*0*' to '*a*' as shown in Fig.3.22(f).

The Switching pattern for single-phase single-leg five level modified diode-clamped inverter is presented for Modes 1 to 5 in Table 3.6. In Fig. 3.22(g) the load voltage waveform for phase is V_{a0} for single phase single-leg five-level modified diode-clamped inverter is given.

The circuit operation is explained for single-phase five-level modified diode-clamped full-bridge topology shown in Fig.3.23, as follows by considering nine-modes.

Table 3.6 Switching pattern for single phase single-leg five-level modified diode-clamped inverter

Modes	V_{a0}	S_{a1}	S_{a2}	S_{a3}	S_{a4}	S_{a4}'	S_{a3}'	S_{a2}'	S_{a1}'	Clamping Diodes	Ia
M1	$+V_{dc}/2$	1	1	1	1	0	0	0	0	----	+ve
M2	$+V_{dc}/4$	0	1	1	1	0	0	0	0	D_{11}	+ve
M3	0	0	0	1	1	0	0	0	0	D_{13}, D_{21}	zero
M4	$-V_{dc}/4$	0	0	0	0	1	1	1	0	D_{16}	-ve
M5	$-V_{dc}/2$	0	0	0	0	1	1	1	1	----	-ve

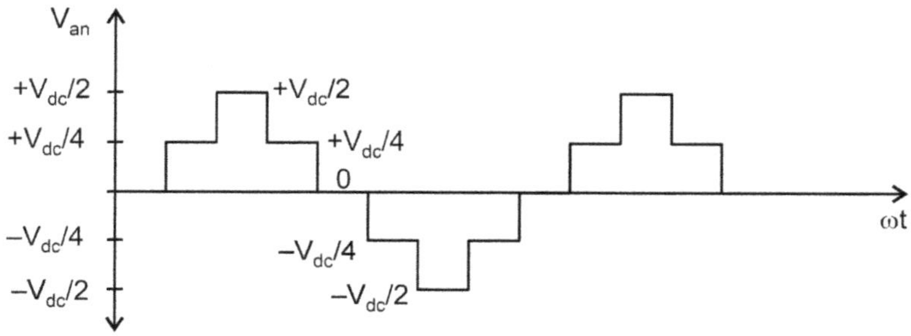

Fig. 3.22 (g) Load phase voltage Va0 for single phase single-leg five-level modified diode-clamped inverter.

Mode 1: For $V_{ab} = +V_{dc}$, with positive load current I_a, the switches S_{a1}, S_{a2}, S_{a3}, S_{a4}, and S_{b4}', S_{b3}', S_{b2}', S_{b1}' are to be turned on to connect source to the load through source terminal V_5, switches S_{a1}, S_{a2}, S_{a3}, S_{a4}, load terminals '*a*' to '*b*', switches S_{b4}', S_{b3}', S_{b2}', S_{b1}', and source terminal V_1, and the load current is positive when it flows from load terminals '*a*' to '*b*' as shown in Fig.3.23 (b).

Mode 2: $V_{ab} = +3V_{dc}/4$, with positive load current I_a, the switches S_{a2}, S_{a3}, S_{a4}, and S_{b4}', S_{b3}', S_{b2}', S_{b1}' are to be turned on to connect source to the load through source terminal V4, clamping diode D_{11}, switches S_{a2}, S_{a3}, S_{a4}, load terminals '*a*' to '*b*', switches S_{b4}', S_{b3}', S_{b2}', S_{b1}', and source terminal V_1, and the load current is positive when it flows from load terminals '*a*' to '*b*' as shown in Fig.3.23 (c).

Mode 3: $V_{ab} = +V_{dc}/2$, with positive load current I_a, the switches S_{a3}, S_{a4}, and S_{b4}', S_{b3}', S_{b2}', S_{b1}' are to be turned on to connect source to the load through source terminal V_3, clamping diodes D_{13}, D_{21}, switches S_{a3}, S_{a4}, load terminals '*a*' to '*b*', switches S_{b4}', S_{b3}', S_{b2}', S_{b1}', and source terminal V_1, and the load current is positive when it flows from load terminals '*a*' to '*b*' as shown in Fig.3.23 (d).

Mode 4: $V_{ab} = +V_{dc}/4$, with positive load current I_a, the switches S_{a4}, and S_{b4}', S_{b3}', S_{b2}', S_{b1}' are to be turned on to connect source to the load through source terminal V_2, clamping diodes D_{15}, D_{23}, D_{31}, switch S_{a4}, load terminals '*a*' to '*b*', switches S_{b4}', S_{b3}', S_{b2}', S_{b1}', and source terminal

V_1, and the load current is positive when it flows from load terminals '*a*' to '*b*' as shown in Fig.3.23 (e).

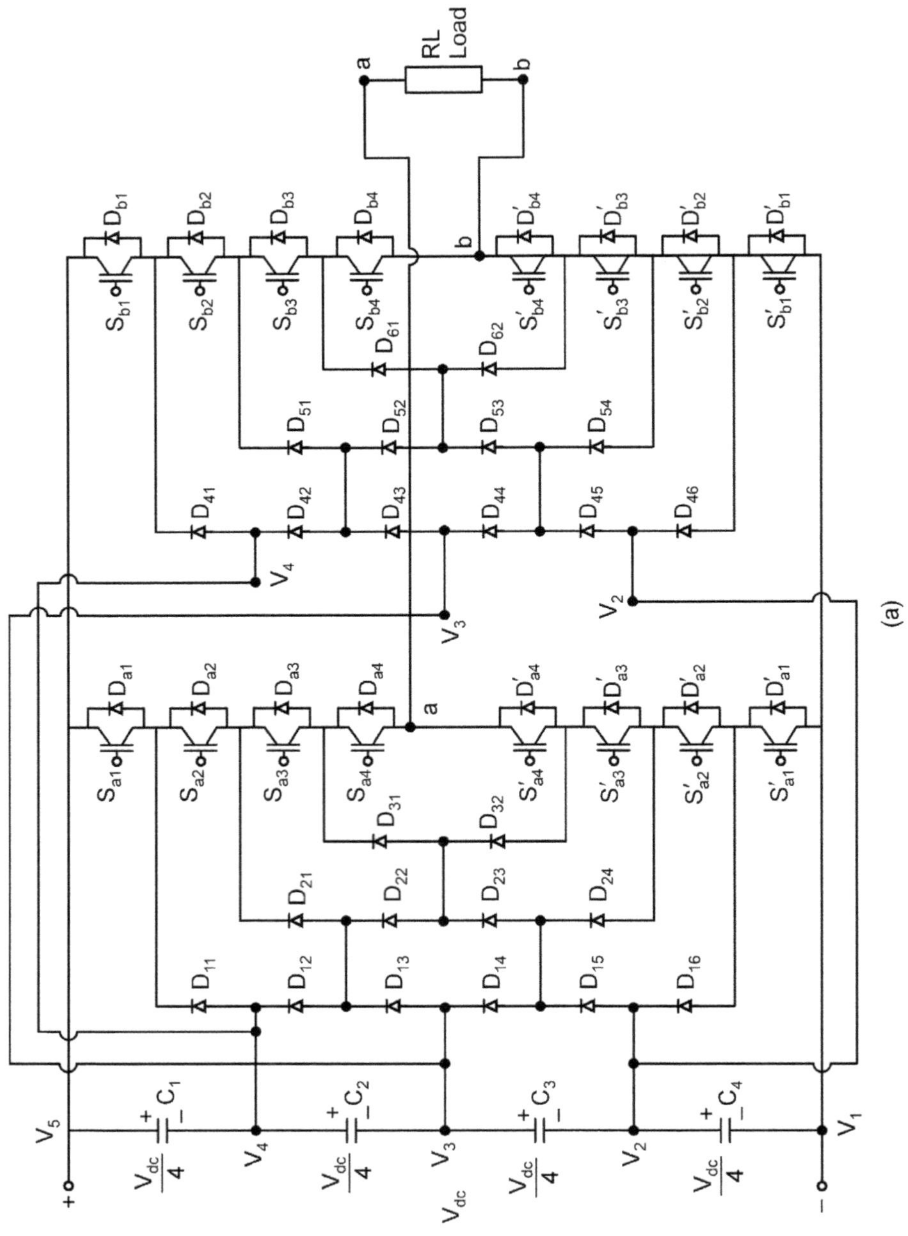

Fig. 3.23 (a) Single phase five-level modified diode-clamped full bridge topology.

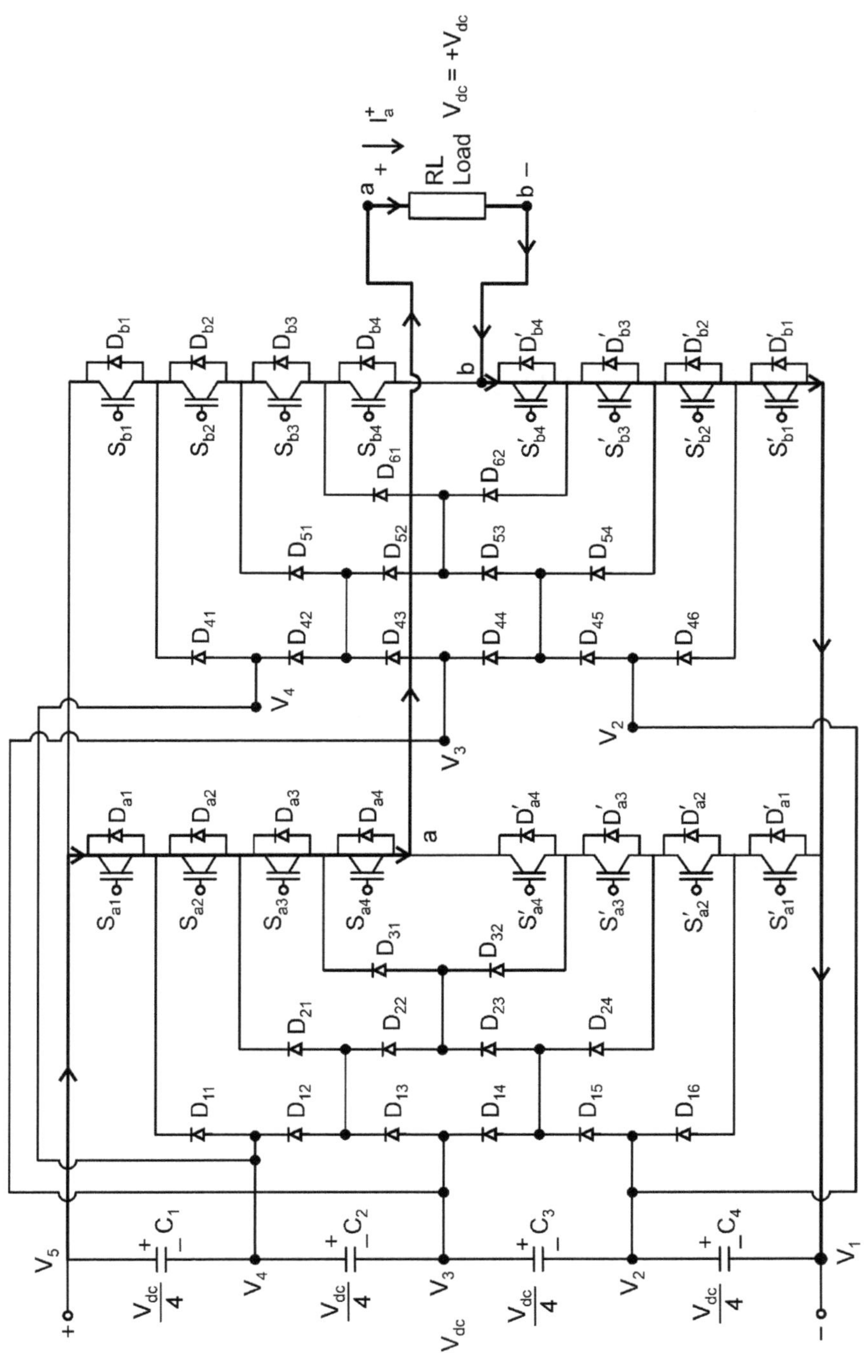

Fig. 3.23 (b) Mode1: V_{ab}= $+V_{dc}$ with positive I_a.

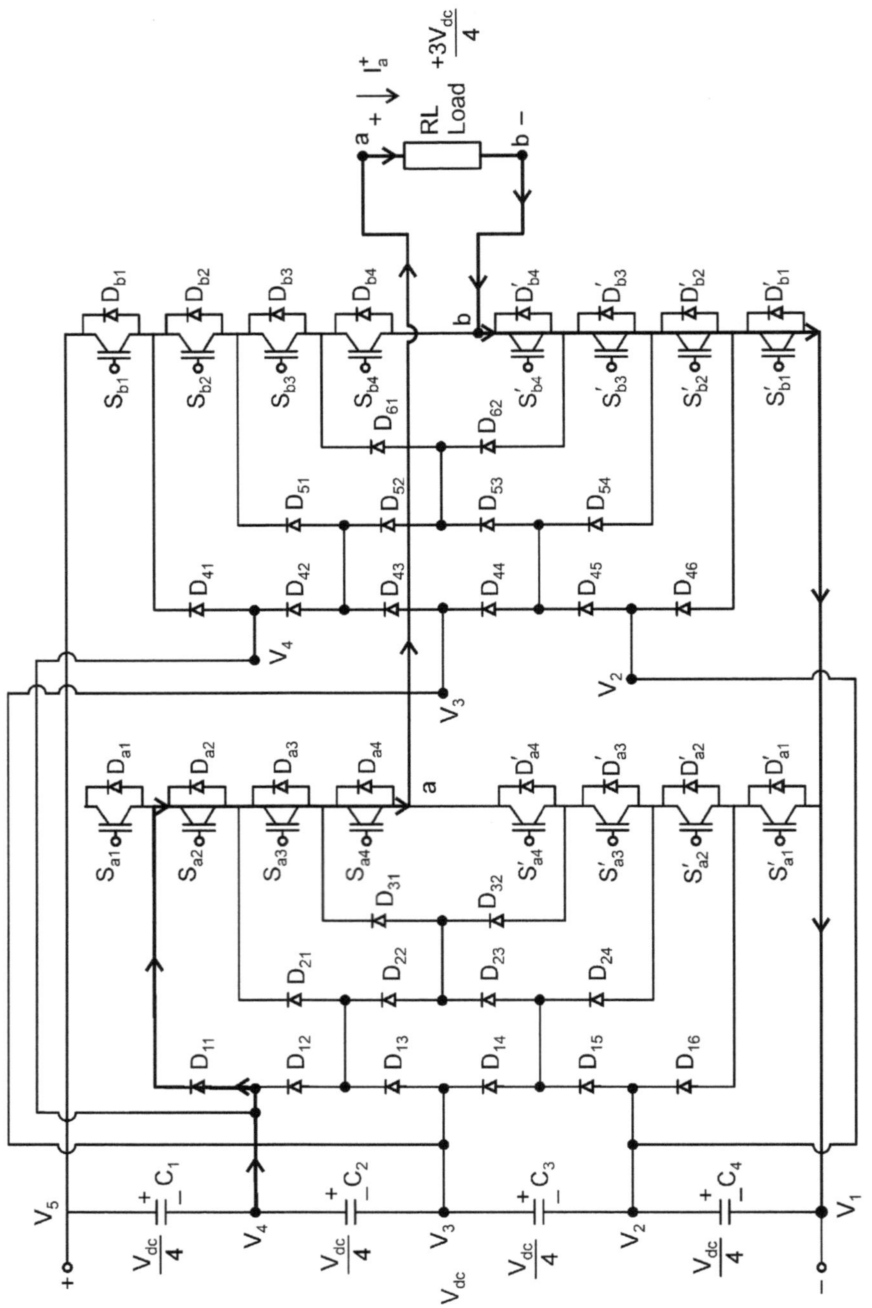

Fig. 3.23 (c) Mode2: V_{ab}=+3V_{dc}/4.

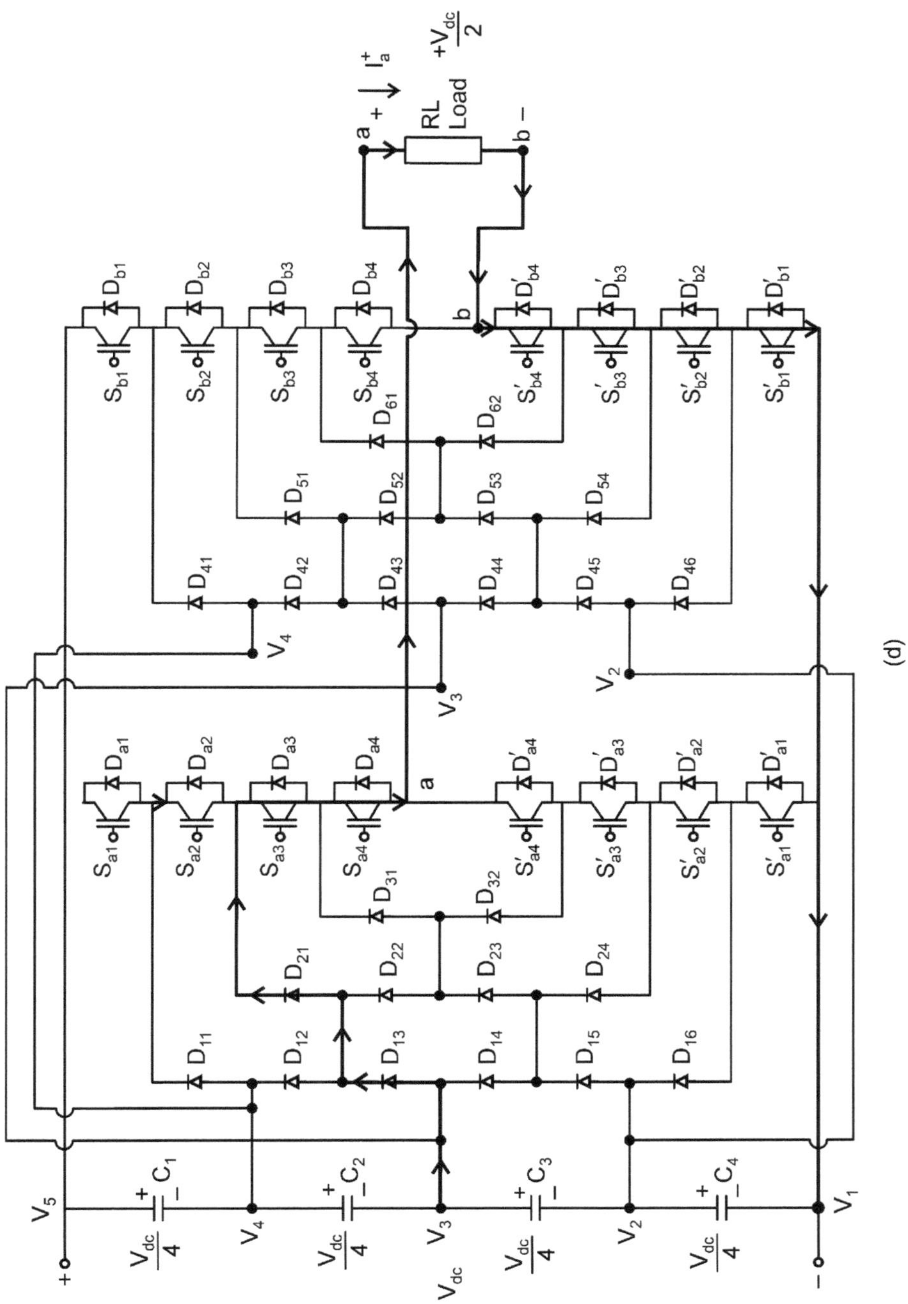

Fig. 3.23 (d) Mode3: V_{ab}=+V_{dc}/2 with positive I_a.

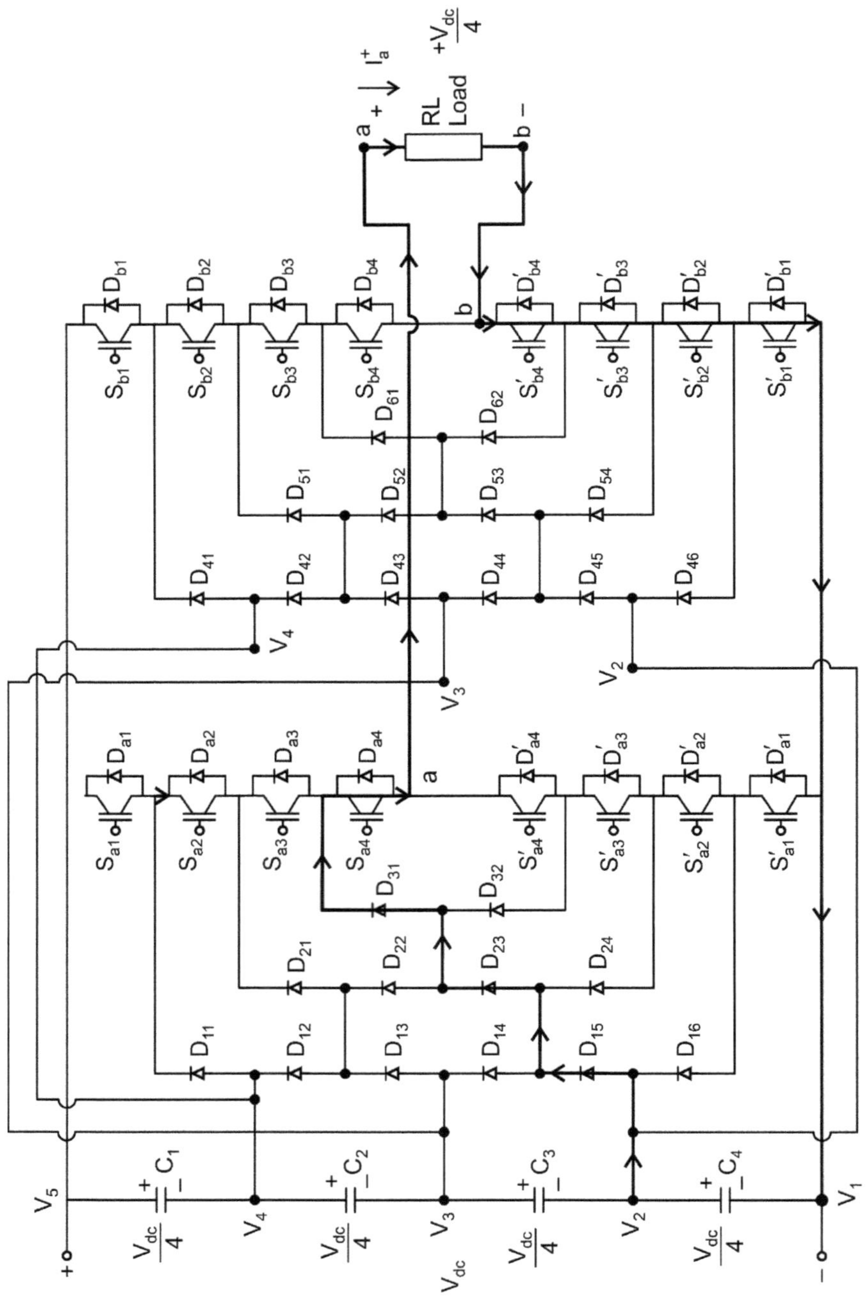

(e)

Fig. 3.23 (e) Mode4: V_{ab}= $+V_{dc}/4$ with positive I_a.

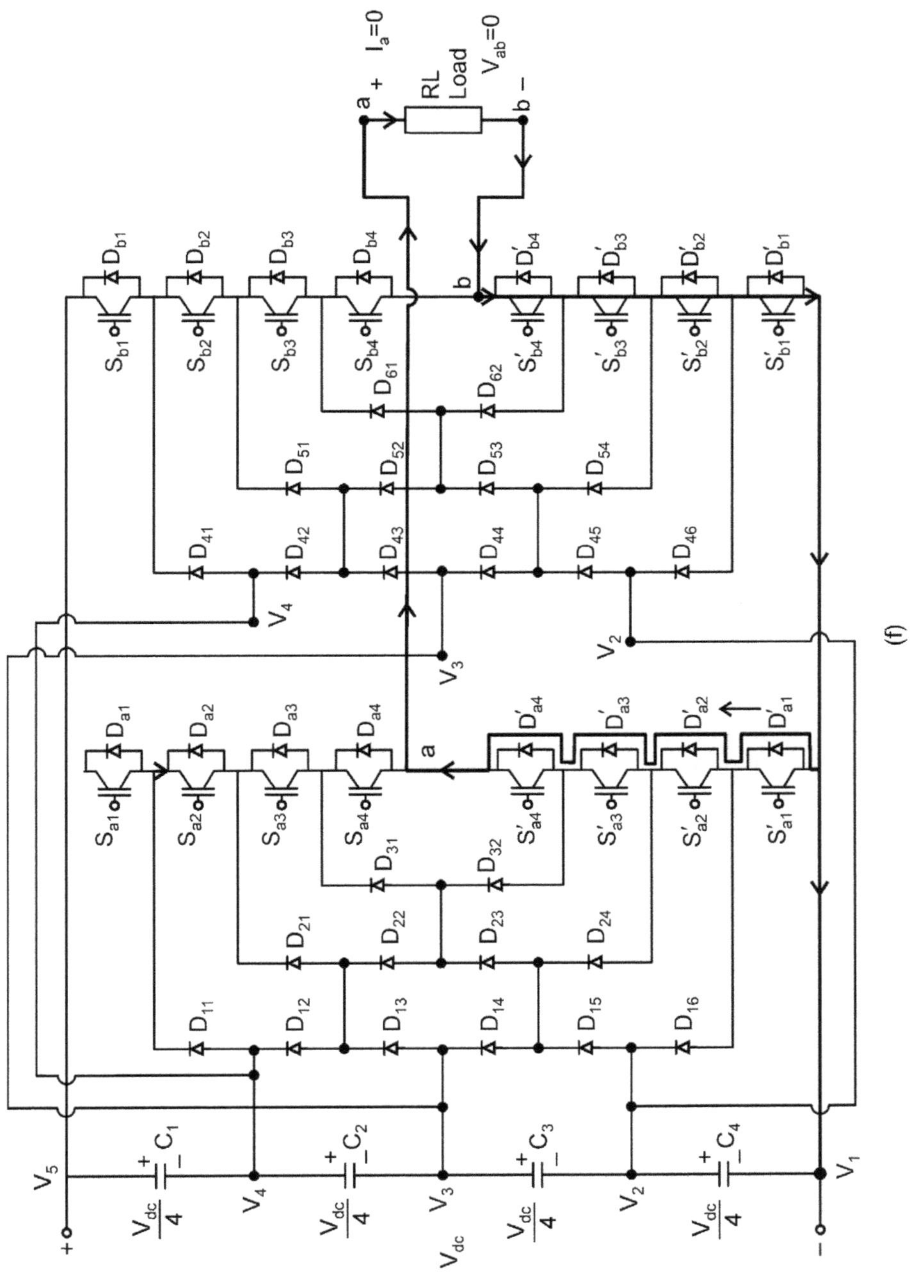

Fig. 3.23 (f) Mode5: V_{ab} = 0 with zero I_a.

Mode 5: $V_{ab} = 0$, with zero load current I_a, the switches S_{b4}', S_{b3}', S_{b2}', S_{b1}' are to be turned on to connect source to the load through source terminal V_1, anti-parallel diodes D_{a1}', D_{a2}', D_{a3}', D_{a4}' switch S_{a4}, load terminals '*a*' to '*b*', switches S_{b4}', S_{b3}', S_{b2}', S_{b1}', and source terminal V_1, and the load current is zero as shown in Fig.3.23 (f).

Mode 6: $V_{ab} = -V_{dc}/4$, with negative load current I_a, the switches S_{b4}, and S_{a4}', S_{a3}', S_{a2}', S_{a1}' are to be turned on to connect source to the load through source terminal V_2, clamping diodes D_{45}, D_{53}, D_{61}, switch S_{b4}, load terminals '*b*' to '*a*', switches S_{a4}', S_{a3}', S_{a2}', S_{a1}', and source terminal V_1, and the load current is negative when it flows from load terminals '*b*' to '*a*' as shown in Fig.3.23 (g).

Mode 7: $V_{ab} = -V_{dc}/2$, with negative load current I_a, the switches S_{b3}, S_{b4}, and S_{a4}', S_{a3}', S_{a2}', S_{a1}' are to be turned on to connect source to the load through source terminal V_3, clamping diodes D_{43}, D_{51}, switches S_{b3}, S_{b4}, load terminals '*b*' to '*a*', switches S_{a4}', S_{a3}', S_{a2}', S_{a1}', and source terminal V_1, and the load current is negative when it flows from load terminals '*b*' to '*a*' as shown in Fig.3.23 (h).

Mode 8: $V_{ab} = -3V_{dc}/4$, with negative load current I_a, the switches S_{b2}, S_{b3}, S_{b4}, and S_{a4}', S_{a3}', S_{a2}', S_{a1}' are to be turned on to connect source to the load through source terminal V_4, clamping diode D_{41}, switches S_{b2}, S_{b3}, S_{b4}, load terminals '*b*' to '*a*', switches S_{a4}', S_{a3}', S_{a2}', S_{a1}', and source terminal V_1, and the load current is negative when it flows from load terminals '*b*' to '*a*' as shown in Fig.3.23 (i).

Mode 9: $V_{ab} = -V_{dc}$, with negative load current I_a, the switches S_{b1}, S_{b2}, S_{b3}, S_{b4}, and S_{a4}', S_{a3}', S_{a2}', S_{a1}' are to be turned on to connect source to the load through source terminal V_5, switches S_{b1}, S_{b2}, S_{b3}, S_{b4}, load terminals '*b*' to '*a*', switches S_{a4}', S_{a3}', S_{a2}', S_{a1}', and source terminal V_1, and the load current is negative when it flows from load terminals '*b*' to '*a*' as shown in Fig.3.23 (j).

The switching pattern for switching devices, clamping diodes and anti-parallel diodes for various modes of operation for single phase five-level modified diode-clamped bridge inverter is given in Table 3.7.

The load line voltage V_{ab} for single phase five-level full bridge modified diode-clamped inverter is presented in Fig. 3.23 (k).

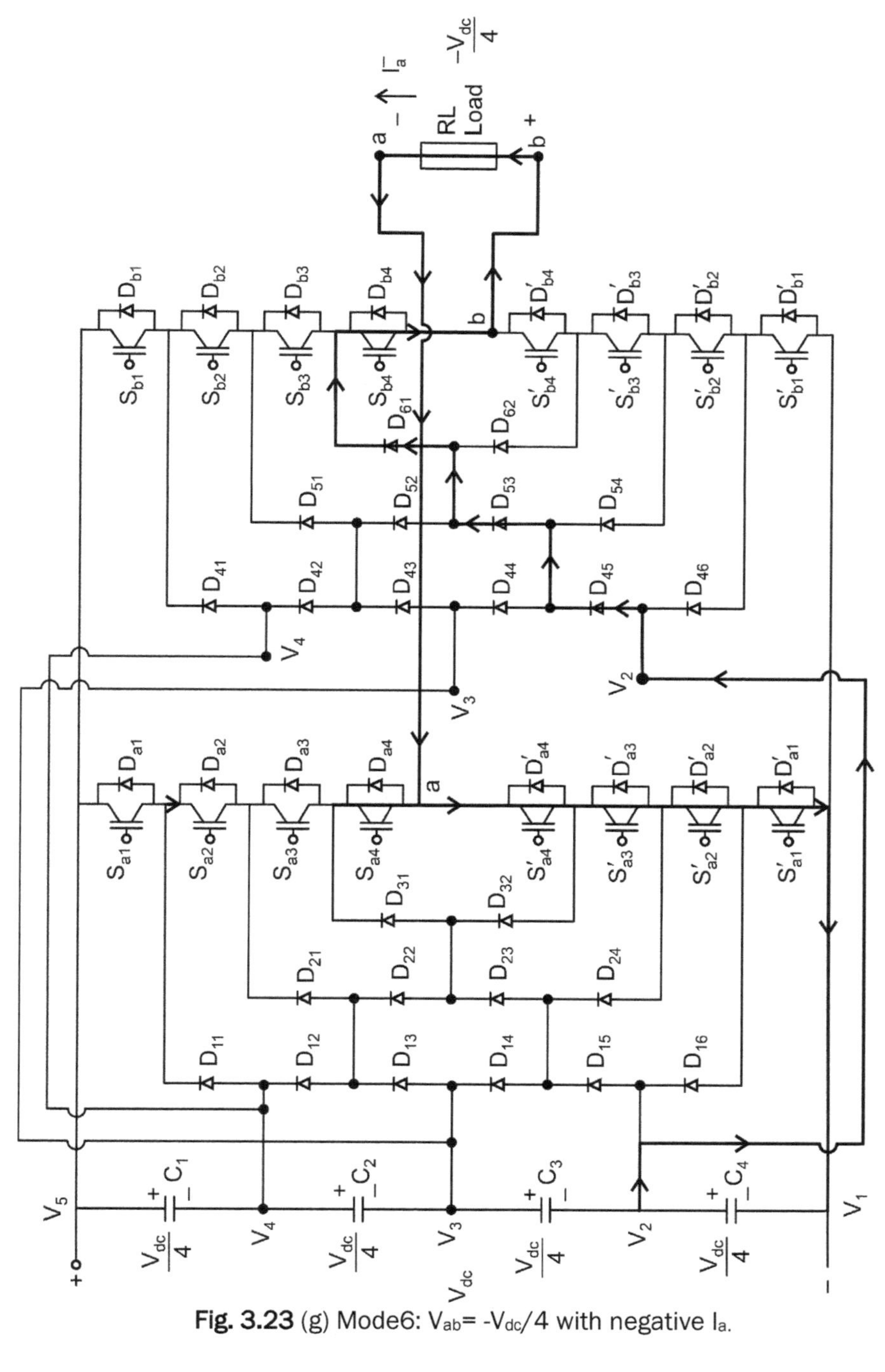

Fig. 3.23 (g) Mode6: $V_{ab} = -V_{dc}/4$ with negative I_a.

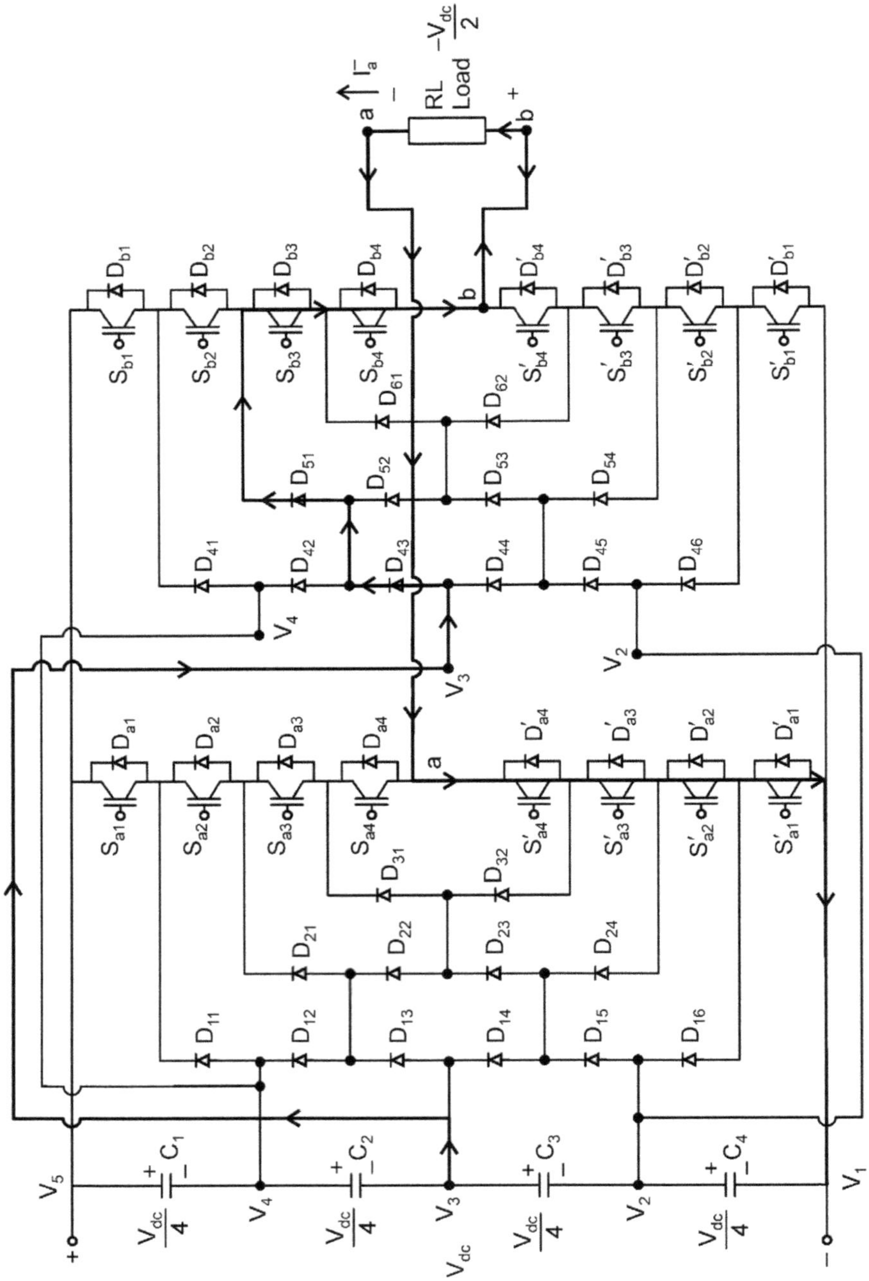

Fig. 3.23 (h) Mode7: V_{ab}= $-V_{dc}/2$ with negative I_a.

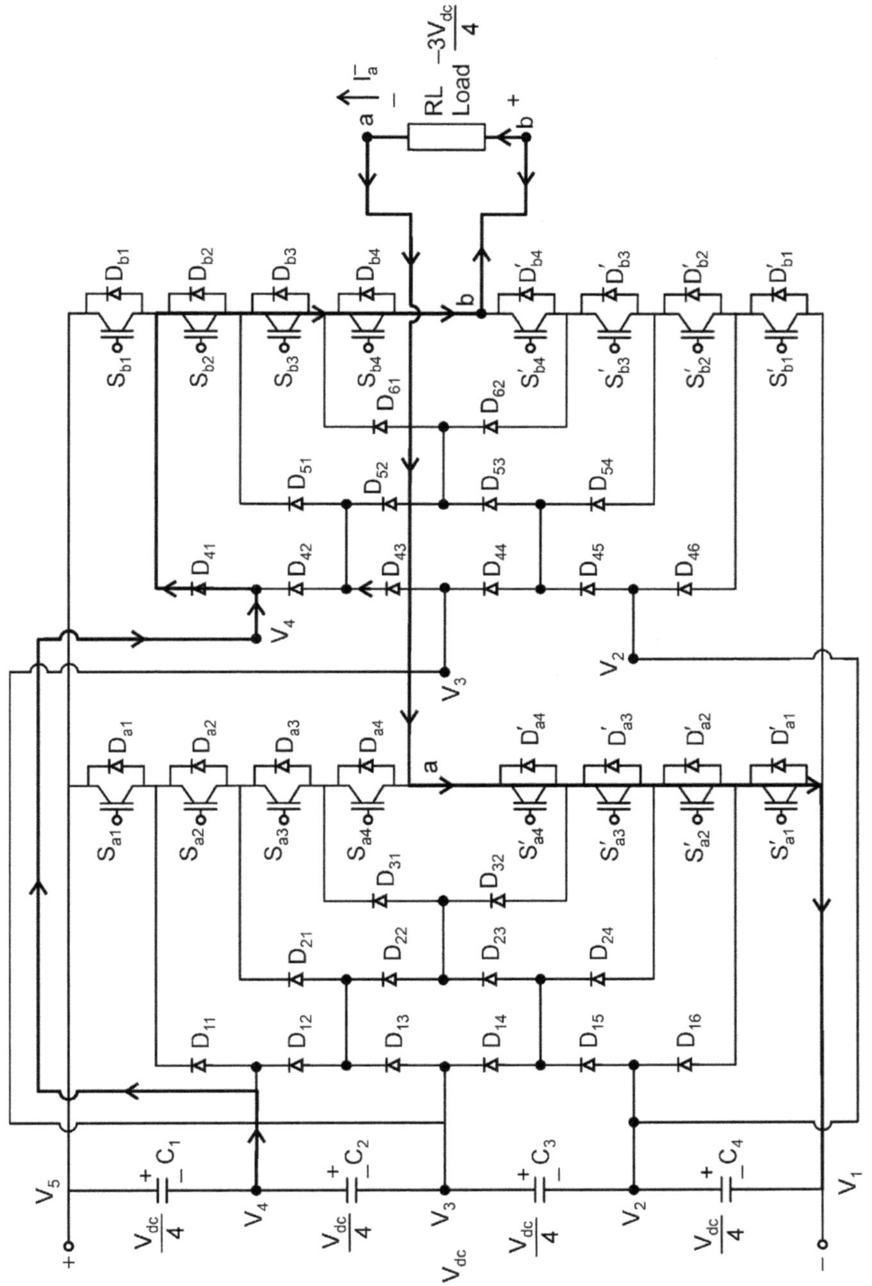

Fig. 3.23 (i) Mode 8: V_{ab}= $-3V_{dc}/4$ with negative I_a.

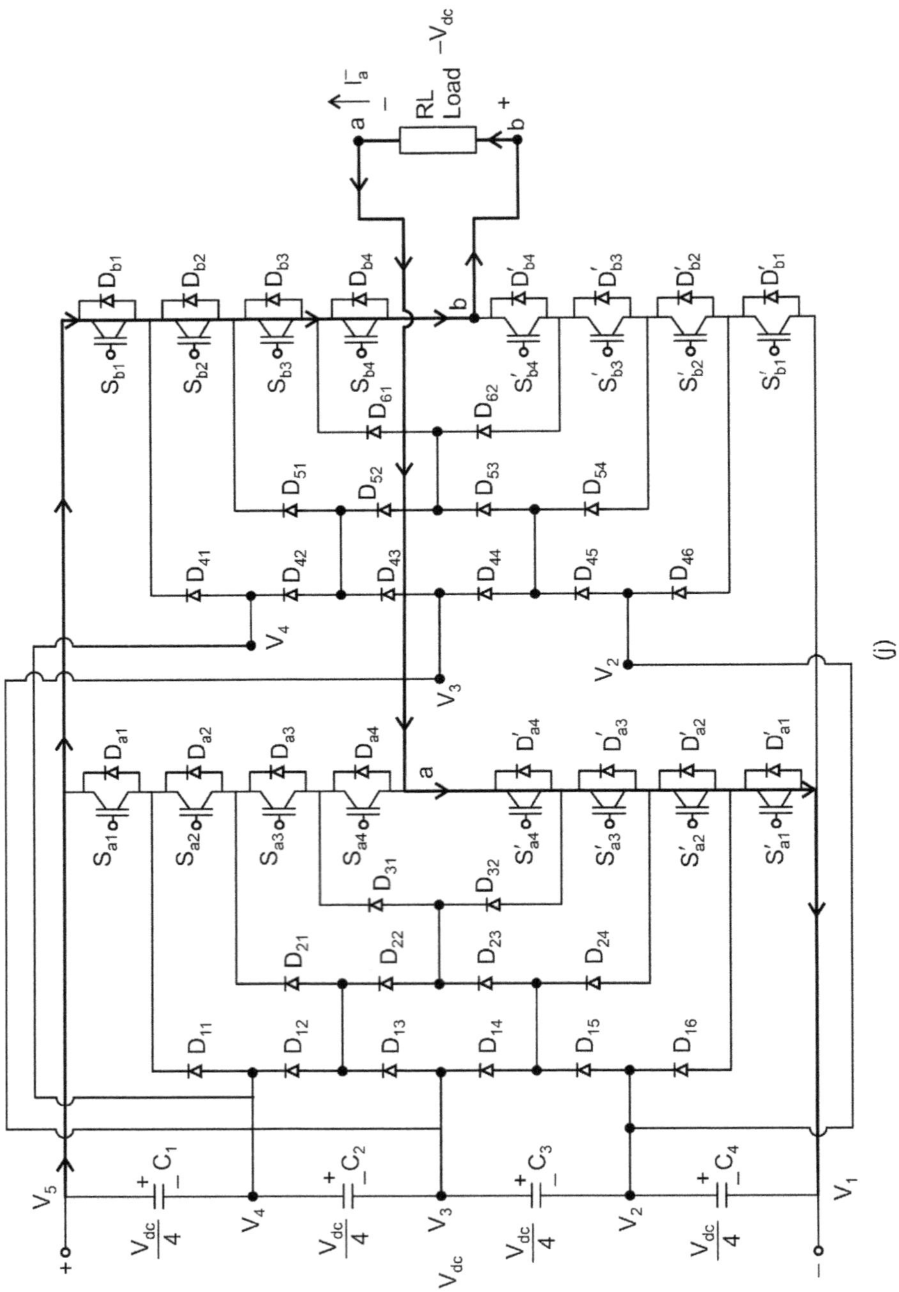

Fig. 3.23 (j) Mode 9: V_{ab}= $-V_{dc}$ with negative I_a.

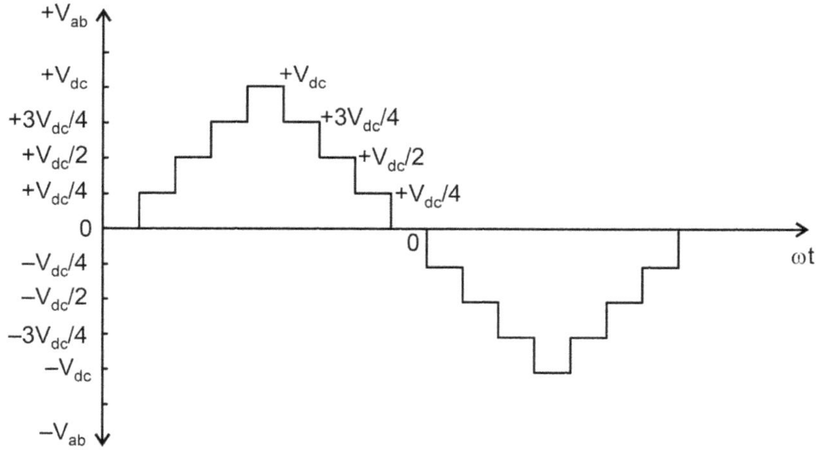

(k)

Fig. 3.23 (k) Load line voltage V_{ab} for single phase five-level full-bridge modified diode-clamped inverter.

Table 3.7 Switching pattern for single phase five-level modified diode-clamped bridge inverter.

Modes	V_{ab}	S_{a1} /S_{a1}'	S_{a2} /S_{a2}'	S_{a3} /S_{a3}'	S_{a4} /S_{a4}'	S_{b1} /S_{b1}'	S_{b2} /S_{b2}'	S_{b3} /S_{b3}'	S_{b4} /S_{b4}'	Clamping Diodes/Anti-parallel Diodes	Ia
M1	$+V_{dc}$	1/0	1/0	1/0	1/0	0/1	0/1	0/1	0/1	----	+ve
M2	$+3V_{dc}/4$	0/0	1/0	1/0	1/0	0/1	0/1	0/1	0/1	D_{11}	+ve
M3	$+V_{dc}/2$	0/0	0/0	1/0	1/0	0/1	0/1	0/1	0/1	D_{13}, D_{21}	+ve
M4	$+V_{dc}/4$	0/0	0/0	0/0	0/0	0/1	0/1	0/1	0/1	D_{15}, D_{23}, D_{31}	+ve
M5	0	0/0	0/0	0/0	0/0	0/1	0/1	0/1	0/1	D_{a1}',D_{a2}', D_{a3}', D_{a4}'	Zero
M6	$-V_{dc}/4$	0/1	0/1	0/1	0/1	0/0	0/0	0/0	1/0	D_{45}, D_{53}, D_{61}	-ve
M7	$-V_{dc}/2$	0/1	0/1	0/1	0/1	0/0	0/0	1/0	1/0	D_{43}, D_{51}	-ve
M8	$-3V_{dc}/4$	0/1	0/1	0/1	0/1	0/0	1/0	1/0	1/0	D_{41}	-ve
M9	$-V_{dc}$	0/1	0/1	0/1	0/1	1/0	1/0	1/0	1/0	----	-ve

3.4 m-Level Diode Clamped Inverter

The diode clamped multilevel inverter employs clamping diodes and cascaded dc capacitors to produce ac voltage waveforms with multiple levels. However, as the number of levels is increased, the number of clamping diodes, switching devices and dc capacitors also increases, thus the circuit configuration becomes more complicated. To increase the inverter voltage rating and improves its waveform quality, high level diode clamped inverter can be employed. The m-level diode clamped inverter is given in Fig. 3.24. In general, for an m-level diode clamped inverter, for each leg 2(m-1) switching devices, (m-1) x (m-2) clamping diodes and (m-1) dc link capacitors are required.

The m-level circuit topology is given and discussed in detail.

In Fig. 3.24, D_{a2} requires two diodes in series because it blocks two capacitor voltages, and the diode D_a(m-2) require (m-2) series connected diodes because it blocks (m-2) capacitor voltages. The voltage across each capacitor for an m-level diode clamped inverter at steady state is V_{dc}/(m-1), where V_{dc} is dc bus voltage. Though switching devices in diode clamped topology are required to block only a voltage level of V_{dc} and they are of equal ratings. On the other hand, the clamping diodes require different ratings to block the reverse voltage and diverse voltage rated clamping diodes need to be chosen. For every additional level of the output voltage, it requires an extra pair of complementary switches in each phase leg and additional capacitors and clamping diodes.

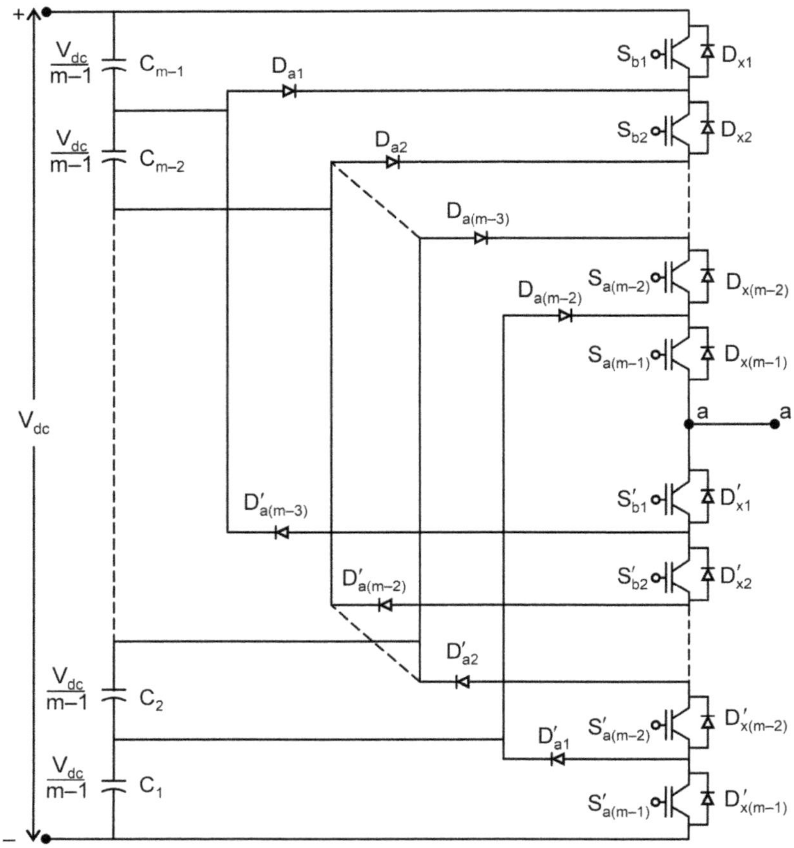

Fig. 3.24 m-level diode clamped inverter.

As the number of levels is increased the capacitor voltage balancing becomes a major problem. The m-level diode clamped inverter has an m-levels in output phase voltage and a (2m-1)-levels in output line voltage waveform.

This topology suffers from unequal load distribution among the switching devices particularly, when the inverter operates with pulse width modulation technique, and the reverse recovery of the clamping diodes is also a major design challenge. Hence, the circuit configurations and overall control algorithms become more complex. The dc-link capacitor voltage imbalance becomes the dominant problem with the increase in inverter voltage levels.

The main features of the m-level diode clamped inverter are

- A common dc bus voltage can be shared by all the phases which minimizes capacitance requirements in the inverter.
- Capacitors can be pre-charged as a group.
- When the number of levels is high, harmonic content is low enough to avoid the need of filters.
- Inverter efficiency is high, if it operates at fundamental switching frequency.
- Control logic is simple.

There are also certain drawbacks of the topology,

- Since, the number of clamping diodes required is quadratically related to the number of levels, which results in more complications in design as the number of levels increases.
- Diode-clamped inverters need high voltage rating diodes to block the reverse voltages and thus increasing the overall cost of diodes used.
- Different current ratings of the switching devices are required due to the difference in conduction periods.
- Possibility of deviation of neutral point voltages as the number of levels are more than three.
- Uneven loss distribution in the devices. In fundamental cycle, the conduction period of inner devices is more than the outer devices. This causes unequal losses in devices in a leg. This problem can be solved by using two additional switches to form Active Neutral Point (ANPC) discussed in chapter 6.

3.5 Case Studies

In this chapter four case studies are considered to demonstrate the working principle of Three phase Diode-clamped inverter for three-level and five-level output load voltages respectively, with resistive-inductive loads. The most popularly known Sinusoidal PWM and advanced Space Vector PWM techniques are used for Diode-clamped inverters. The simulated results are shown for the load phase and line voltages and Total Harmonic Distortion (%THD) of load voltage waveform.

3.5.1 Case Study 1

Simulation of Three phase Three-level Diode-clamped Inverter with RL loads using SPWM techniques is presented. The resistive-inductive loads are R=20 ohms per phase and L= 47.59 mH per phase. The dc source voltage of 500 V is applied to the circuit. The carrier frequency of 2KHz is used for SPWM firing pulse generation.

Solution: Three phase three-level diode-clamped inverter with RL load:

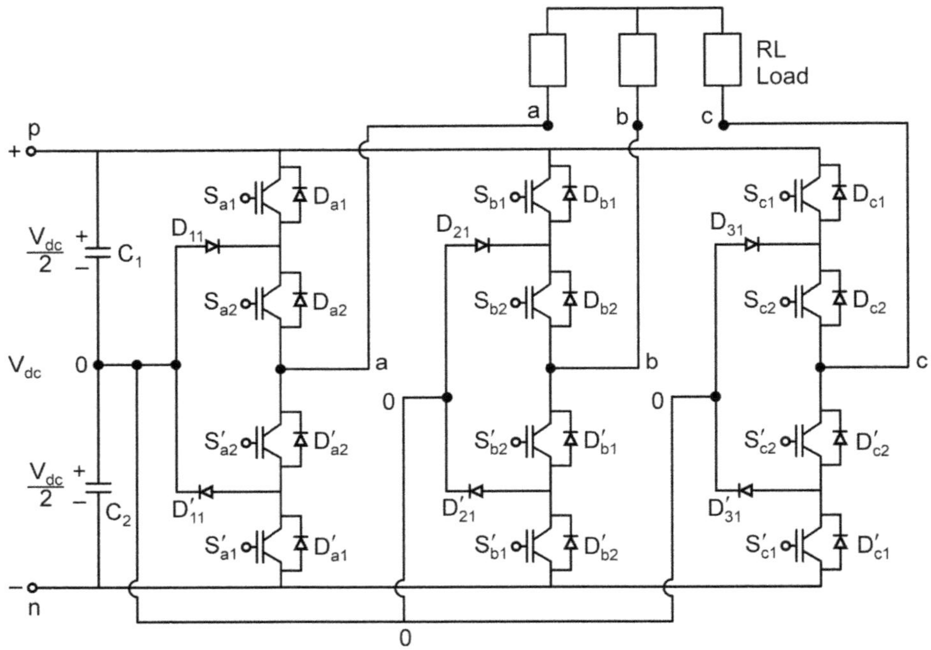

Fig. 3.5.1 (a) Three phase three-level diode-clamped Inverter for RL load.

Simulink block diagram of the above circuit:

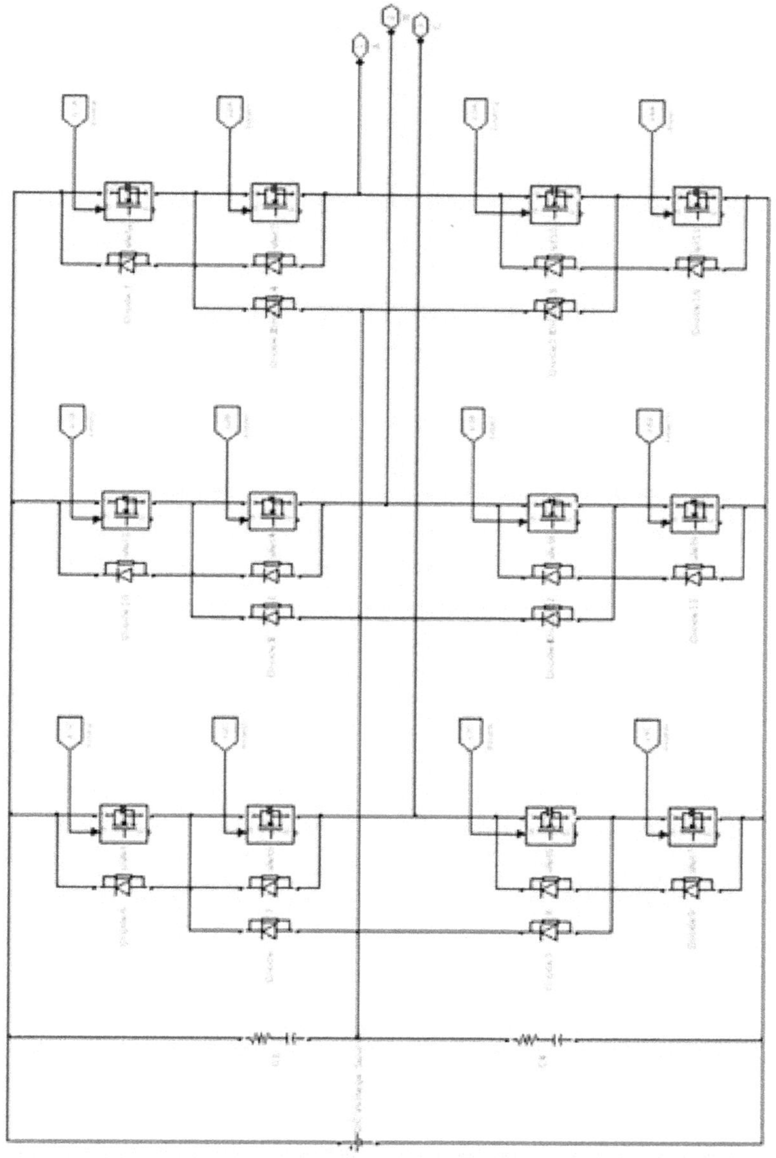

Fig. 3.5.1 (b) Simulink block diagram representation for Three-level Diode-Clamped Inverter for RL load using SPWM.

***abc* to alpha, beta Transformation using Simulink block diagram:**

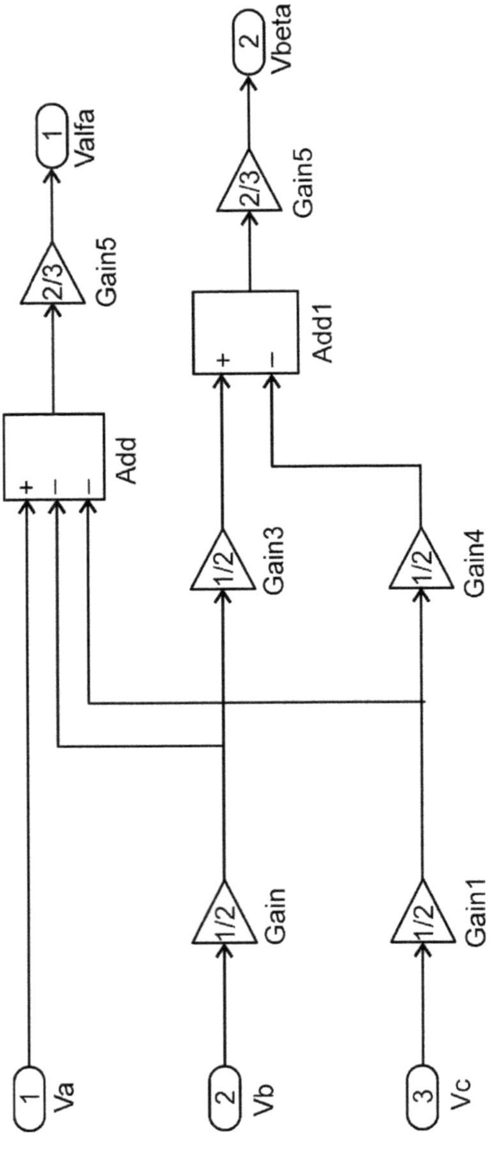

Fig. 3.5.1 (c) Transformation block diagram from *abc* to α, β.

Sinusoidal PWM for three phase three level NPC inverter:

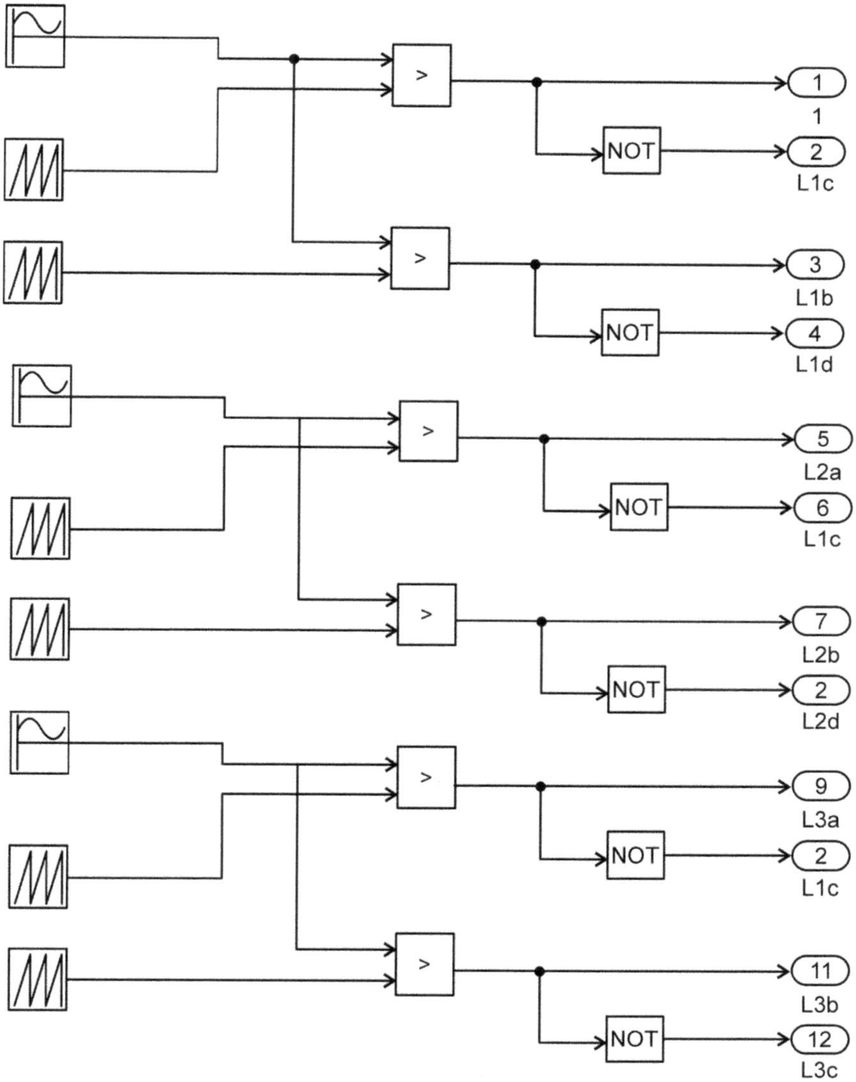

Fig. 3.5.1 (d) The firing pulse generation with SPWM for three phase three level diode clamped inverter.

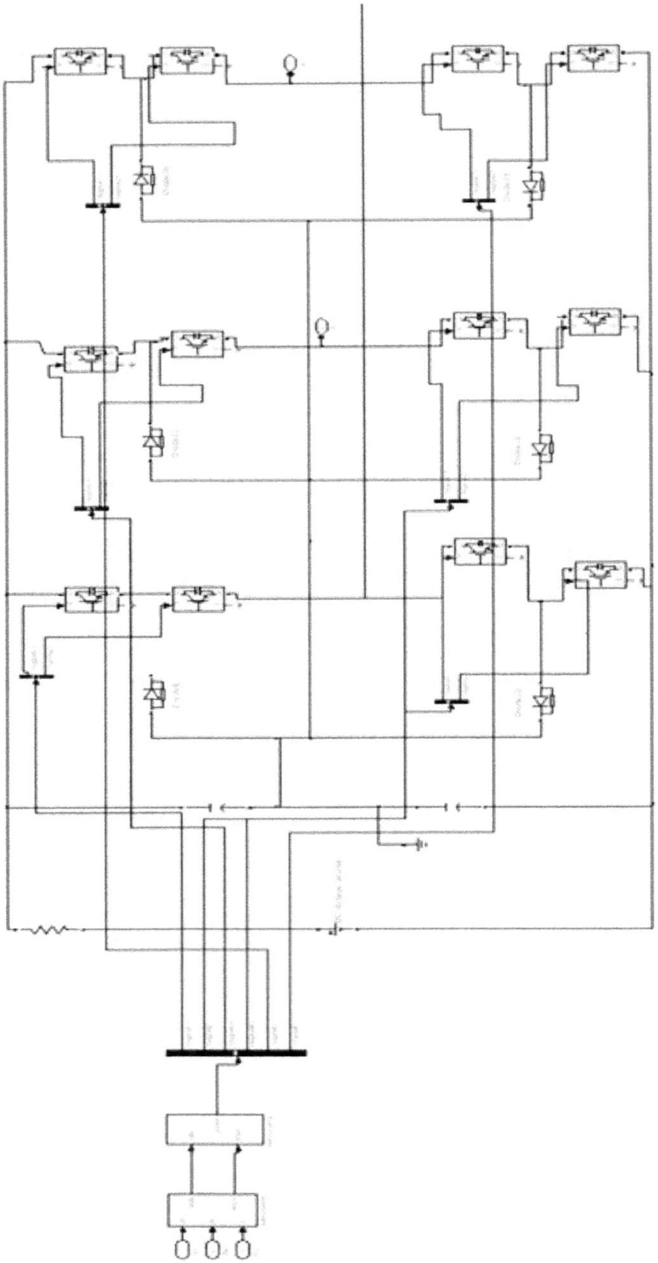

Fig. 3.5.1 (e) Simulink block diagram for three phase three level diode clamped inverter with SPWM firing signals.

The load voltages for three phase three level diode-clamped inverter with Sinusoidal PWM:

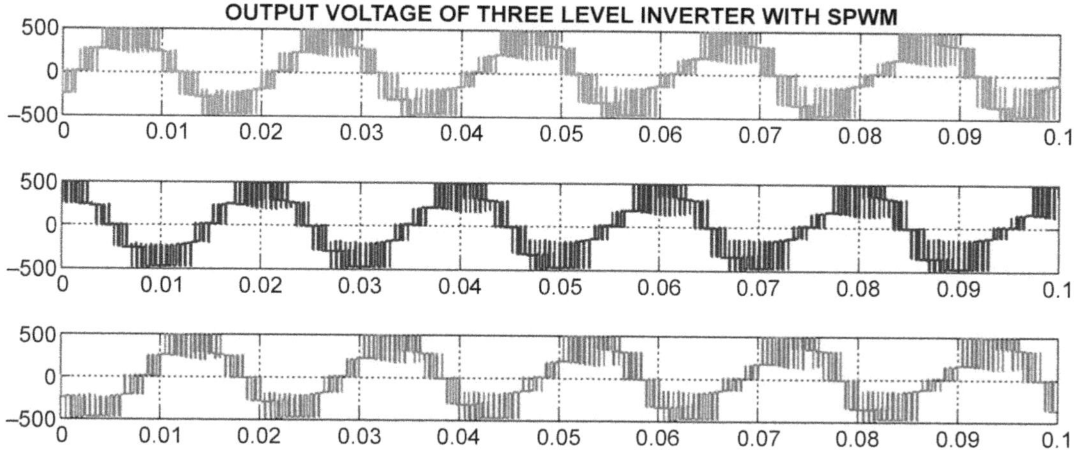

(f)

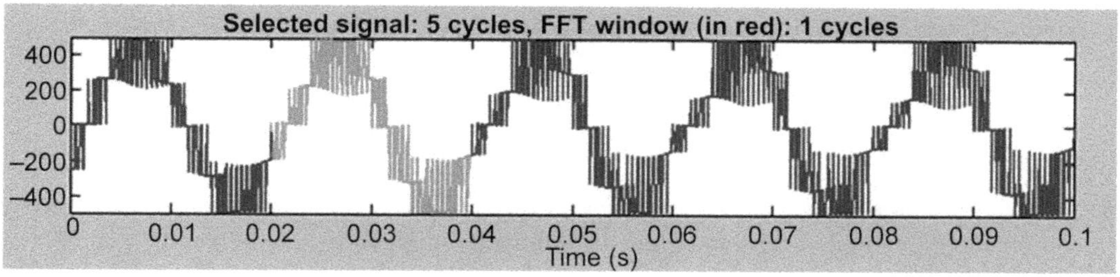

(g)

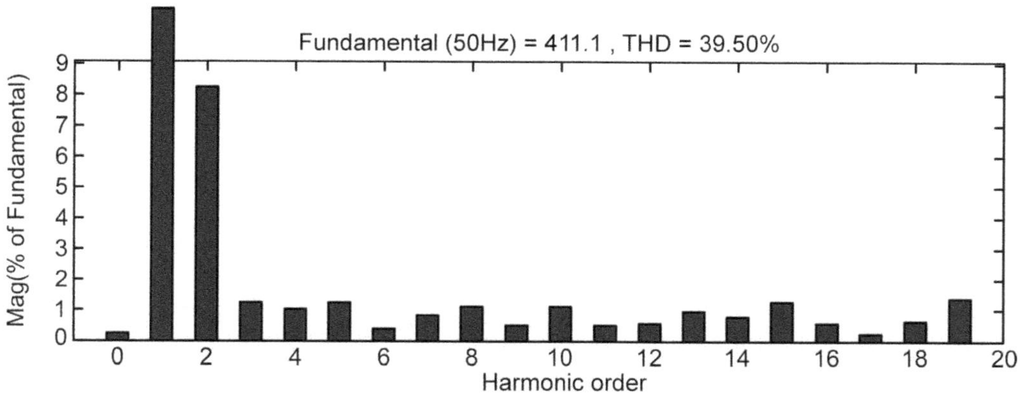

(h)

Fig. 3.5.1 (f) Three phase three-level load line voltage V_{ab}, for RL loads, (g) *a*-phase load voltage waveform and (h) the % total harmonic distortion for diode-clamped inverter with SPWM.

3.5.2 Case Study 2

Simulation of Three phase Three-level Diode-clamped Inverter with RL loads using SVPWM techniques is presented. The resistive-inductive loads are R=20 ohms per phase and L= 47.59 mH per phase. The dc source voltage of 500 V is applied to the circuit. The carrier frequency of 2.5 kHz is used for SVPWM firing pulse generation.

Solution:

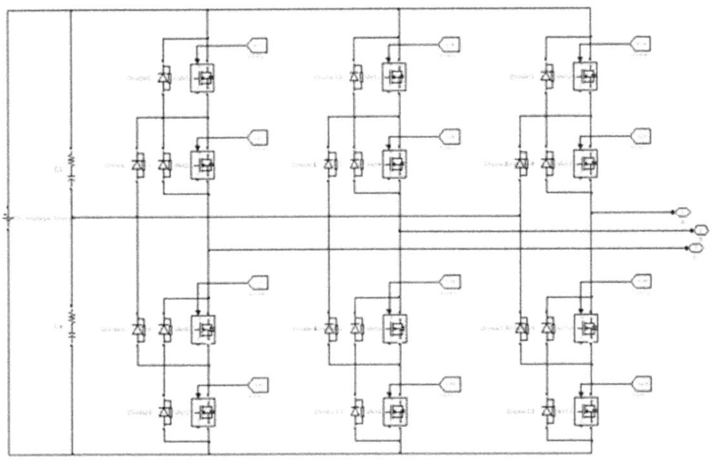

Fig. 3.5.2 (a) Simulink block diagram representation for three phase three-level Diode-Clamped Inverter for RL load using SVPWM.

***abc* to alpha, beta transformation:**

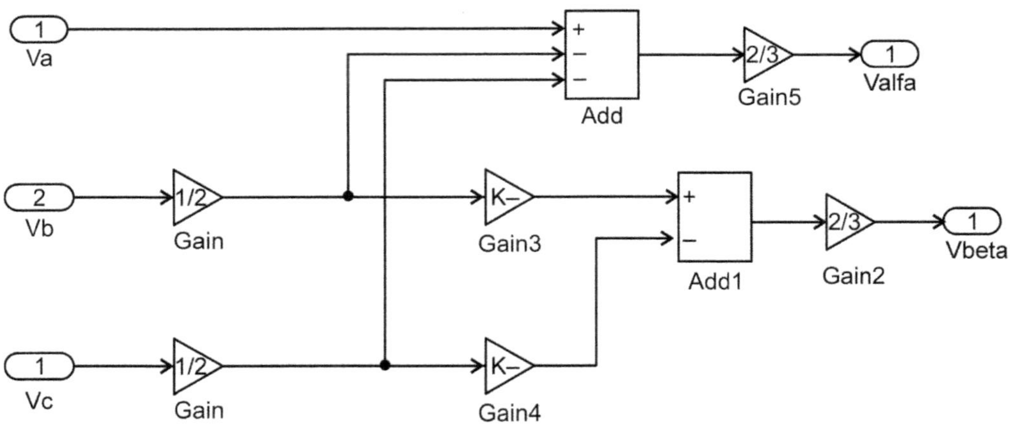

Fig. 3.5.2 (b) Transformation block diagram from *abc* to α, β.

Pulse generation using SVPWM Technique in Simulink:

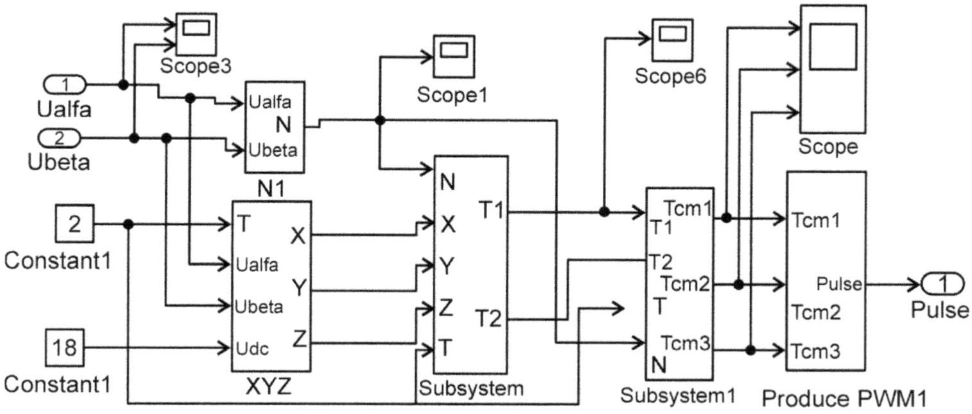

Fig. 3.5.2 (c) The firing pulse generation for SVPWM for three phase three-level diode clamped inverter.

Three phase three level Inverter diode-clamped with firing pulses using SVPWM:

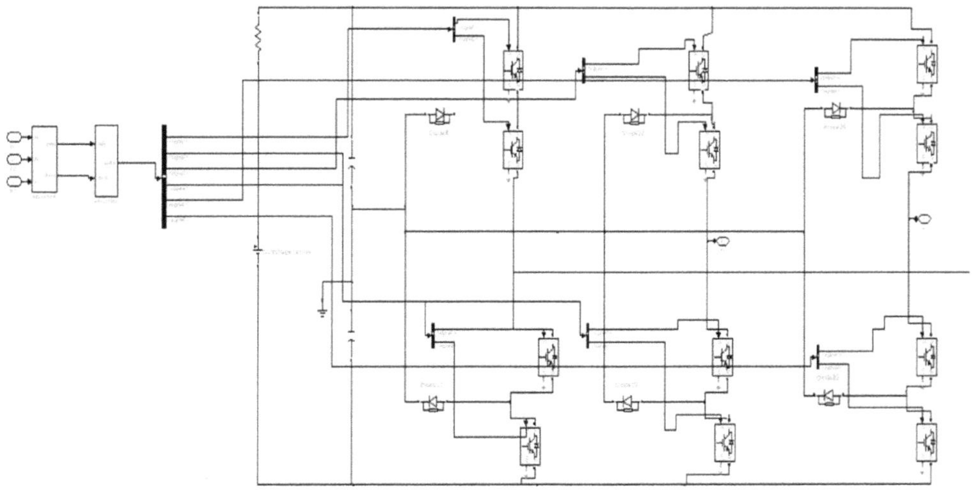

Fig. 3.5.2 (d) Simulink block diagram for three phase three level diode clamped inverter with SVPWM firing signals.

The load voltages for three phase three level diode-clamped inverter with Space Vector PWM:

Three phase three level diode-clamped Inverter with SVPWM:

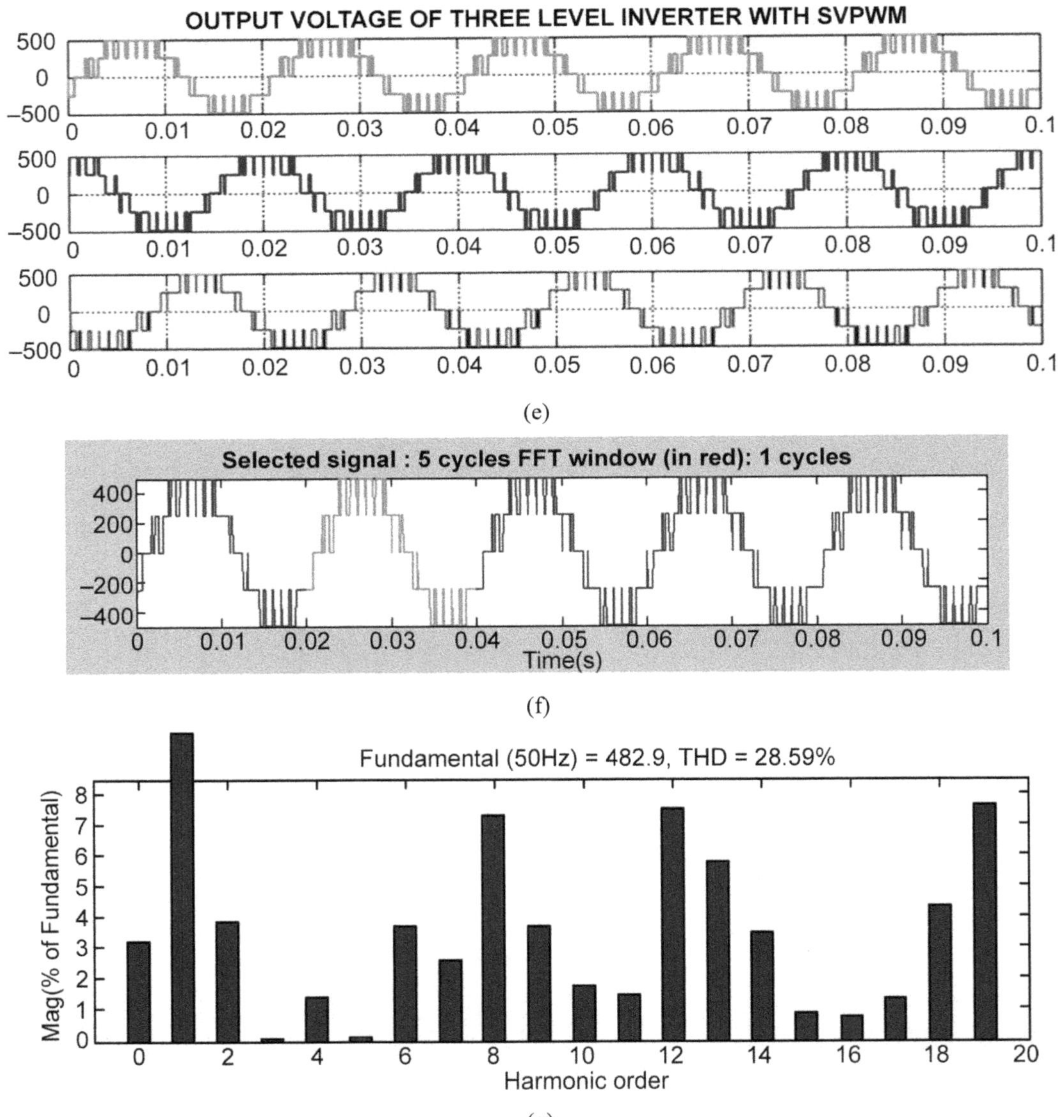

Fig. 3.5.2 (e) Three phase three-level load line voltage V_{ab}, for RL loads, (f) *a*-phase load voltage waveform and (g) the % total harmonic distortion for diode-clamped inverter with SVPWM.

3.5.3 Case Study 3

Simulation of Three phase Five-level Diode-clamped Inverter with RL loads using SPWM techniques is presented. The resistive-inductive loads are R=20 ohms per phase and L= 47.59 mH per phase. The dc source voltage of 500 V is applied to the circuit. The carrier frequency of 2 kHz is used for SPWM firing pulse generation.

Solution: **Three phase Five level diode-clamped inverter:**

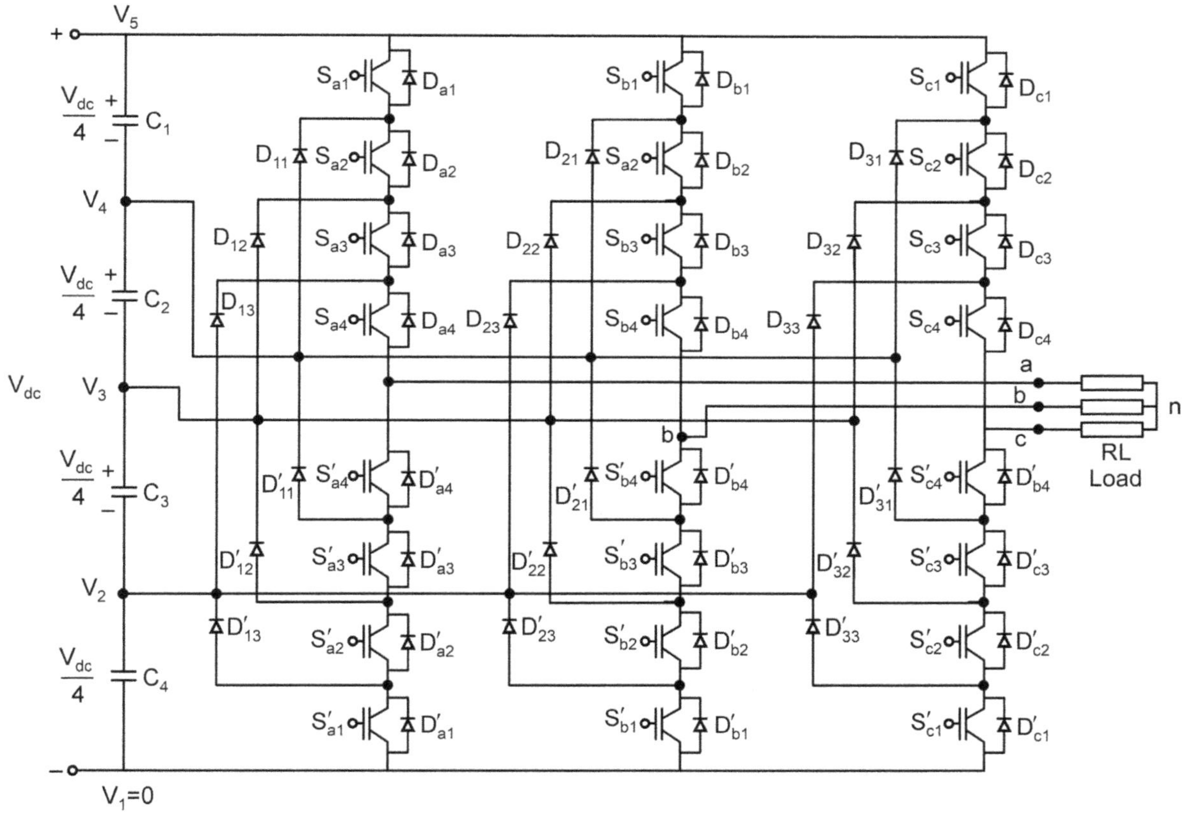

Fig. 3.5.3 (a) Three phase Five-level diode-clamped Inverter for RL load.

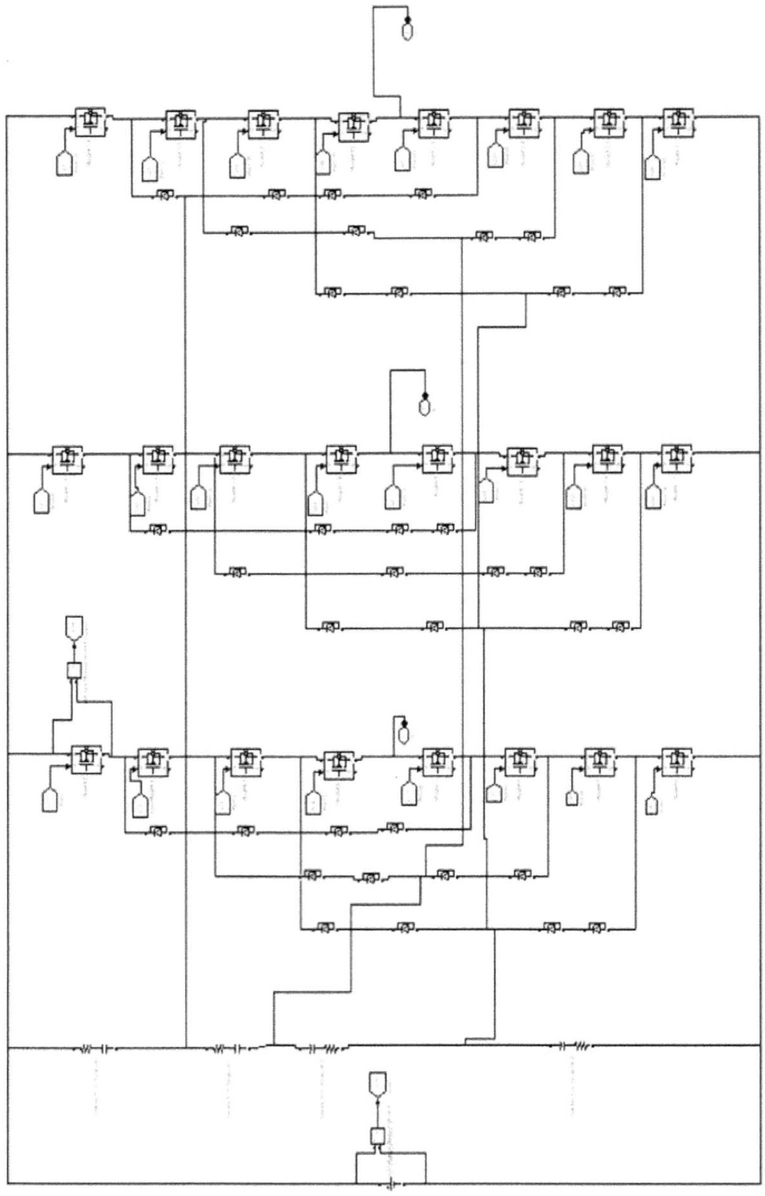

Fig. 3.5.3 (b) Simulink block diagram representation for three phase five-level Diode-Clamped Inverter for RL load using SPWM.

Sinusoidal PWM for three phase five-level diode-clamped inverter (shown for one-phase and similar diagram for other two phases are to be drawn)

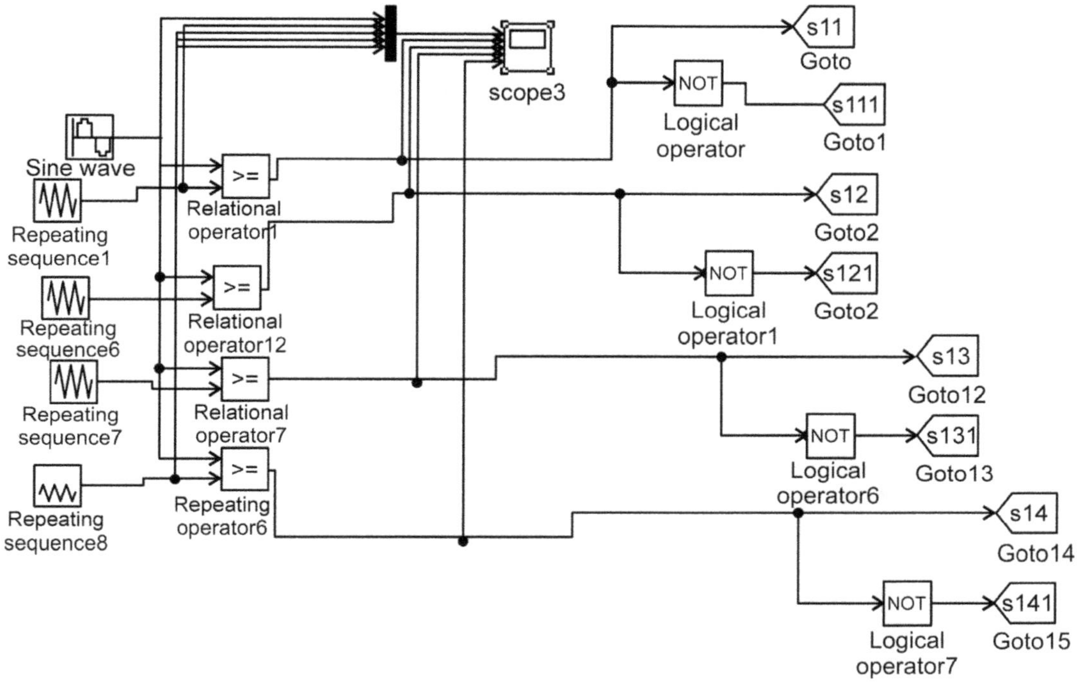

Fig. 3.5.3 (c) Firing pulses generation for three phase five-level diode-clamped inverter with SPWM.

The load voltages for three phase five-level diode-clamped inverter with Sinusoidal PWM:

Three phase five-level diode-clamped Inverter with SPWM:

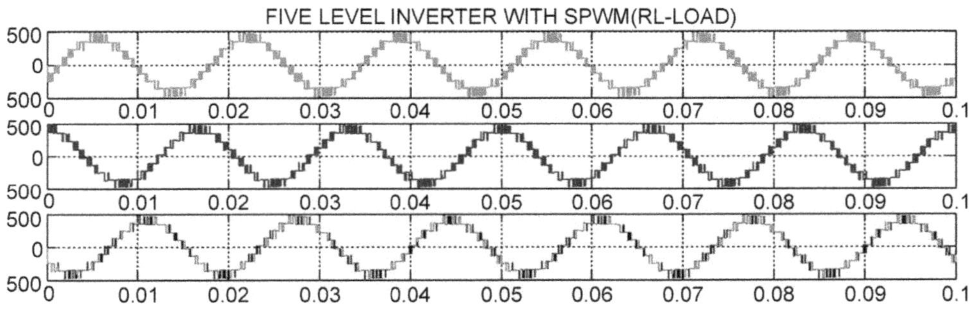

(d)

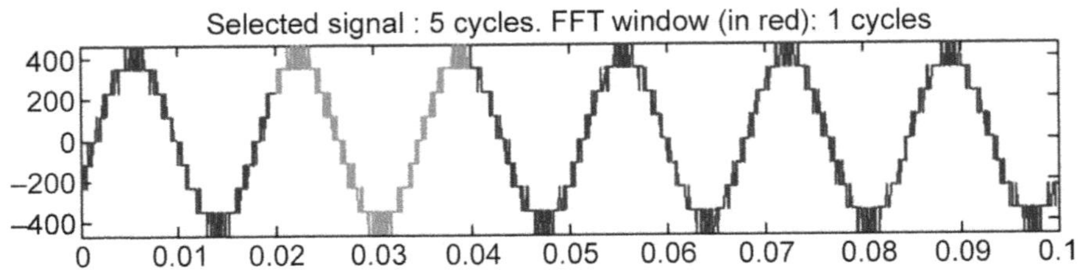

(e)

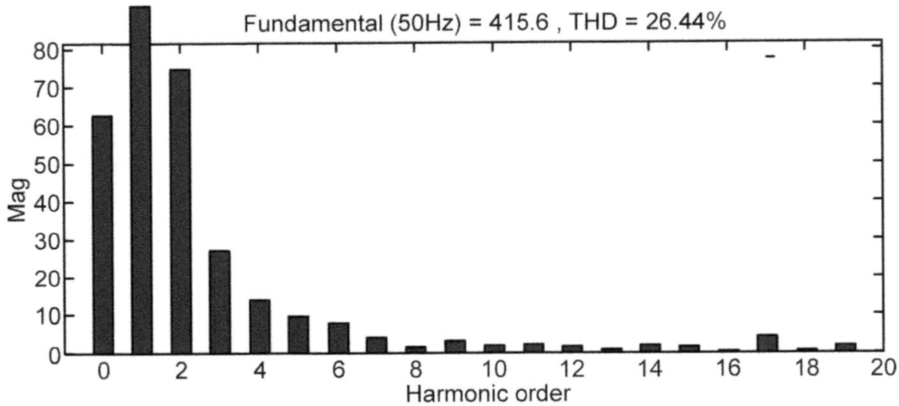

(f)

Fig. 3.5.3 (d) Three phase five-level load line voltage V_{ab}, for RL loads, (e) *a*-phase load voltage waveform and (f) the % total harmonic distortion for diode-clamped inverter with SPWM.

3.5.4 Case Study 4

Simulation of Three phase Five-level Diode-clamped Inverter with RL loads using SVPWM techniques is presented. The resistive-inductive loads, R=20 ohms per phase and L= 47.59 mH per phase. The dc source voltage of 500 V is applied to the circuit. The carrier frequency of 2.5 kHz is used.

Solution:

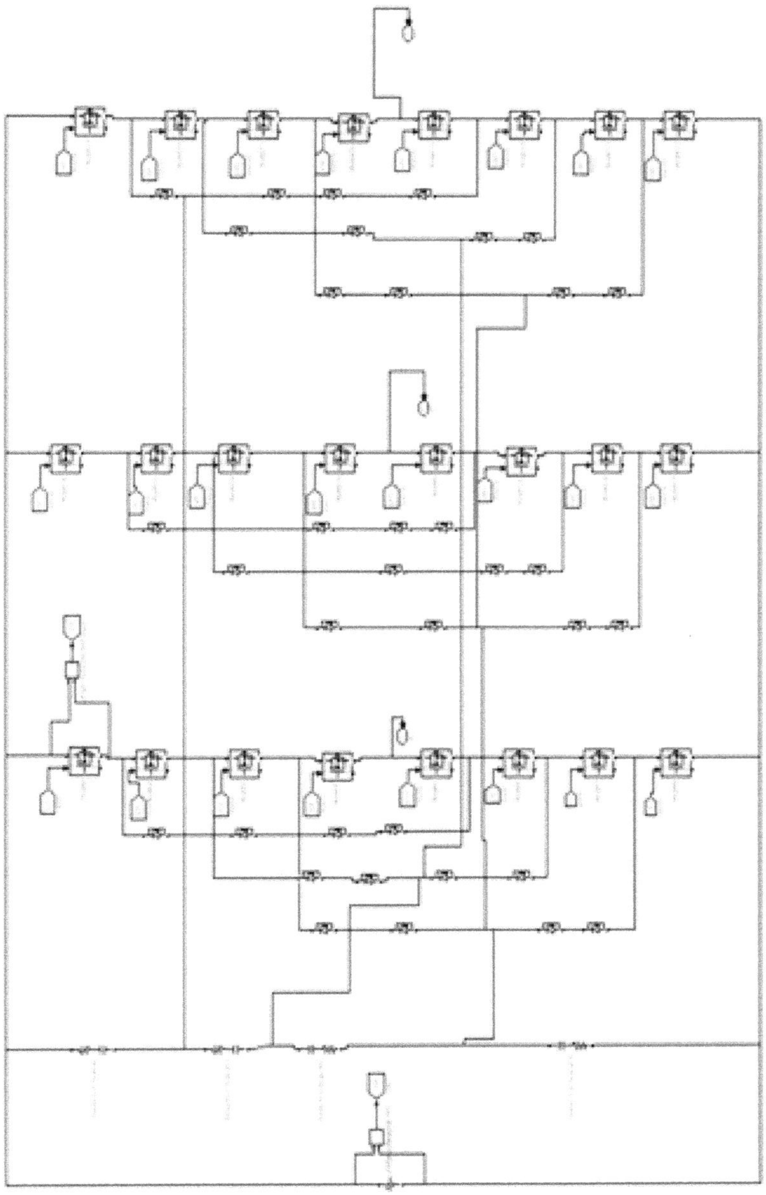

Fig. 3.5.4 (a) Simulink block diagram representation for three phase five-level diode-clamped inverter for RL load using SVPWM.

Firing pulse generation circuit for SVPWM for three phase five-level inverter:

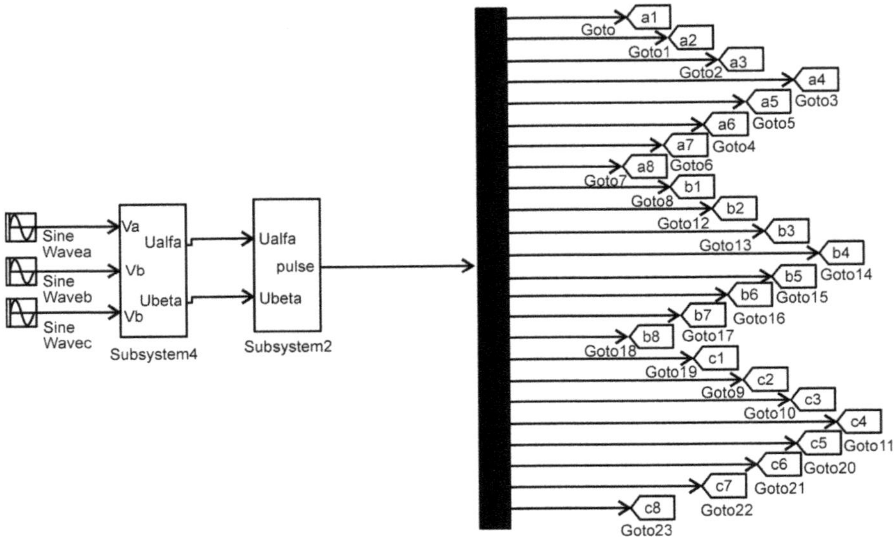

Fig. 3.5.4 (b) The firing pulse generation for SVPWM for three phase five-level diode clamped inverter

Three phase Five-level inverter with SVPWM:

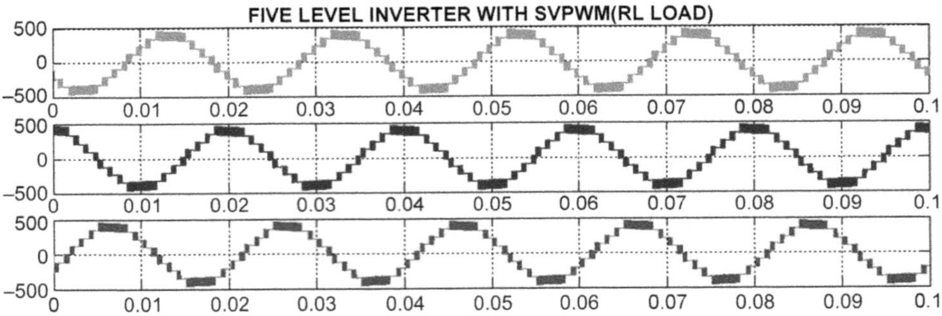

(c)

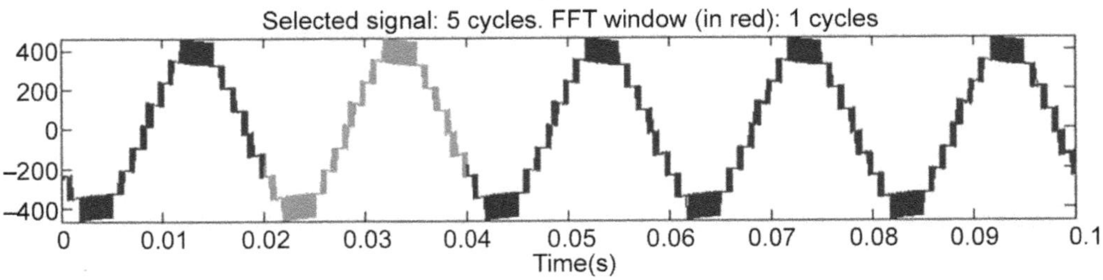

(d)

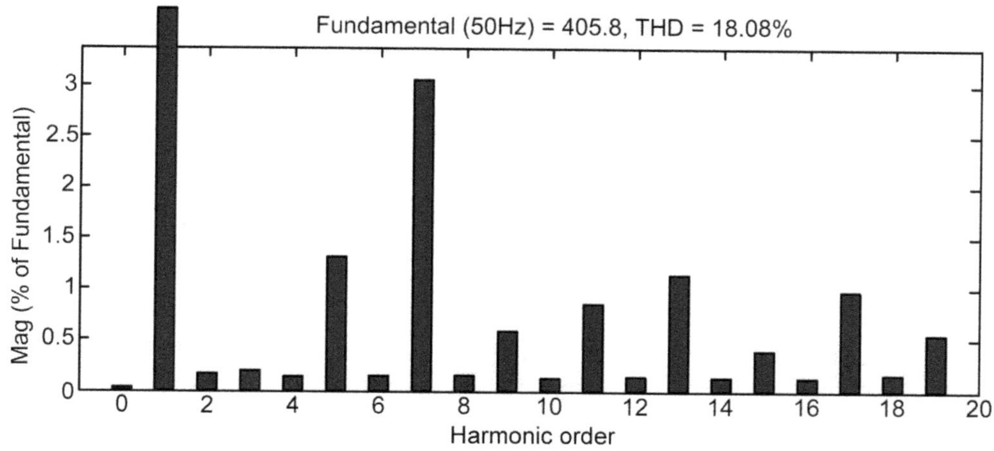

(e)

Fig. 3.5.4 (c) three phase five-level load line voltage, V_{ab}, (d) a-phase load voltage, and (e) the % total harmonic distortion with SVPWM.

Comparison of load voltage % total harmonic distortion for various three phase three- and five-level diode-clamped inverter considered in above 4 case studies are presented in table.3.5.1.

Table 3.5.1 Comparison of % THD of Load voltage.

Type of inverter and PWM method	% Total harmonic distortion of load voltage
Five-level with SVPWM	18.08
Five-level with SPWM	26.44
Three-level with SVPWM	28.59
Three-level with SPWM	39.57

3.6 Summary

The diode-clamped inverter is used in many industrial and power systems applications, due to their inherent advantages. The basic construction of diode-clamped inverter and its functioning is explained elaborately. They are classified into single-leg configuration and full-bridge configuration. They are also available in single-phase and three-phase inverter category.

The higher levels of output voltages for the diode-clamped inverters depends on the number of dc source capacitors. For producing m-level output voltage from diode-clamped inverter, (m-1) dc source capacitors are required. Besides, an m-level diode clamped inverter, for each leg 2(m-1) switching devices, (m-1) x (m-2) clamping diodes and (m-1) dc source capacitors are required.

For example, for m=5, 8 switching devices, 4 dc source capacitors and 12 clamping-diodes are required for the five-level inverter.

The single-phase single-leg and single-phase full-bridge diode-clamped inverters are presented and discussed their operation by considering various modes and switching patterns for three-level, five-level and seven-level output voltages for resistive-inductive loads. Similarly, the three-phase full-bridge diode-clamped inverter topologies with their output line voltages are presented for three-level, five-level and seven-level output voltages for wye-connected resistive-inductive loads. The modified diode-clamped inverter is also introduced for five-level output. The load current depends on the nature of the loads.

The m-level configuration is also presented for diode-clamped inverter. The advantages, disadvantages and applications of all the above discussed multilevel inverters are also given.

The four case studies and exercise problems are given in all varieties of diode-clamped multilevel inverters for practice by modeling and simulation.

Learning Outcomes

- Acquire knowledge about the construction of diode-clamped inverter of basic three-level operation. (CO1) :: (PO1)
- Ability to analyze the operation of each type of diode-clamped inverter of different levels of output voltages. (CO4) :: (PO2)
- To be able to apply suitable pulse width modulation techniques for diode-clamped inverters. (CO3) :: (PO1)
- To be able to develop the simulation model using software tools for diode-clamped inverters. (CO4) :: (PO5)
- To be able to evaluate the performance of the diode-clamped inverters for various loads by simulation. (three-level, five-level, seven-level inverters). (CO5) :: (PO3)
- To be able to extend the single-phase diode-clamped inverter concepts to three-phase diode-clamped inverter and investigate the performance of the inverter, PWM schemes for three-phase loads. (CO6) :: (PO4).

Questions

3.1 Construct the circuit for a single-phase three-level diode-clamped bridge Inverter using single-pole triple throw switches for RL loads, and explain the operation of the circuit and draw output voltage waveform?

3.2 Realize the basic circuit of single-phase three-level diode-clamped by using bidirectional IGBTs and draw the final topology for single-leg arrangement?

3.3 Analyze single-phase single-leg three-level diode-clamped inverter topology and present the working modes for positive load current with the help of output voltage waveform?

3.4 Explain the importance of blanking time interval during commutation for a single-phase single-leg three-level diode-clamped inverter?

3.5 Discuss the advantages and applications of single-phase three-level diode-clamped inverter?

3.6 Analyze the topology of single-phase five-level diode-clamped inverter with single-leg arrangement and bring out detailed operation for both positive and negative load current?

3.7 Analyze the five-level diode-clamped inverter circuit given in question no. 3.6 and bring out the switching states for generating five-level output voltages by considering single-leg arrangement?

3.8 Analyze the single-phase five-level diode-clamped bridge inverter by considering various modes of operations to obtain five-level output voltages?

3.9 Analyze the single-phase five-level diode-clamped bridge inverter by considering various modes of operations with the influence of induced emf in load inductance (stored energy) of RL load, to obtain five-level output voltages?

3.10 Analyze the topology of single-phase seven-level diode-clamped inverter with single-leg arrangement and bring out various modes of operation for both positive and negative load current?

3.11 Analyze the seven-level diode-clamped inverter circuit given in question no. 3.10 and bring out the switching states for generating seven-level output voltages by considering single-leg arrangement and sketch output voltage waveform?

3.12 Analyze the single-phase seven-level diode-clamped inverter with single-leg arrangement by considering various modes of operations with the influence of induced emf in load inductance (stored energy) of RL load, to obtain seven-level output voltages?

3.13 Analyze the single-phase seven-level diode-clamped bridge inverter by considering various modes of operations to obtain seven-level output voltages?

3.14 Analyze the seven-level diode-clamped inverter bridge circuit given in Question no. 3.13 and bring out the switching states for generating seven-level output voltages and sketch output voltage waveform?

3.15 Draw the circuit topology for three-phase three-level diode-clamped bridge inverter for star-connected load and sketch the three-phase output load voltages?

3.16 Draw the circuit topology for three-phase five-level diode-clamped bridge inverter for star-connected load and sketch the three-phase output load voltages?

3.17 Draw the circuit topology for three-phase seven-level diode-clamped bridge inverter for star-connected load and sketch the three-phase output load voltages?

3.18 Analyze the modified single-phase three-level diode-clamped inverter with distributed clamped diodes for single-leg arrangement and sketch the output load voltage?

3.19 Analyze the modified single-phase five-level diode-clamped bridge inverter with distributed clamped diodes for single-leg arrangement and sketch the output load voltage?

3.20 Draw the m-level generalized diode-clamped inverter with single-leg arrangement? Bring out main features and drawbacks?

Multiple Choice Questions

3.1 What is the purpose of diodes in diode-clamped inverter other than anti-parallel diodes?

(a) for feed back (b) for clamping

(c) both (a) and (b) (d) none of the above

3.2 Which of the following is also called Neutral Point Clamped inverter?

(a) Cascaded H-Bridge inverter (b) flying capacitor inverter

(c) diode-clamped inverter (d) transistor clamped inverter

3.3 How many dc source capacitors are needed for three-level diode-clamped inverter?

(a) three (b) one

(c) zero (d) two

3.4 How many power semiconductor devices are necessary for three-level diode-clamped inverter per leg?

(a) four PSC devices (b) three PSC devices

(c) two PSC devices (d) six PSC devices

3.5 What is the purpose of anti-parallel diodes connected across each power semiconductor switching device in multilevel inverter?

(a) source energy transfer to load

(b) load energy transfer to source

(c) to short-circuit the respective conducting PSC devices

(d) none of the above

3.6 How many dc source capacitors are needed for five-level diode-clamped inverter?

(a) three (b) one

(c) four (d) two

3.7 How many power semiconductor devices are necessary for five-level diode-clamped inverter per leg?

(a) four PSC devices (b) six PSC devices

(c) two PSC devices (d) eight PSC devices

3.8 How many clamping diodes are required for five-level diode-clamped Inverter per leg?

(a) six clamping diodes (b) four clamping diodes

(c) eight clamping diodes (d) two clamping diodes

3.9 How many dc source capacitors are needed for seven-level diode-clamped inverter?

(a) three (b) six

(c) four (d) five

3.10 How many power semiconductor devices are necessary for seven-level diode-clamped inverter per leg?

(a) ten power semiconductor switching devices

(b) six power semiconductor switching devices

(c) twelve power semiconductor switching devices

(d) eight power semi-conductor switching devices

3.11 How many clamping diodes are required for seven-level diode-clamped Inverter per leg?

(a) sixteen clamping diodes

(b) fourteen clamping diodes

(c) eighteen clamping diodes

(d) thirty clamping diodes

3.12 How many clamping diodes are required for modified five-level diode-clamped Inverter per leg with distributed clamping diodes topology?

(a) sixteen clamping diodes

(b) twelve clamping diodes

(c) eighteen clamping diodes

(d) fourteen clamping diodes

3.13 For m-level diode-clamped inverter, how many switching devices are required per each leg?

(a) (m-1) x (m-2) switching devices

(b) (m-1) + (m-2) switching devices

(c) 2(m-1) switching devices

(d) (m-1) switching devices

3.14 For m-level diode-clamped inverter, how many clamping-diodes are needed per each leg?

(a) (m-1) x (m-2) clamping diodes

(b) (m-1) clamping diodes

(c) 2(m-1) clamping diodes

(d) (m-1) + (m-2) clamping diodes

3.15 For m-level diode-clamped inverter, how many dc link source capacitors are required for the circuit?

(a) (m-2) dc link source capacitors

(b) (m-1) x (m-2) dc link source capacitors

(c) 2(m-1) dc link source capacitors

(d) (m-1) dc link source capacitors

3.16 For five-level diode-clamped inverter (m=5), how many power semiconductors switching devices, dc link source capacitors, and clamping diodes are required per each leg?
   (a) 4 switching devices, 8 dc source capacitors, 12 clamping diodes
   (b) 8 switching devices, 4 dc source capacitors, 12 clamping diodes
   (c) 6 switching devices, 4 dc source capacitors, 6 clamping diodes
   (d) 8 switching devices, 6 dc source capacitors, 4 clamping diodes

3.17 For three-level diode-clamped inverter (m=3), how many power semiconductors switching devices, dc link source capacitors, and clamping diodes are required per each leg?
   (a) 4 switching devices, 8 dc source capacitors, 2 clamping diodes
   (b) 6 switching devices, 4 dc source capacitors, 4 clamping diodes
   (c) 4 switching devices, 2 dc source capacitors, 2 clamping diodes
   (d) 8 switching devices, 6 dc source capacitors, 4 clamping diodes

3.18 For seven-level diode-clamped inverter (m=7), how many power semiconductors switching devices, dc link source capacitors, and clamping diodes are required per each leg?
   (a) 12 switching devices, 6 dc source capacitors, 30 clamping diodes
   (b) 8 switching devices, 4 dc source capacitors, 16 clamping diodes
   (c) 10 switching devices, 8 dc source capacitors, 12 clamping diodes
   (d) 8 switching devices, 6 dc source capacitors, 24 clamping diodes

Exercise Problems

Ex.3.1: Simulate single phase diode-clamped inverter with single-leg circuit, when dc source voltage V_{dc}=200V, R=20 ohms and L= 10 mH and obtain gating signals, output phase voltage and load current by using any software tools available. Use Sinusoidal PWM modulation technique for three-level output.

Ex.3.2: Evaluate the performance of single-phase diode-clamped inverter with single-leg circuit, when dc source voltage V_{dc}=200V, R=20 ohms and L=10 mH and obtain gating signals, output phase voltage and load current by simulation using any software tools available. Use Carrier-based PWM modulation technique for three-level output.

Ex.3.3: Simulate single phase diode-clamped inverter with single-leg circuit, when dc source voltage V_{dc}=200V, R=20 ohms and L= 10 mH and obtain gating signals, output phase voltage and load current by using any software tools available. Use SVPWM modulation technique for three-level output.

Ex.3.4: Evaluate the performance of single-phase diode-clamped full-bridge inverter when dc source voltage V_{dc}=200V, R=20 ohms and L=10 mH and obtain gating signals, output line voltage and load current by simulation using any software tools available. Use Sinusoidal PWM modulation technique for three-level output.

Ex.3.5: Evaluate the performance of single-phase diode-clamped full-bridge inverter when dc source voltage V_{dc}=200V, R=20 ohms and L=10 mH and obtain gating signals, output line voltage and load current by simulation using any software tools available. Use Space-vector PWM modulation technique for three-level output.

Ex.3.6: Simulate single phase diode-clamped full-bridge inverter when dc source voltage V_{dc}=100V, R=20 ohms and L=10 mH and obtain gating signals, output line voltage and load current by using any software tools available. Use Carrier-based PWM modulation technique for three-level output.

Ex.3.7: Simulate three phase flying-capacitor full-bridge inverter when dc source voltage V_{dc}=500V, R=20 ohms per phase and L=47.59 mH per phase and obtain output line voltages and load current by using any software tools available. Use Sinusoidal PWM modulation technique for three-level output.

Ex.3.8: Simulate single phase diode-clamped inverter with single-leg circuit, when dc source voltage V_{dc}=200V, R=20 ohms and L=10 mH and obtain gating signals, output phase voltage and load current by using any software tools available. Use Sinusoidal PWM modulation technique for five-level output.

Ex.3.9: Evaluate the performance of single-phase diode-clamped inverter with single-leg circuit, when dc source voltage V_{dc}=200V, R=20 ohms and L=10 mH and obtain gating signals, output phase voltage and load current by simulation using any software tools available. Use Carrier-based PWM modulation technique for five-level output.

Ex.3.10: Simulate single phase diode-clamped inverter with single-leg circuit, when dc source voltage V_{dc}=200V, R=20 ohms and L=10 mH and obtain gating signals, output phase voltage and load current by using any software tools available. Use Space-vector PWM modulation technique for five-level output.

Ex.3.11: Evaluate the performance of single-phase diode-clamped full-bridge inverter when dc source voltage V_{dc}=200V, R=20 ohms and L=10 mH and obtain gating signals, output line voltage and load current by simulation using any software tools available. Use Sinusoidal PWM modulation technique for five-level output.

Ex.3.12: Evaluate the performance of single-phase diode-clamped full-bridge inverter when dc source voltage V_{dc}=200V, R=20 ohms and L=10 mH and obtain gating signals, output line voltage and load current by simulation using any software tools available. Use Space-vector PWM modulation technique for five-level output.

Ex.3.13: Simulate single phase diode-clamped full-bridge inverter when dc source voltage V_{dc}=200V, R=20 ohms and L=10 mH and obtain gating signals, output line voltage and load current by using any software tools available. Use Carrier-based PWM modulation technique for five-level output.

Ex.3.14: Evaluate the performance of three phase diode-clamped full-bridge inverter when dc source voltage V_{dc}=500V, R=20 ohms per phase and L=47.59 mH per phase and obtain output line voltages and load current by using any software tools available. Use Sinusoidal PWM modulation technique for five-level output.

Ex.3.15: Simulate single phase diode-clamped inverter with single-leg circuit, when dc source voltage V_{dc}=200V, R=20 ohms and L=10 mH and obtain gating signals, output phase voltage and load current by using any software tools available. Use Sinusoidal PWM modulation technique for seven-level output.

Ex.3.16: Evaluate the performance of single-phase diode-clamped inverter with single-leg circuit, when dc source voltage V_{dc}=200V, R=20 ohms and L=10 mH and obtain gating signals, output phase voltage and load current by simulation using any software tools available. Use Carrier-based PWM modulation technique for seven-level output.

Ex.3.17: Simulate single phase diode-clamped inverter with single-leg circuit, when dc source voltage V_{dc}=200V, R=20 ohms and L=10 mH and obtain gating signals, output phase voltage and load current by using any software tools available. Use Space-vector PWM modulation technique for seven-level output.

Ex.3.18: Evaluate the performance of single-phase diode-clamped full-bridge inverter when dc source voltage V_{dc}=200V, R=20 ohms and L=10 mH and obtain gating signals, output line voltage and load current by simulation using any software tools available. Use Sinusoidal PWM modulation technique for seven-level output.

Ex.3.19: Evaluate the performance of single-phase diode-clamped full-bridge inverter when dc source voltage V_{dc}=200V, R=20 ohms and L=10 mH and obtain gating signals, output line voltage and load current by simulation using any software tools available. Use Space-vector PWM modulation technique for seven-level output.

Ex.3.20: Simulate single phase diode-clamped full-bridge inverter when dc source voltage V_{dc}=200V, R=20 ohms and L=10 mH and obtain gating signals, output line voltage and load current by using any software tools available. Use Carrier-based PWM modulation technique for seven-level output.

Ex.3.21: Evaluate the performance of three phase diode-clamped full-bridge inverter when dc source voltage V_{dc}=500V, R=20 ohms per phase and L=47.59 mH per phase and obtain output line voltages and load current by using any software tools available. Use Sinusoidal PWM modulation technique for seven-level output.

Ex.3.22: Simulate single phase modified diode-clamped full-bridge inverter when dc source voltage V_{dc}=200V, R=20 ohms and L=10 mH and obtain gating signals, output line voltage and load current by using any software tools available. Use Sinusoidal PWM modulation technique for three-level output.

Ex.3.23: Simulate single phase modified diode-clamped full-bridge inverter when dc source voltage V_{dc}=200V, R=20 ohms and L=10 mH and obtain gating signals, output line voltage and load current by using any software tools available. Use SVPWM modulation technique for five-level output.

Ex.3.24: Simulate single phase modified diode-clamped full-bridge inverter when dc source voltage V_{dc}=200V, R=20 ohms and L=10 mH and obtain gating signals, output line voltage and load current by using any software tools available. Use Sinusoidal PWM modulation technique for seven-level output.

Calculation of Load parameters and other specifications:

The Following are the Values used in the Simulation.

Vdc --- 500 Volts for 'Five Level Inverter with both SPWM and SVPWM'

500 Volts for 'Three Level Inverter with both SPWM and SVPWM'

Carrier Frequency: 2000 Hz for Three and Five Level Inverter with SPWM

2500 Hz for Three and Five Level Inverter with SVPWM

Series "RL" Load is used in the Simulation:

The Active Power is 8000 Watts and the Reactive Power is 6000 Watts. Voltage = 500 V.

The Calculation for R and L Values is done as under.

VI Cos φ = 8000 W VI sin φ = 6000 W

Tan φ = 0.75 and Cos φ = 0.8

$500 \times I \times 0.8 = 8000$, therefore Current, I = 20 A

Z = 500 / 20 = 25 Ohms

$R = Z \cos \varphi = 25 \times 0.8 = 20\ \Omega$

$X = 25 \times 0.6 = 15\ \Omega$ L = 47.59 mH

Answers to Multiple Choice Questions

3.1	(b)	3.2	(c)	3.3	(d)	3.4	(a)	3.5	(b)
3.6	(c)	3.7	(d)	3.8	(a)	3.9	(b)	3.10	(c)
3.11	(d)	3.12	(b)	3.13	(c)	3.14	(a)	3.15	(d)
3.16.	(b)	3.17.	(c)	3.18.	(a)				

CHAPTER

4 Flying Capacitor Multilevel Inverters

LEARNING OBJECTIVES

- ✓ *To understand and acquire knowledge about construction of flying-capacitor multilevel inverters.*
- ✓ *To prepare learners to analyze the operation of three-level, five-level, seven-level, and m-level flying-capacitor inverters.*
- ✓ *To prepare learners to practice through case studies on flying-capacitor multilevel inverters.*
- ✓ *To prepare learners to simulate the performance of three-level, five-level, seven-level, and m-level flying-capacitor inverters with suitable pulse width modulation techniques.*
- ✓ *To understand the applications of various levels of flying-capacitor inverters.*

4.1 Construction of Three-level Flying Capacitor Inverter

A single-phase three-level voltage source inverter equivalent circuit is drawn with one single-pole double throw switch S_A and two single pole single throw switches (S_1 and S_4) for RL load and as shown in Fig.4.1. The operation of the circuit is explained as follows for single-leg arrangement for simplicity and for ease of understanding the various modes of operation are depicted in Fig.4.1 (b) – (d).

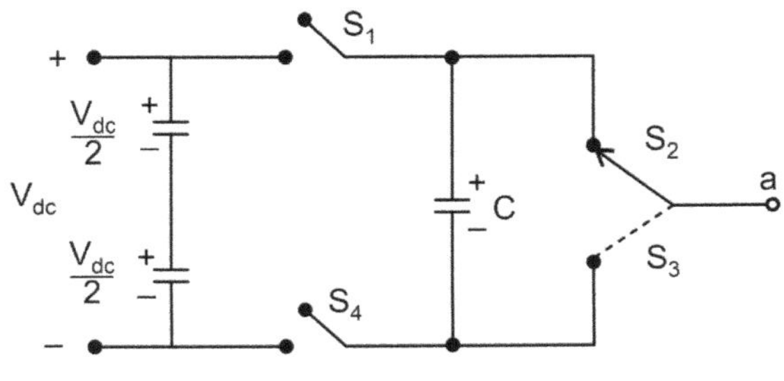

(a)

If switch S_A is connected to terminal 2 and switch S_1 is ON which connects positive dc source terminal 1 (and S_4 is OFF), then $V_{a0} = +V_{dc}/2$ and current will be positive flowing from terminal 'a' to terminal '0' of RL load as shown in Fig.4.1(b). When switch S_A is connected to terminal 3 and Switch S_4 is ON which connects negative dc source terminal 4 (and S_1 is OFF), then $V_{a0} = -V_{dc}/2$ is given in Fig.4.1(c). There are two possibilities to connect load terminal 'a' to terminal '0' to get V_{a0} as zero. (i) If switch S_A is connected to terminal 3 and switch S_1 is ON (and S_4 is OFF), then $V_{A0} = 0$. And (ii) When switch S_A is connected to terminal 2 and switch S_4 is ON (and S_1 is OFF), then $V_{a0} = 0$. When $V_{dc}/2$ of source capacitor and $V_{dc}/2$ of Flying capacitor gets cancelled then the resultant voltage V_{a0} will be Zero as shown in Fig.4.1(d).

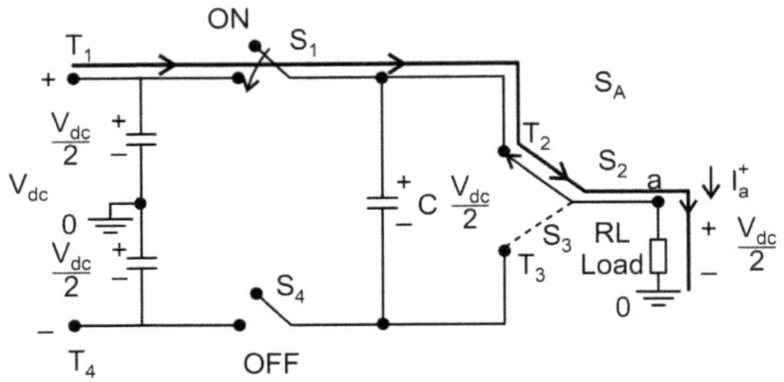

(b) $V_{ao} = \frac{+V_{dc}}{2}$ and I_a is positive

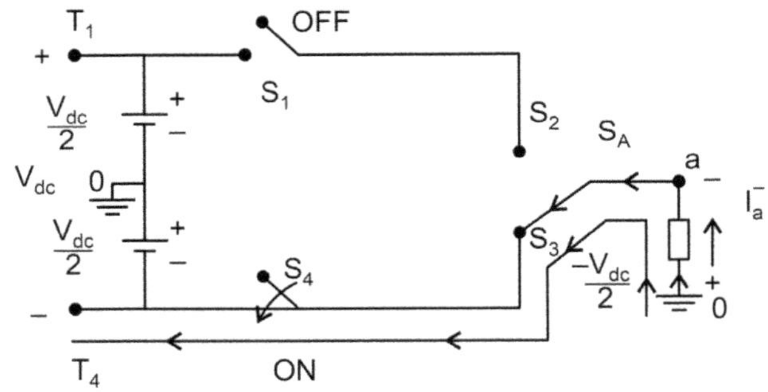

(c) $V_{ao} = \frac{-V_{dc}}{2}$ and I_a is negative

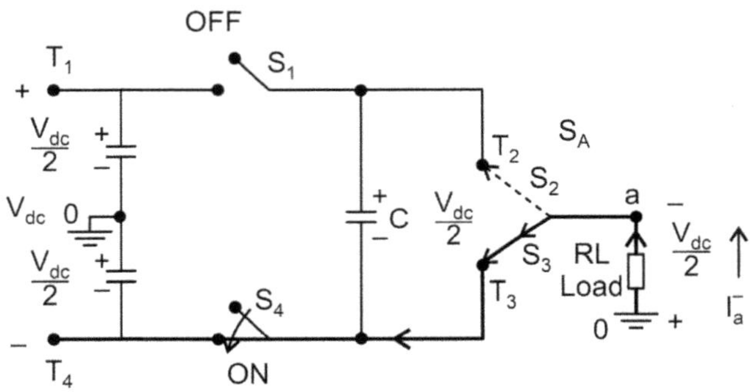

(d) V_{ao}=0 and I_a=0

Fig. 4.1 Basic circuit operation of Single phase three-level flying-capacitor, (a) basic circuit, (b) mode1: +V_{dc}/2 operation, (c) mode2: -V_{dc}/2 operation, and (d) Mode3: 0 operation.

The working of the circuit is summarized in the Table 4.1.

Table 4.1 Single phase three level flying capacitor for R L-loads with various modes of operation.

Mode	Switch S_A	Switch S_1	Switch S_4	Output Voltage, V_{AO}	Load current, I_o
M1	Terminal 2	ON	OFF	$+V_{dc}/2$	Positive (a to 0)
M2	Terminal 3	OFF	ON	$-V_{dc}/2$	Negative (0 to a)
M3	Terminal 3	ON	OFF	0	Zero
M3	Terminal 2	OFF	ON	0	Zero

The basic circuit is given in Fig.4.2(a). The switches S_1 and S_4 can be realized by using two bidirectional semiconductor devices. An IGBT device with anti-parallel diode arrangement can be used here to explain the realization by considering only single leg for simple and ease of understanding. This is shown in Fig.4.2 (b).

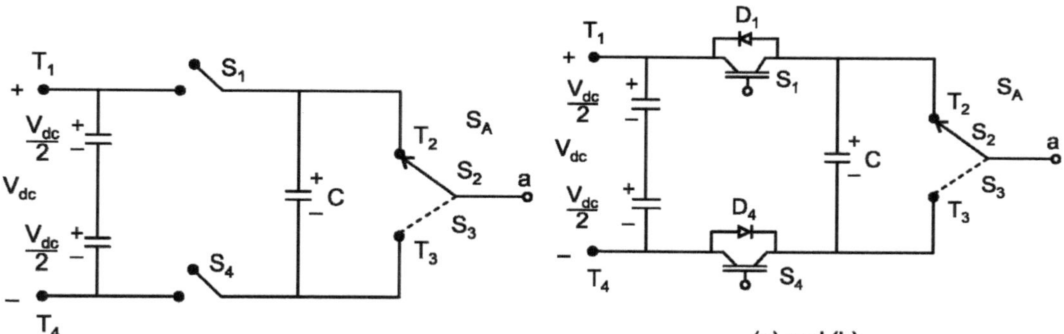

(a) and (b)

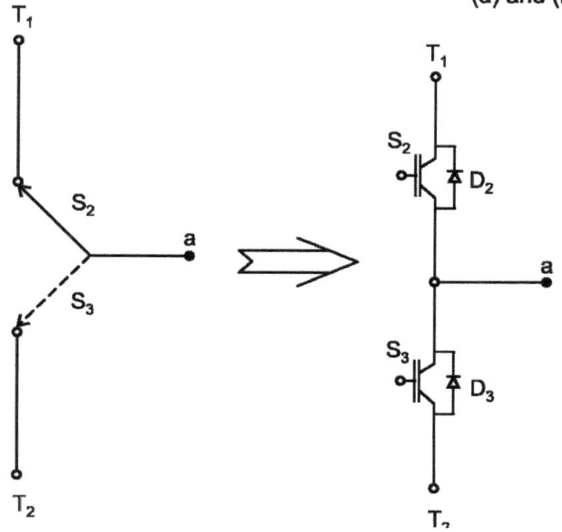

(c) and (d)

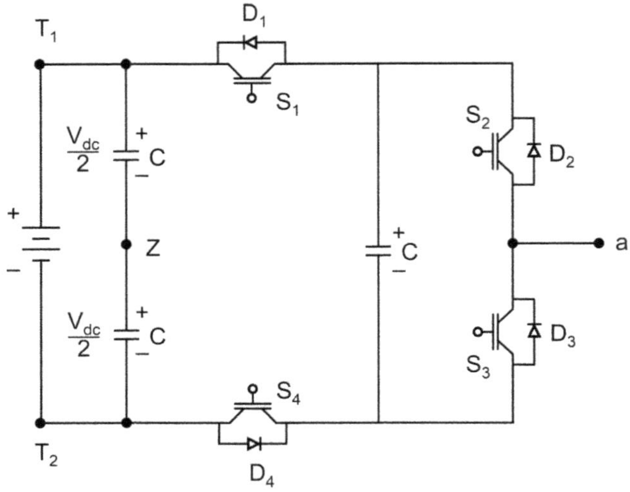

(e)

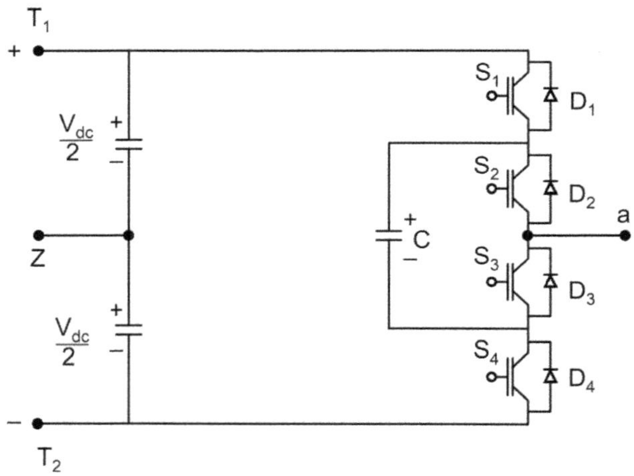

(f)

Fig. 4.2 Construction of Single phase three-level flying-capacitor from basic circuit, (a) basic circuit, (b) switches S_1 and S_4 realization by IGBTs, (c) and (d) switch S_A for T_2 and T_3 connections realization by IGBTs as S_2 and S_3, (e) realized complete circuit and (f) standard arrangement of circuit topology.

Similarly, the three-level operation of switch S_A can be realized using a two more bidirectional semiconductor device, one for terminal 2 and another for terminal 3 operations, which also capable to conduct in both directions for inductive loads and to block voltages across off switches to $\pm V_{dc}/2$. This circuit is shown in Fig.4.2 (c) and (d). The complete circuit realization is shown in Fig.4.2 (e). The standard arrangement of the same circuit topology as per the literature is given in Fig.4.2 (f).

The IGBT based single phase three-level Flying capacitor inverter with single-leg arrangement with proper labeling the switches are shown in the Fig.4.3. The detailed operation of three-level flying capacitor inverter is explained in section 4.2.

4.2 Three level Flying Capacitor Inverter

A. Single phase three-level Flying-capacitor inverter with single-leg arrangement: Single phase flying capacitor inverter topology requires four switching devices, two at the

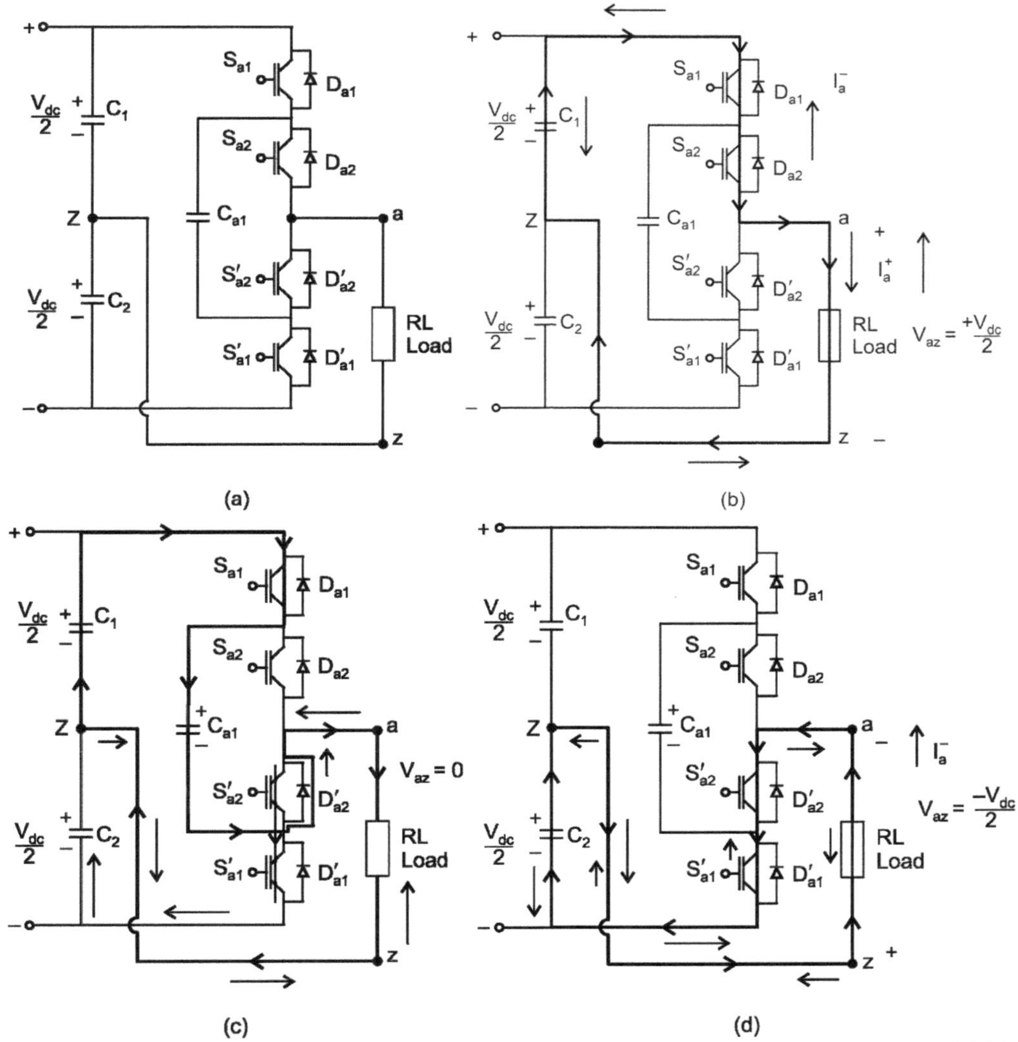

Fig. 4.3 Single phase flying capacitor three-level inverter with single-leg arrangement, (a) Main circuit, (b) Mode1: $+V_{dc}/2$; (c) Mode2: 0; and (d) Mode3: $-V_{dc}/2$.

upper and two at the bottom of a leg and four anti-parallel diodes, one clamping capacitor between the junction of Sa_1, Sa_2 and Sa_2', Sa_1' and two input capacitors for splitting the dc input voltage V_{dc} into two halves. A neutral is connected at the mid-point of two input capacitors. Fig. 4.3. Shows the single-phase flying capacitor three-level inverter. The detailed operation is explained with respect to Fig. 4.3. (a), (b), (c) and (d).

From Fig. 4.3 (a), the three-modes of operation of the circuit is given as follows:

Mode 1: S_{a1} and S_{a2} are fired on, so that positive potential of upper capacitor C_1 ($+V_{dc}/2$) is connected to the load terminal 'a' through S_{a1} and S_{a2}, then V_{az} is $+V_{dc}/2$. Positive current Ia flows from 'a' to (z) as shown in thick lines (Fig. 4.3(b)).

Mode 2: For $V_{az} = 0$, S_{a1} and S_{a2}' are fired on, then mid-point neutral terminal 'z' is connected through upper capacitor C_1, switching device Sa_1, Capacitor C_{a1} and anti-parallel diode D_{a2}' to the load terminal 'a', voltages across the capacitors C_1 and C_{a1} get cancelled, resulting in zero voltage.

Mode 3: For $V_{az} = -V_{dc}/2$, S_{a1}' and S_{a2}' are fired ON and negative potential of lower capacitor C_2 is connected to the load terminal 'a' through S_{a2}' and S_{a1}'. Negative current Ia flows from terminals 'z' to 'a' as shown in thick lines (Fig. 4.3(d)).

Thus, three levels of voltages are made available at the output terminal $V_{az} = +V_{dc}/2$, 0, $-V_{dc}/2$ and voltage waveform is given in Fig.4.4.

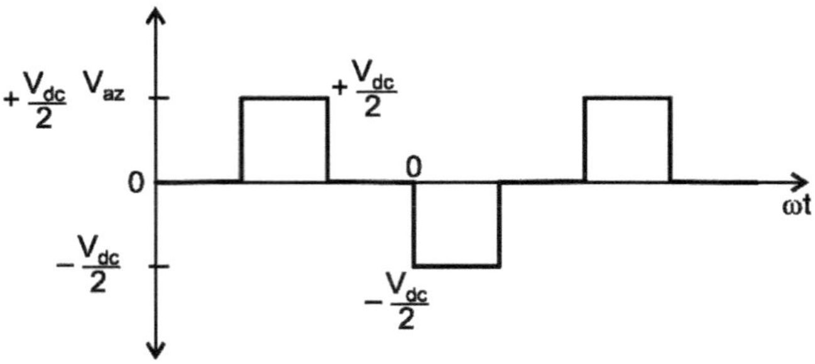

Fig. 4.4 Output voltage waveform for single phase three-level circuit with single-leg arrangement, for V_{az}.

B. Single phase three-level Flying-capacitor full-bridge inverter: If the same operation is extended for second-leg also to have a single-phase complete bridge circuit and the output voltage across load terminals '*a*' and '*b*' respectively, namely line-voltage V_{ab}, will be 0, $\pm V_{dc}/2$, $\pm V_{dc}$ and gives a three-level voltage as shown in Fig.4.5. The detailed operation of the single-phase full bridge flying-capacitor inverter for RL load is explained as follows:

Mode 1: (i) To obtain $V_{ab} = +V_{dc}$, the switches S_{a1}, S_{a2}, S_{b2}' and S_{b1}' are turned ON and the positive load current flows from upper source capacitor C_1, S_{a1}, S_{a2}, load *a* to *b*, S_{b2}', S_{b1}' lower source capacitor C_2 as shown in thick lines (ii) Since the load is inductive, during the mode1 operation the inductor stored energy and after the period of operation the inductive load acts as source and forces a current in the reverse direction shown with arrows to feed the source until the stored energy is exhausted/dissipated, (Fig.4.5(b)).

Mode 2: For $V_{ab} = +V_{dc}/2$, the switches S_{a1}, S_{b2}' and S_{b1}' are turned ON and diode D_{a2}' made to conduct so that the positive load current flows from upper source capacitor C_1, S_{a1}, flying- capacitor C_{a1} which is re-charged to voltage $+V_{dc}/2$, D_{a2}', load *a* to *b*, S_{b2}', S_{b1}' and lower source capacitor C_2 as shown in thick line. The total source voltage is $+V_{dc}$ minus the charged flying capacitor C_{a1} voltage $+V_{dc}/2$ and remaining $+V_{dc}/2$ will appear as V_{ab}.

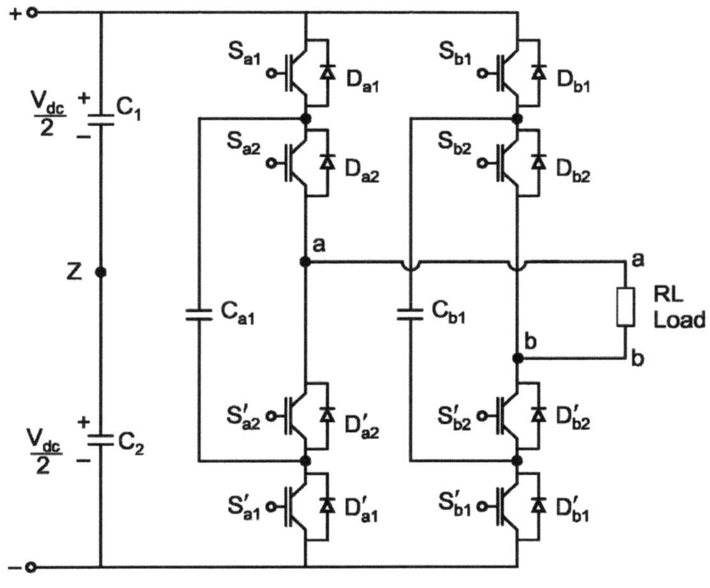

(a)

Fig. 4.5 Single-phase Flying-capacitor full bridge inverter circuit for RL-loads with various modes of operation, (a) main circuit, (b) Mode1: $+V_{dc}$, (c) Mode2: $+V_{dc}/2$, (d) Mode3: 0, (e) Mode4: $-V_{dc}/2$, and (f) Mode5: $-V_{dc}$.

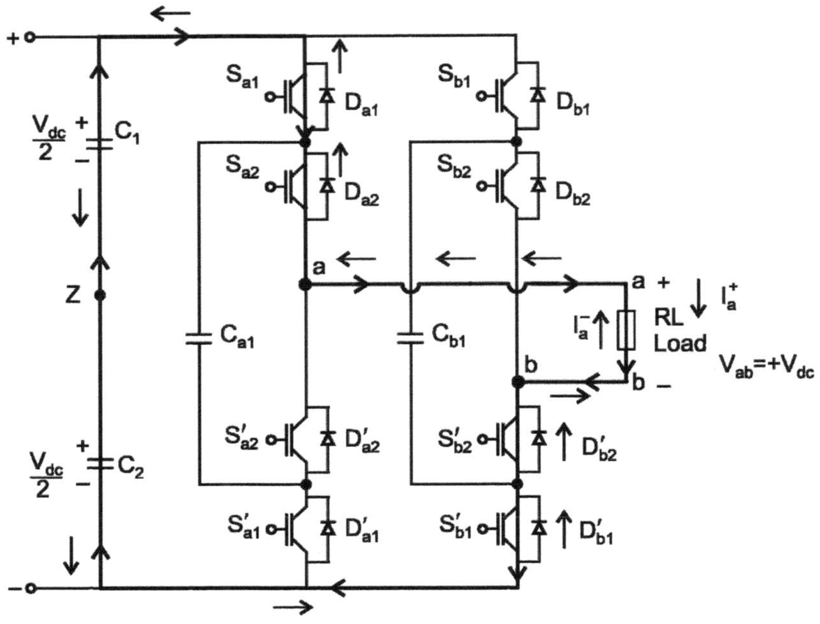
+
S$_{a1}$
D$_{a1}$
S$_{b1}$
D$_{b1}$
$\frac{V_{dc}}{2}$
C$_1$
S$_{a2}$
D$_{a2}$
S$_{b2}$
D$_{b2}$
a
Z
C$_{a1}$
C$_{b1}$
a +
I$_a^+$
I$_a^-$
RL
Load
b
b –
V$_{ab}$=+V$_{dc}$
S′$_{a2}$
D′$_{a2}$
S′$_{b2}$
D′$_{b2}$
$\frac{V_{dc}}{2}$
C$_2$
S′$_{a1}$
D′$_{a1}$
S′$_{b1}$
D′$_{b1}$
–

(b)

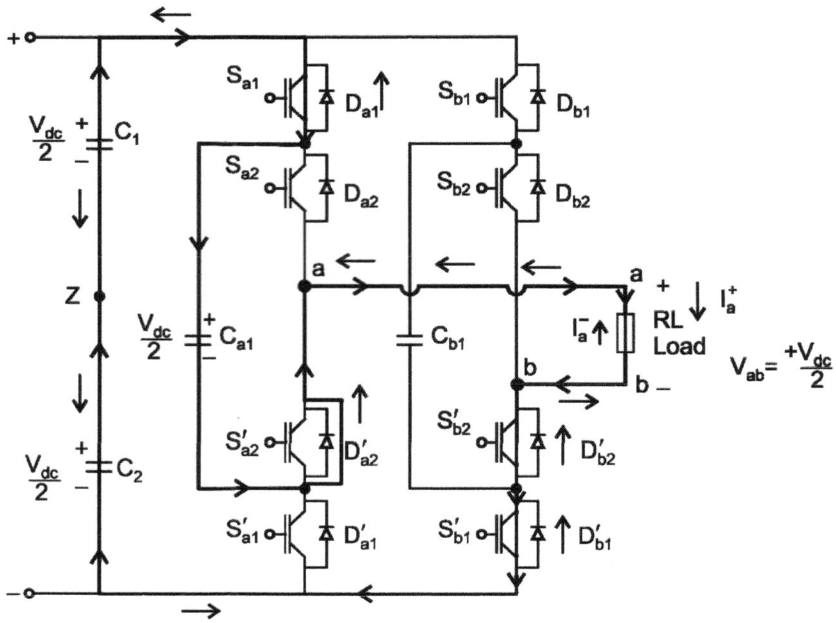
+
S$_{a1}$
D$_{a1}$
S$_{b1}$
D$_{b1}$
$\frac{V_{dc}}{2}$
C$_1$
S$_{a2}$
D$_{a2}$
S$_{b2}$
D$_{b2}$
a
Z
$\frac{V_{dc}}{2}$
C$_{a1}$
C$_{b1}$
a +
I$_a^+$
I$_a^-$
RL
Load
b
b –
V$_{ab}$= $\frac{+V_{dc}}{2}$
S′$_{a2}$
D′$_{a2}$
S′$_{b2}$
D′$_{b2}$
$\frac{V_{dc}}{2}$
C$_2$
S′$_{a1}$
D′$_{a1}$
S′$_{b1}$
D′$_{b1}$
–

(c)

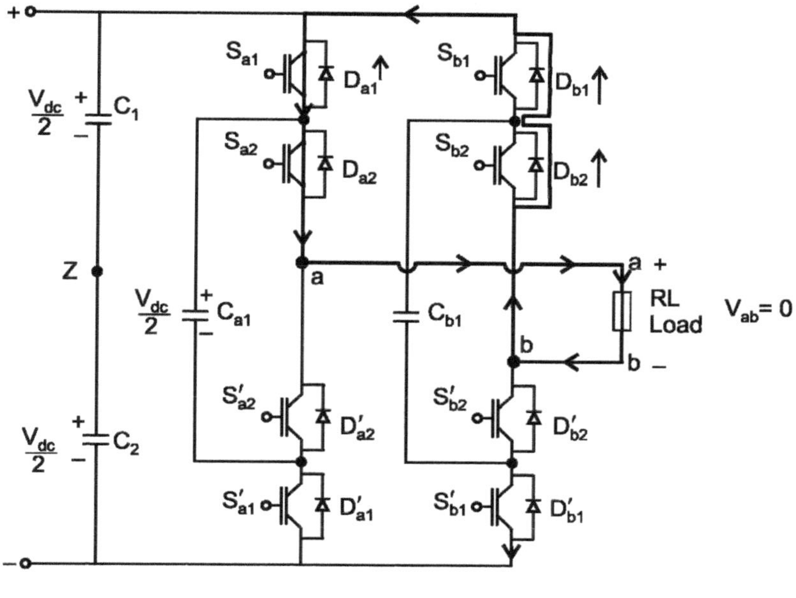

S_{a1}
D_{a1}
S_{b1}
D_{b1}
$\frac{V_{dc}}{2}$
C_1
S_{a2}
D_{a2}
S_{b2}
D_{b2}
Z
a
$\frac{V_{dc}}{2}$
C_{a1}
C_{b1}
RL Load
$V_{ab}= 0$
b
S'_{a2}
D'_{a2}
S'_{b2}
D'_{b2}
$\frac{V_{dc}}{2}$
C_2
S'_{a1}
D'_{a1}
S'_{b1}
D'_{b1}

(d)

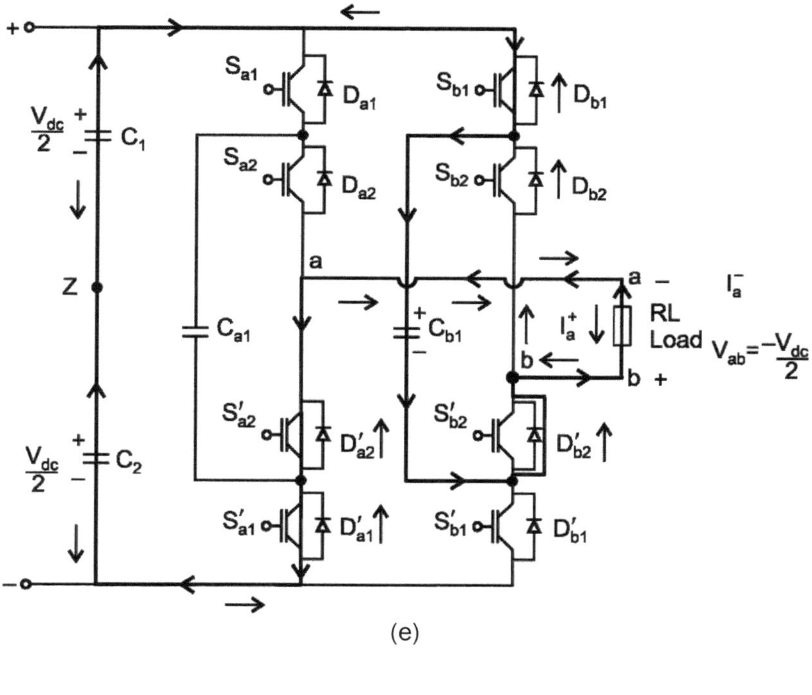

S_{a1}
D_{a1}
S_{b1}
D_{b1}
$\frac{V_{dc}}{2}$
C_1
S_{a2}
D_{a2}
S_{b2}
D_{b2}
a
Z
I_a^-
C_{a1}
C_{b1}
I_a^+
RL Load
$V_{ab}=-\frac{V_{dc}}{2}$
b
S'_{a2}
D'_{a2}
S'_{b2}
D'_{b2}
$\frac{V_{dc}}{2}$
C_2
S'_{a1}
D'_{a1}
S'_{b1}
D'_{b1}

(e)

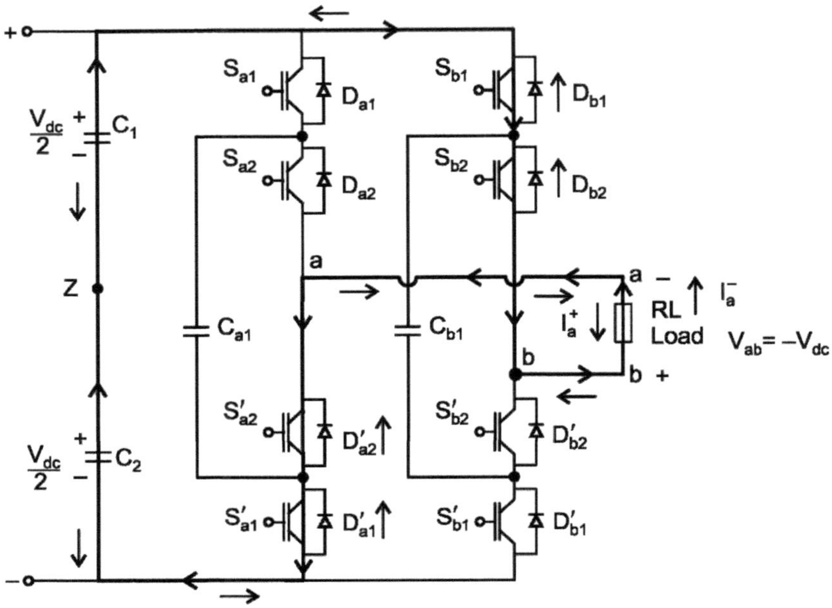

(f)

Fig. 4.5 Single-phase Flying-capacitor full bridge inverter circuit for RL-loads with various modes of operation, (a) main circuit, (b) Mode1: $+V_{dc}$, (c) Mode2: $+V_{dc}/2$, (d) Mode3: 0, (e) Mode4: $-V_{dc}/2$, and (f) Mode5: $-V_{dc}$.

Table 4.2 Single-phase Flying-capacitor full bridge inverter circuit for RL-loads with various modes of operation.

Modes	V_{ab}	S_{a1}	S_{a2}	S_{a2}'	S_{a1}'	S_{b1}	S_{b2}	S_{b2}'	S_{b1}'	Diodes
M1	$+V_{dc}$	1	1	0	0	0	0	1	1	-----
M2	$+V_{dc}/2$	1	0	0	0	0	0	1	1	D_{a2}'
M3	0	1	1	0	0	0	0	0	0	D_{b2}, D_{b1}
M4	$-V_{dc}/2$	0	0	1	1	1	0	0	0	D_{b2}'
M5	$-V_{dc}$	0	0	1	1	1	1	0	0	-----

The stored energy due to inductive load is fed back to dc source with reverse current pathway shown by broken arrows, (Fig.4.5(c)).

Mode 3: To obtain V_{ab}=0, the switches S_{a1}, S_{a2}, are turned ON, the conduction takes place from S_{a1}, S_{a2}, load *a* to *b*, diodes D_{b2}, D_{b1}, forming a short-circuit resulting in V_{ab} as zero voltage, (Fig.4.5(d)) as shown in thick line.

Mode 4: For V_{ab}= $-V_{dc}/2$, the switches S_{b1}, S_{a2}', and S_{a1}' are turned ON and diode D_{b2}' made to conduct so that the negative load current flows from upper source capacitor C_1, S_{b1}, flying- capacitor C_{b1} which is charged to voltage $+V_{dc}/2$, D_{b2}', load *b* to *a*, S_{a2}', S_{a1}' and lower source capacitor C_2. The total source voltage is $+V_{dc}$ minus the charged flying capacitor C_{b1} voltage $+V_{dc}/2$ and the remaining $+V_{dc}/2$ will appear as V_{ba} and it will be -

$V_{dc}/2$ across V_{ab}. The stored energy due to inductive load is fed back to dc source with reverse current pathway shown by broken arrows, (Fig.4.5(e)).

Mode 5: To obtain V_{ab}= $-V_{dc}$, the switches S_{b1}, S_{b2}, S_{a2}' and S_{a1}' are turned ON and the negative load current flows from upper source capacitor C_1, S_{b1}, S_{b2}, load *b* to *a*, S_{a2}', S_{a1}' lower source capacitor C_2. The stored energy due to inductive load is fed back to dc source with reverse current pathway shown by broken arrows, (Fig.4.5(f)).

The expected output voltage, V_{ab} across RL load is a stepped voltage giving a three-level voltage (0, $\pm V_{dc}/2$ and $\pm V_{dc}$) for single phase full bridge flying-capacitor inverter as given in Fig. 4.6.

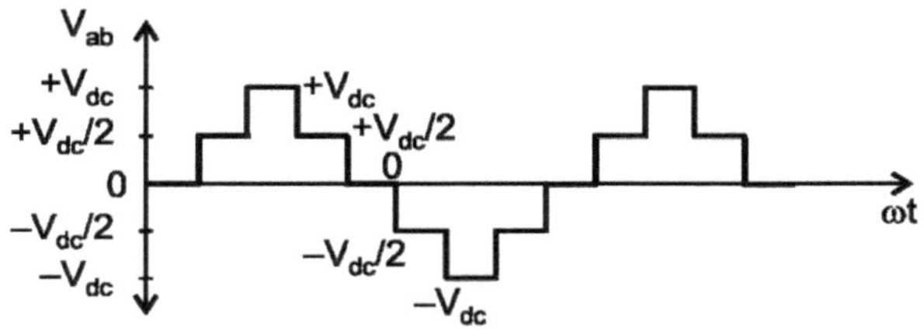

Fig. 4.6 Output line voltage waveform for single phase three-level full-bridge flying-capacitor inverter, for V_{ab}.

The switching pattern for single-phase flying capacitor full bridges inverter for various nodes of operation is given in Table 4.2.

C. **Three phase three-level Flying-capacitor full-bridge inverter:** The three single-phase flying-capacitor single-leg inverter circuits are connected to form a three-phase three-level flying-capacitor full-bridge inverter as shown in the Fig. 4.7. The output voltage of three-phase inverter is feeding power to the balanced star-connected load. The phase voltages are V_{an}, V_{bn}, V_{cn}, and the line voltages are V_{ab}, V_{bc}, and V_{ca} of the three-phase load. The phase voltage waveform V_{an} is similar to the load voltage obtained for single-phase single-leg arrangement, the only difference is the V_{bn}, and V_{cn} are phase shifted with respect to V_{an} by 120° and 240° respectively. The line voltage waveforms are only shown in the Fig.4.7 (b).

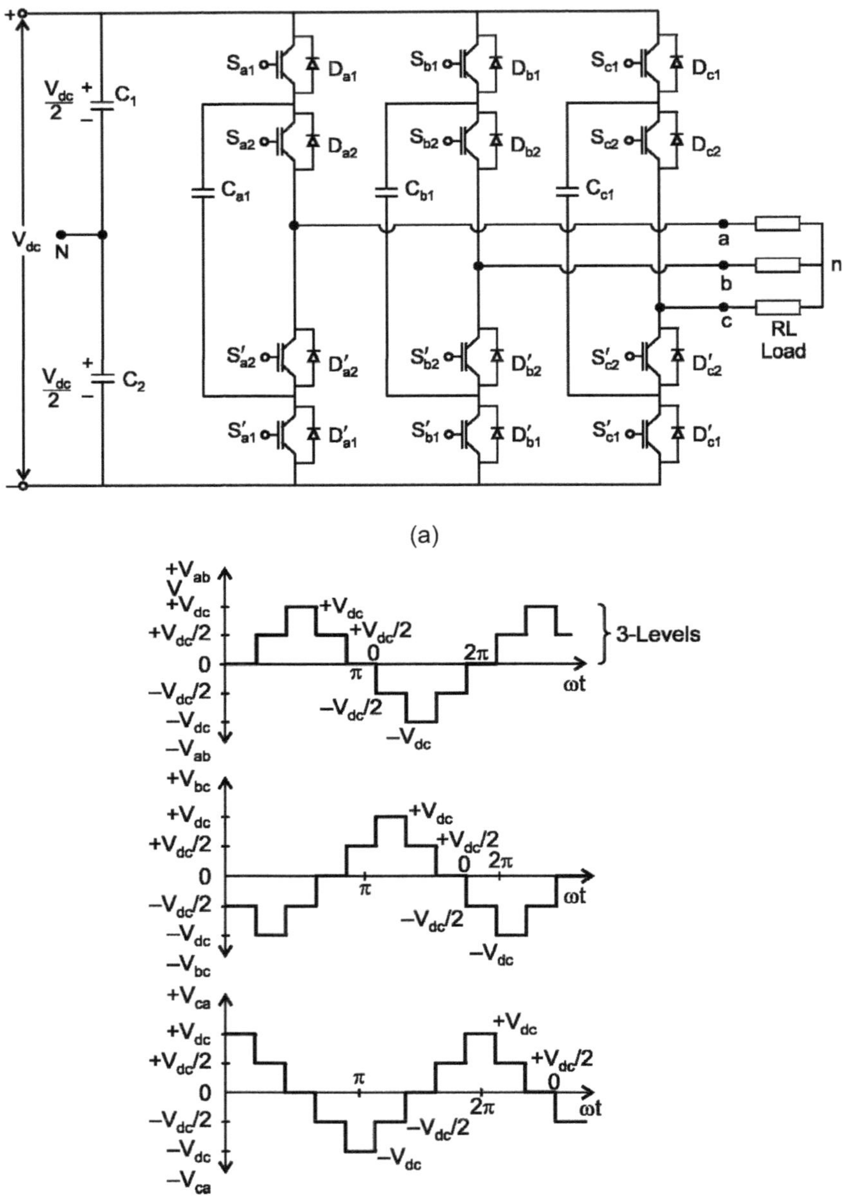

(b)

Fig. 4.7 (a) Three-phase three-level flying-capacitor full-bridge inverter circuit (b) Line Voltages of the load, V_{ab}, V_{bc}, V_{ca} for three-level output.

4.3 Five Level Flying Capacitor Inverter

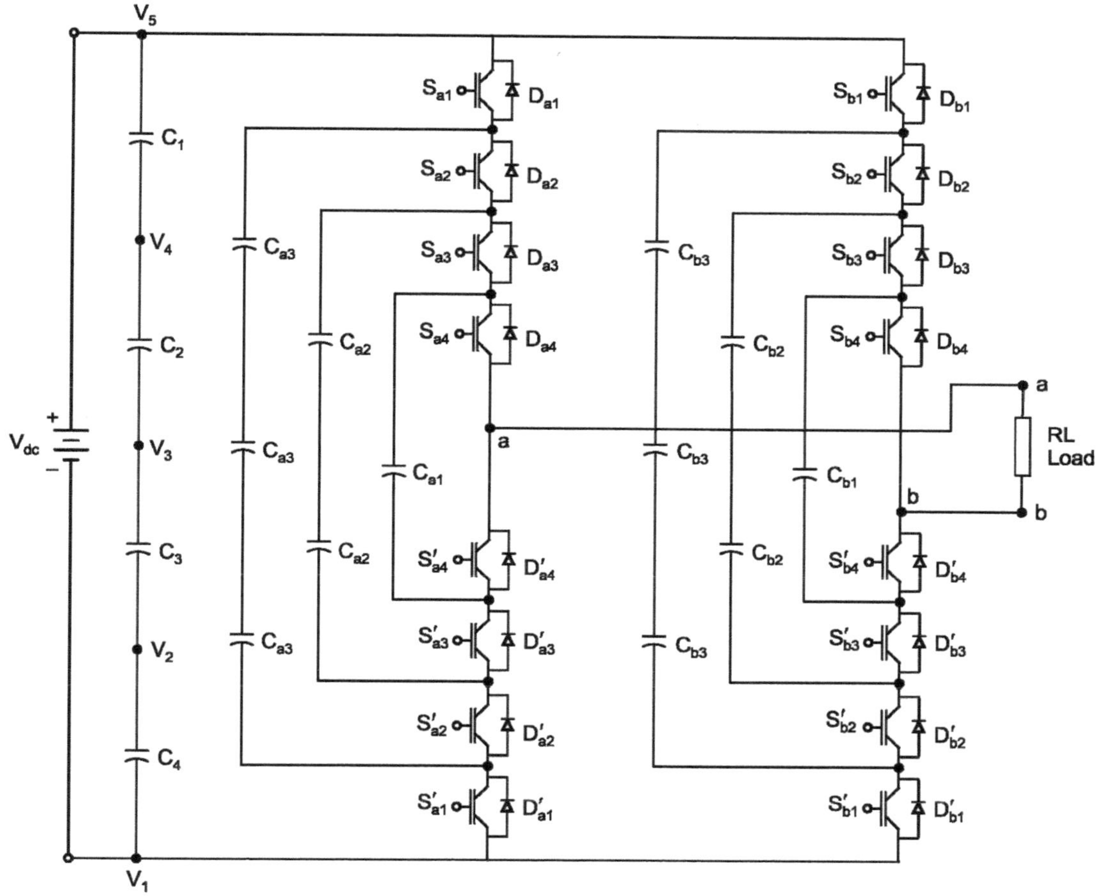

(a)

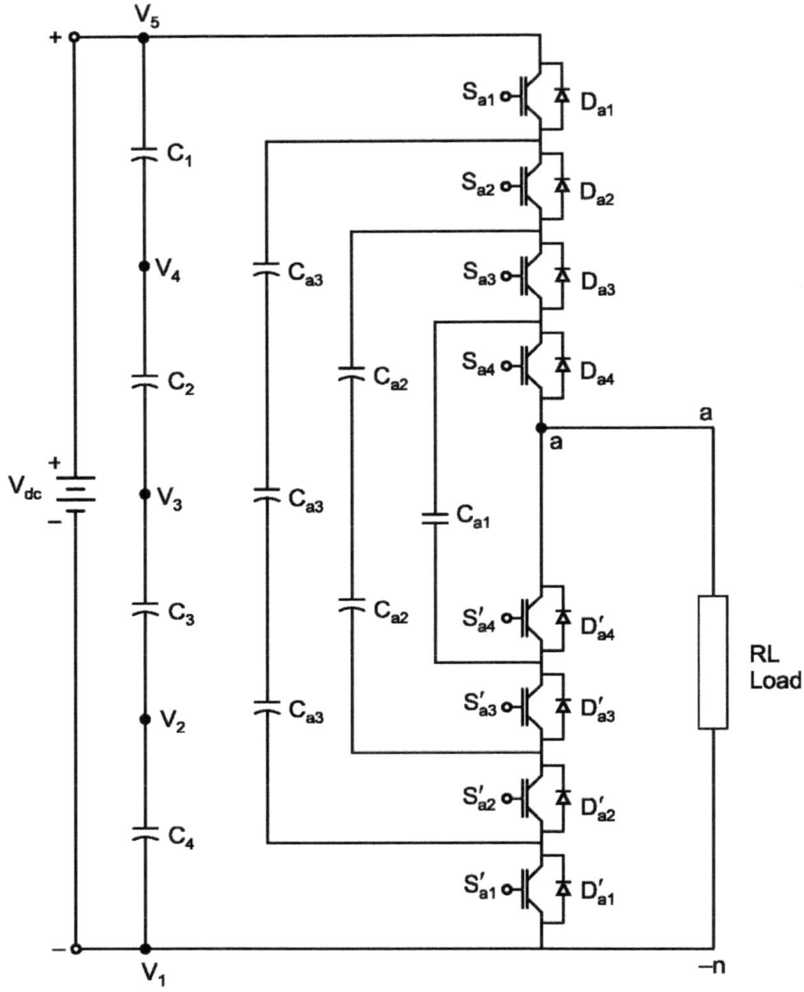

(b)

Fig. 4.8 (a) Single phase five-level full-bridge flying capacitor inverter with inductive load (b) Single-leg arrangement for better understanding the operation.

The single phase five-level flying capacitor full-bridge inverter with inductive load is shown in Fig. 4.8 (a). This circuit requires eight switching devices along with eight anti-parallel diodes per leg, and six flying-capacitors distributed as shown in the figure as one C_{a1}, two C_{a2} and three C_{a3} capacitors per leg. Four input capacitors are required for five-level inverter (as per (m-1) source capacitors), namely C_1, C_2, C_3, C_4 dividing dc input voltage into five-levels as $V_{c4}=V_{dc}/4$, $V_{c3}=V_{dc}/4$, $V_{c2}=V_{dc}/4$, and $V_{c1}=V_{dc}/4$. The single-leg arrangement of single-phase five-level flying-capacitor inverter circuit topology as given in Fig. 4.8. (b)

A. Single-phase five-level flying-capacitor inverter with single-leg arrangement: The circuit topology shown in Fig.4.9 (a) with neutral is arranged at the mid-point of input dc voltages and one single leg of circuit for single phase five-level flying-capacitor inverter is considered for explanation to obtain five-level ac output voltage V_{an} as $+V_{dc}/2$, $+V_{dc}/4$, 0, $-V_{dc}/4$, $-V_{dc}/2$. There are four switching devices S_{a1}, S_{a2}, S_{a3}, S_{a4} along with four anti-parallel diodes for upper leg and four switching devices S_{a1}', S_{a2}', S_{a3}', S_{a4}' along with four anti-parallel diodes for lower leg and four input dc capacitors $C_1=C_2=C_3=C_4$ each has $V_{dc}/4$ voltages with mid-point neutral and flying capacitors $C_{a1}=C_{a2}=C_{a2}=C_{a3}=C_{a3}=C_{a3}=V_{dc}/4$ with upper plate at positive potential.

The operation of the above circuit for five-levels of output ac voltage are explained with Fig. 4.9. (b) – (f). There are many alternative pathways to obtain certain level of output voltage and only one possible operation is considered and provided in the table 4.3.

Table 4.3 Shows two possibilities for $+V_{dc}/4$ and the rest one possibility though there are several possibilities to obtain those voltages for single-phase single-leg flying-capacitor inverter.

Output ac voltage V_{an}	Switching States								Diodes
	S_{a1}	S_{a2}	S_{a3}	S_{a4}	S_{a4}'	S_{a3}'	S_{a2}'	S_{a1}'	
$+V_{dc}/2$	1	1	1	1	0	0	0	0	-----
$+V_{dc}/4$ Option 1	1	1	1	0	0	0	0	0	D_{a4}'
$+V_{dc}/4$ Option 2	1	0	1	1	0	0	0	0	D_{a2}'
0	1	1	0	0	0	0	0	0	D_{a3}', D_{a4}'
$-V_{dc}/4$	0	0	0	0	0	1	1	1	D_{a4}
$-V_{dc}/2$	0	0	0	0	1	1	1	1	-----

Mode 1: For $V_{an}=+V_{dc}/2$, the switching devices S_{a1}, S_{a2}, S_{a3}, and S_{a4} are turned-on so that, $+V_{dc}/2$ from upper input capacitors C_1 and C_2 ($V_{dc}/4+V_{dc}/4$), is connected to the load terminal 'a' through S_{a1}, S_{a2}, S_{a3}, S_{a4} to get $V_{an}= + V_{dc}/2$ as shown in Fig.4.9 (b), with I_a positive.

Mode 2: $V_{an}=+Vdc/4$, (i) the First option of connecting upper most C_1 and C_2 ($+V_{dc}/2$) is through switching devices S_{a1}, S_{a2}, S_{a3}, flying capacitor C_{a1} ($+V_{dc}/4$), anti-parallel diode D_{a4}' and output voltage terminal '*a*' to '*n*', the output voltage is obtained as $V_{an}= (+V_{dc}/2) - (+V_{dc}/4) = +V_{dc}/4$, as shown in Fig. 4.9 (c), for I_a positive.

And (ii) the Second option is by firing S_{a1}, S_{a3}, and S_{a4} as shown in Fig 4.9 (d), and now the upper dc source capacitors C_2 and C_1 ($+V_{dc}/2$ together) is connected to output terminal '*a*' $V_{an}= +V_{dc}/4$, through switching devices S_{a1}, flying capacitors C_{a3} ($+V_{dc}/4$), C_{a3} ($+V_{dc}/4$), C_{ca3} ($+V_{dc}/4$) and Da_2' and C_{a2} ($+V_{dc}/4$), C_{a2} ($+V_{dc}/4$), S_{a3}, S_{a4}, load terminal '*a*' to '*n*', in this way V_{an} is obtained as summation of voltages across each capacitor ($C_1 + C_2 - C_{a3} - C_{a3} - C_{a3} + C_{a2} + C_{a2} = V_{dc}/4 + V_{dc}/4 - V_{dc}/4 - V_{dc}/4 - V_{dc}/4 + V_{dc}/4 + V_{dc}/4 = +V_{dc}/4$, for I_a positive.

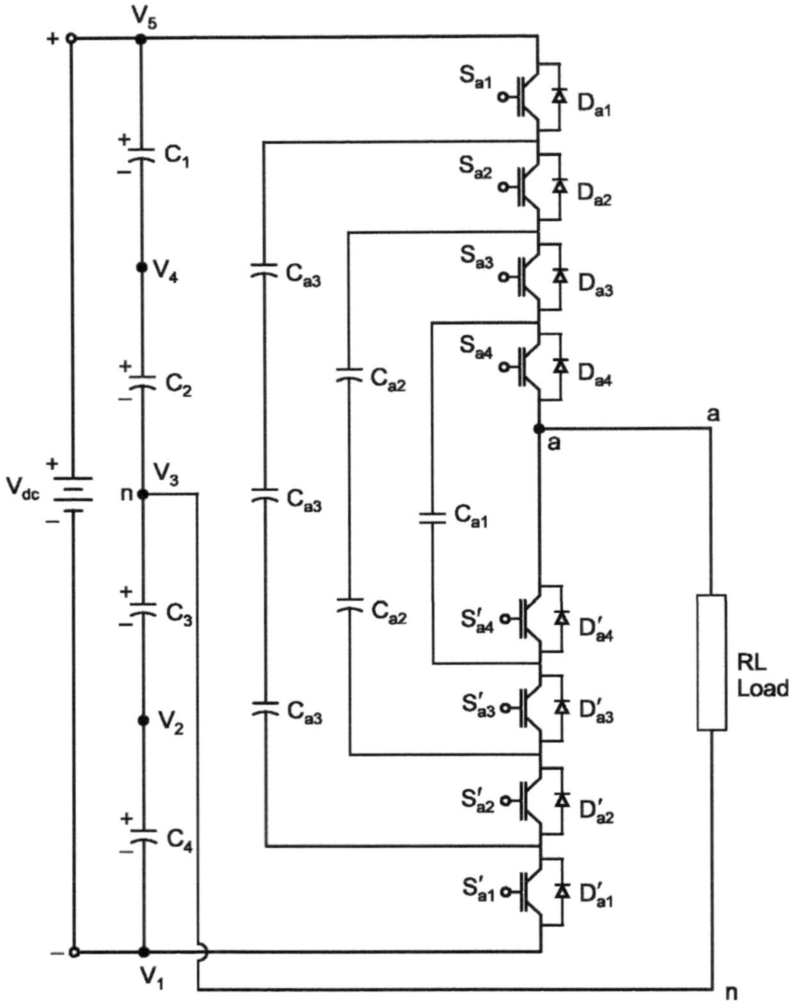
V5
+
Sa1
Da1
C1
Sa2
Da2
V4
Ca3
Sa3
Da3
Sa4
Da4
C2
Ca2
a
a
Vdc
n
V3
Ca3
Ca1
C3
Ca2
S'a4
D'a4
RL
Load
V2
Ca3
S'a3
D'a3
S'a2
D'a2
C4
S'a1
D'a1
V1
n

(a)

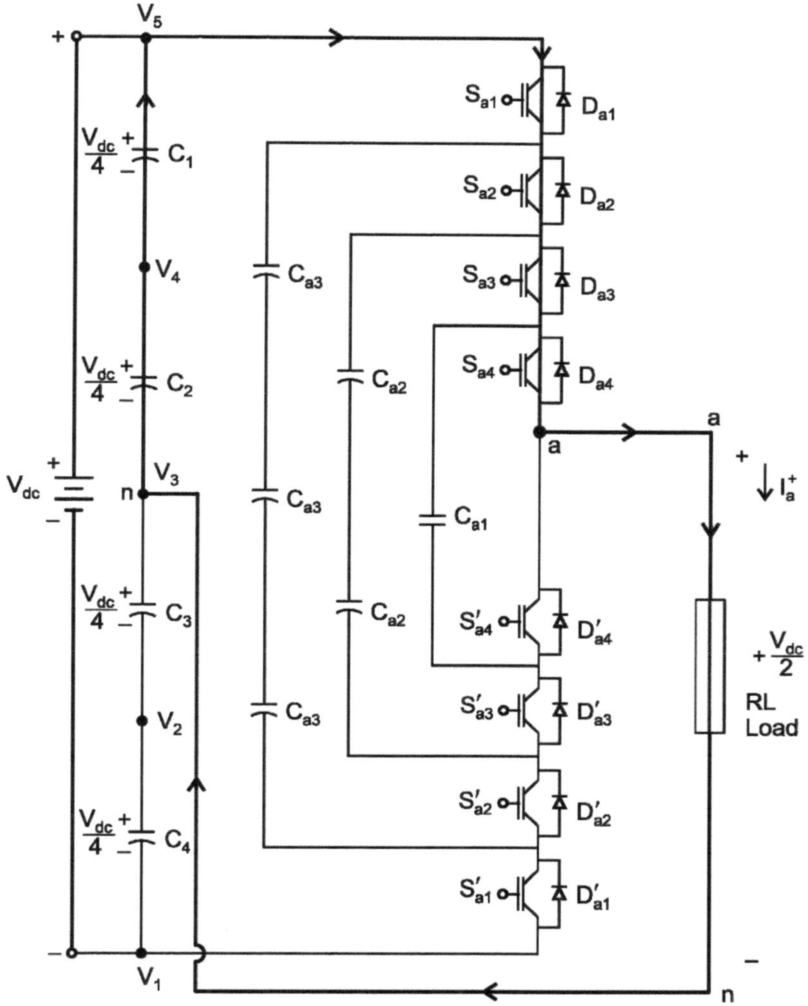

(b)

Fig. 4.9 Single phase five-level flying capacitor topology shown for one leg of circuit with neutral at the mid-point of the input dc voltages, (a) Main Circuit and (b) Mode1: $V_{an}= +V_{dc}/2$.

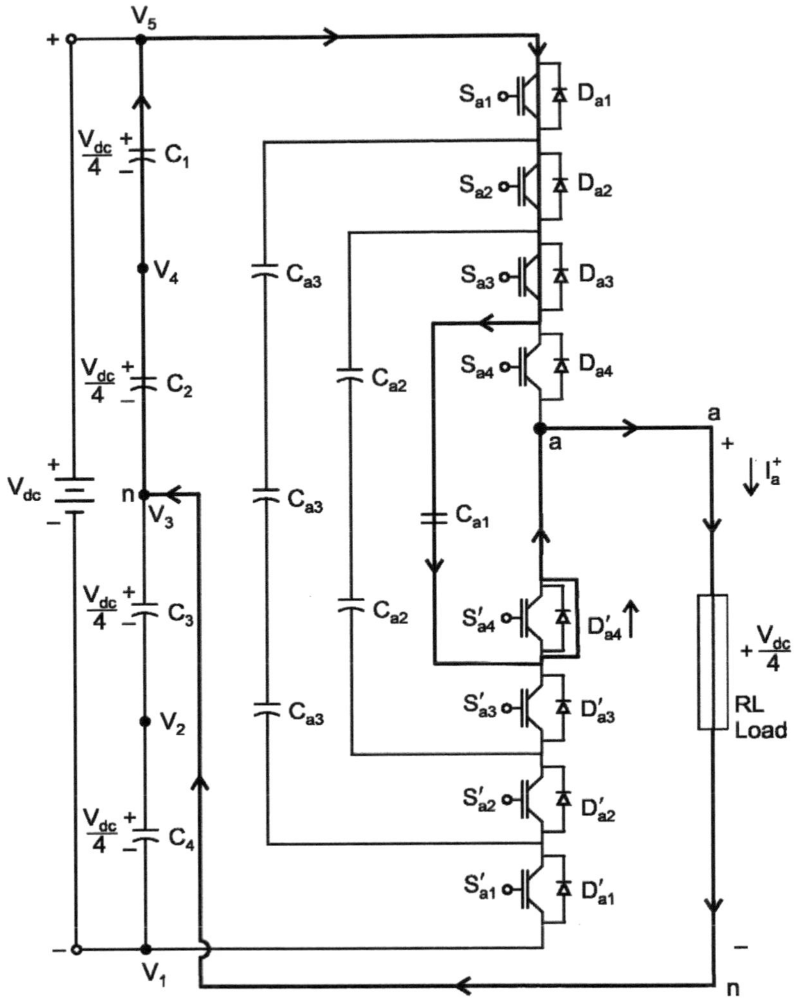
V5
+
Vdc/4
C1
V4
C2
Vdc
n
V3
C3
V2
C4
V1
−
Sa1
Da1
Sa2
Da2
Sa3
Da3
Sa4
Da4
S'a4
D'a4
S'a3
D'a3
S'a2
D'a2
S'a1
D'a1
Ca3
Ca2
Ca1
a
Ia+
+ Vdc/4
RL
Load
n

(c)

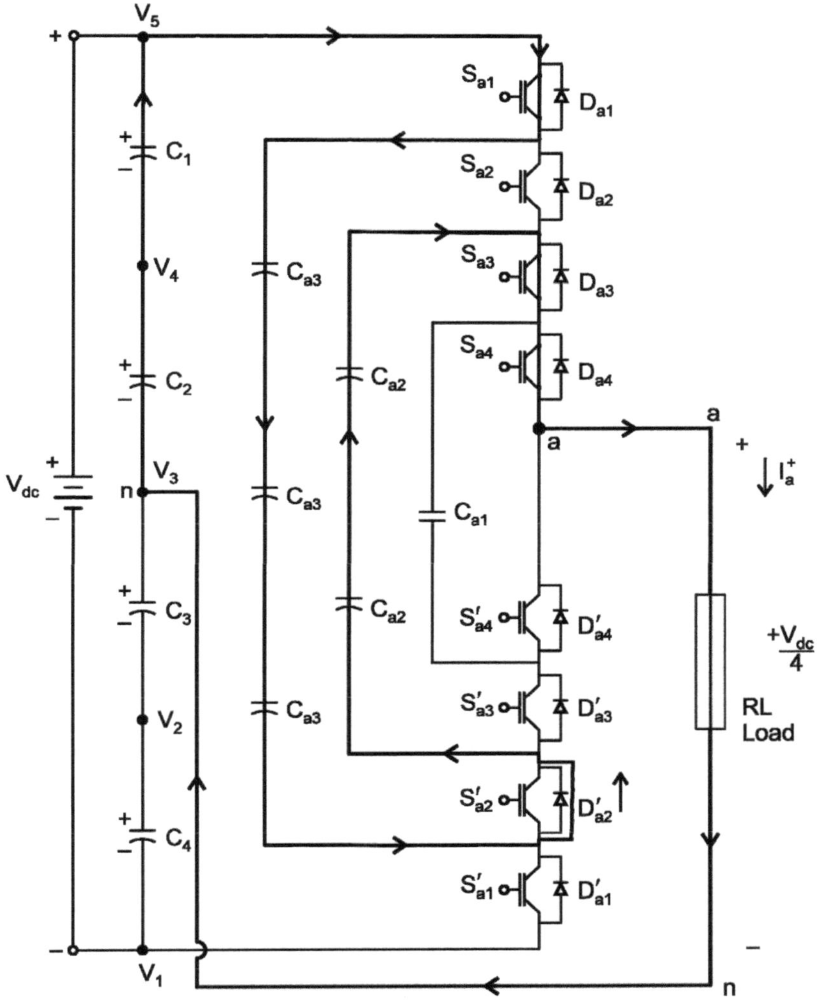

(d)

Fig. 4.9 Mode 2: (c) For Van= $+V_{dc}/4$ when S_{a1}, S_{a2}, S_{a3} are fired on, $\{V_{dc}/2 - V_{dc}/4\}$, (d) Other possibility for $V_{an}=+V_{dc}/4$ when S_{a1}, S_{a3} and S_{a4} are fired on, $\{V_{dc}/2 - 3V_{dc}/4 + V_{dc}/2\}$.

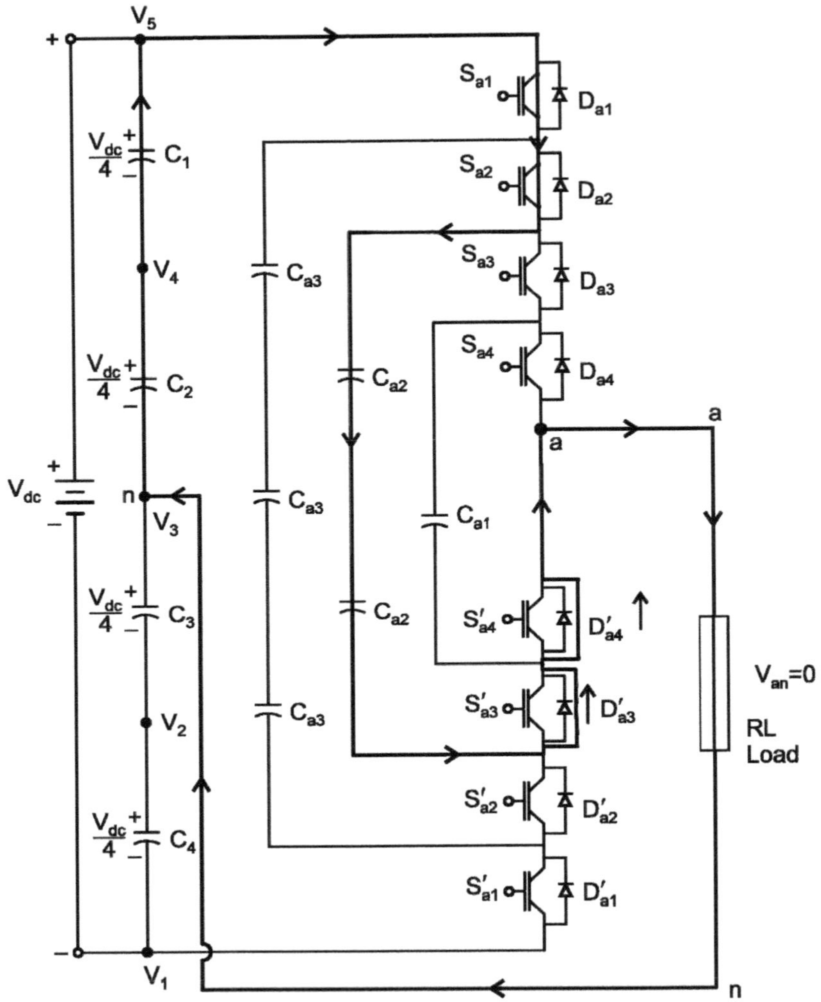

(e)

Fig. 4.9 Mode 3: (e) V_{an}= 0, when S_{a1} and S_{a2} are fired on and to conduct D_{a3}', D_{a4}'.

Mode 3: V_{an}=0 also obtained in many different pathways but only one possibility is considered. The switches S_{a1}, S_{a2}, are turned-on, and the conduction takes place from the upper dc source capacitors C_1, C_2, S_{a1}, S_{a2}, flying capacitors C_{a2}, C_{a2}, anti-parallel diodes D_{a3}' and D_{a4}', load terminal '*a*' to '*n*'. The net voltage V_{an} = Voltages of $(C_1 + C_2 - C_{a2} - C_{a2}) = (+V_{dc}/4) + (+Vdc/4) - (+V_{dc}/4) - (+V_{dc}/4) = 0$ voltage, as shown in Fig.4.9 (e).

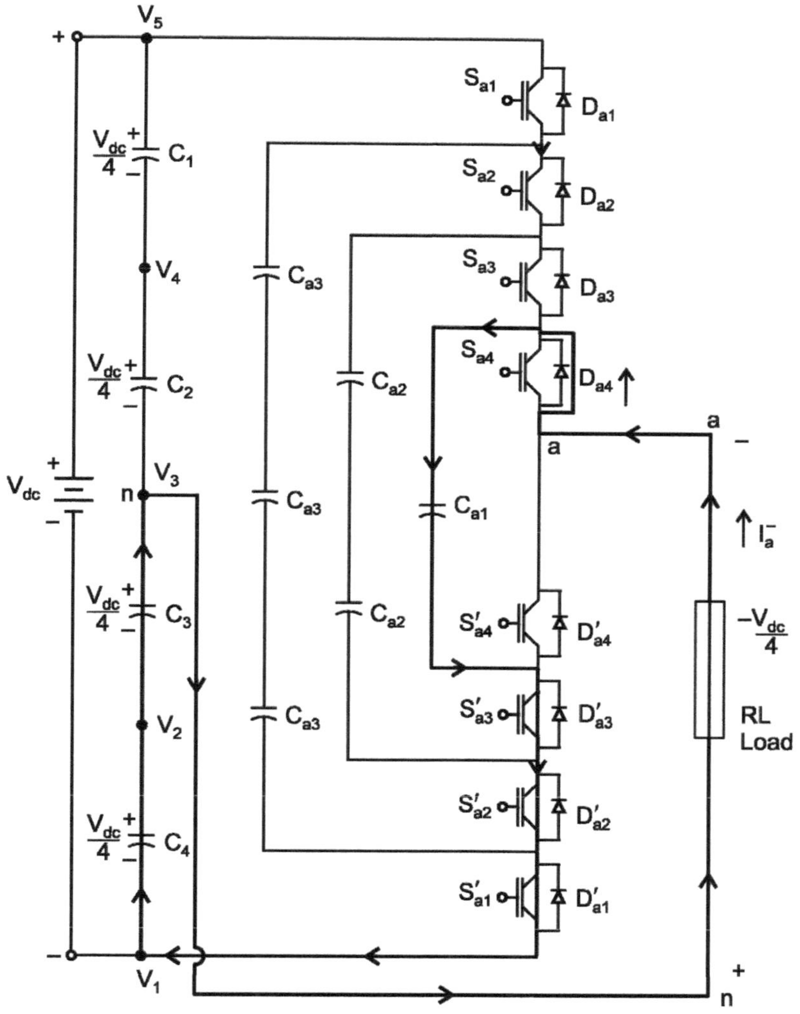

(f)

Fig. 4.9 Mode 4: (f) $V_{an}=\frac{-V_{dc}}{4}$, when S'_{a3}, S'_{a2}, S'_{a1} are fired and D_{a4} conducts.

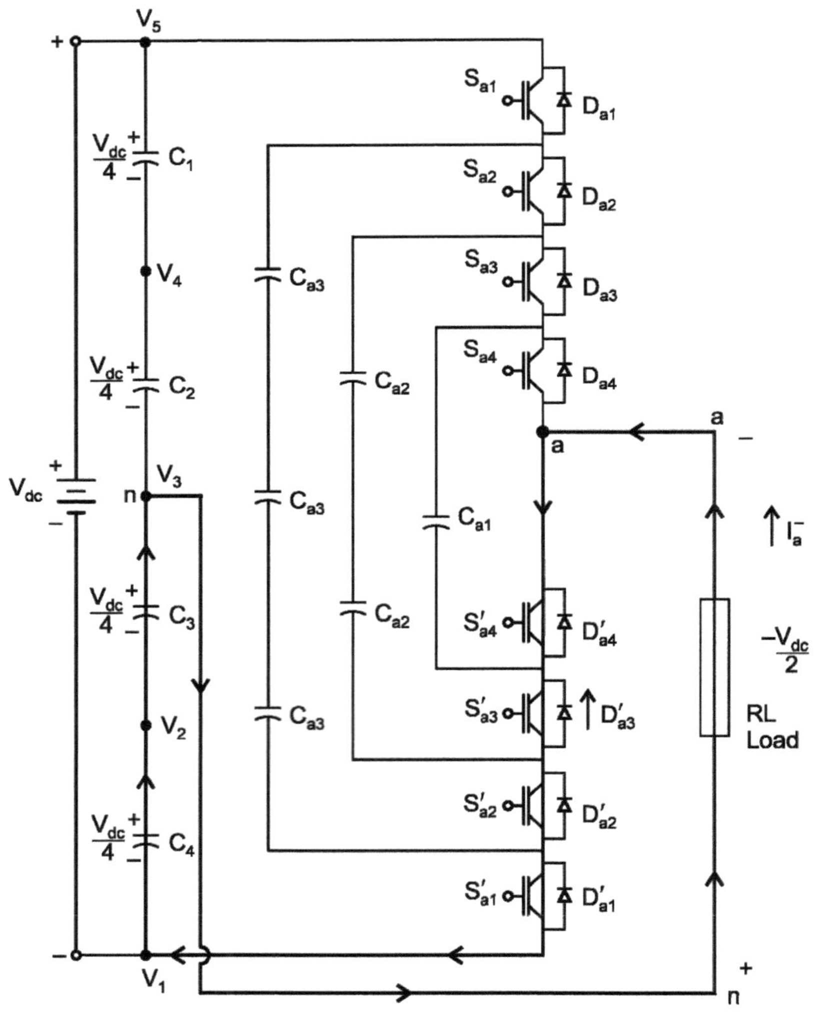

(g)

Fig. 4.9 Mode 5: (g) $V_{an} = -V_{dc}/2$, when S_{a4}', S_{a3}', S_{a2}' and S_{a1}' are fired on.

Mode 4: For output voltage V_{an} = - $V_{dc}/4$, the circuit operation is shown in Fig. 4.9 (f) when S_{a3}', S_{a2}', S_{a1}' are fired on, and lower dc source capacitors C_3, C_4 are connected to load terminal '*n*' to '*a*' anti-parallel diode D_{a4}, flying capacitor C_{a1}, through switching device S_{a3}', S_{a2}' S_{a1}', back to dc source terminal C_4. $V_{na} = +V_{dc}/4$ is obtained as Voltages of $(C_4 + C_3 - C_{a1}) = (+V_{dc}/4) + (+V_{dc}/4) - (+V_{dc}/4)) = +V_{dc}/4$, and voltage across V_{an} is - $V_{dc}/4$, for I_a negative.

Mode 5: For output voltage V_{an}= -$V_{dc}/2$, is obtained in a single straight path by firing switching devices S_{a4}', S_{a3}', S_{a2}', and S_{a1}' on, and conduction takes place through lower dc source capacitors C_4,C_3, load terminal '*n*' to '*a*' and switches S_{a4}', S_{a3}' S_{a2}', S_{a1}' and back to dc source terminal C_4. The -$V_{dc}/2$ is connected to output terminal '*a*' as V_{an}, as shown in Fig.4.9 (g).

Thus, all five level of voltages are obtained at V_{an} as $+V_{dc}/2$, $+V_{dc}/4$, 0, $-V_{dc}/4$, $-V_{dc}/2$ from single phase five-level flying capacitor inverter topology shown in Fig. 4.9. The Five-level voltage waveform is given in Fig. 4.10.

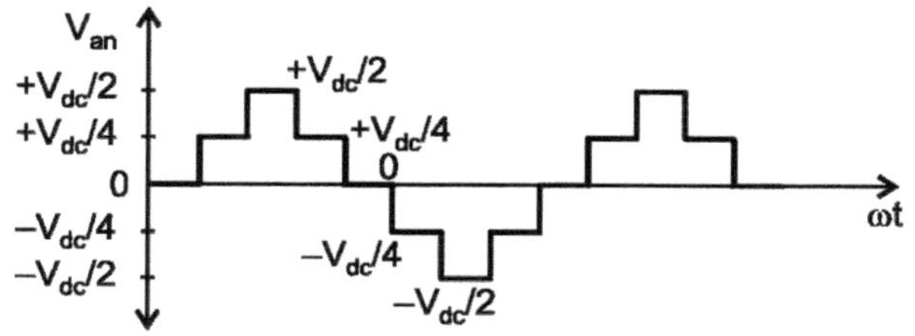

Fig. 4.10 Five-level output voltage waveform for single phase flying capacitor inverter with single-leg arrangement.

B. Single-phase Five-level Flying Capacitor full-bridge inverter: The single phase five-level flying capacitor full-bridge inverter with inductive load is shown in Fig. 4.11 (a).

The circuit topology shown in Fig.4.11. (a) – (j), explained the operation of the bridge circuit by dividing it into nine-modes to give a five-level line voltage for the inductive-resistive load. Eight power IGBTs with anti-parallel diodes are connected per leg and for complete bridge sixteen devices are required. To get five-level voltages at the load, the number of dc source capacitors required are (m-1), i.e. four equal magnitude voltage capacitors. The detailed operation is presented as follows.

Mode 1: For $V_{ab} = +V_{dc}$, the dc source terminal node V_5 is connected to the load terminal '*a*' through switches S_{a1}, S_{a2}, S_{a3}, S_{a4}, to load terminal '*b*' and S_{b4}', S_{b3}', S_{b2}', S_{b1}', to the dc source terminal node V_1. The load current I_a is positive, flowing from load terminal '*a*' to '*b*', as given in Fig.4.11 (b).

Mode 2: To obtain V_{ab}=+3$V_{dc}/4$, the dc source terminal node V_5 is connected to the load terminal '*a*' through switches S_{a1}, S_{a2}, S_{a3}, S_{a4}, to load terminal '*b*' and diode D_{b4}, flying capacitor C_{b1} which is pre-charged to $+V_{dc}/4$, S_{b3}', S_{b2}', S_{b1}', to the dc source terminal node V_1. The net load voltage appears as $+3V_{dc}/4$ from $+V_{dc}$ (V_5) minus $+V_{dc}/4$ (C_{b1}). The load current I_a is positive, flowing from load terminal '*a*' to '*b*' as given in Fig.4.11 (c).

Mode 3: For V_{ab}=$+V_{dc}/2$, the dc source terminal node V_5 is connected to the load terminal '*a*' through switches S_{a1}, S_{a2}, flying capacitors C_{a2}, C_{a2}, anti-parallel diodes D_{a3}', D_{a4}', to load terminal '*b*' and to turn-on switches S_{b4}', S_{b3}', S_{b2}', S_{b1}' to connect to the dc source terminal node V_1. The net load voltage appears as $+V_{dc}/2$ from $+V_{dc}$ (V_5) minus $+V_{dc}/4$ (C_{a2}) minus $+V_{dc}/4$ (C_{a2}). The load current I_a is positive, flowing from load terminal '*a*' to '*b*', as depicted in Fig.4.11 (d).

Mode 4: To get V_{ab}=$+V_{dc}/4$, the conduction takes place from the dc source terminal node V_5, through switch S_{a1}, flying capacitors C_{a3}, C_{a3}, C_{a3}, anti-parallel diodes D_{a2}', D_{a3}', D_{a4}', load terminal '*a*' to '*b*' and to turn-on switches S_{b4}', S_{b3}', S_{b2}', S_{b1}' through which it can connect to the dc source terminal node V_1. The net load voltage appears as $+V_{dc}/4$ from $+V_{dc}$ (V_5) minus $+V_{dc}/4$ (C_{a3}) minus $+V_{dc}/4$ (C_{a3}) minus $+V_{dc}/4$ (C_{a3}). The load current I_a is positive, flowing from load terminal '*a*' to '*b*', as depicted in Fig.4.11 (e).

Mode 5: To obtain V_{ab}=0, the conduction takes place from the dc source terminal node V_5, through switch S_{a1}, S_{a2}, flying capacitors C_{a2}, C_{a2}, anti-parallel diodes D_{a3}', D_{a4}' of first-leg, load terminal '*a*' to '*b*' and diodes D_{b4}, D_{b3}, flying capacitors C_{b2}, C_{b2}, and to turn-on switches S_{b2}', S_{b1} ' of second-leg, through which it can connect to the dc source terminal node V_1. The net load voltage appears as Zero (0) from $+V_{dc}$ (V_5) minus $+V_{dc}/4$ (C_{a2}) minus $+V_{dc}/4$ (C_{a2}) minus $+V_{dc}/4$ (C_{b2}) minus $+V_{dc}/4$ (C_{b2}). The load current I_a is zero since V_{ab} is zero, as depicted in Fig.4.11 (f).

Mode 6: For V_{ab}=$-V_{dc}/4$, the conduction takes place from the dc source terminal node V_5, through switch S_{b1},, flying capacitors C_{b3}, C_{b3}, C_{b3}, anti-parallel diodes D_{b2}', D_{b3}', D_{b4}', load terminal '*b*' to '*a*' and to turn-on switches S_{a4}', S_{a3}', S_{a2}', S_{a1}' through which it can connect to the dc source terminal node V_1. The net load voltage appears for V_{ab} as $-V_{dc}/4$ from $+V_{dc}$ (V_5) minus $+V_{dc}/4$ (C_{b3}) minus $+V_{dc}/4$ (C_{b3}) minus $+V_{dc}/4$ (C_{b3}) {V_{ba}=+Vdc/4}. The load current I_a is negative, flowing from load terminal '*b*' to '*a*', as depicted in Fig.4.11 (g).

Mode 7: To get V_{ab}=$-V_{dc}/2$, the dc source terminal node V_5 is connected to the load terminal '*b*' through switches S_{b1}, S_{b2}, flying capacitors C_{b2}, C_{b2}, anti-parallel diodes D_{b3}', D_{b4}', load terminal '*b*' to '*a*' and to turn-on switches S_{a4}', S_{a3}', S_{a2}', S_{a1}' to connect to the dc source terminal node V_1. The net load voltage V_{ab} appears as $-V_{dc}/2$ from $+V_{dc}$ (V_5) minus $+V_{dc}/4$ (C_{b2}) minus $+V_{dc}/4$ (C_{b2}). The load current I_a is negative, flowing from load terminal '*b*' to '*a*', as depicted in Fig.4.11 (h).

Mode 8: To obtain V_{ab} =$-3V_{dc}/4$, the dc source terminal node V_5 is connected to the load terminal '*b*' through switches S_{b1}, S_{b2}, S_{b3}, S_{b4}, load terminal '*b*' to '*a*' and diode D_{a4}, flying capacitor C_{a1} which is pre-charged to $+V_{dc}/4$, and turn-on switches S_{a3}', S_{a2}', S_{a1}', through which it can connect to the dc source terminal node V_1. The net load voltage V_{ab}, appears as $-3V_{dc}/4$ from $+V_{dc}$ (V_5) minus +Vdc/4 (C_{a1}). The load current I_a is negative, flowing from load terminal '*b*' to '*a*' as given in Fig.4.11 (i).

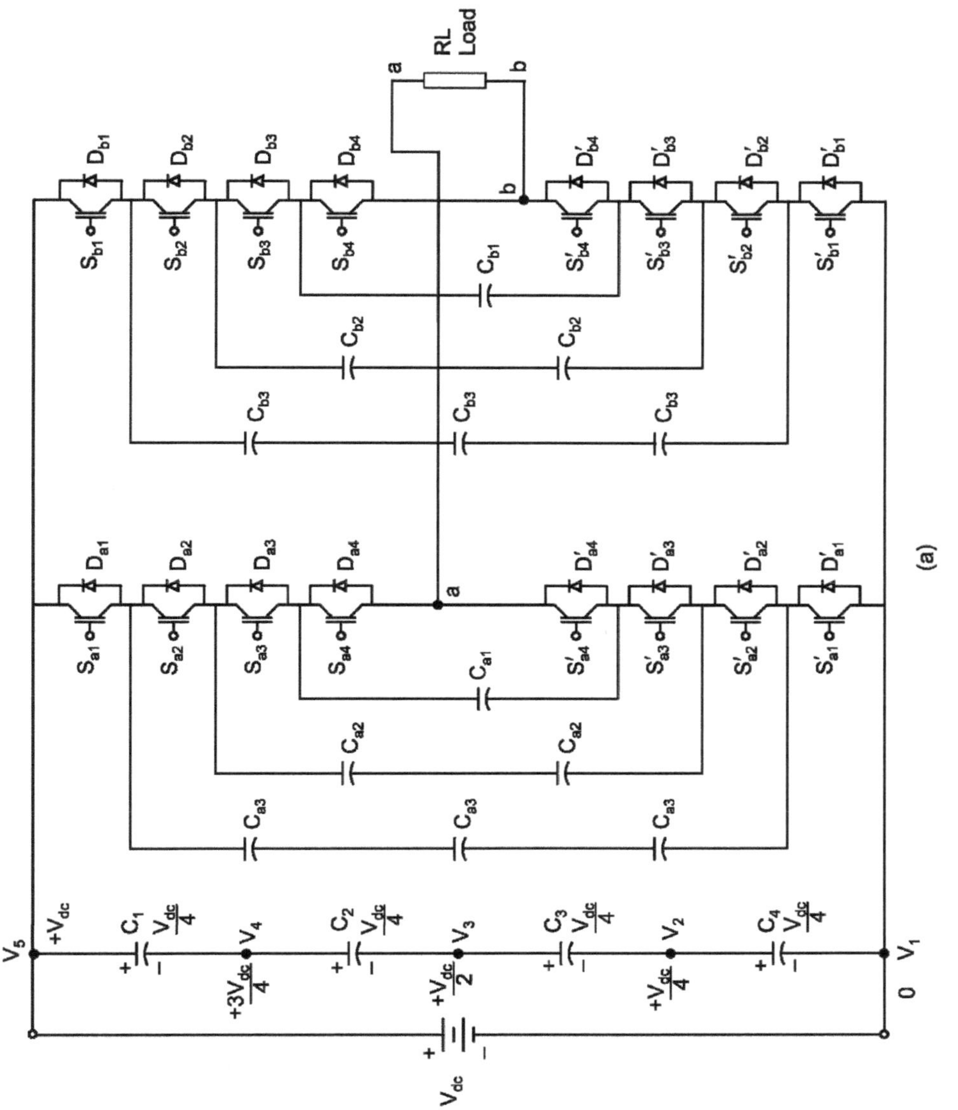

Fig. 4.11 (a) Main circuit.

Fig. 4.11 Single-Phase five-level flying-capacitor full-bridge inverter, (a) Main circuit, (b) Mode1: $+V_{dc}$, (c) Mode2: $+3V_{dc}/4$, (d) Mode3: $+V_{dc}/2$, (e) Mode4: $+V_{dc}/4$ for positive half-wave of line voltage across the load, V_{ab} and (f) Mode 5: $V_{ab} = 0$, (g) Mode 6: $-V_{dc}/4$, (h) Mode 7: $-V_{dc}/2$, (i) Mode 8: $-3V_{dc}/4$, (j) Mode 9: $-V_{dc}$ for negative half-wave of the voltage across the load, V_{ab}.

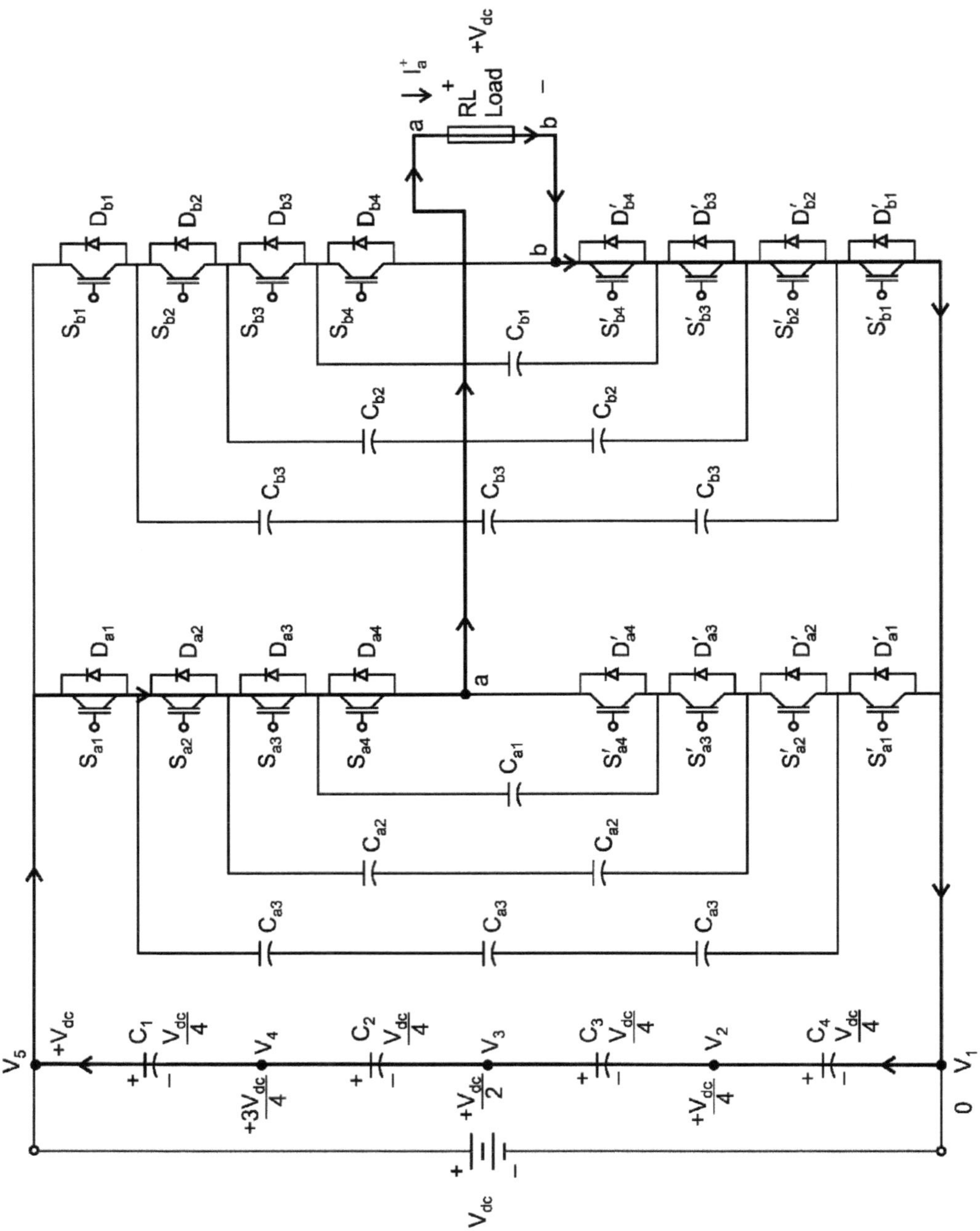

Fig. 4.11 (b) Mode 1: $V_{ab} = +V_{dc}$ with positive current.

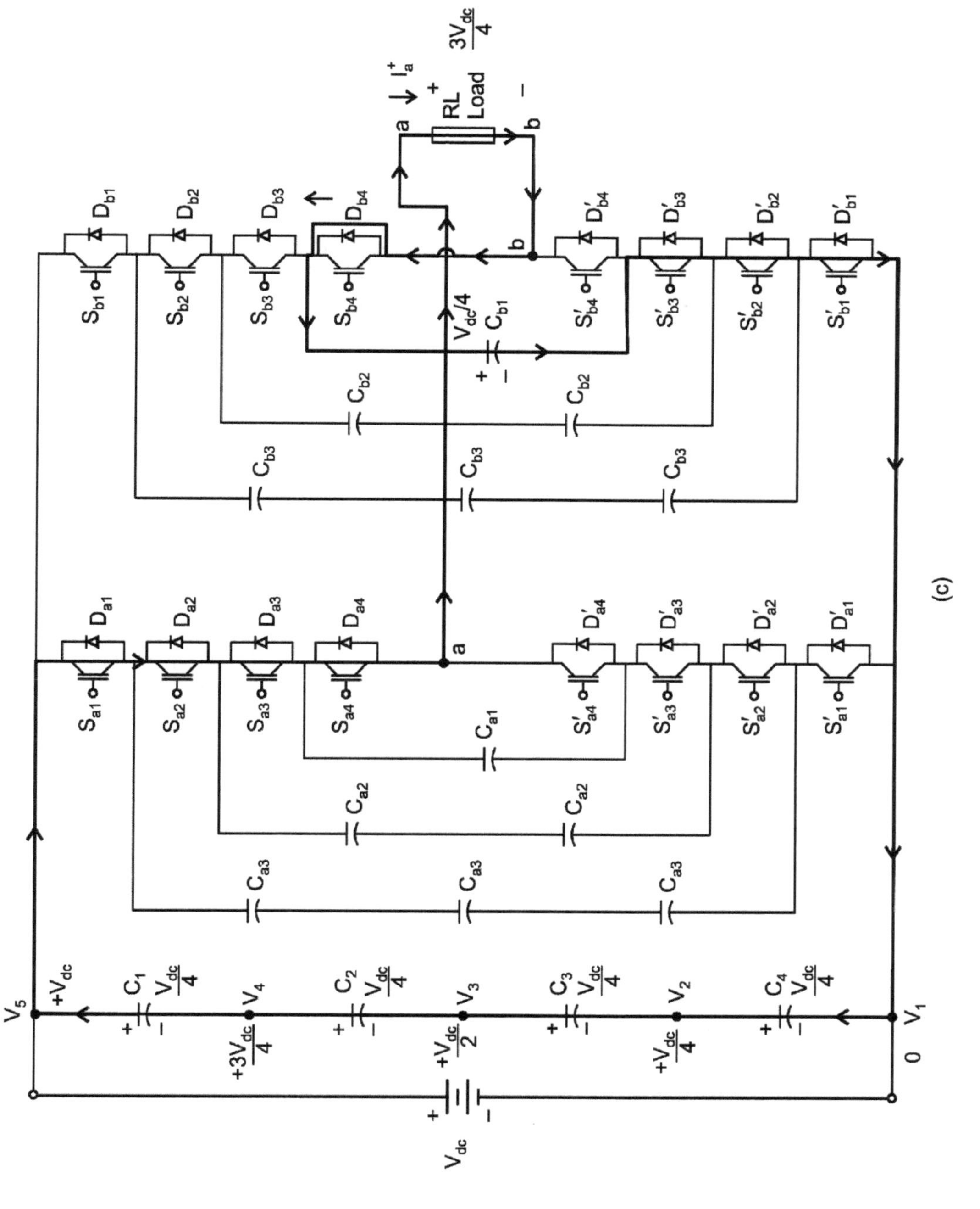

Fig. 4.11 (c) Mode 2: $V_{ab} = +3V_{dc}/4$, with positive I_a.

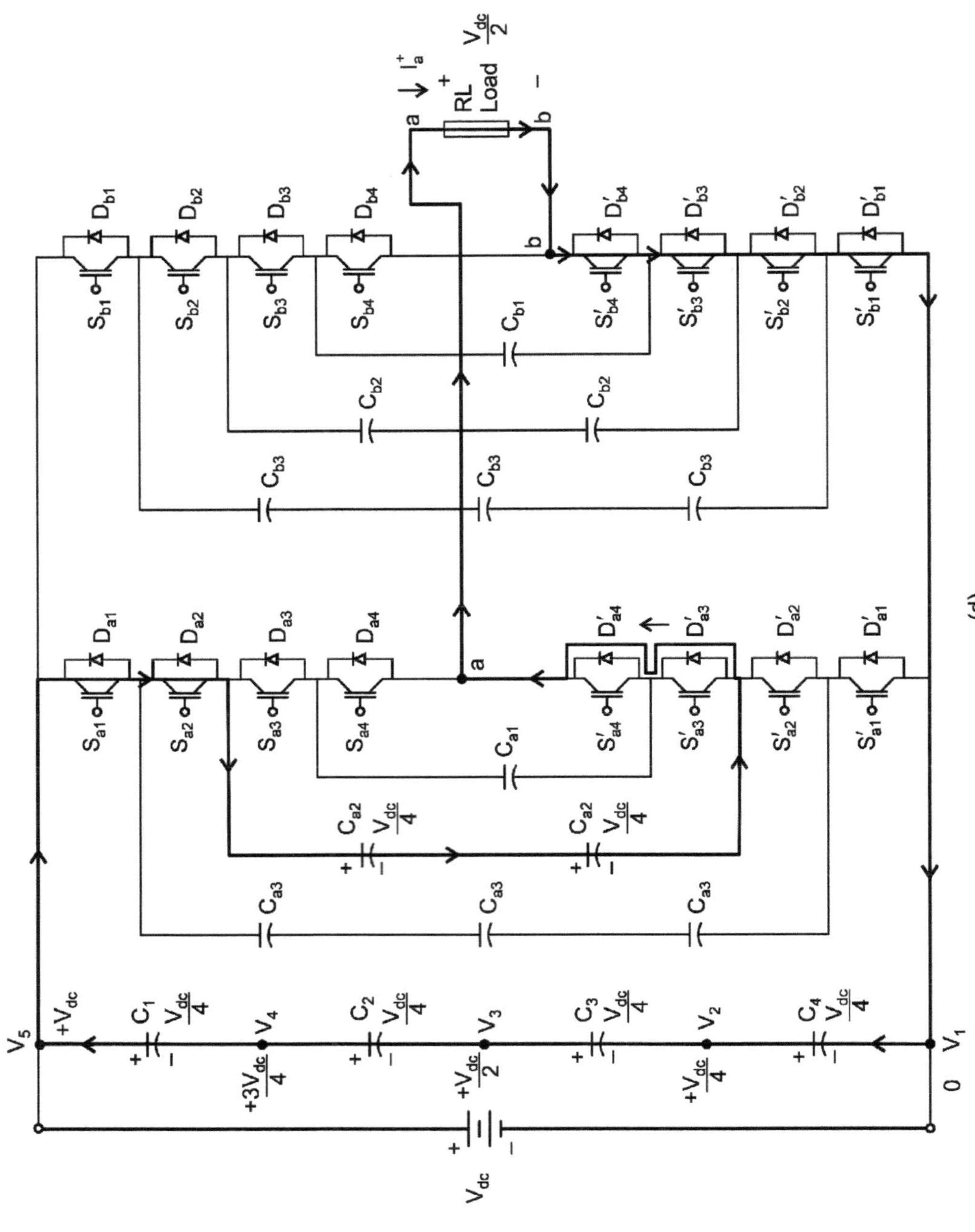

Fig. 4.11 (d) Mode 3: V_{ab} = $+V_{dc}/2$ with positive I_a.

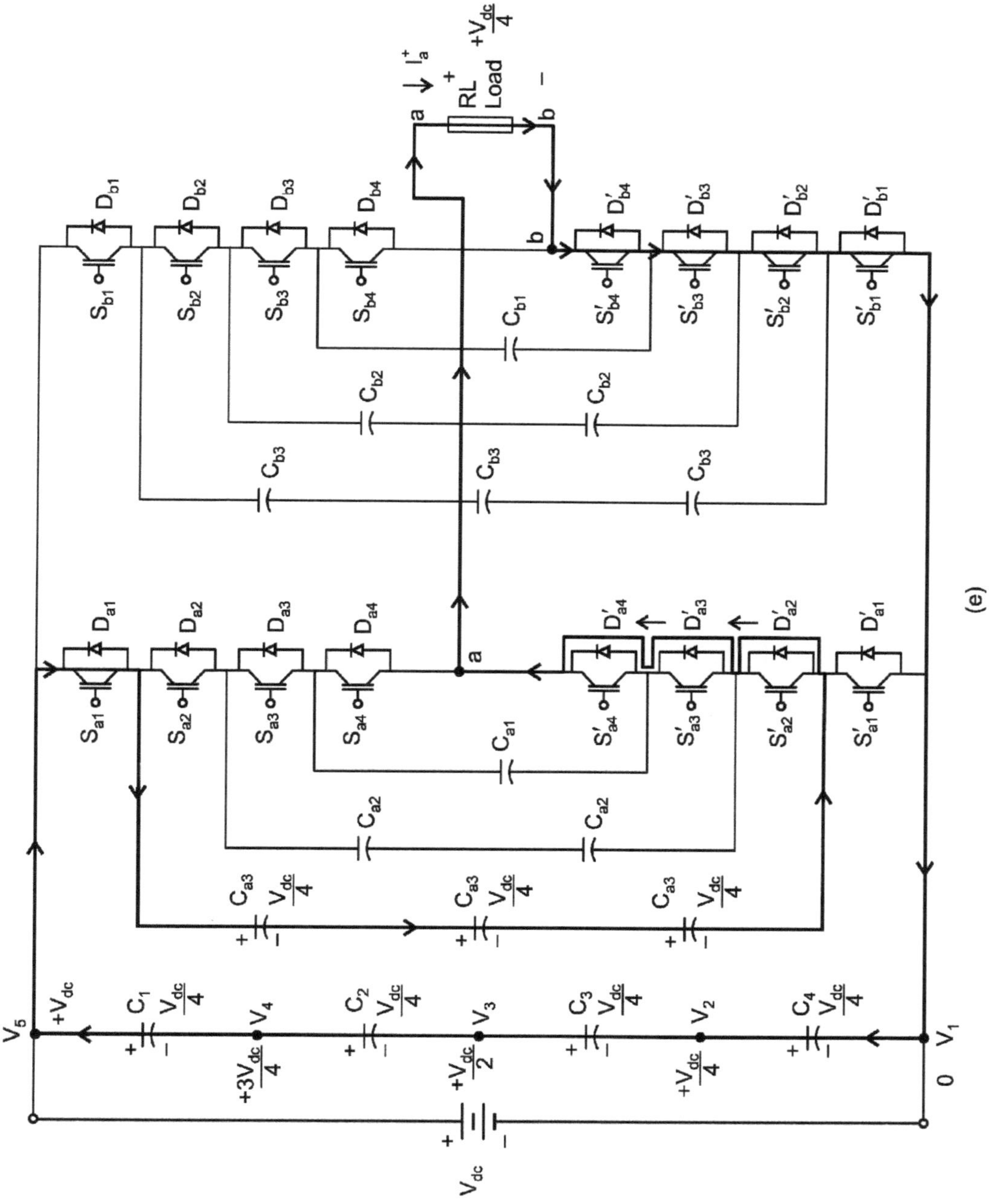

Fig. 4.11 (e) Mode 4: $V_{ab} = +V_{dc}/4$, with positive I_a.

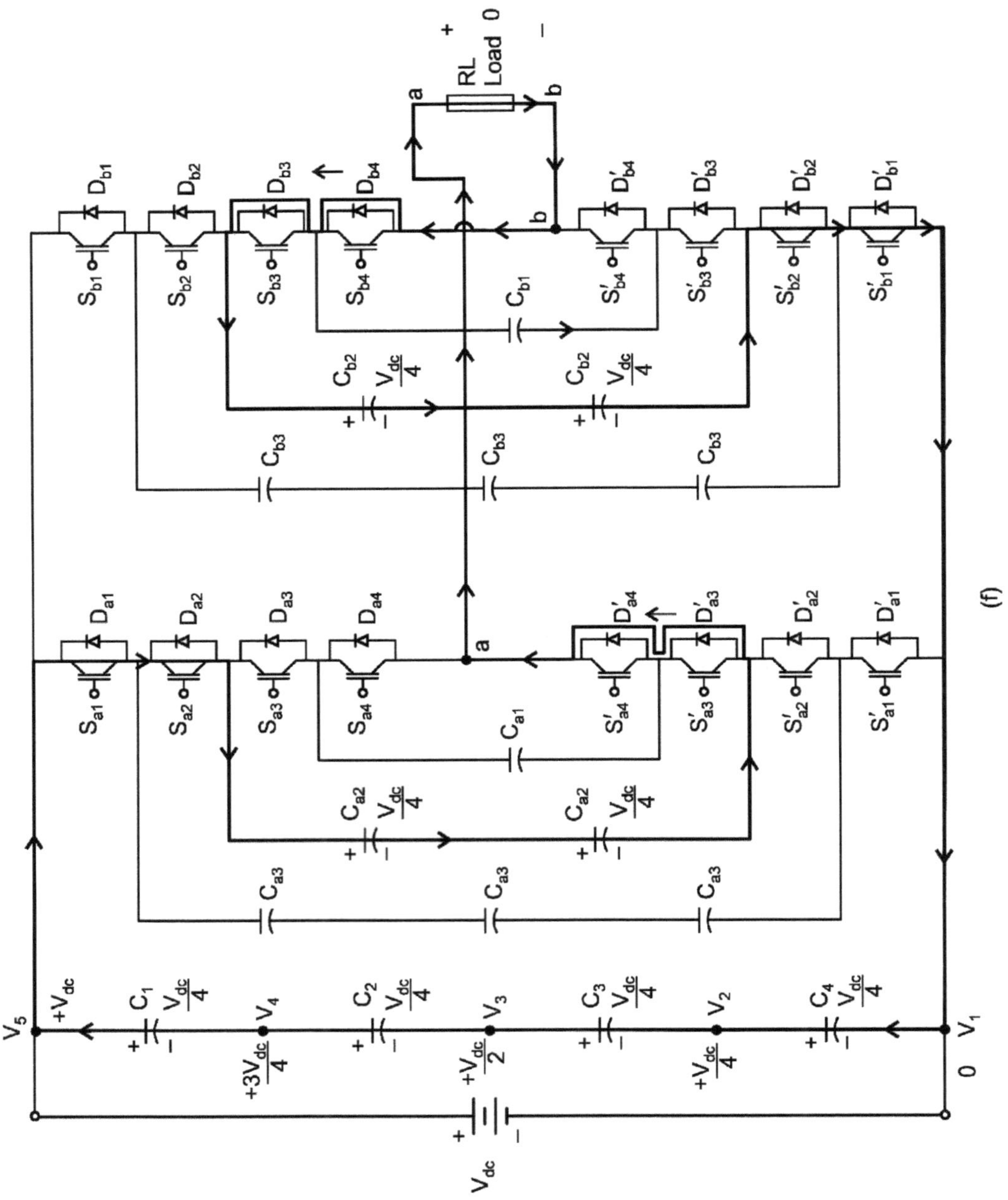

Fig. 4.11 (f) Mode 5: $V_{ab} = 0$ with $I_a = 0$.

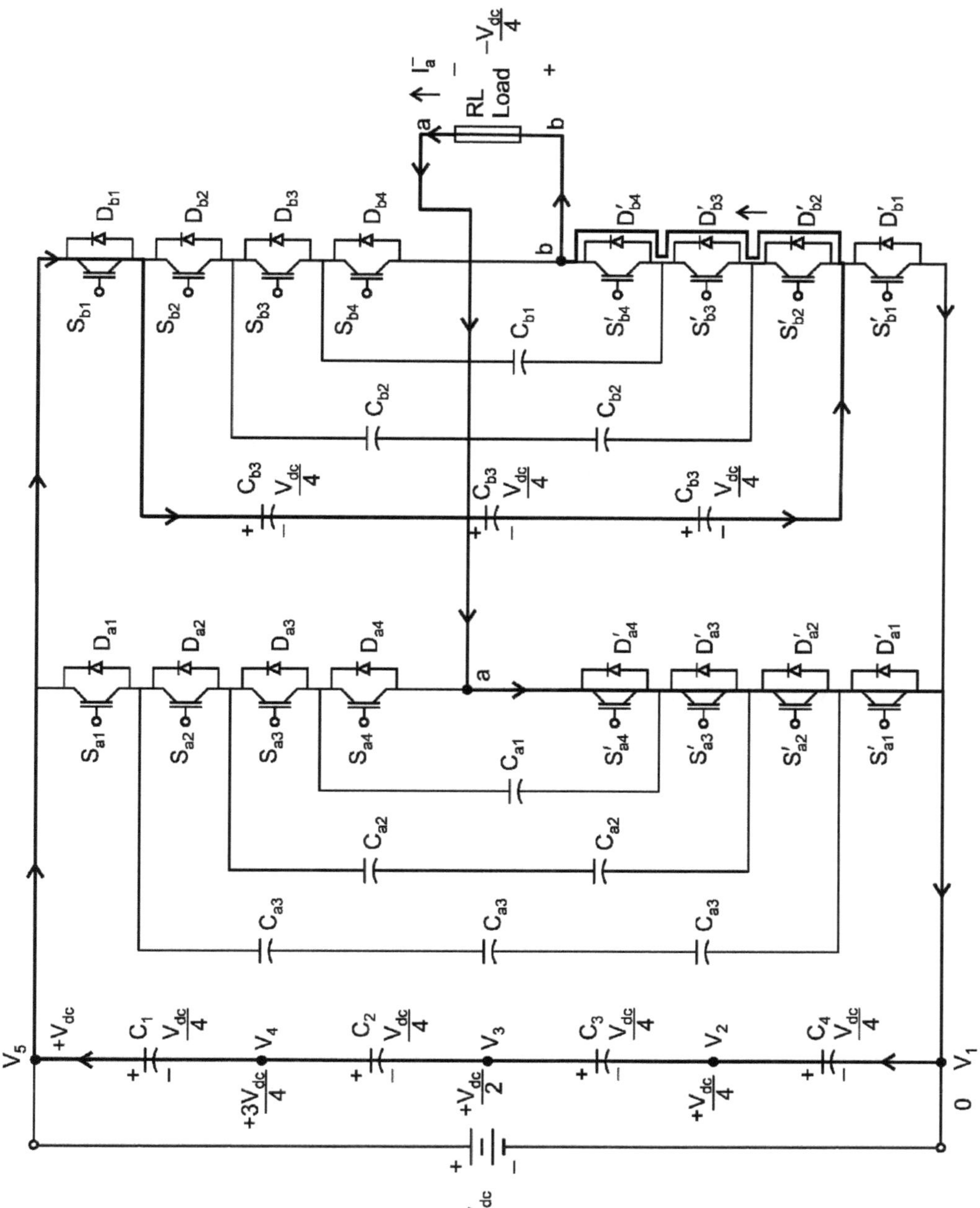

(g)

Fig. 4.11 (g) Mode 6: $V_{ab} = -V_{dc}/4$ with negative I_a.

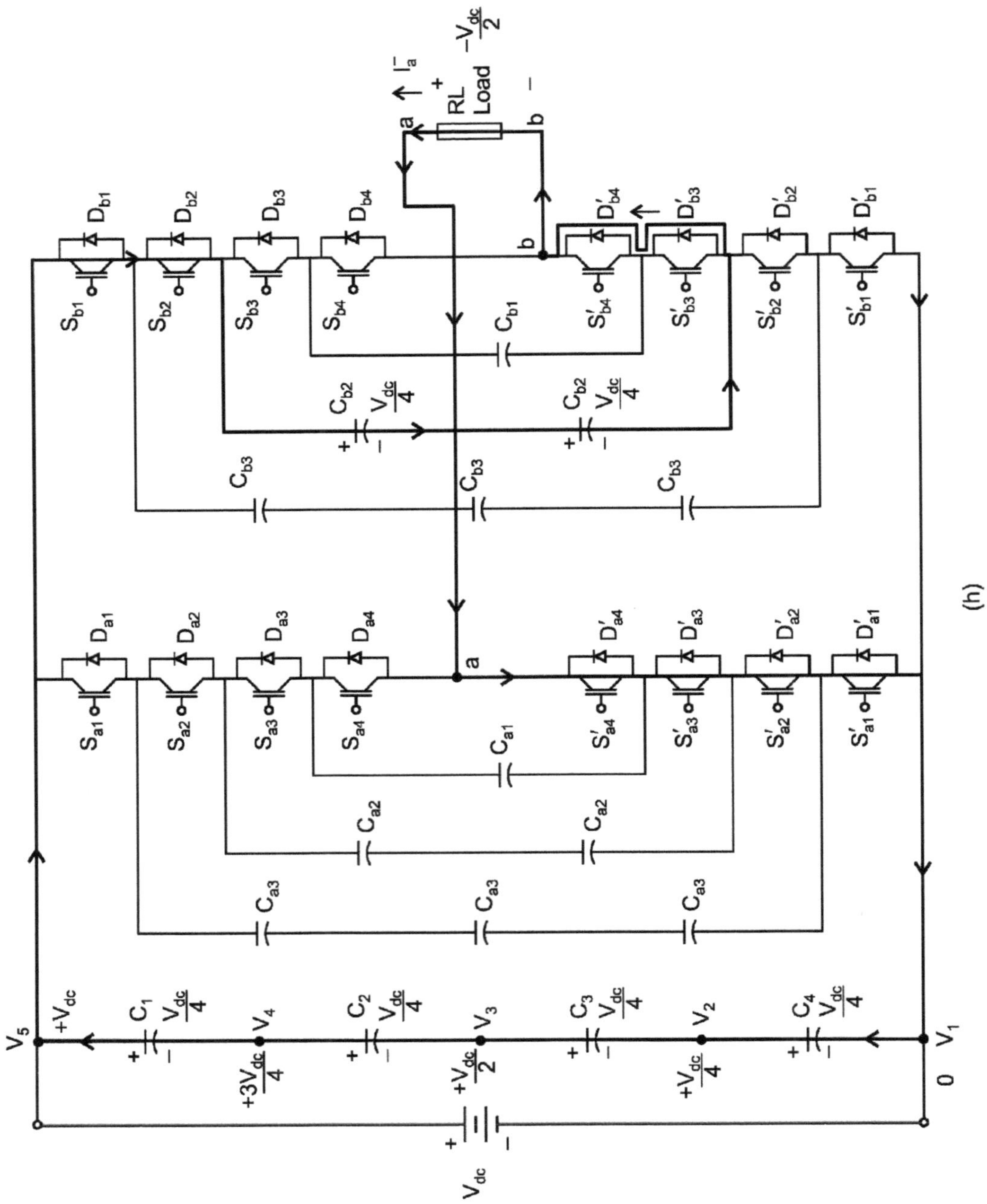

Fig. 4.11 (h) Mode 7: $V_{ab} = -V_{dc}/2$ with negative I_a.

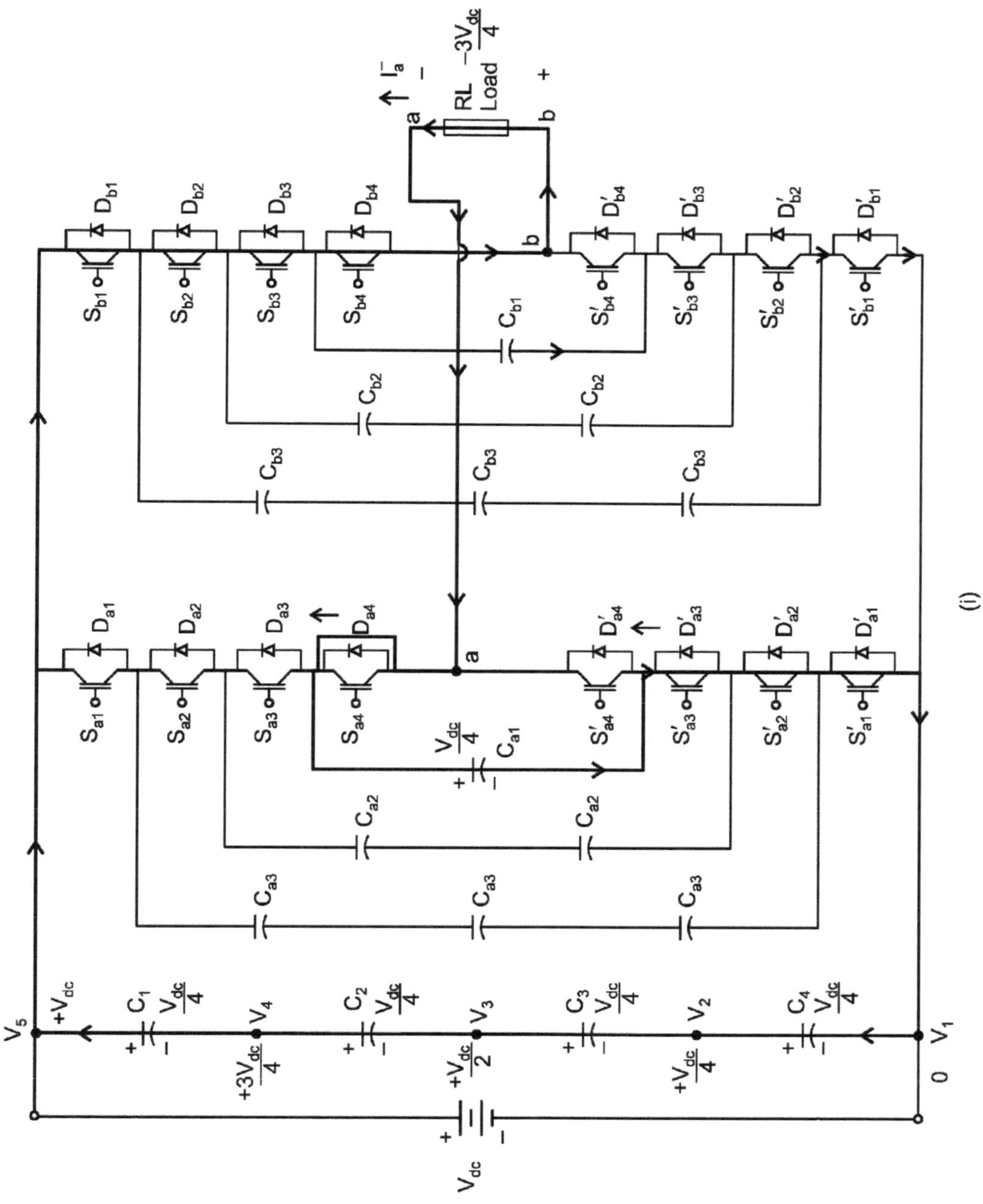

Fig. 4.11 (i) Mode 8: $V_{ab} = -3V_{dc}/4$ with negative I_a.

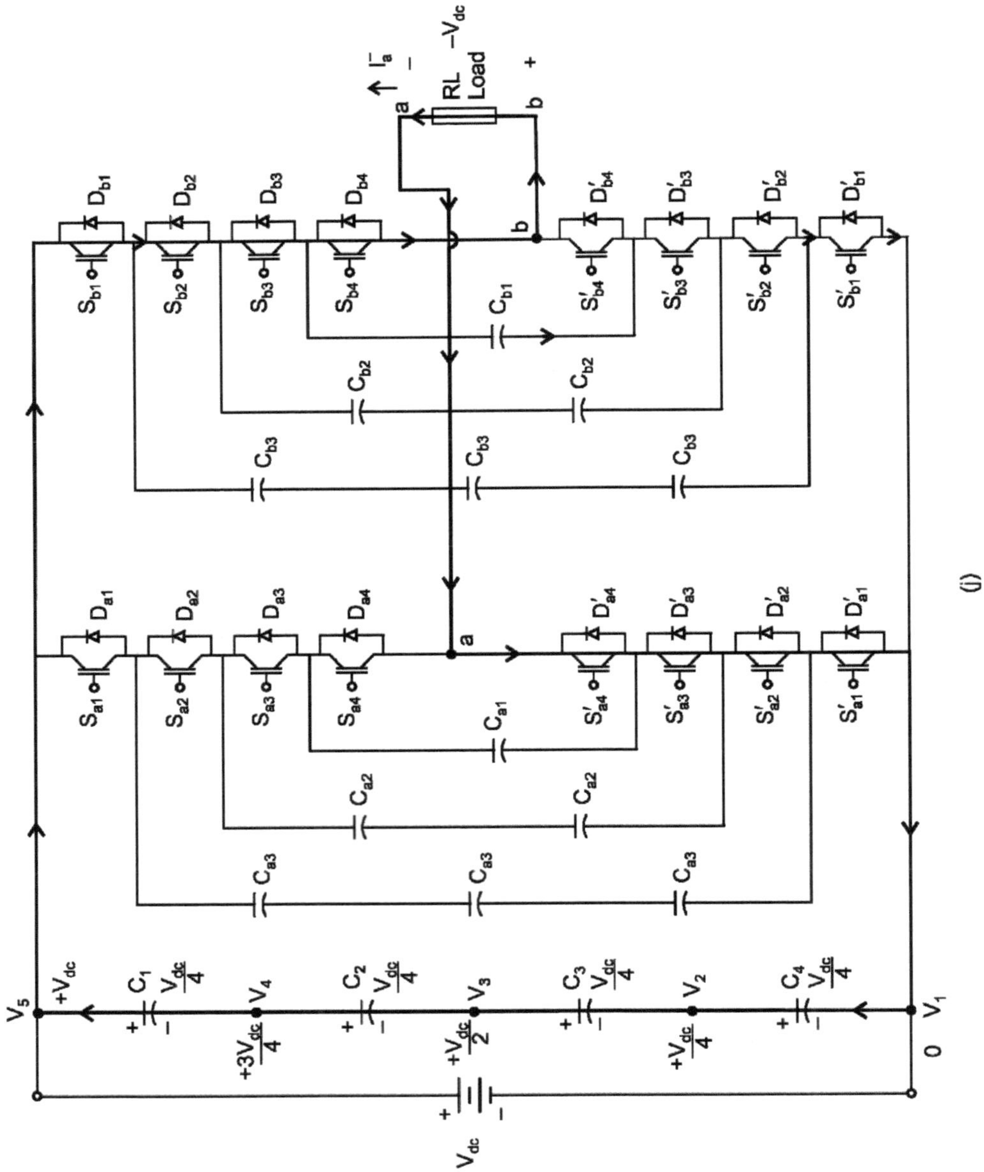

Fig. 4.11 (j) Mode 9: $V_{ab} = -V_{dc}$ with negative I_a.

Fig. 4.11 Single-Phase five-level flying-capacitor full-bridge inverter, (a) Main circuit, (b) Mode1: $+V_{dc}$, (c) Mode2: $+3V_{dc}/4$, (d) Mode3: $+V_{dc}/2$, (e) Mode4: $+V_{dc}/4$ for positive half-wave of line voltage across the load, V_{ab} and (f) Mode 5: $V_{ab} = 0$, (g) Mode 6: $-V_{dc}/4$, (h) Mode 7: $-V_{dc}/2$, (i) Mode 8: $-3V_{dc}/4$, (j) Mode 9: $-V_{dc}$ for negative half-wave of the voltage across the load, V_{ab}.

Mode 9: For V_{ab}=-V_{dc}, the dc source terminal node V_5 is connected to the load terminal '*b*' through switches S_{b1}, S_{b2}, S_{b3}, S_{b4}, load terminal '*b*' to '*a*' and S_{a4}', S_{a3}', S_{a2}', S_{a1}', through which it can connect to the dc source terminal node V_1. The load current I_a is negative, flowing from load terminal '*b*' to '*a*', as given in Fig.4.11 (j). The Five-level output voltage waveform for single-phase flying-capacitor full-bridge inverter is given in Fig. 4.12.

The Five-level output line voltage waveform for single-phase flying-capacitor full bridge inverter is shown in Fig. 4.12.

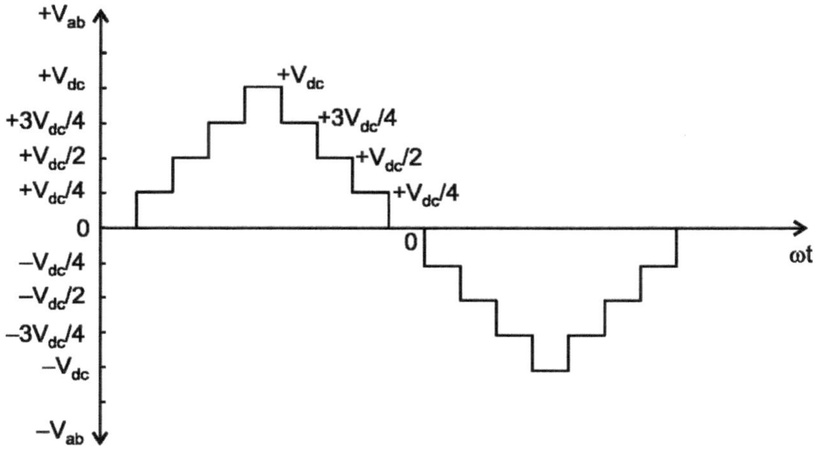

Fig. 4.12 Five-level output line voltage waveform for single phase flying-capacitor full-bridge inverter.

C. **Three phase Five-level Flying Capacitor full-bridge inverter:** The three single-phase five-level flying-capacitor single-leg inverter circuits are connected to form a three-phase five-level flying-capacitor full-bridge inverter as shown in the Fig. 4.13(a). The output voltage of three-phase inverter is feeding power to the balanced star-connected load. The phase voltages are V_{an}, V_{bn}, V_{cn}, and the line voltages are V_{ab}, V_{bc}, and V_{ca} of the three-phase load. The phase voltage waveform V_{an} is similar to the load voltage obtained for single-phase single-leg arrangement, the only difference is the V_{bn}, and V_{cn} are phase shifted with respect to V_{an} by 120º and 240º respectively. The line voltage waveforms are only shown in the Fig.4.13 (b).

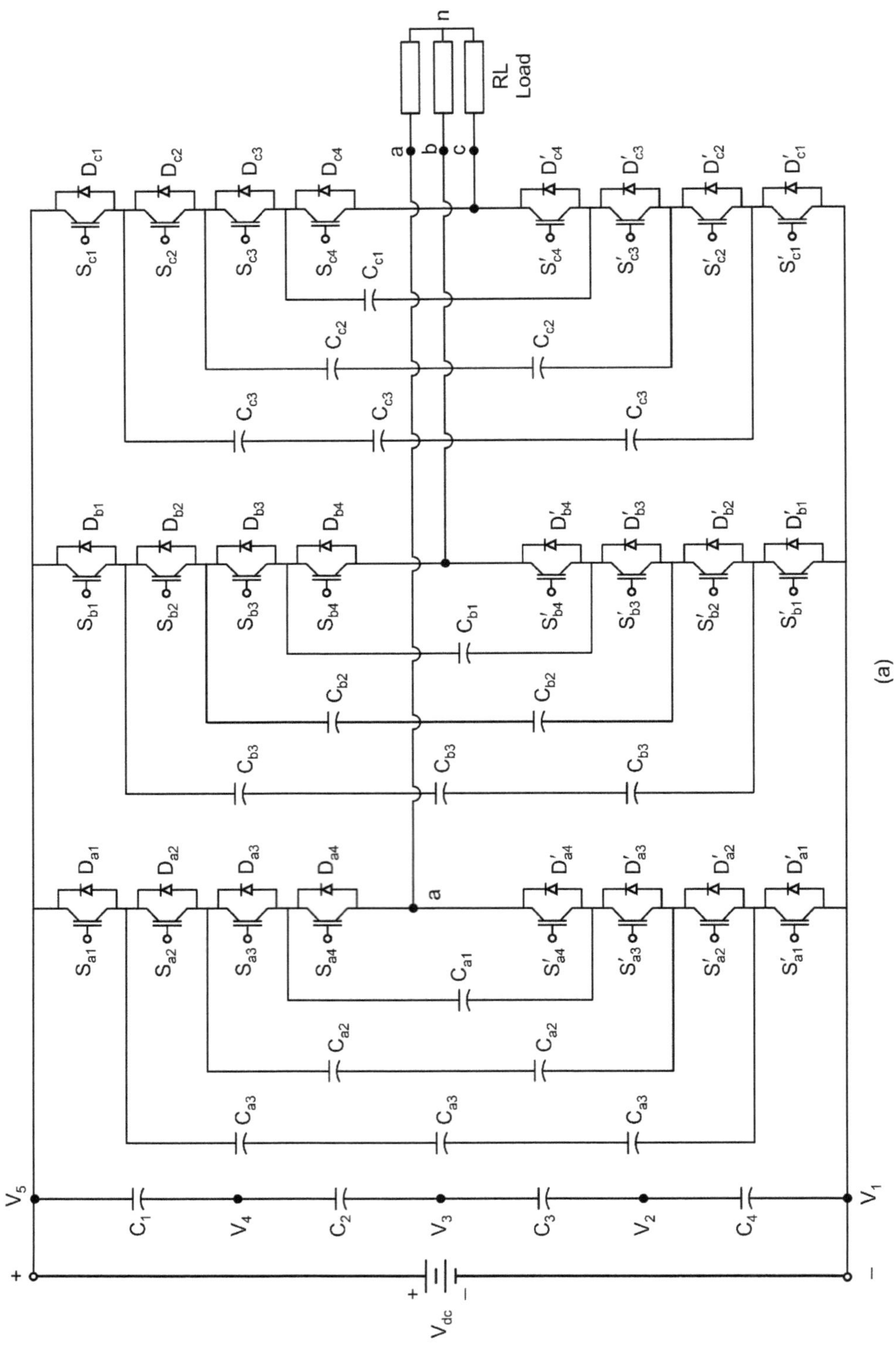
n
RL
Load
a
b
c
D_{c1}
D_{c2}
D_{c3}
D_{c4}
D'_{c4}
D'_{c3}
D'_{c2}
D'_{c1}
S_{c1}
S_{c2}
S_{c3}
S_{c4}
C_{c1}
S'_{c4}
S'_{c3}
S'_{c2}
S'_{c1}
C_{c2}
C_{c2}
C_{c3}
C_{c3}
C_{c3}
D_{b1}
D_{b2}
D_{b3}
D_{b4}
D'_{b4}
D'_{b3}
D'_{b2}
D'_{b1}
S_{b1}
S_{b2}
S_{b3}
S_{b4}
C_{b1}
S'_{b4}
S'_{b3}
S'_{b2}
S'_{b1}
C_{b2}
C_{b2}
C_{b3}
C_{b3}
C_{b3}
D_{a1}
D_{a2}
D_{a3}
D_{a4}
a
D'_{a4}
D'_{a3}
D'_{a2}
D'_{a1}
S_{a1}
S_{a2}
S_{a3}
S_{a4}
C_{a1}
S'_{a4}
S'_{a3}
S'_{a2}
S'_{a1}
C_{a2}
C_{a2}
C_{a3}
C_{a3}
C_{a3}
V_5
C_1
V_4
C_2
V_3
C_3
V_2
C_4
V_1
+
−
V_{dc}

(a)

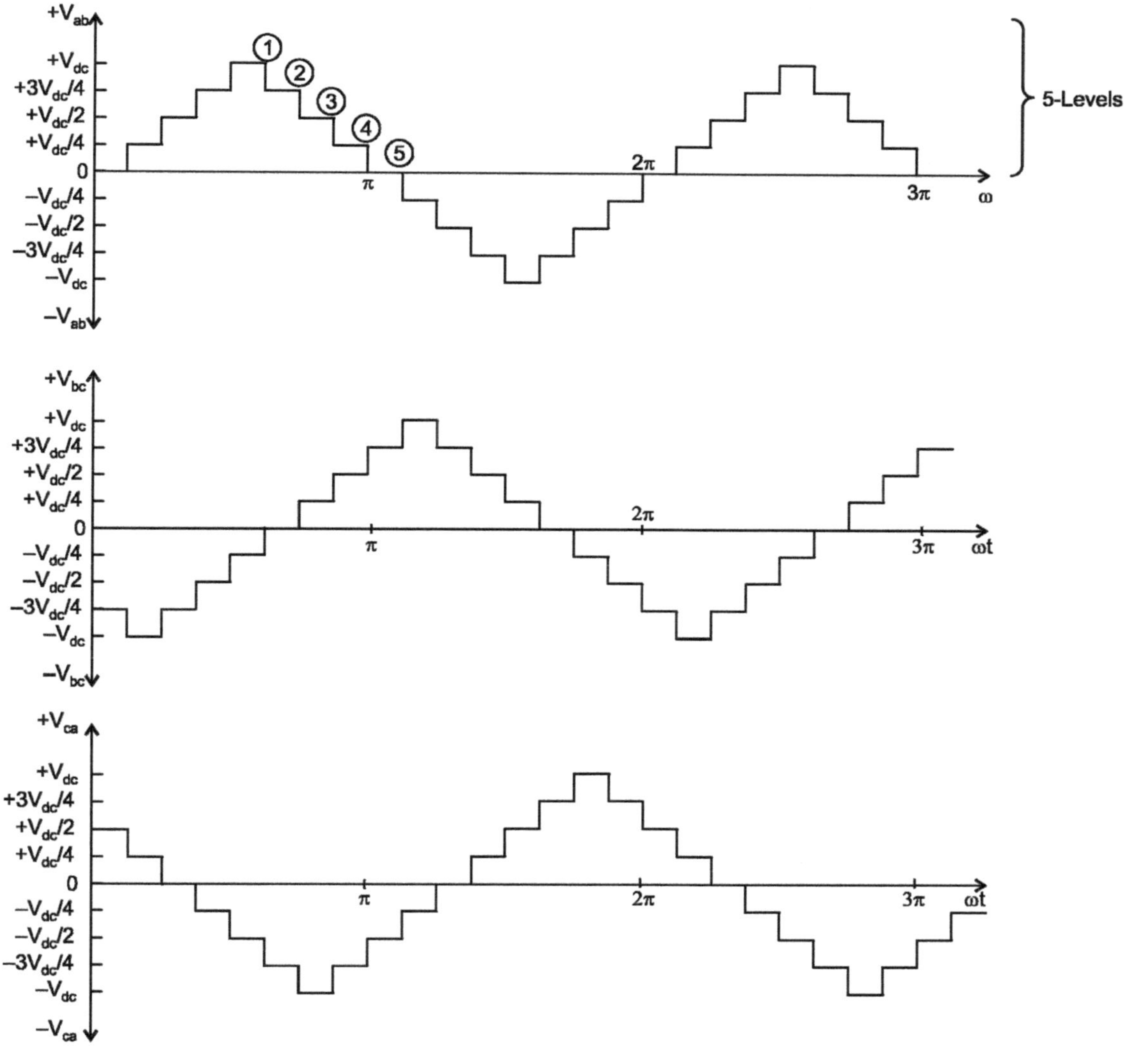

(b)

Fig. 4.13 (a) Three phase Five-level Flying Capacitor full-bridge inverter circuit. (b) Line Voltage Waveforms of the load for five-level output, V_{ab}, V_{bc}, V_{ca}.

1. For an output voltage level $v_{ao} = V_{dc}$, turn on all upper-half switches S_{a1} through S_{a4}.
2. For an output voltage level $v_{ao} = 3V_{dc}/4$, there are four combinations:
   (a) $v_{ao} = V_{dc} - V_{dc}/4$ by turning on devices S_{a1}, S_{a2}, S_{a3}, and S'_{a4}.
   (b) $v_{ao} = 3V_{dc}/4$ by turning on devices S_{a2}, S_{a3}, S_{a4}, and S'_{a1}.
   (c) $v_{ao} = -3V_{dc}/4 + V_{dc}/2$ by turning on devices S_{a1}, S_{a3}, S_{a4}, and S'_{a2}.
   (d) $v_{ao} = V_{dc} - V_{dc}/2 + V_{dc}/4$ by turning on devices S_{a1}, S_{a2}, S_{a4}, and S'_{a3}.
3. For an output voltage level $v_{ao} = V_{dc}/2$, there are six combinations:
   (a) $v_{ao} = V_{dc} - V_{dc}/2$ by turning on devices S_{a1}, S_{a2}, S'_{a3}, and S'_{a4}.
   (b) $v_{ao} = V_{dc}/2$ by turning on devices S_{a3}, S_{a4}, S'_{a1}, and S'_{a2}.
   (c) $v_{ao} = V_{dc} - 3V_{dc}/4 + V_{dc}/2 - V_{dc}/4$ by turning on devices S_{a1}, S_{a3}, S'_{a2}, and S'_{a4}.
   (d) $v_{ao} = V_{dc} - 3V_{dc}/4 + V_{dc}/4$ by turning on devices S_{a1}, S_{a4}, S'_{a2}, and S'_{a3}.
   (e) $v_{ao} = 3V_{dc}/4 - V_{dc}/2 + V_{dc}/4$ by turning on devices S_{a2}, S_{a4}, S'_{a1}, and S'_{a3}.
   (f) $v_{ao} = 3V_{dc}/4 + V_{dc}/4$ by turning on devices S_{a2}, S_{a3}, S'_{a1}, and S'_{a4}.
4. For an output voltage level $v_{ao} = V_{dc}/4$, there are four combinations:
   (a) $v_{ao} = V_{dc} - 3V_{dc}/4$ by turning on devices S_{a1}, S'_{a2}, S'_{a3}, and S'_{a4}.
   (b) $v_{ao} = V_{dc}/4$ by turning on devices S_{a4}, S'_{a1}, S'_{a2}, and S'_{a3}.
   (c) $v_{ao} = V_{dc}/2 - V_{dc}/4$ by turning on devices S_{a3}, S'_{a1}, S'_{a2}, and S'_{a4}.

4.4 Seven-Level Flying Capacitor Inverter

The seven-level flying-capacitor inverter with single-leg arrangement is presented in Fig. 4.14 (a) for ease of understanding its operation. Each leg comprises of six power semiconductor switches with anti-parallel diodes in the upper leg and six similar devices in the lower leg and five-layers of flying-capacitors are arranged in the fashion shown in the figure totaling 15 capacitors. In addition, six dc source capacitors are connected and each sharing $V_{dc}/6$ volts with neutral at the mid-point of the source. The Inductive-resistive load is connected between phase '*a*' and neutral '*n*', which can give a seven-level phase voltage, V_{an}. The load current I_a when it flows from phase '*a*' to '*n*, it is positive and when it flows from neutral '*n*' to phase '*a*', then it is negative. In the process of conduction, load inductor stores energy and for which a reverse direction current path needs to be provided to feed the stored energy to the dc source till it is completely dissipated/exhausted. This current path is not shown here for this topology and left as the assignment for the readers, based on the knowledge acquired in the case of three-level flying capacitor inverter explained earlier in this chapter.

A. **Single-phase Seven-level flying-capacitor inverter with single-leg arrangement:** The seven-level flying-capacitor single-leg inverter is elaborated by considering seven modes of operation and as shown in the Fig.4.14 (b) – (h), conduction path is given with thick line with arrows and switching states for each mode is also given in the Table.4.4.

Mode 1: For V_{an}=+V_{dc}/2, the upper-leg switches S_{a1}, S_{a2}, S_{a3}, S_{a4}, S_{a5}, S_{a6} are turned-on, then the conduction takes place from upper dc source capacitors C_1, C_2, C_3, (total voltage of 3V_{dc}/6) through switches S_{a1}, S_{a2}, S_{a3}, S_{a4}, S_{a5}, S_{a6}, load terminal '*a*' to '*n*', as shown in the Fig.4.14 (b). The load current I_a is positive when it flows from load terminal '*a*' to '*n*'.

Mode 2: To obtain V_{an}=+V_{dc}/3, the upper-leg switches S_{a1}, S_{a2}, S_{a3}, S_{a4}, S_{a5} are turned-on, then the positive load current flows from dc source capacitors C_1, C_2, C_3, (total voltage of 3V_{dc}/6) through switches S_{a1}, S_{a2}, S_{a3}, S_{a4}, S_{a5}, flying-capacitor C_{a1} (pre-charged to +V_{dc}/6), diode D_{a6}', load terminal '*a*' to '*n*', completes that path, as shown in Fig.4.14 (c).

Mode 3: To get V_{an}=+V_{dc}/6, the upper-leg switches S_{a1}, S_{a2}, S_{a3}, S_{a4} are turned-on, then the conduction takes place from upper dc source capacitors C_1, C_2, C_3 (total voltage of 3V_{dc}/6) through switches S_{a1}, S_{a2}, S_{a3}, S_{a4}, through series connected flying-capacitors C_{a2}, C_{a2} (each pre-charged with +V_{dc}/6), diodes D_{a5}', D_{a6}', load terminal '*a*' to '*n*', as shown in Fig.4.14 (d).

Mode 4: For V_{an}= 0, the upper-leg switches S_{a1}, S_{a2}, S_{a3}, are turned-on, then the conduction takes place from upper dc source capacitors C_1, C_2, C_3 (total voltage of 3V_{dc}/6) through switches S_{a1}, S_{a2}, S_{a3}, and series connected flying-capacitors C_{a3}, C_{a3}, C_{a3} (each pre-charges with +V_{dc}/6), diodes D_{a4}', D_{a5}', D_{a6}', load terminal '*a*' to '*n*', as given in Fig.4.14 (e). Since load voltage is zero, the load current I*a* is also zero.

Mode 5: To get V_{an}= -V_{dc}/6, the lower-leg switches S_{a4}', S_{a3}', S_{a2}', S_{a1}' are turned-on, then the negative load current flows from lower-leg dc source capacitors C_4, C_5, C_6, (total voltage of 3V_{dc}/6), load terminal '*n*' to '*a*', through diodes D_{a6}, D_{a5}, series connected flying-capacitor C_{a2}, C_{a2} (each pre-charged with +V_{dc}/6) and switches S_{a4}', S_{a3}', S_{a2}', S_{a1}', completes that path, as shown in Fig.4.14 (f).

Mode 6: To obtain V_{an}= -V_{dc}/3, the lower-leg switches S_{a5}', S_{a4}', S_{a3}', S_{a2}', S_{a1}' are turned-on, then the negative load current flows from lower-leg dc source capacitors C_4, C_5, C_6, (total voltage of 3V_{dc}/6), load terminal '*n*' to '*a*', through diode D_{a6}, flying-capacitor C_{a1} (pre-charged with +V_{dc}/6) and switches S_{a5}', S_{a4}', S_{a3}', S_{a2}', S_{a1}', completes that path, as shown in Fig.4.14 (g).

Mode 7: For V_{an}= -V_{dc}/2, the lower-leg switches S_{a6}', S_{a5}', S_{a4}', S_{a3}', S_{a2}', S_{a1}' are turned-on, then the conduction takes place from lower dc source capacitors C_4, C_5, C_6, (total voltage of 3V_{dc}/6) through load terminal '*n*' to '*a*', and through switches S_{a6}', S_{a5}', S_{a4}', S_{a3}', S_{a2}', S_{a1}',as shown in the Fig.4.14 (h). The load current I_a is negative when it flows from load terminal '*n*' to '*a*'.

The load voltage V_{an} for seven-level flying-capacitor inverter with single-leg is given in Fig.4.15.

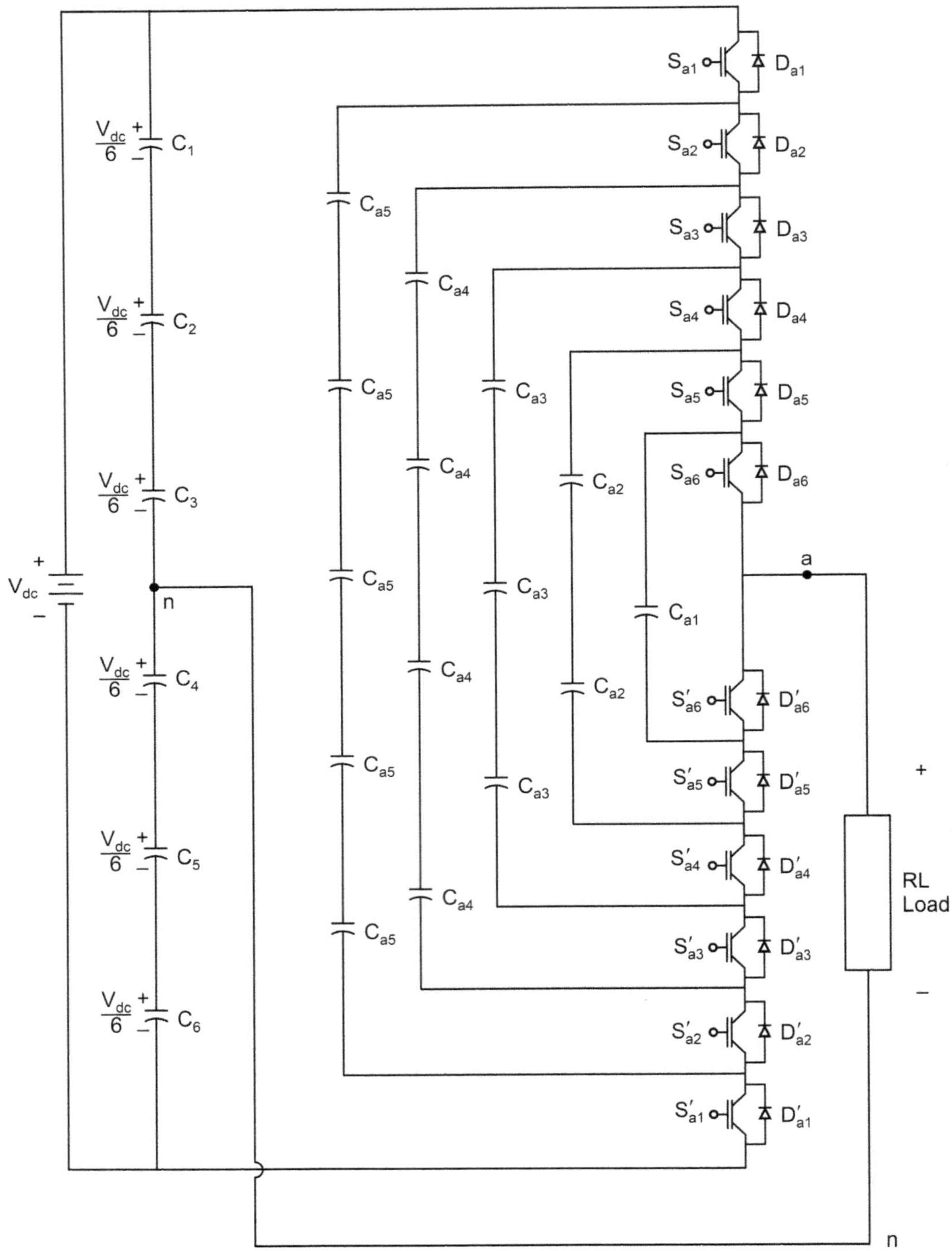
Vdc
Vdc/6 C1
Vdc/6 C2
Vdc/6 C3
n
Vdc/6 C4
Vdc/6 C5
Vdc/6 C6
Ca5
Ca4
Ca3
Ca2
Ca1
Sa1 Da1
Sa2 Da2
Sa3 Da3
Sa4 Da4
Sa5 Da5
Sa6 Da6
a
S′a6 D′a6
S′a5 D′a5
S′a4 D′a4
S′a3 D′a3
S′a2 D′a2
S′a1 D′a1
RL Load
n

(a)

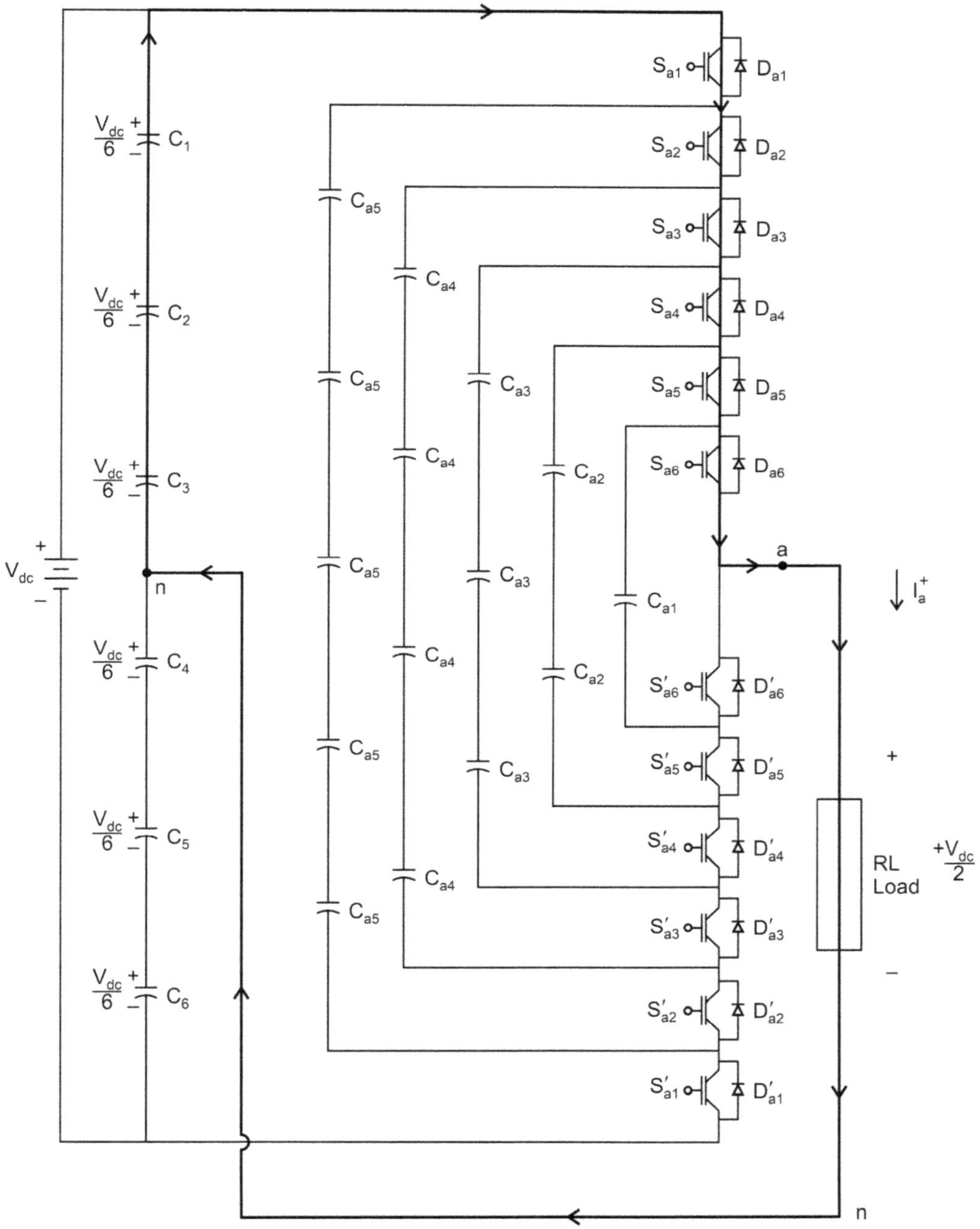

(b)

Fig. 4.14 Seven-level flying-capacitor inverter with single-leg arrangement (a) Main circuit, (b) Mode1: V_{an}=+V_{dc}/2.

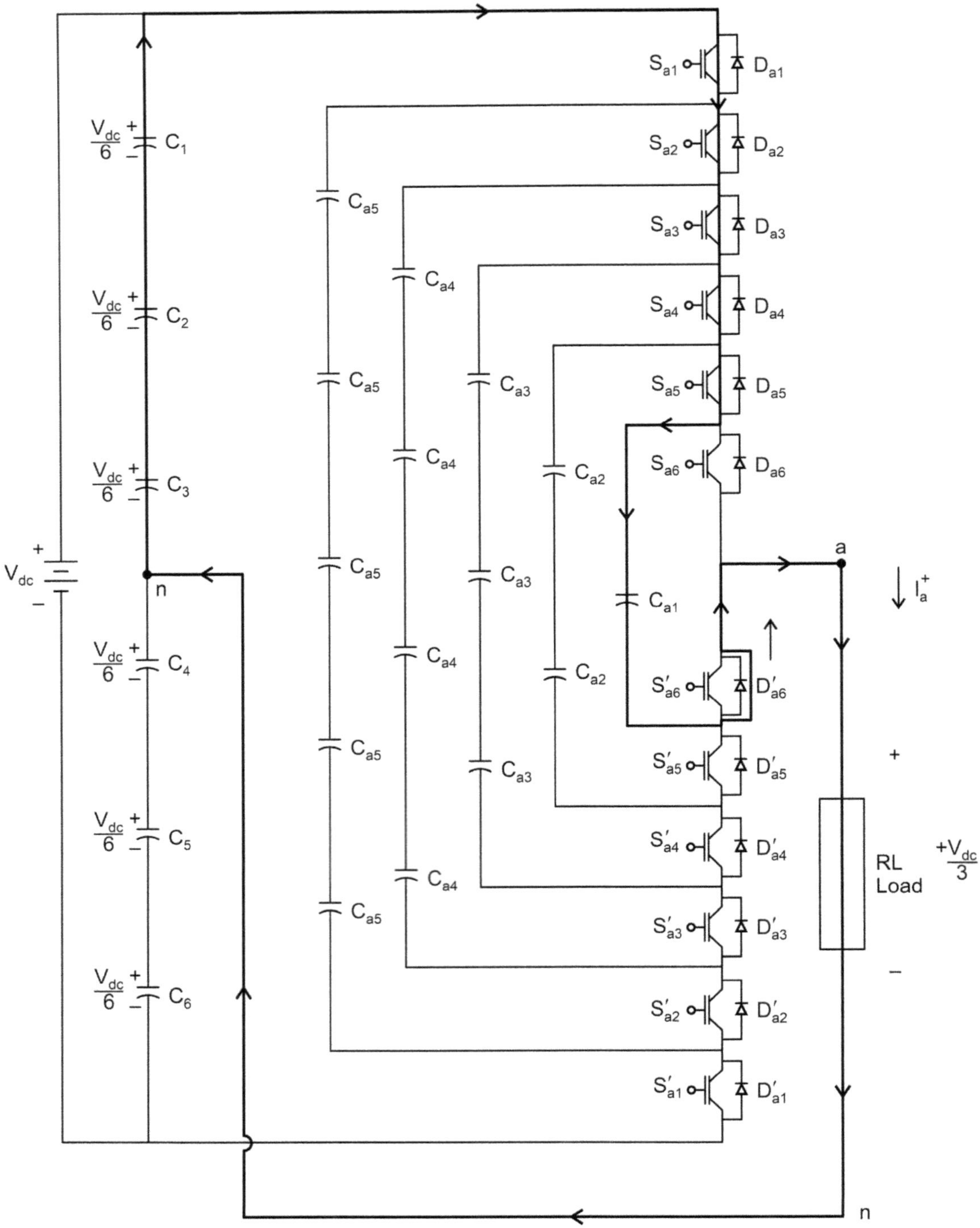

S_{a1} D_{a1}
S_{a2} D_{a2}
S_{a3} D_{a3}
S_{a4} D_{a4}
S_{a5} D_{a5}
S_{a6} D_{a6}
S'_{a6} D'_{a6}
S'_{a5} D'_{a5}
S'_{a4} D'_{a4}
S'_{a3} D'_{a3}
S'_{a2} D'_{a2}
S'_{a1} D'_{a1}
$\frac{V_{dc}}{6}$ C_1
$\frac{V_{dc}}{6}$ C_2
$\frac{V_{dc}}{6}$ C_3
$\frac{V_{dc}}{6}$ C_4
$\frac{V_{dc}}{6}$ C_5
$\frac{V_{dc}}{6}$ C_6
V_{dc}
n
C_{a5}
C_{a4}
C_{a3}
C_{a2}
C_{a1}
a
I_a^+
RL Load
$+\frac{V_{dc}}{3}$
n

(c)

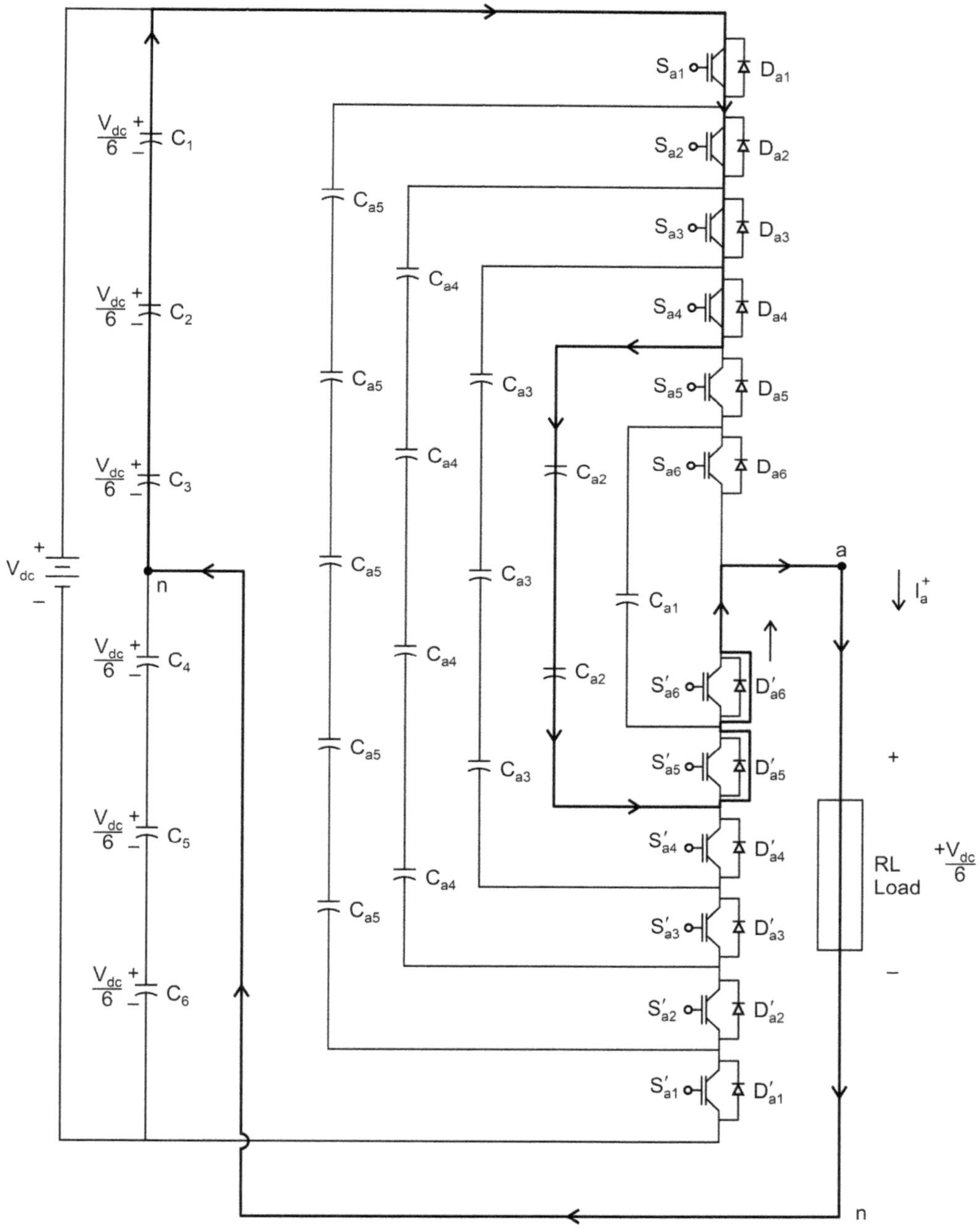

(d)

Fig. 4.14 Seven-level flying-capacitor inverter with single-leg arrangement (c) Mode2: $V_{an}=+V_{dc}/3$, (d) Mode3: $V_{an}=+V_{dc}/6$.

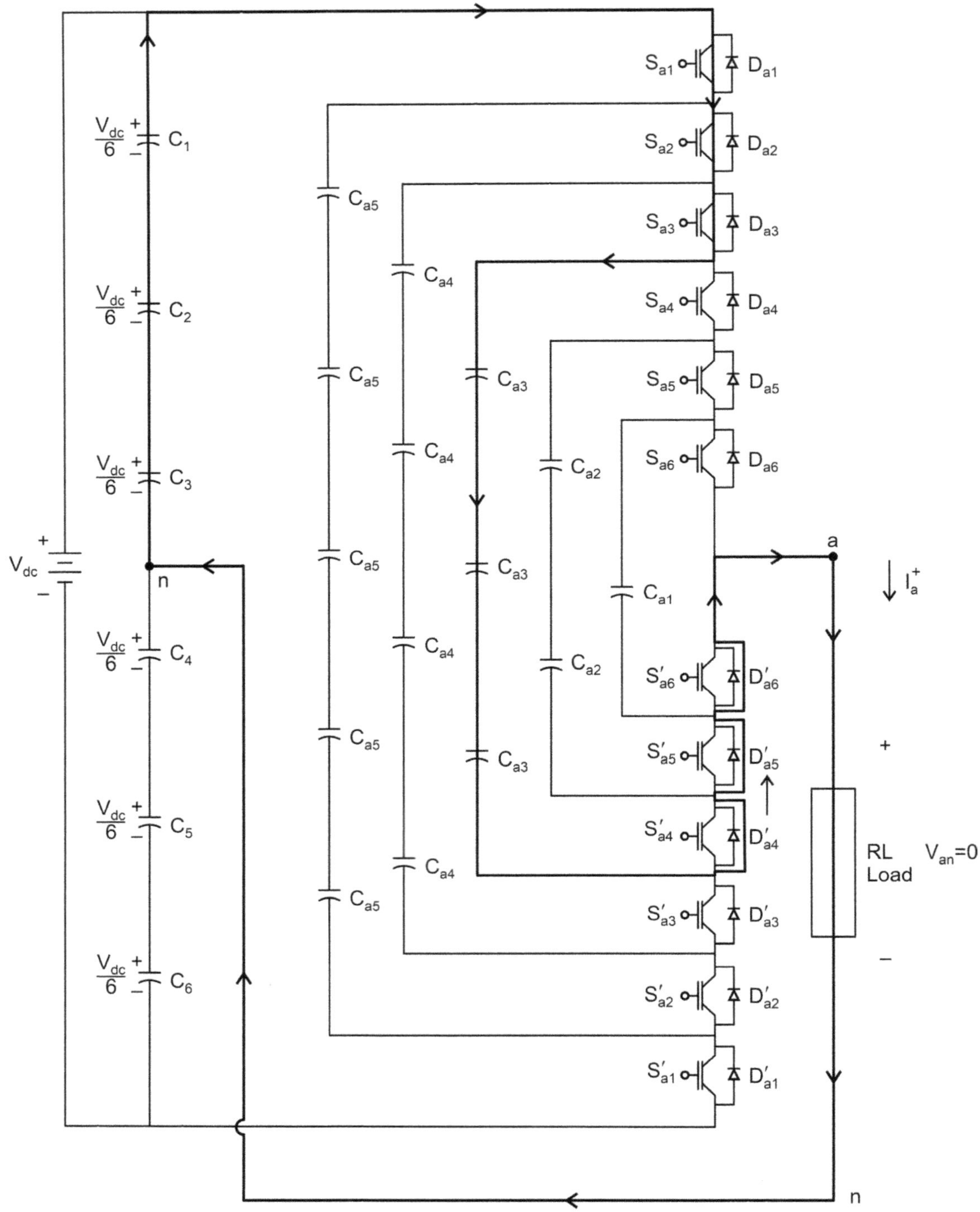
S_{a1} D_{a1}
S_{a2} D_{a2}
S_{a3} D_{a3}
S_{a4} D_{a4}
S_{a5} D_{a5}
S_{a6} D_{a6}
S'_{a6} D'_{a6}
S'_{a5} D'_{a5}
S'_{a4} D'_{a4}
S'_{a3} D'_{a3}
S'_{a2} D'_{a2}
S'_{a1} D'_{a1}
$\frac{V_{dc}}{6}$ C_1
$\frac{V_{dc}}{6}$ C_2
$\frac{V_{dc}}{6}$ C_3
$\frac{V_{dc}}{6}$ C_4
$\frac{V_{dc}}{6}$ C_5
$\frac{V_{dc}}{6}$ C_6
V_{dc}
n
C_{a5}
C_{a4}
C_{a3}
C_{a2}
C_{a1}
a
I_a^+
RL Load
$V_{an}=0$
n

(e)

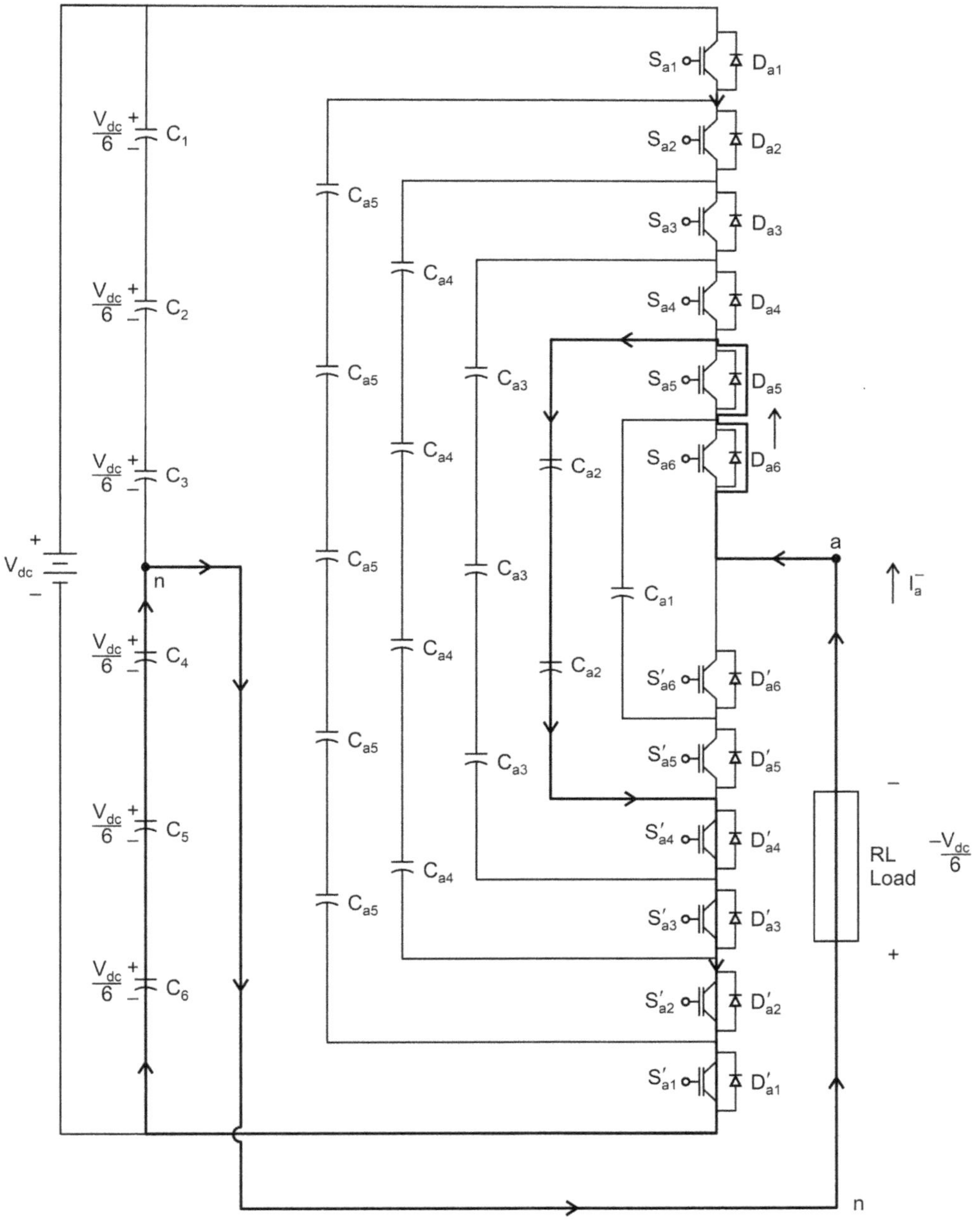

(f)

Fig. 4.14 Seven-level flying-capacitor inverter with single-leg arrangement (e) Mode4: V_{an}= 0, (f) Mode5: V_{an}= $-V_{dc}/6$.

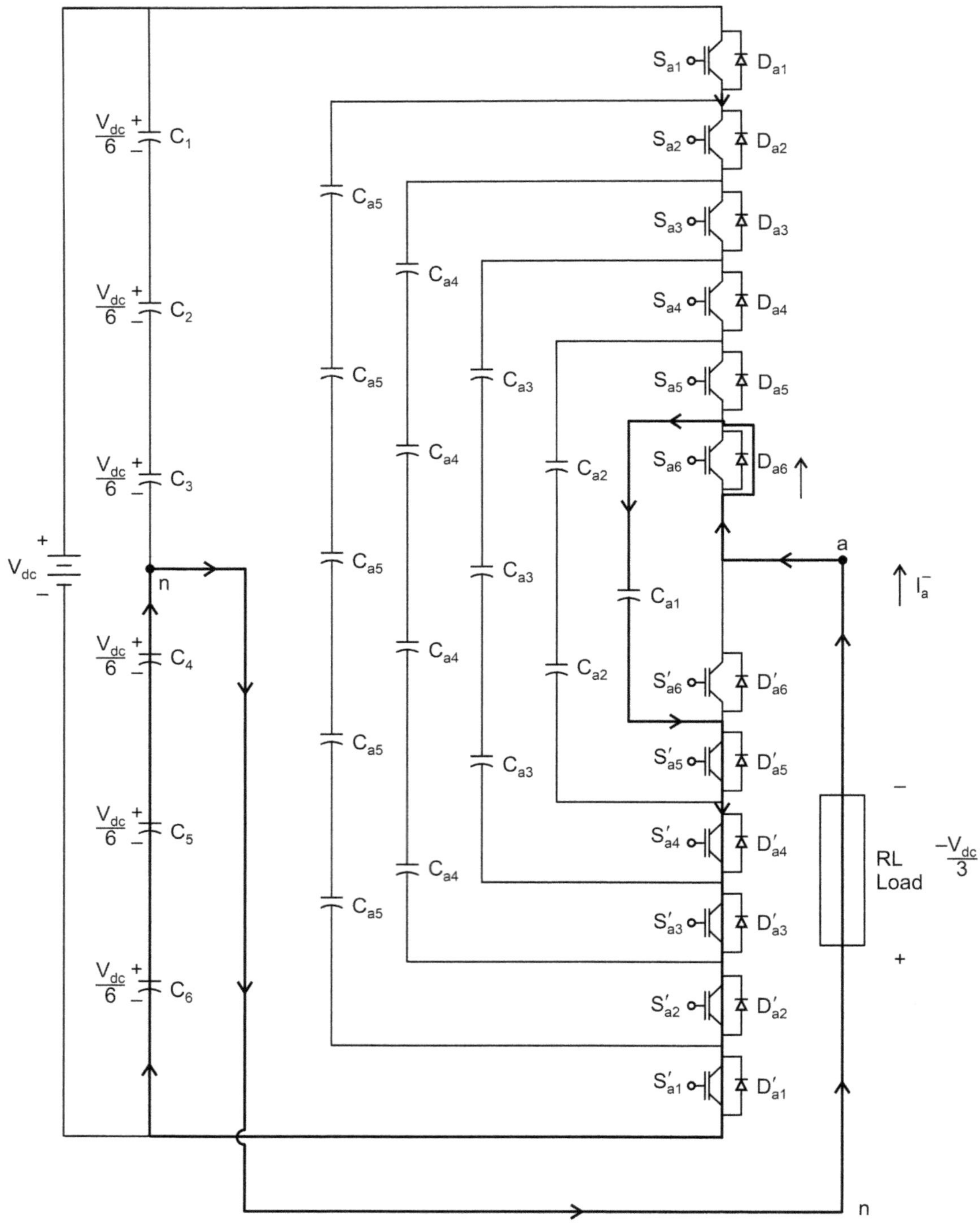

S_{a1} D_{a1}
S_{a2} D_{a2}
S_{a3} D_{a3}
S_{a4} D_{a4}
S_{a5} D_{a5}
S_{a6} D_{a6}
S'_{a6} D'_{a6}
S'_{a5} D'_{a5}
S'_{a4} D'_{a4}
S'_{a3} D'_{a3}
S'_{a2} D'_{a2}
S'_{a1} D'_{a1}
$\frac{V_{dc}}{6}$ C_1
$\frac{V_{dc}}{6}$ C_2
$\frac{V_{dc}}{6}$ C_3
$\frac{V_{dc}}{6}$ C_4
$\frac{V_{dc}}{6}$ C_5
$\frac{V_{dc}}{6}$ C_6
V_{dc}
n
C_{a5}
C_{a4}
C_{a3}
C_{a2}
C_{a1}
a
I_a^-
RL Load
$-\frac{V_{dc}}{3}$
n

(g)

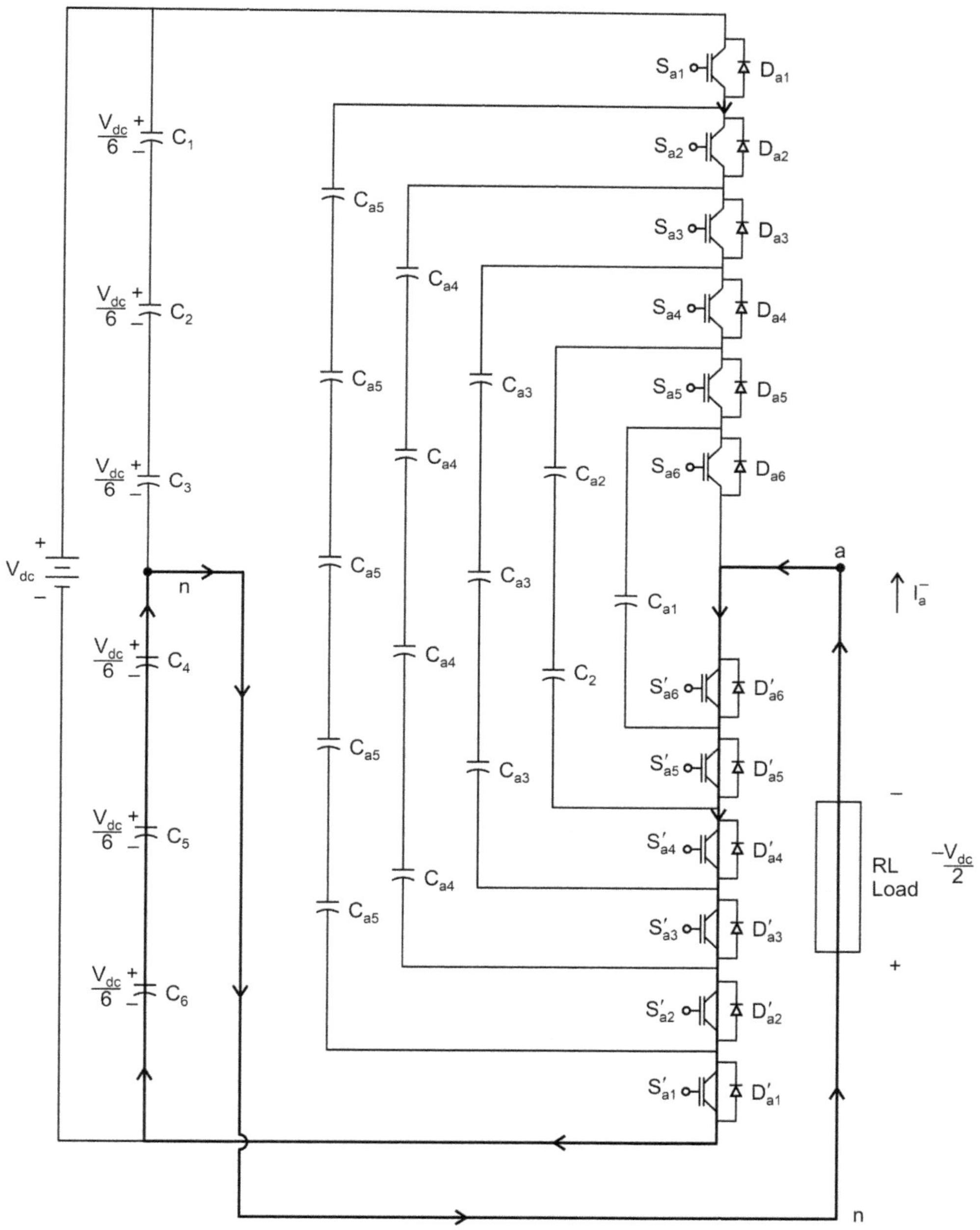

(h)

Fig. 4.14 Seven-level flying-capacitor inverter with single-leg arrangement (g) Mode6: Van= -Vdc/3, (h) Mode7: V_{an}= -Vdc/2.

Table 4.4 Switching States for Seven-level flying-capacitor inverter with single-leg.

Modes	V_{an}	S_{a1}	S_{a2}	S_{a3}	S_{a4}	S_{a5}	S_{a6}	$S_{a6'}$	$S_{a5'}$	$S_{a4'}$	$S_{a3'}$	$S_{a2'}$	$S_{a1'}$	Diodes
M1	$+V_{dc}/2$	1	1	1	1	1	1	0	0	0	0	0	0	----
M2	$+V_{dc}/3$	1	1	1	1	1	0	0	0	0	0	0	0	D_{a6}'
M3	$+V_{dc}/6$	1	1	1	1	0	0	0	0	0	0	0	0	D_{a5}' D_{a6}'
M4	0	1	1	1	0	0	0	0	0	0	0	0	0	D_{a4}', D_{a5}', D_{a6}'
M5	$-V_{dc}/6$	0	0	0	0	0	0	0	0	1	1	1	1	D_{a6}, D_{a5}
M6	$-V_{dc}/3$	0	0	0	0	0	0	0	1	1	1	1	1	D_{a6}
M7	$-V_{dc}/2$	0	0	0	0	0	0	1	1	1	1	1	1	----

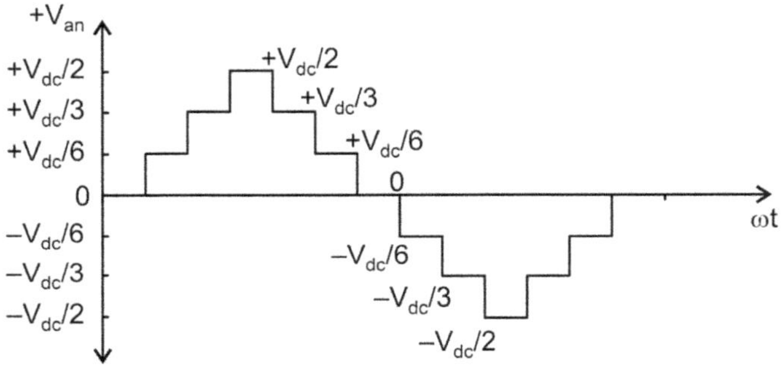

Fig. 4.15 The load voltage Van for seven-level flying-capacitor inverter with single-leg.

B. Single phase Seven-level flying-capacitor full-bridge inverter:

Mode 1: For $V_{ab} = +V_{dc}$, the switches S_{a1}, S_{a2}, S_{a3}, S_{a4}, S_{a5}, S_{a6} and S_{b6}', S_{b5}', S_{b4}', S_{b3}', S_{b2}', S_{b1}' are turned-on, then the conduction takes place from dc source node V_6 through switches S_{a1}, S_{a2}, S_{a3}, S_{a4}, S_{a5}, S_{a6}, load terminal '*a*' to '*b*' and switches S_{b6}', S_{b5}', S_{b4}', S_{b3}', S_{b2}', S_{b1}' to dc source node V_0, thus $+V_{dc}$ appears across the load as shown in Fig.4.16 (b). The load current is positive since it is flowing from load terminal '*a*' to '*b*'. The load current path is indicated with thick line and arrows.

Mode 2: To obtain $V_{ab} = +5V_{dc}/6$, the switches S_{a1}, S_{a2}, S_{a3}, S_{a4}, S_{a5}, and S_{b6}', S_{b5}', S_{b4}', S_{b3}', S_{b2}', S_{b1}' are turned-on, then the conduction takes place from dc source node V_6 through switches S_{a1}, S_{a2}, S_{a3}, S_{a4}, S_{a5}, flying capacitor C_{a1} (pre-charged to $+V_{dc}/6$), diode D_{a6}', load terminal '*a*' to '*b*' and switches S_{b6}', S_{b5}', S_{b4}', S_{b3}', S_{b2}', S_{b1}' to dc source node V_0, thus $+5V_{dc}/6$ appears across the load as given in Fig. 4.16 (c). The load current is positive since it is flowing from load terminal '*a*' to '*b*'.

Mode 3: To get $V_{ab} = +2V_{dc}/3$, the switches S_{a1}, S_{a2}, S_{a3}, S_{a4}, and S_{b6}', S_{b5}', S_{b4}', S_{b3}', S_{b2}', S_{b1}' are turned-on, then the conduction takes place from dc source node V_6 through switches S_{a1}, S_{a2}, S_{a3}, S_{a4}, flying capacitors C_{a2}, C_{a2} (each pre-charged to $+V_{dc}/6$), diodes D_{a5}',D_{a6}', load terminal '*a*' to '*b*' and switches S_{b6}', S_{b5}', S_{b4}', S_{b3}', S_{b2}', S_{b1}' to dc source node V_0, thus $+2V_{dc}/3$ appears across the load as given in Fig. 4.16 (d). The load current is positive since it is flowing from load terminal '*a*' to '*b*'.

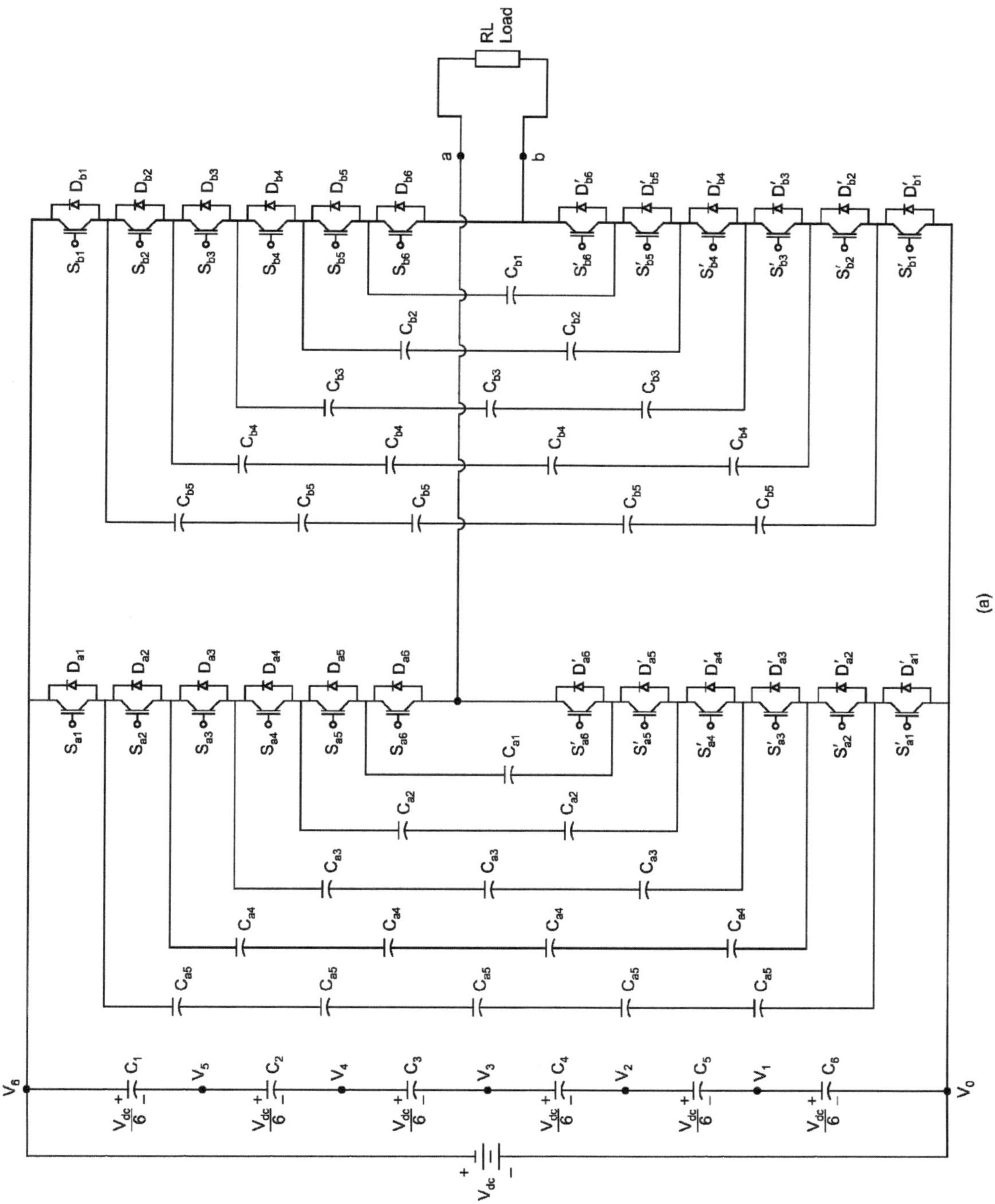

Fig. 4.16 (a) Main circuit.

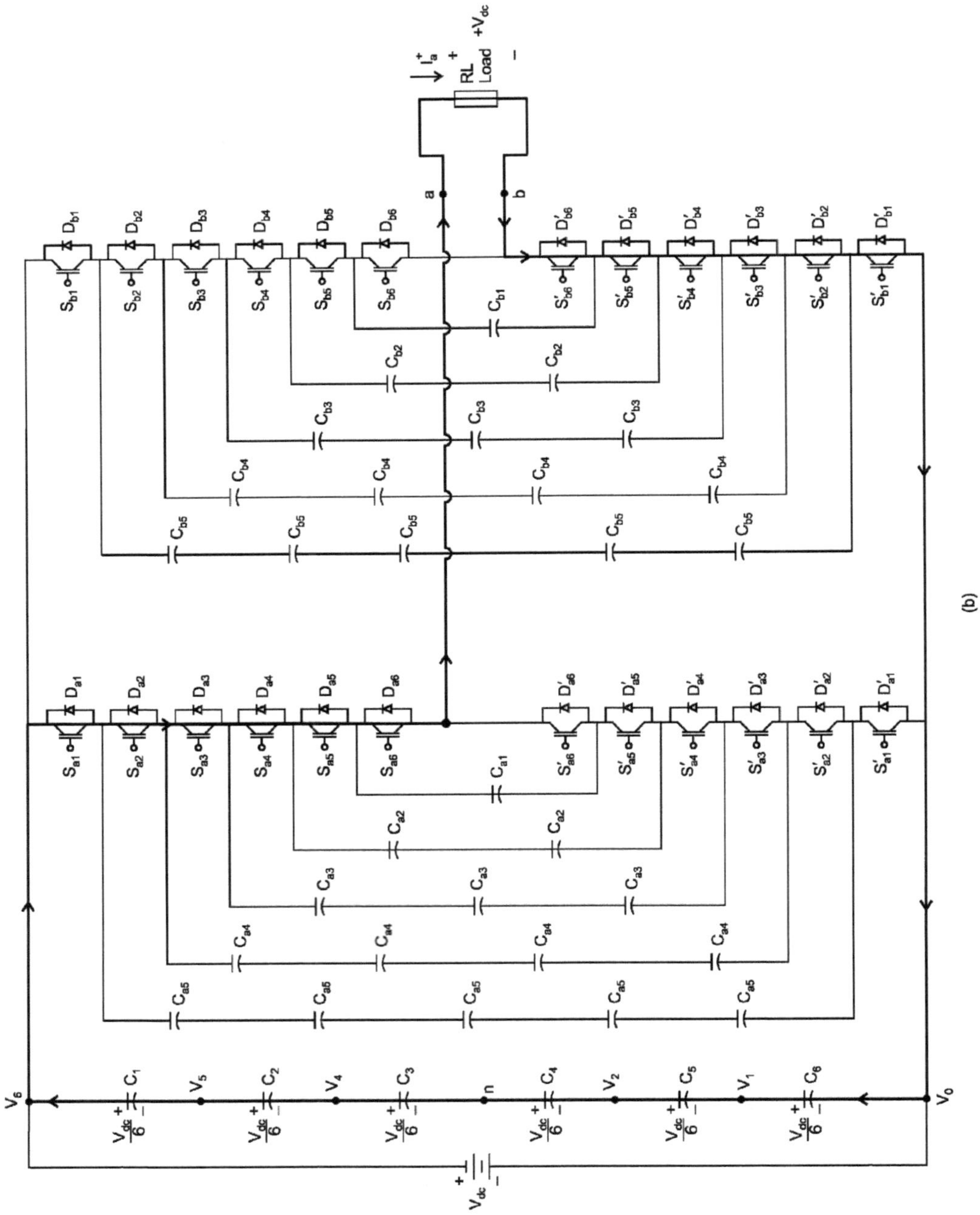

Fig. 4.16 (a) Main circuit, and (b) Mode1: V_{ab}= +V_{dc} with positive I_a.

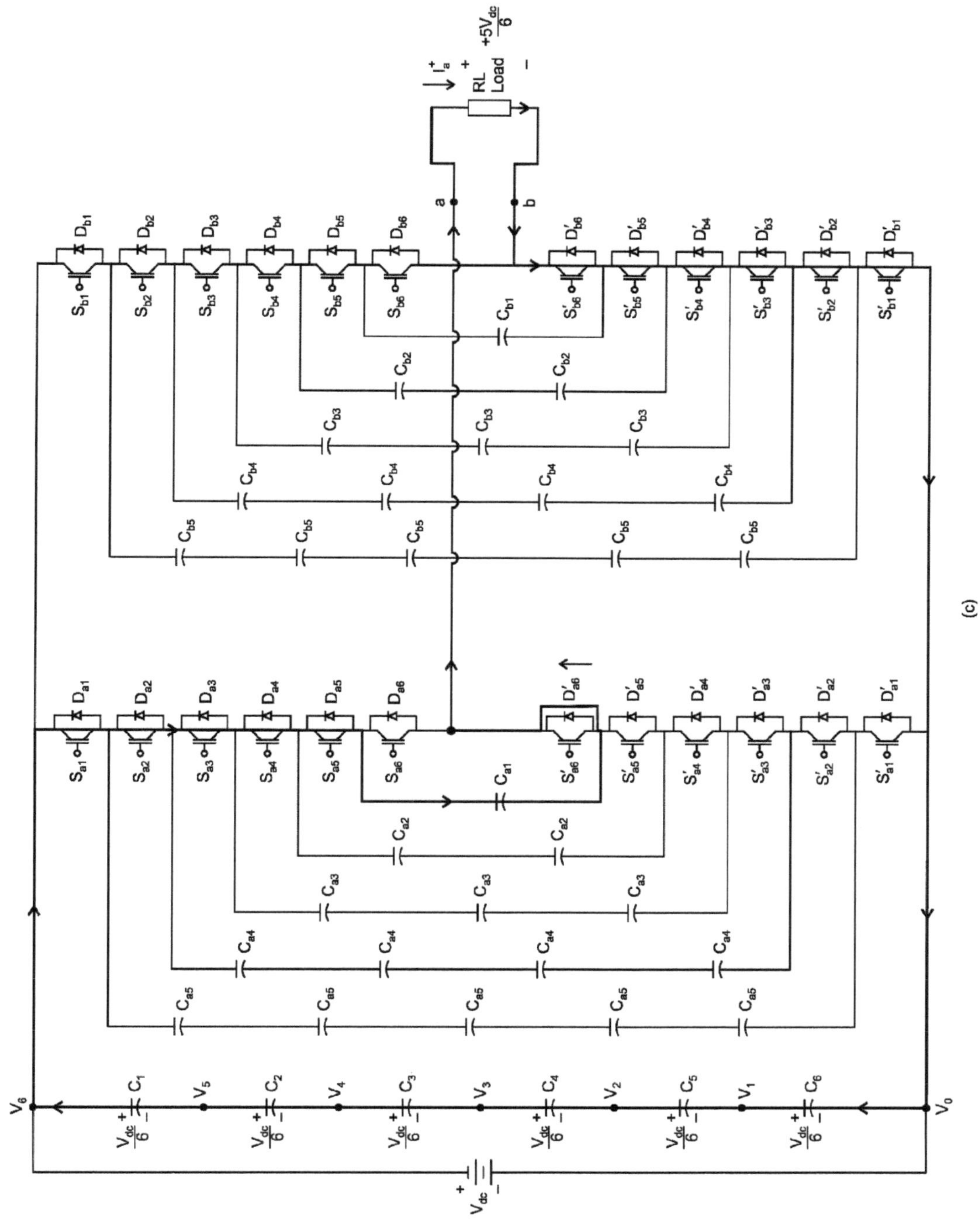

Fig. 4.16 (c) Mode 2: $V_{ab} = +5V_{dc}/6$ with positive I_a.

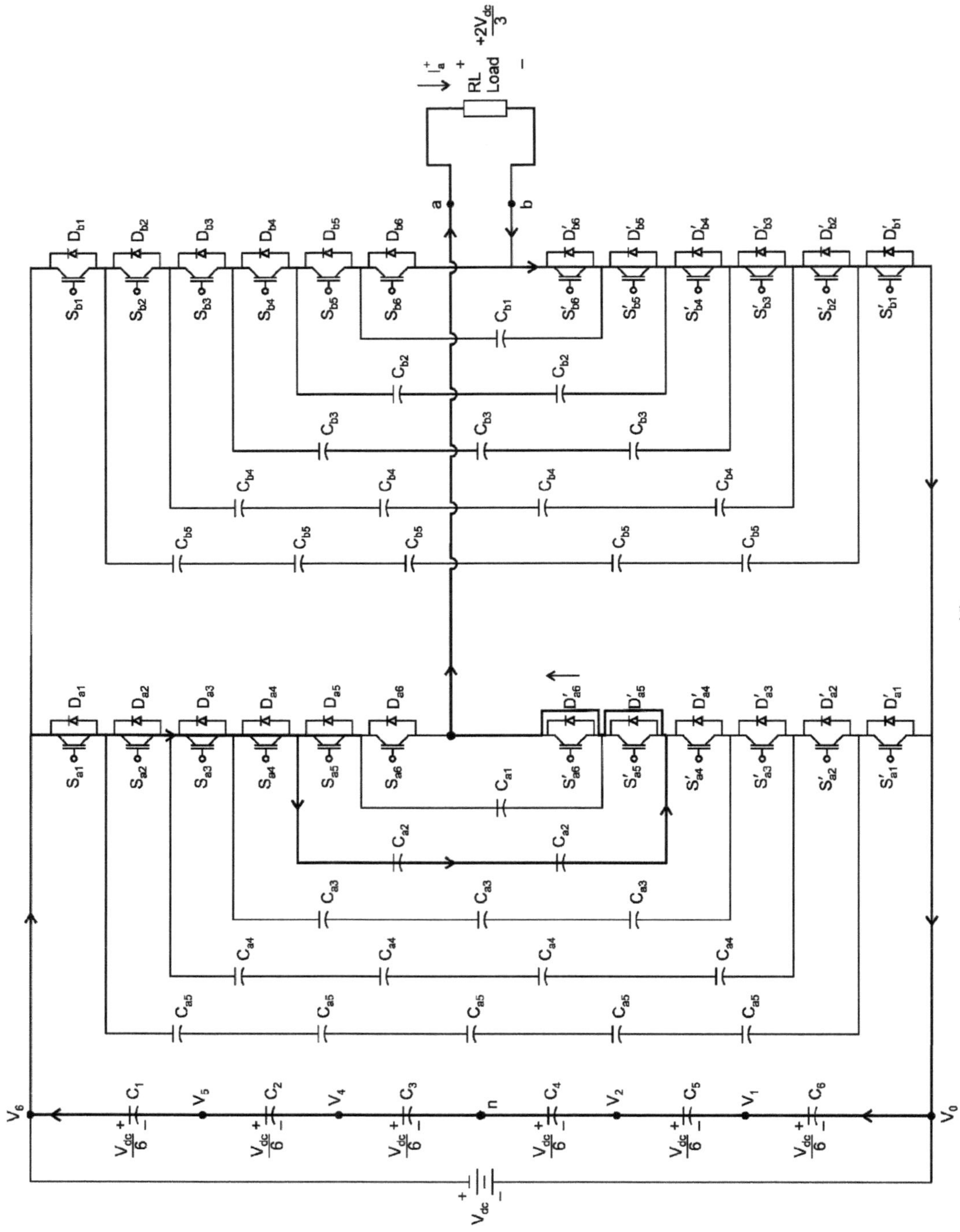

Fig. 4.16 (d) Mode 3: $V_{ab} = +2V_{dc}/3$ with positive I_a.

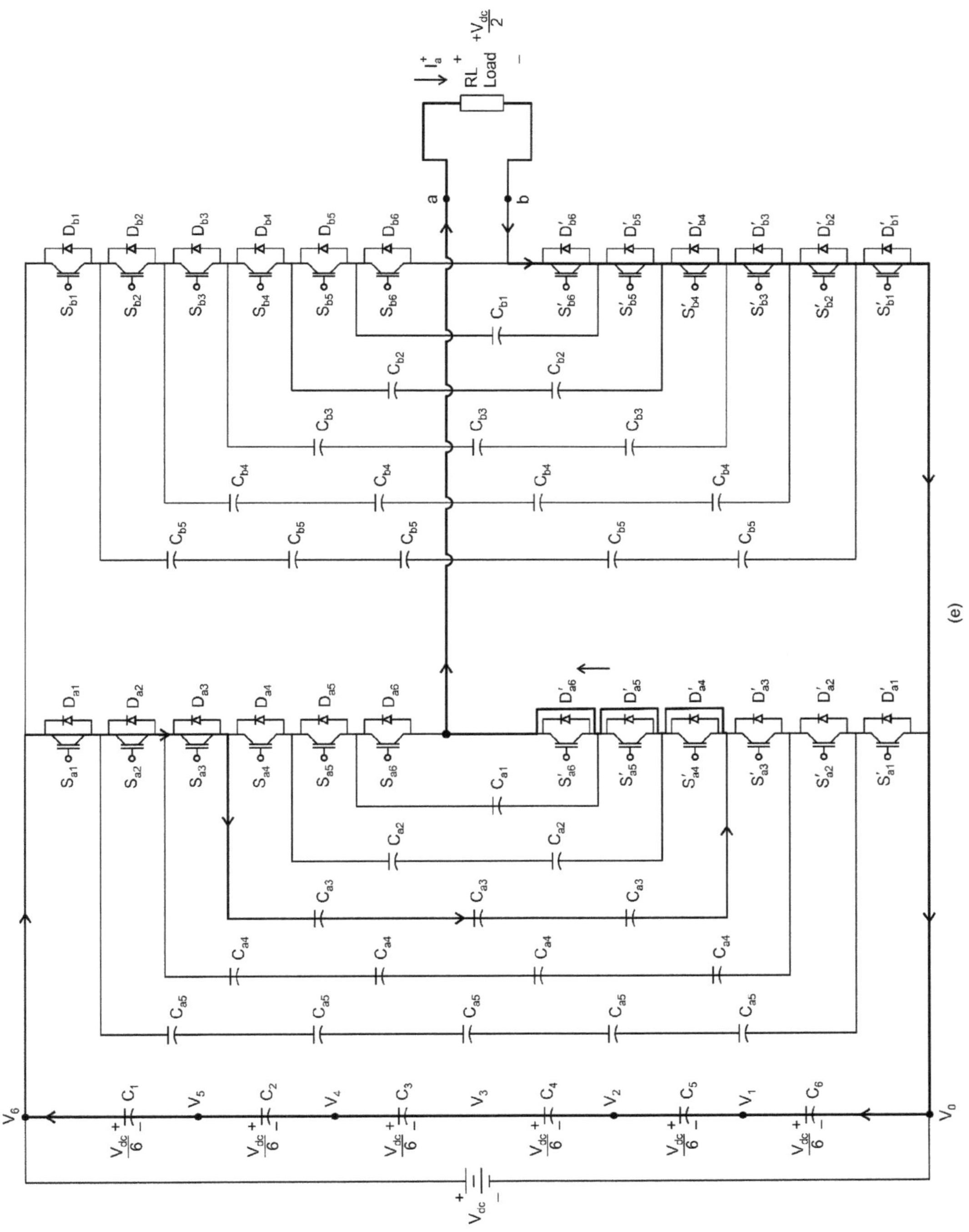

Fig. 4.16 (e) Mode 4: $V_{ab} = +V_{dc}/2$ with positive I_a.

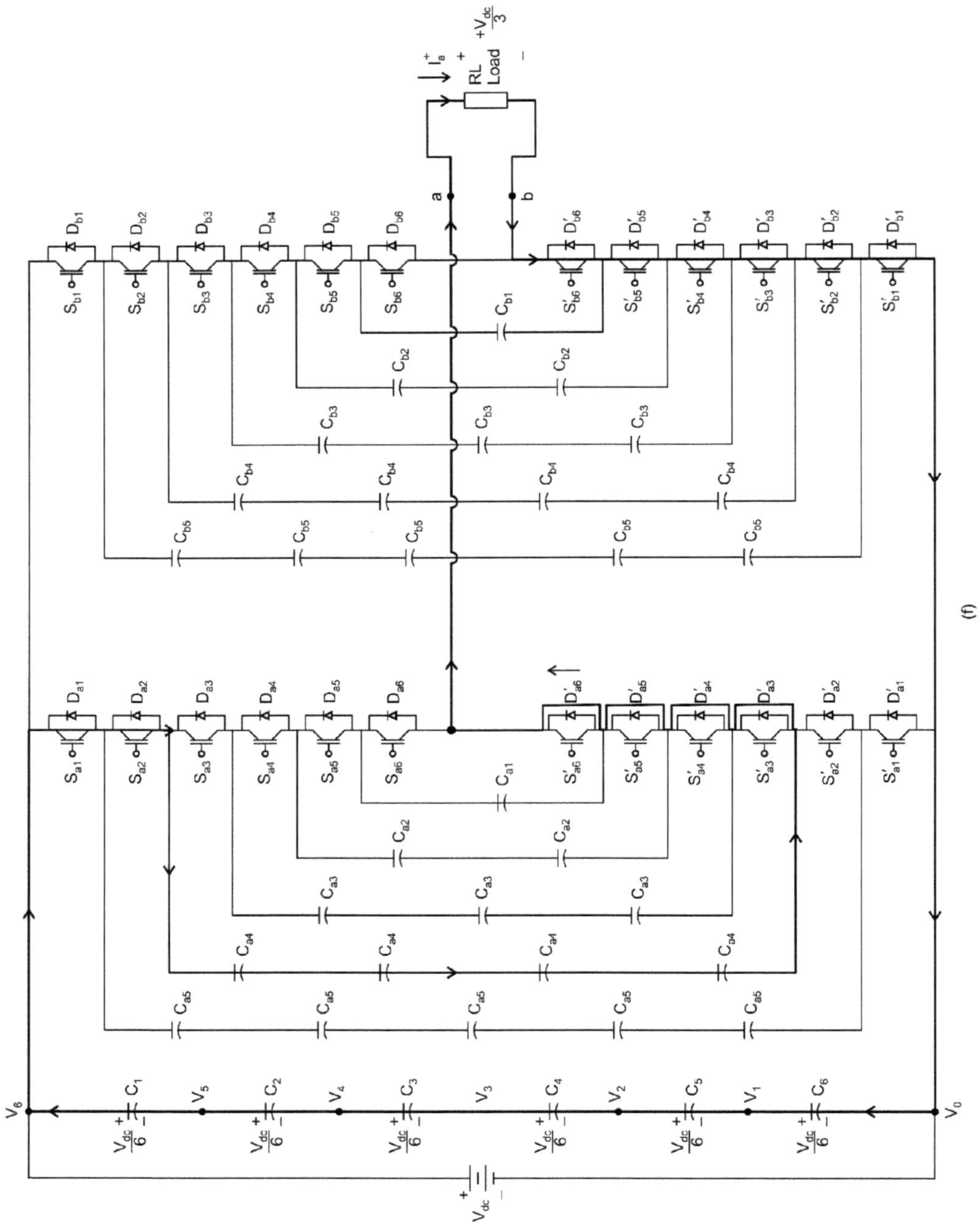

Fig. 4.16 (f) Mode5: $V_{ab} = +V_{dc}/3$ with positive I_a.

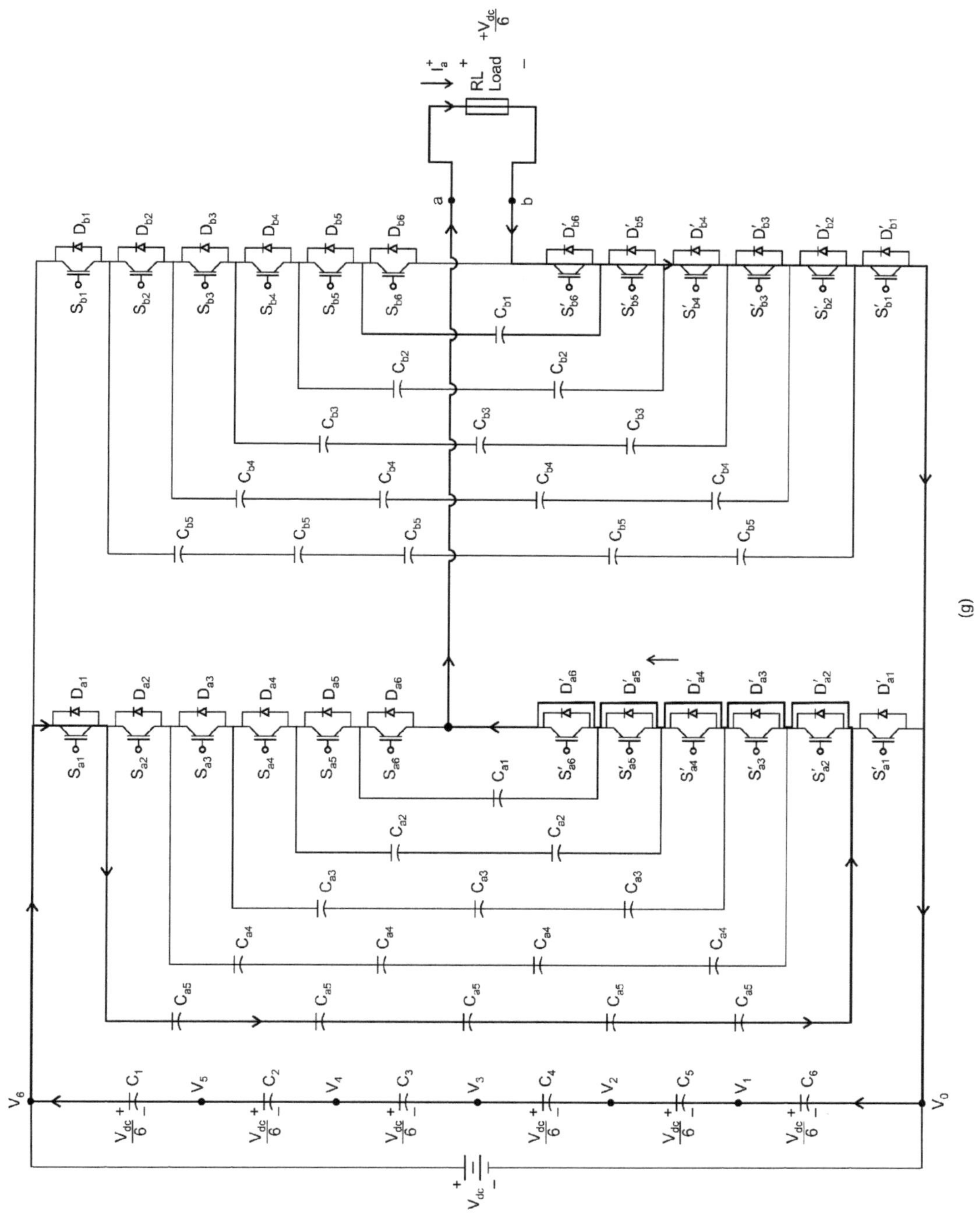

Fig. 4.16 (g) Mode 6: $V_{ab} = +V_{dc}/6$ with positive I_a.

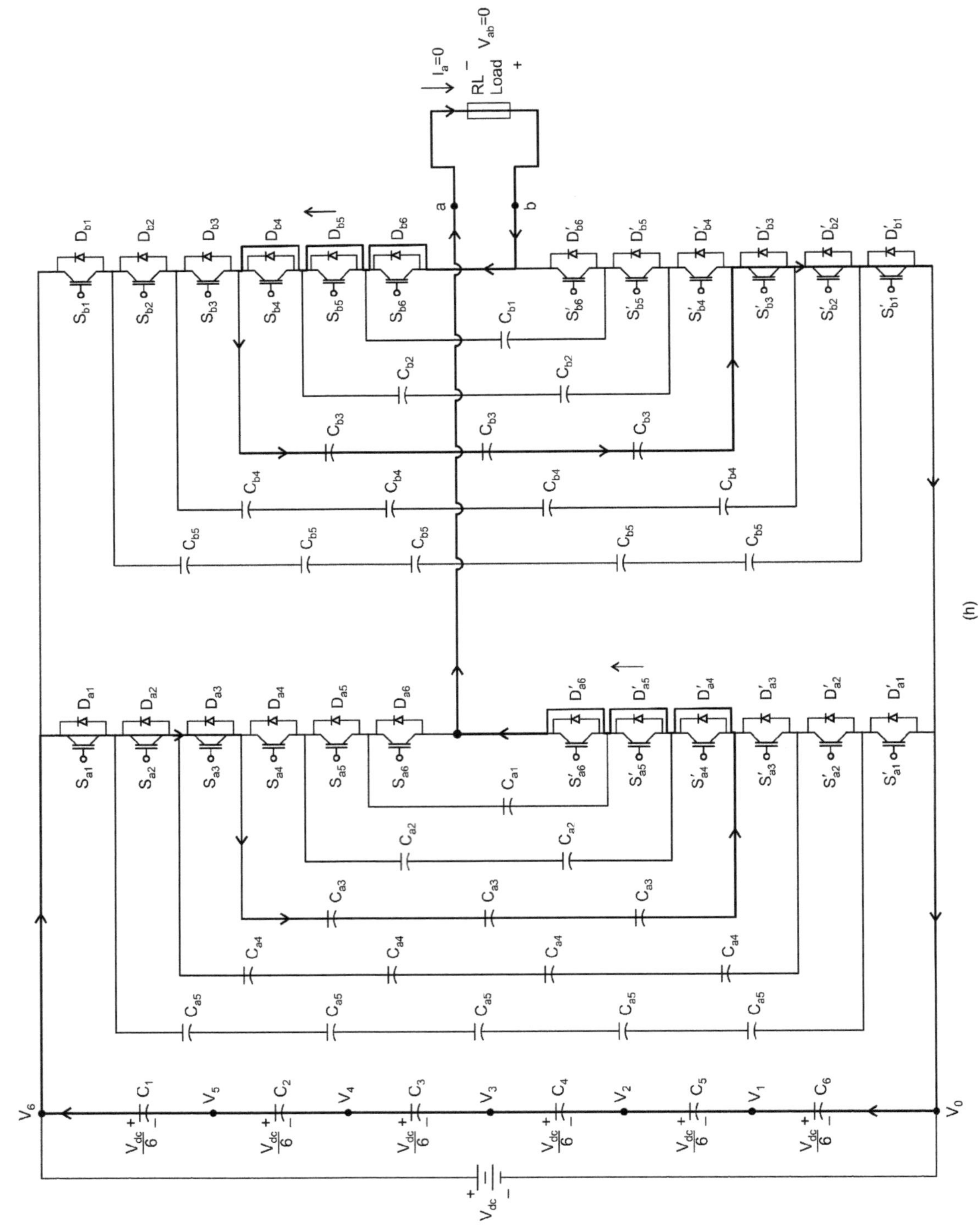

Fig. 4.16 (h) Mode 7: $V_{ab} = 0$, with $I_a = 0$.

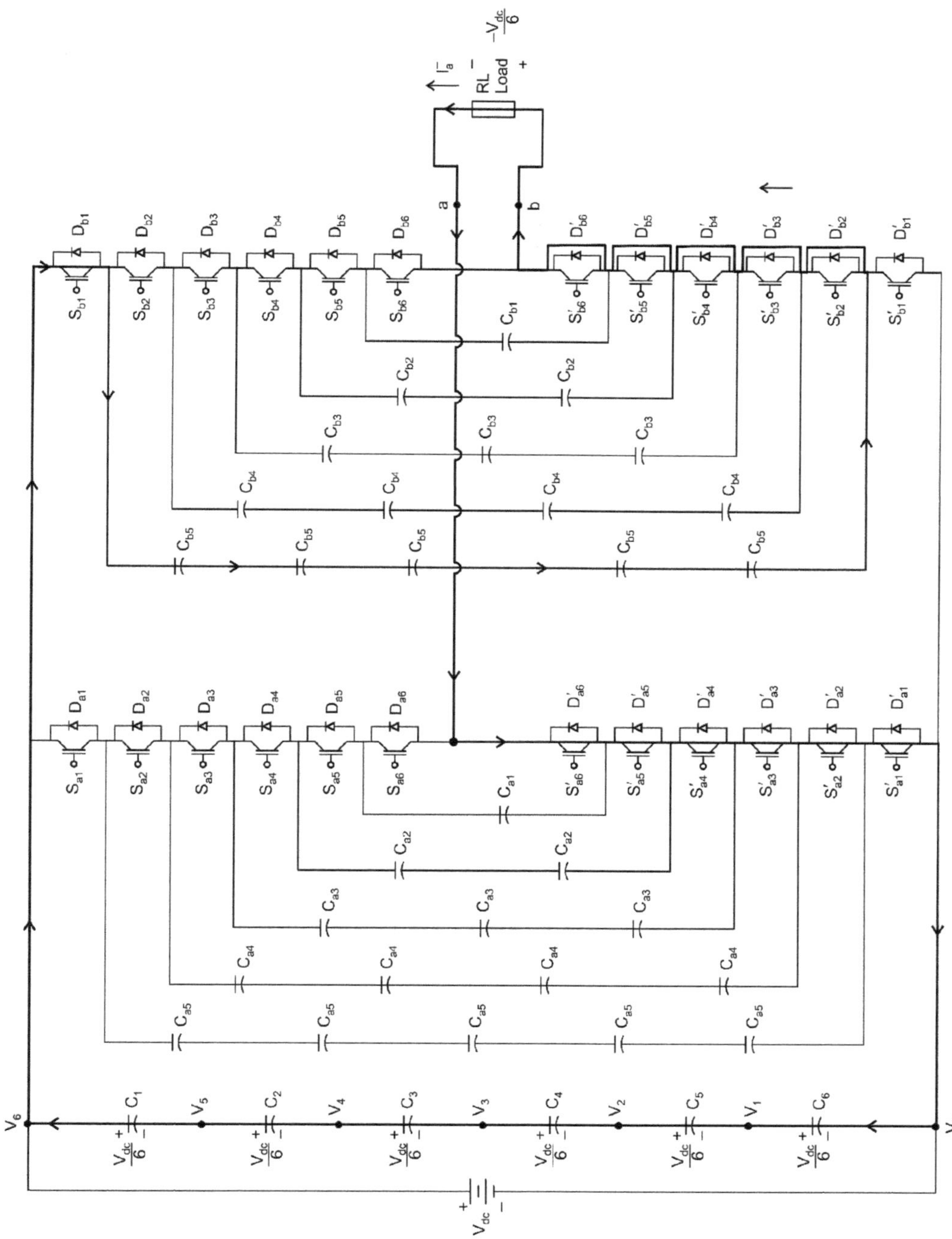

Fig. 4.16 (i) Mode 8: $V_{ab} = -V_{dc}/6$ with negative I_a.

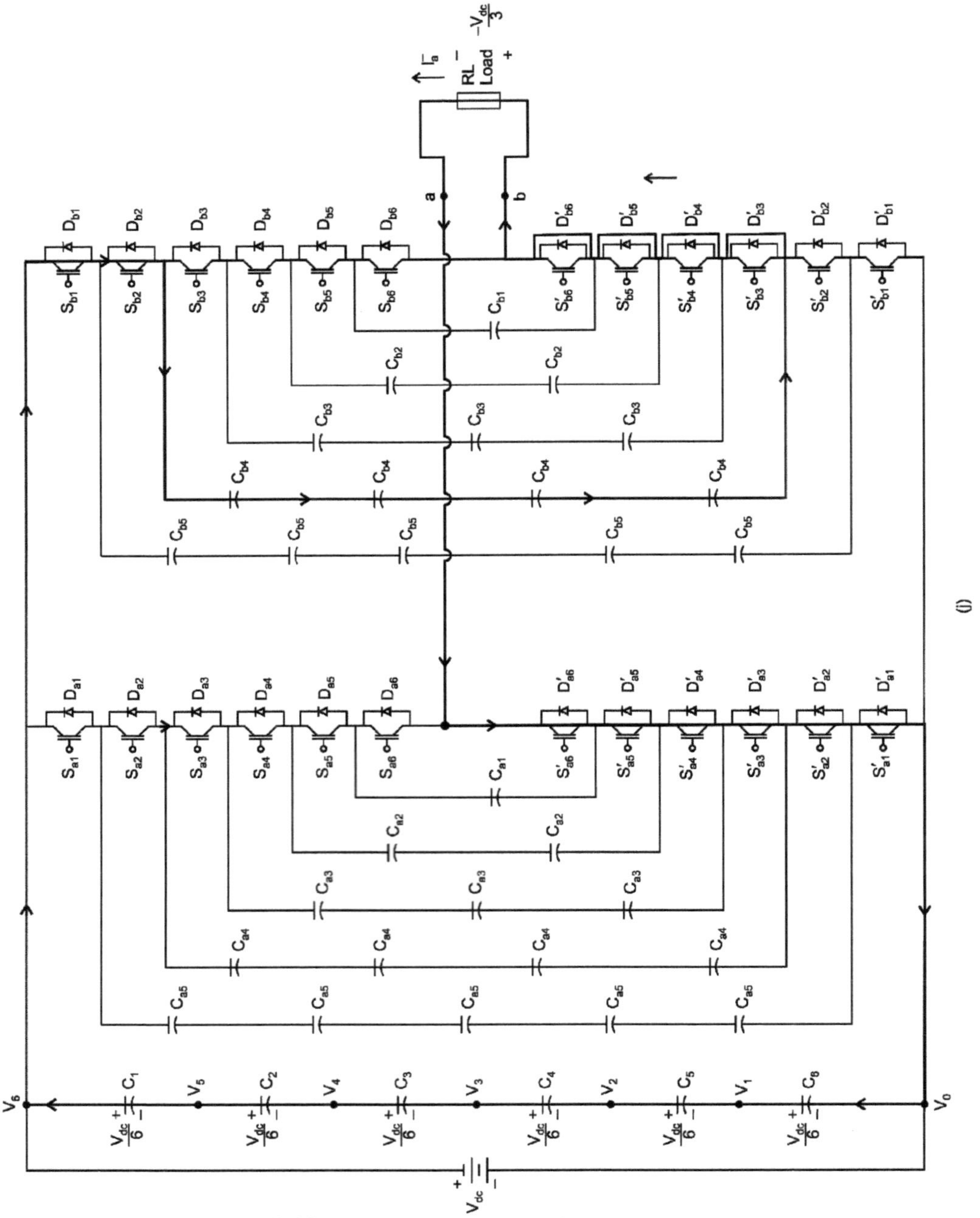

Fig. 4.16 (j) Mode9: $V_{ab} = -V_{dc}/3$, with negative I_a.

Mode 4: To get V_{ab} = +V_{dc}/2, the switches S_{a1}, S_{a2}, S_{a3}, and S_{b6}', S_{b5}', S_{b4}', S_{b3}', S_{b2}', S_{b1}' are turned-on, then the conduction takes place from dc source node V_6 through switches S_{a1}, S_{a2}, S_{a3}, flying capacitors C_{a3}, C_{a3}, C_{a3} (each pre-charged to +V_{dc}/6), diodes D_{a4}',D_{a5}',D_{a6}', load terminal '*a*' to '*b*' and switches S_{b6}', S_{b5}', S_{b4}', S_{b3}', S_{b2}', S_{b1}' to dc source node V_0, thus +V_{dc}/2 appears across the load as given in Fig. 4.16 (e). The load current is positive since it is flowing from load terminal '*a*' to '*b*'.

Mode 5: For V_{ab} = +V_{dc}/3, the switches S_{a1}, S_{a2}, and S_{b6}', S_{b5}', S_{b4}', S_{b3}', S_{b2}', S_{b1}' are turned-on, then the conduction takes place from dc source node V_6 through switches S_{a1}, S_{a2}, flying capacitors C_{a4}, C_{a4}, C_{a4}, C_{a4} (each pre-charged to +V_{dc}/6), diodes D_{a3}',D_{a4}',D_{a5}',D_{a6}', load terminal '*a*' to '*b*' and switches S_{b6}', S_{b5}', S_{b4}', S_{b3}', S_{b2}', S_{b1}' to dc source node V_0, thus +V_{dc}/3 appears across the load as given in Fig. 4.16 (f). The load current is positive since it is flowing from load terminal '*a*' to '*b*'.

Mode 6: To obtain V_{ab}= +V_{dc}/6, the switches S_{a1}, and S_{b6}', S_{b5}', S_{b4}', S_{b3}', S_{b2}', S_{b1}' are turned-on, then the conduction takes place from dc source node V_6 through switch S_{a1}, flying capacitors C_{a5}, C_{a5}, C_{a5}, C_{a5}, C_{a5} (each pre-charged to +V_{dc}/6), diodes D_{a2}',D_{a3}',D_{a4}',D_{a5}',D_{a6}', load terminal '*a*' to '*b*' and switches S_{b6}', S_{b5}', S_{b4}', S_{b3}', S_{b2}', S_{b1}' to dc source node V_0, thus +V_{dc}/6 appears across the load as given in Fig. 4.16 (g). The load current is positive since it is flowing from load terminal '*a*' to '*b*'.

Mode 7: For V_{ab}= 0, the switches S_{a1}, S_{a2}, S_{a3}, and S_{b3}', S_{b2}', S_{b1}' are turned-on, then the conduction takes place from dc source node V_6 through switches S_{a1}, S_{a2}, S_{a3}, flying capacitors C_{a3}, C_{a3}, C_{a3} (each pre-charged to +V_{dc}/6), diodes D_{a4}',D_{a5}',D_{a6}', load terminal '*a*' to '*b*' and diodes D_{b6}, D_{b5}, D_{b4}, flying capacitors C_{b3}, C_{b3}, C_{b3} (each pre-charged to +Vdc/6), switches S_{b3}', S_{b2}', S_{b1}' to dc source node V_0, thus zero voltage appears across the load as given in Fig. 4.16 (h). The load current is also zero. The source dc voltages and flying capacitor voltages get cancelled leads to zero load voltage.

Mode 8: To get V_{ab}= -V_{dc}/6, the switches S_{b1}, and S_{a6}', S_{a5}', S_{a4}', S_{a3}', S_{a2}', S_{a1}' are turned-on, then the conduction takes place from dc source node V_6 through switch S_{b1}, flying capacitors C_{b5}, C_{b5}, C_{b5}, C_{b5}, C_{b5} (each pre-charged to +V_{dc}/6), diodes D_{b2}',D_{b3}',D_{b4}',D_{b5}',D_{b6}', load terminal '*b*' to '*a*' and switches S_{a6}', S_{a5}', S_{a4}', S_{a3}', S_{a2}', S_{a1}' to dc source node V_0, thus -V_{dc}/6 appears across the load as given in Fig. 4.16 (i). The load current is negative since it is flowing from load terminal '*b*' to '*a*'.

Mode 9: For V_{ab}= -V_{dc}/3, the switches S_{b1}, S_{b2}, and S_{a6}', S_{a5}', S_{a4}', S_{a3}', S_{a2}', S_{a1}' are turned-on, then the conduction takes place from dc source node V_6 through switches S_{b1}, S_{b2}, flying capacitors C_{b4}, C_{b4}, C_{b4}, C_{b4} (each pre-charged to +V_{dc}/6), diodes D_{b3}',D_{b4}',D_{b5}',D_{b6}', load terminal '*b*' to '*a*' and switches S_{a6}', S_{a5}', S_{a4}', S_{a3}', S_{a2}', S_{a1}' to dc source node V_0, thus -V_{dc}/3 appears across the load as given in Fig. 4.16 (j). The load current is negative since it is flowing from load terminal '*b*' to '*a*'.

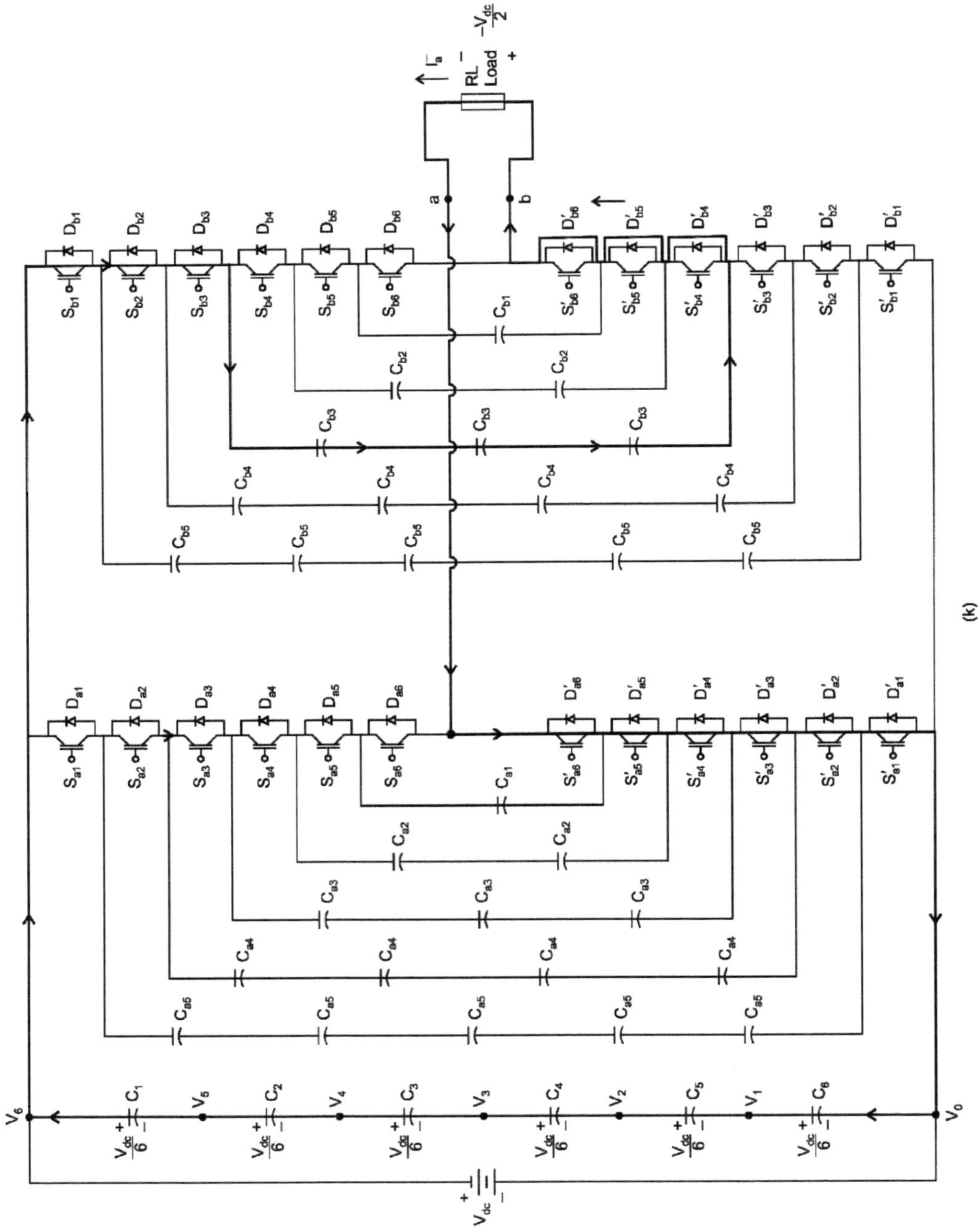

Fig. 4.16 (k) Mode 10: $V_{ab} = -V_{dc}/2$ with negative I_a.

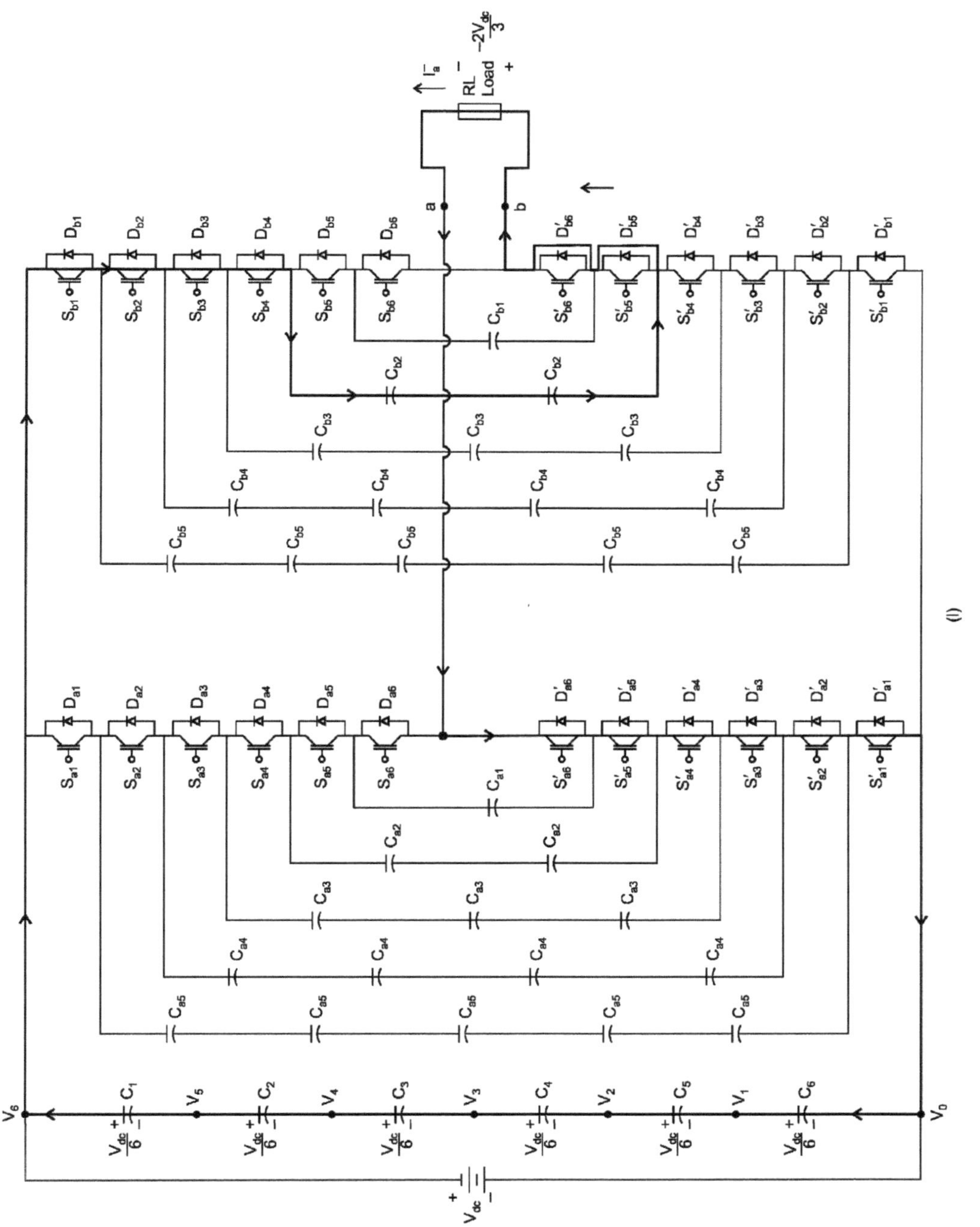

Fig. 4.16 (l) Mode 11: $V_{ab} = -2V_{dc}/3$ with negative I_a.

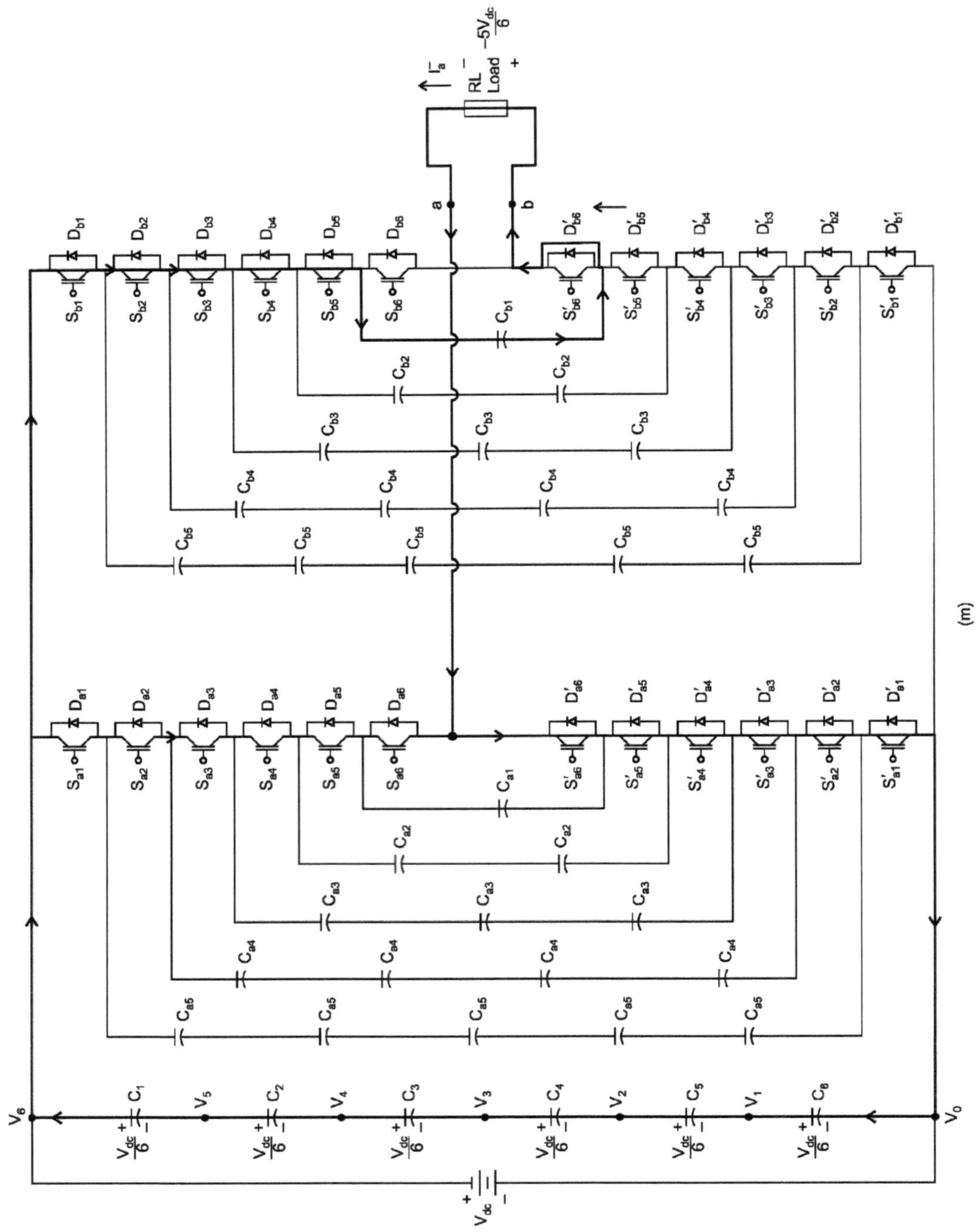

Fig. 4.16 (m) Mode 12: $V_{ab} = -5V_{dc}/6$ with negative I_a.

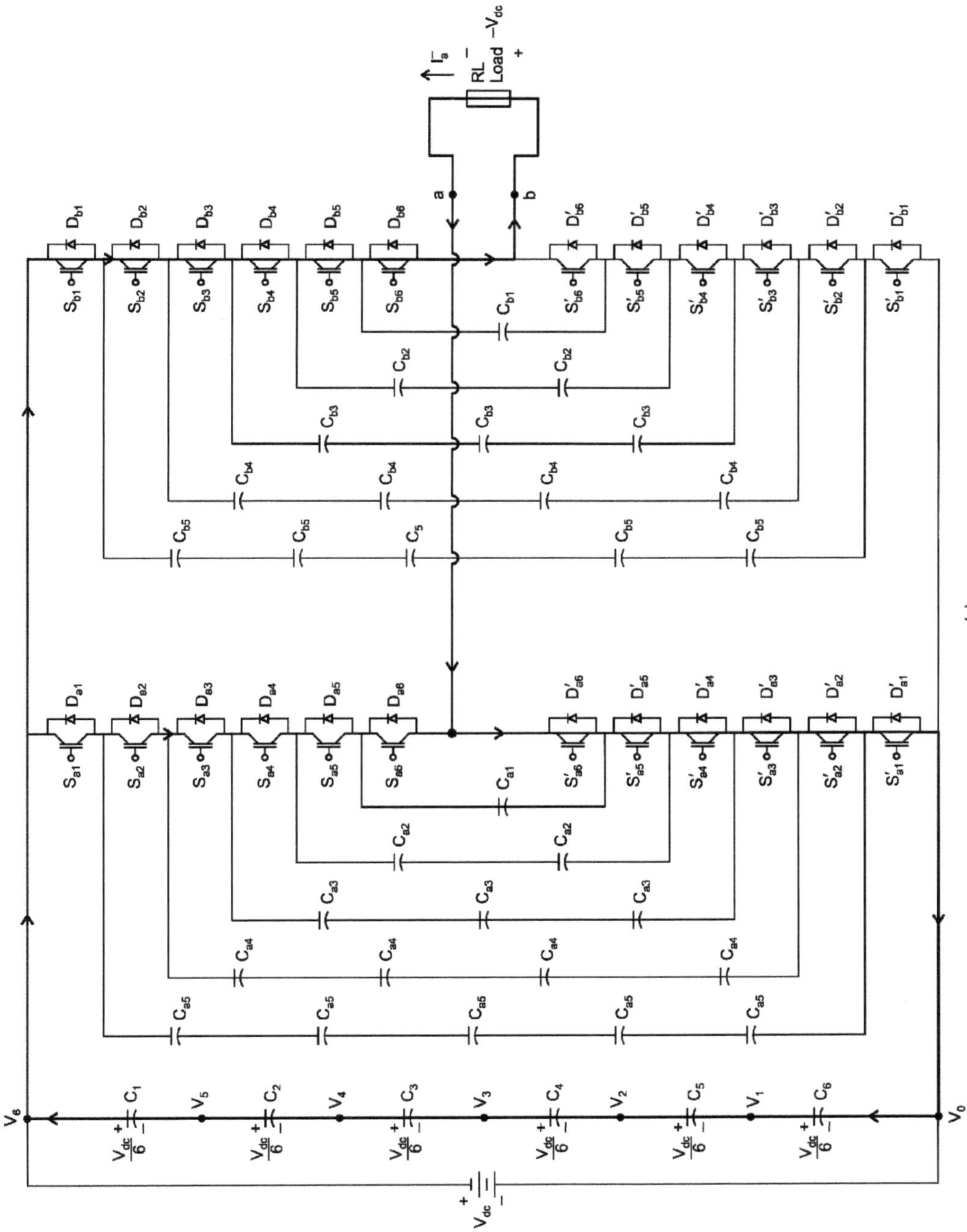

Fig. 4.16 (n) Mode13: V_{ab}= -V_{dc} with negative I_a.

Mode 10: To obtain V_{ab}= $-V_{dc}/2$, the switches S_{b1}, S_{b2}, S_{b3}, and S_{a6}', S_{a5}', S_{a4}', S_{a3}', S_{a2}', S_{a1}' are turned-on, then the conduction takes place from dc source node V_6 through switches S_{b1}, S_{b2}, S_{b3}, flying capacitors C_{b3}, C_{b3}, C_{b3} (each pre-charged to $+V_{dc}/6$), diodes D_{b4}',D_{b5}',D_{b6}', load terminal '*b*' to '*a*' and switches S_{a6}', S_{a5}', S_{a4}', S_{a3}', S_{a2}', S_{a1}' to dc source node V_0, thus $-V_{dc}/2$ appears across the load as given in Fig. 4.16 (k). The load current is negative since it is flowing from load terminal '*b*' to '*a*'.

Mode 11: For V_{ab}= $-2V_{dc}/3$, the switches S_{b1}, S_{b2}, S_{b3}, S_{b4}, and S_{a6}', S_{a5}', S_{a4}', S_{a3}', S_{a2}', S_{a1}' are turned-on, then the conduction takes place from dc source node V_6 through switches S_{b1}, S_{b2}, S_{b3}, S_{b4}, flying capacitors C_{b2}, C_{b2} (each pre-charged to $+V_{dc}/6$), diodes D_{b5}',D_{b6}', load terminal '*b*' to '*a*' and switches S_{a6}', S_{a5}', S_{a4}', S_{a3}', S_{a2}', S_{a1}' to dc source node V_0, thus $-2V_{dc}/3$ appears across the load as given in Fig. 4.16 (l). The load current is negative since it is flowing from load terminal '*b*' to '*a*'.

Mode 12: To get V_{ab}= $-5V_{dc}/6$, the switches S_{b1}, S_{b2}, S_{b3}, S_{b4}, S_{b5}, and S_{a6}', S_{a5}', S_{a4}', S_{a3}', S_{a2}', S_{a1}' are turned-on, then the conduction takes place from dc source node V_6 through switches S_{b1}, S_{b2}, S_{b3}, S_{b4}, S_{b5}, flying capacitor C_{b1} (pre-charged to $+V_{dc}/6$), diode D_{b6}', load terminal '*b*' to '*a*' and switches S_{a6}', S_{a5}', S_{a4}', S_{a3}', S_{a2}', S_{a1}' to dc source node V_0, thus $-5V_{dc}/6$ appears across the load as given in Fig. 4.16 (m). The load current is negative since it is flowing from load terminal '*b*' to '*a*'.

Mode 13: To obtain V_{ab}= $-V_{dc}$, the switches S_{b1}, S_{b2}, S_{b3}, S_{b4}, S_{b5}, S_{b6} and S_{a6}', S_{a5}', S_{a4}', S_{a3}', S_{a2}', S_{a1}' are turned-on, then the conduction takes place from dc source node V_6 through switches S_{b1}, S_{b2}, S_{b3}, S_{b4}, S_{b5}, S_{b6}, load terminal '*b*' to '*a*' and switches S_{a6}', S_{a5}', S_{a4}', S_{a3}', S_{a2}', S_{a1}' to dc source node V_0, thus $-V_{dc}$ appears across the load as shown in Fig.4.16 (n). The load current is negative since it is flowing from load terminal '*b*' to '*a*'.

Table 4.5 Switching states for Single phase seven-level flying-capacitor full-bridge Inverter.

Modes:	M1	M2	M3	M4	M5	M6	M7	M8	M9	M10	M11	M12	M13
V_{ab}	$+V_{dc}$	$+\frac{5}{6}V_{dc}$	$+\frac{2}{3}V_{dc}$	$+V_{dc}/2$	$+\frac{1}{3}V_{dc}$	$+\frac{1}{6}V_{dc}$	0	$-\frac{1}{6}V_{dc}$	$-\frac{1}{3}V_{dc}$	$-V_{dc}/2$	$-\frac{2}{3}V_{dc}$	$-\frac{5}{6}V_{dc}$	$-V_{dc}$
S_{a1}/S_{a1}'	1/0	1/0	1/0	1/0	1/0	1/0	1/0	0/1	0/1	0/1	0/1	0/1	0/1
S_{a2}/S_{a2}'	1/0	1/0	1/0	1/0	1/0	—	1/0	0/1	0/1	0/1	0/1	0/1	0/1
S_{a3}/S_{a3}'	1/0	1/0	1/0	1/0	—	—	1/0	0/1	0/1	0/1	0/1	0/1	0/1
S_{a4}/S_{a4}'	1/0	1/0	1/0	—	—	—	—	0/1	0/1	0/1	0/1	0/1	0/1
S_{a5}/S_{a5}'	1/0	1/0	—	—	—	—	—	0/1	0/1	0/1	0/1	0/1	0/1
S_{a6}/S_{a6}'	1/0	—	—	—	—	—	—	0/1	0/1	0/1	0/1	0/1	0/1
S_{b1}/S_{b1}'	0/1	0/1	0/1	0/1	0/1	0/1	0/1	1/0	1/0	1/0	1/0	1/0	1/0
S_{b2}/S_{b2}'	0/1	0/1	0/1	0/1	0/1	0/1	0/1	—	1/0	1/0	1/0	1/0	1/0
S_{b3}/S_{b3}'	0/1	0/1	0/1	0/1	0/1	0/1	0/1	—	—	1/0	1/0	1/0	1/0
S_{b4}/S_{b4}'	0/1	0/1	0/1	0/1	0/1	0/1	—	—	—	—	1/0	1/0	1/0
S_{b5}/S_{b5}'	0/1	0/1	0/1	0/1	0/1	0/1	—	—	—	—	—	1/0	1/0
S_{b6}/S_{b6}'	0/1	0/1	0/1	0/1	0/1	0/1	—	—	—	—	—	—	1/0
D_{a1}/D_{a1}'	—	—	—	—	—	—	—	—	—	—	—	—	—
D_{a2}/D_{a2}'	—	—	—	—	—	0/1	—	—	—	—	—	—	—
D_{a3}/D_{a3}'	—	—	—	—	0/1	0/1	—	—	—	—	—	—	—
D_{a4}/D_{a4}'	—	—	—	0/1	0/1	0/1	0/1	—	—	—	—	—	—

Contd...

Modes:	M1	M2	M3	M4	M5	M6	M7	M8	M9	M10	M11	M12	M13
D_{a5}/D_{a5}'	---	---	0/1	0/1	0/1	0/1	0/1	---	---	---	---	---	---
D_{a6}/D_{a6}'	---	0/1	0/1	0/1	0/1	0/1	0/1	---	---	---	---	---	---
D_{b1}/D_{b1}'	---	---	---	---	---	---	---	---	---	---	---	---	---
D_{b2}/D_{b2}'	---	---	---	---	---	---	---	0/1	---	---	---	---	---
D_{b3}/D_{b3}'	---	---	---	---	---	---	---	0/1	0/1	---	---	---	---
D_{b4}/D_{b4}'	---	---	---	---	---	---	1/0	0/1	0/1	0/1	---	---	---
D_{b5}/D_{b5}'	---	---	---	---	---	---	1/0	0/1	0/1	0/1	0/1	---	---
D_{b6}/D_{b6}'	---	---	---	---	---	---	1/0	0/1	0/1	0/1	0/1	0/1	---
I_a	$+I_a$	$+I_a$	$+I_a$	$+I_a$	$+I_a$	$+I_a$	0	$-I_a$	$-I_a$	$-I_a$	$-I_a$	$-I_a$	$-I_a$
Levels	L_1	L_2	L_3	L_4	L_5	L_6	L_7	---	---	---	---	---	---

The switching states for seven level operation for single phase seven-level flying capacitor full-bridge inverter for various modes are presented in Table 4.5 and the seven level load line voltage waveform is given in Fig. 4.17.

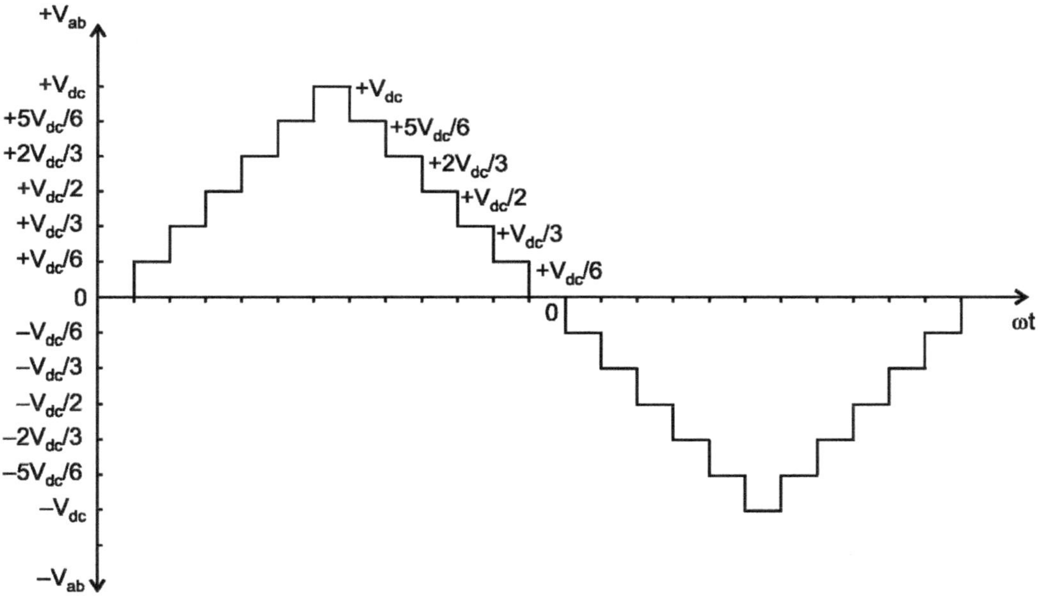

Fig. 4.17 Load line voltage for single phase seven-level flying-capacitor full-bridge inverter.

C. **Three phase Seven-level flying-capacitor full-bridge inverter:** The three single-phase seven-level flying-capacitor single-leg inverter circuits are connected to form a three-phase seven-level flying-capacitor full-bridge inverter as shown in the Fig. 4.18 (a). The output voltage of three-phase inverter is feeding power to the balanced star-connected load. The phase voltages are V_{an}, V_{bn}, V_{cn}, and the line voltages are V_{ab}, V_{bc}, and V_{ca} of the three-phase load. The phase voltage waveform V_{an} is similar to the load voltage obtained for single-phase single-leg arrangement, the only difference is the V_{bn}, and V_{cn} are phase shifted with respect to V_{an} by 120° and 240° respectively. The line voltage waveforms are only shown in the Fig.4.18 (b).

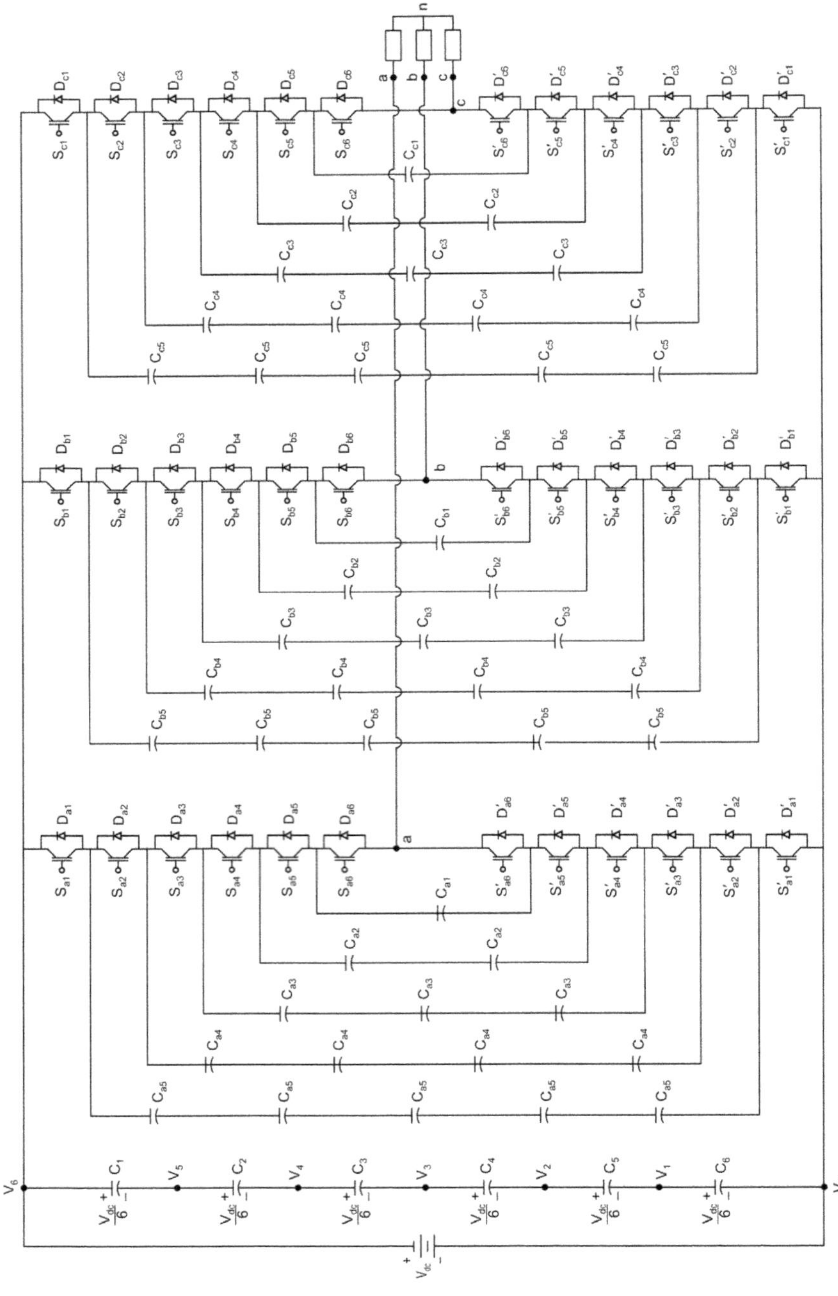

(a)

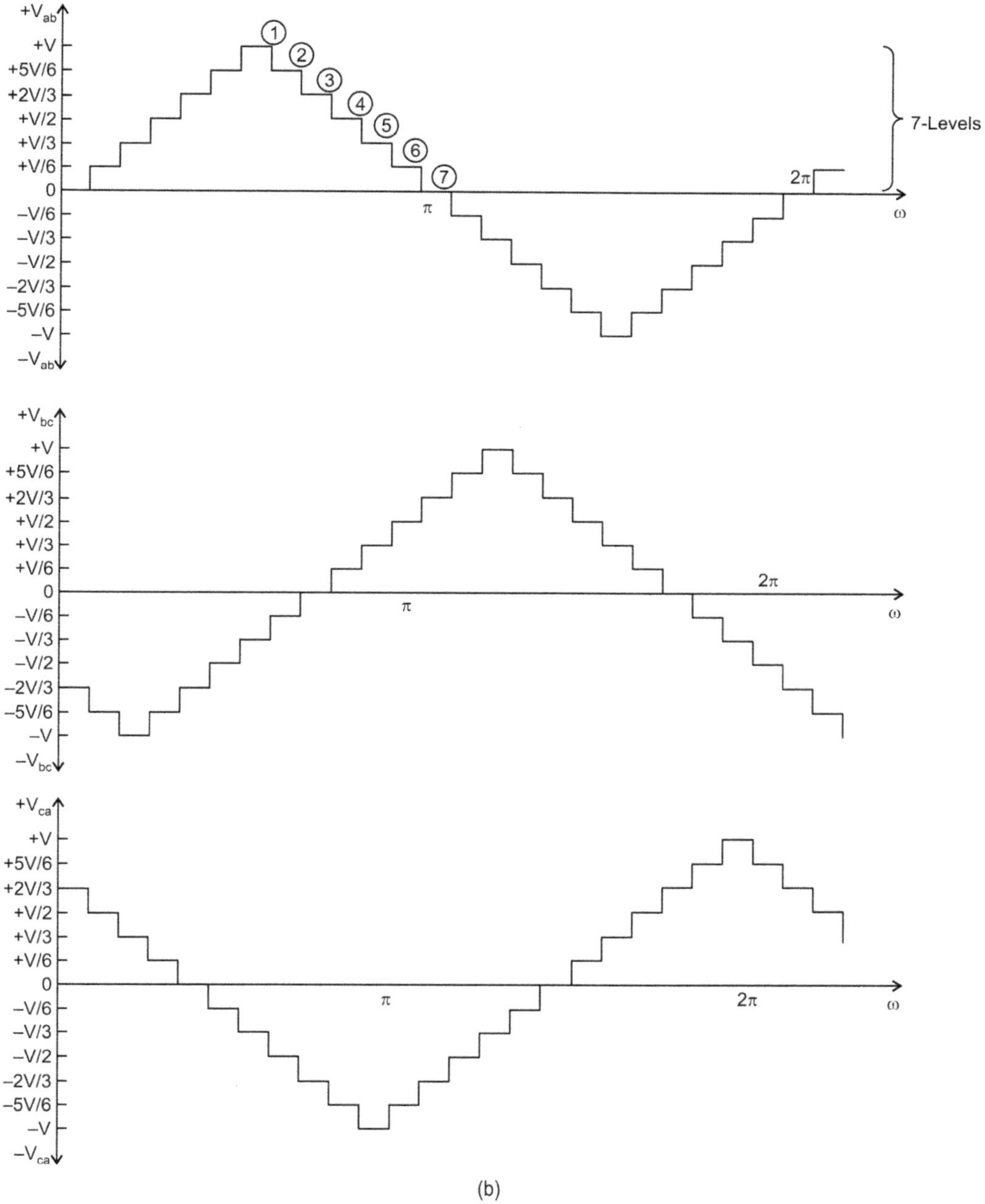

Fig. 4.18 (a) Three phase Seven-level flying-capacitor full-bridge inverter circuit (b) Line Voltage waveforms of the load for seven-level output, V_{ab}, V_{bc}, V_{ca}.

4.5 m-level Flying Capacitor Inverter

The flying capacitor inverter topology is similar to that of NPC topology which uses capacitors instead of clamping diodes. The number of levels in the output voltage waveforms is the voltage increment between two adjacent capacitor legs. One important advantage of the flying capacitor topology is that two or more valid switch combinations are possible to synthesize an output voltage.

The m-level flying capacitor multilevel inverter requires (m-1) x (m-2)/2 auxiliary capacitors per phase in addition to (m-1) main dc link input capacitors under assumption that the voltage ratings of the capacitors are identical to that of the main switches.

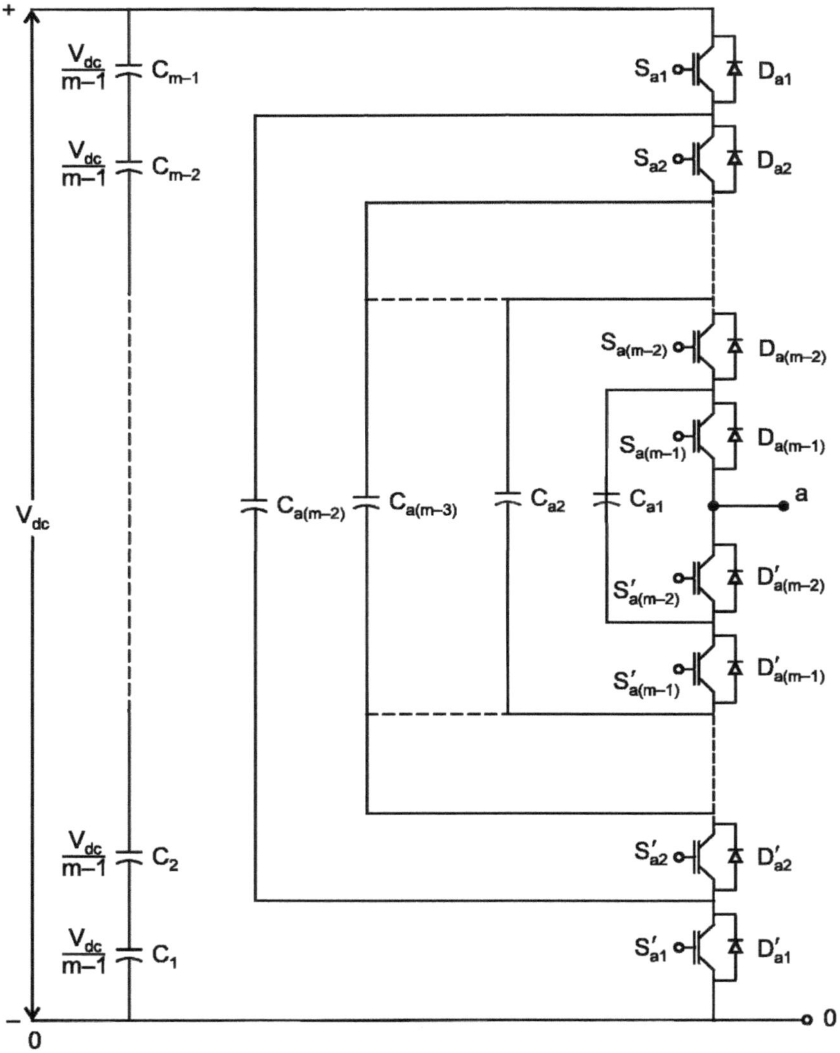

Fig. 4.19 m-level flying capacitor multilevel inverter.

Thus, as the number of levels is increased a greater number of storage capacitors are required resulting in a bulky and expensive compared to NPC topology.

Advantages are

Balancing the voltage levels of the capacitors is possible due to flexible phase redundancy.

- The large number of storage capacitors enable the inverter to ride through short duration outages and deep voltage sags.
- Real and reactive power flows can be controlled.
- Like diode-clamped inverter with more levels harmonic content is low enough to avoid need for filters.

Disadvantages:

- The requirement of a greater number of storage capacitors makes the inverter bulky and expensive.
- Complex control system is required to maintain the capacitor voltage balance.
- Inverter control becomes complicated as number of levels are higher,
- Packaging is difficult for higher number of levels for the inverter.

Applications:

- This inverter is used for static var generation in power system.

4.6 Case Studies for Flying-capacitor Multilevel Inverter

In this chapter two case studies are considered to demonstrate the working principle of Three phase Flying-capacitor inverter for three-level and five-level output load voltages respectively, with resistive-inductive loads. The most popularly known Sinusoidal PWM technique for Flying-capacitor inverters are used. The simulated results are shown for the load phase and line voltages and Total Harmonic Distortion (%THD) of load voltage waveform.

4.6.1 Case Study 1

Simulation of Three phase Three-level Flying-capacitor Inverter with RL loads using SPWM techniques is presented. The resistive-inductive loads, R=100 ohms per phase and L=1 mH per phase. The dc source voltage of 600 V is applied to the circuit.

Three Phase Three Level Flying Capacitor Multilevel Inverter Topology

The circuit diagram of three level flying capacitor MLI is shown below. It can produce an inverter phase voltage with THREE voltage levels.

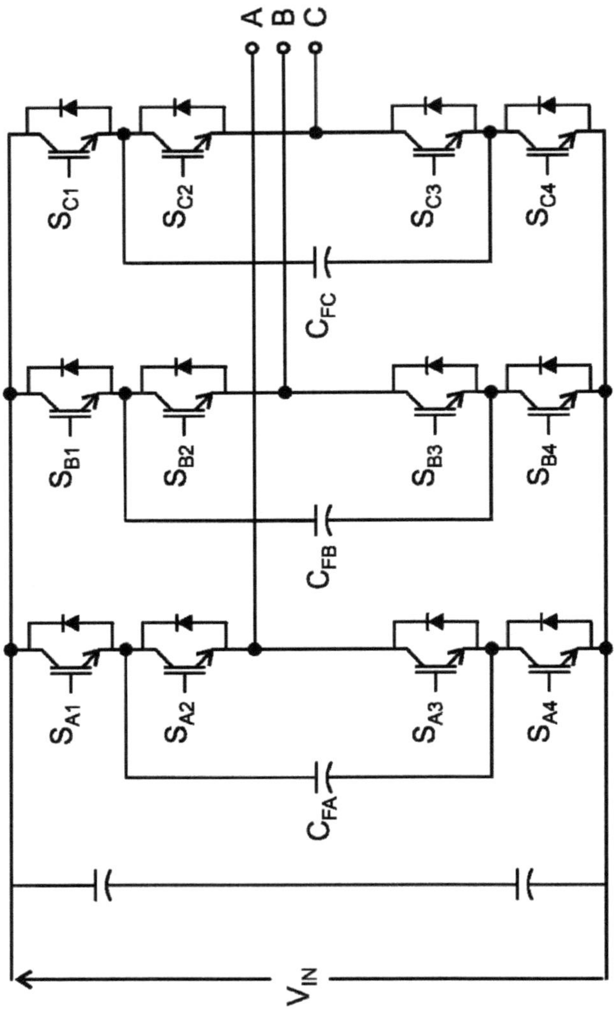

Fig. 4.6.1 (a) Main circuit topology of Three phase three-level flying-capacitor Inverter.

The simulation diagram of the circuit is shown below. SPWM is used to generate the PWM signals to the switches of the circuit.

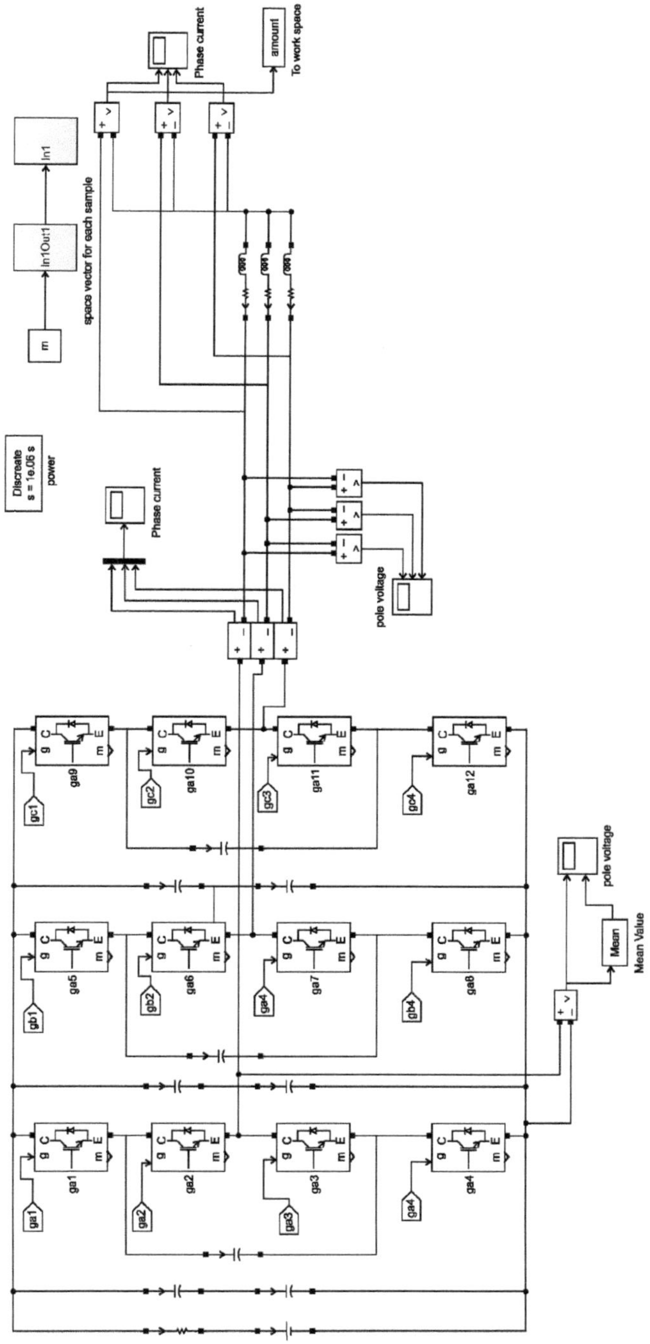

Fig. 4.6.1 (b) The Simulink block diagram of three-phase three-level flying-capacitor Inverter.

Vab wave form:

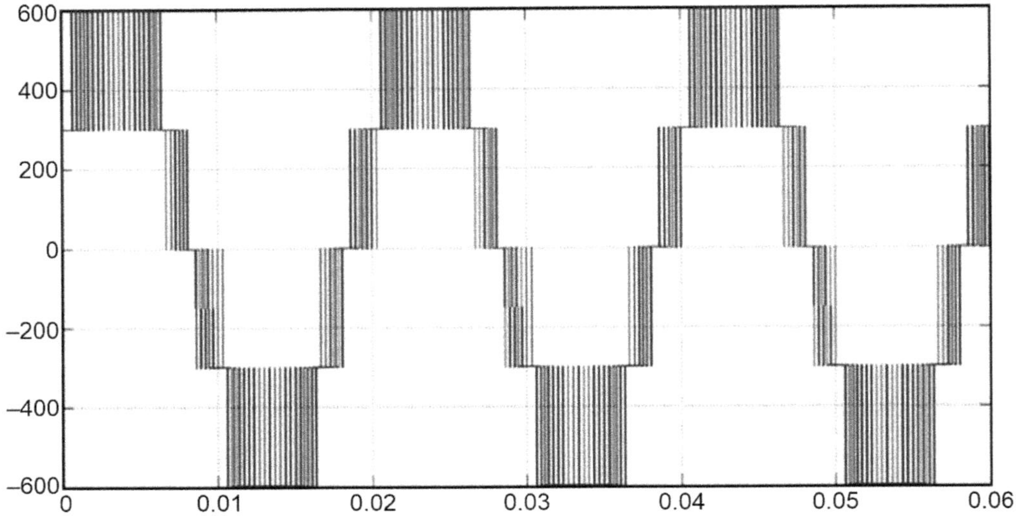

Fig. 4.6.1 (c) Line voltage waveform across resistive-inductive load, V_{ab} for three level output.

Van wave form

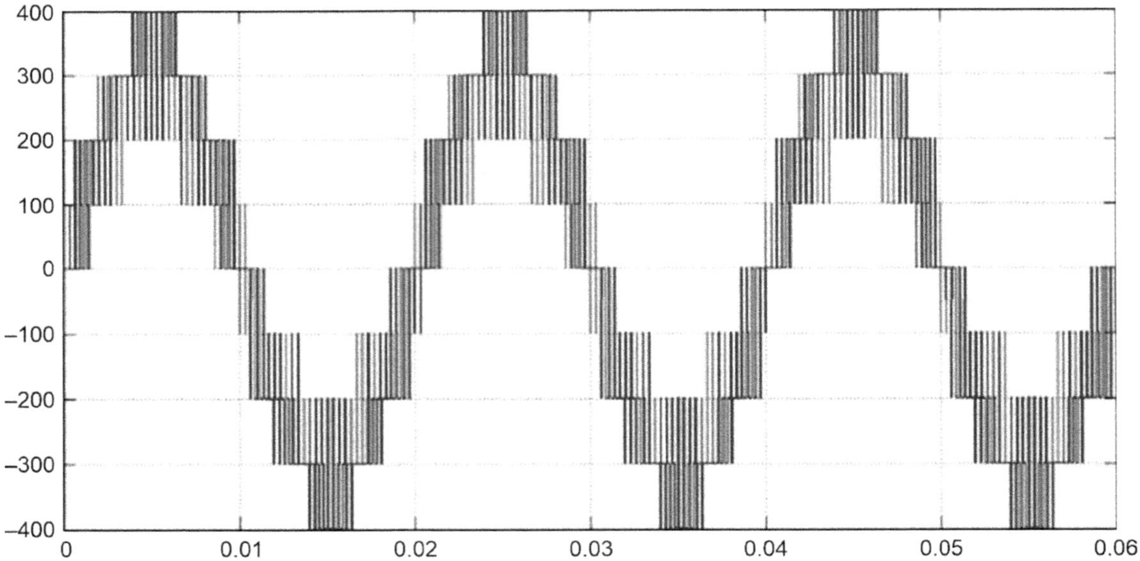

Fig. 4.6.1 (d) Phase voltage waveform across resistive-inductive load, Van for three level output.

The total harmonic distortion is carried out for load line voltage waveform, V_{ab} and THD of $V_{ab} = 2.01\%$ is obtained.

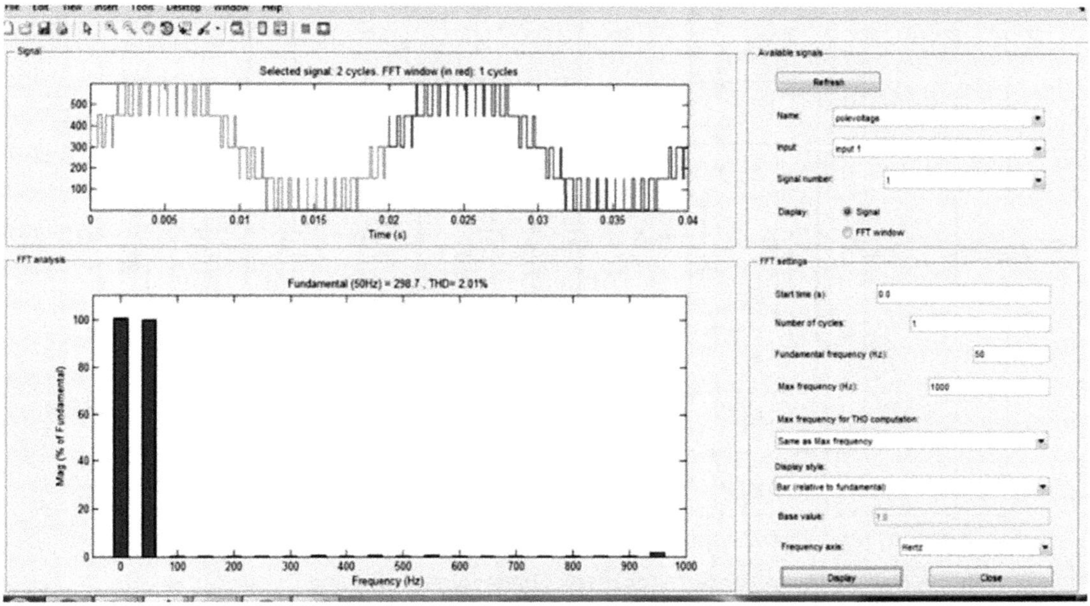

Fig. 4.6.1 (e) Total Harmonic Distortion spectrum of load line voltage waveform.

4.6.2 Case Study 2

Simulation of Three phase Five-level Flying-capacitor Inverter with RL loads using SPWM techniques is presented. The resistive-inductive loads, R=1.0 ohms per phase and L=1 mH per phase. The dc source voltage of 600 V (Vdc=4E) is applied to the circuit. Amplitude and frequency modulation considered for this example is m_a=1, m_f=33, f=50 Hz.

Five Level Flying Capacitor Multilevel Inverter

The circuit diagram of five level flying capacitor MLI is shown below. It can produce an inverter phase voltage with five voltage levels. When switches S_1, S_2, S_3, and S_4 conduct, the inverter phase voltage V_{AN} is 4E, which is the voltage at the inverter terminal A with respect to the negative dc bus N. Similarly, with S_1, S_2, and S_3 switched on, V_{AN} = 3E. Table lists all the voltage levels and their corresponding switching states. It is noted that some voltage levels can be obtained by more than one switching state. The voltage level 2E, for instance, can be produced by six sets of different switching states. The switching state redundancy is a common phenomenon in multilevel converters, which provides a great flexibility for the switching pattern design.

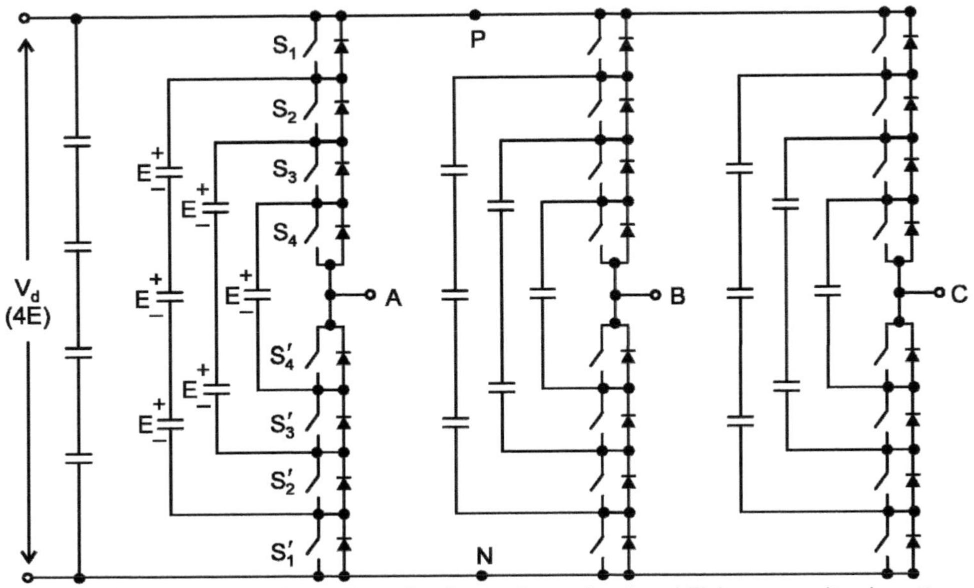

Fig. 4.6.2 (a) Main circuit topology for three-phase five-level Flying-capacitor Inverter.

Table 4.6.2 Switching pattern for three-phase five-level Flying-Capacitor Inverter.

Inverter Phase Voltage V_{AN}	Switching State			
	S_1	S_2	S_3	S_4
4E	1	1	1	1
3E	1	1	1	0
	0	1	1	1
	1	0	1	1
	1	1	0	1
	1	1	1	0
2E	1	1	0	0
	0	0	1	1
	1	0	0	1
	0	1	1	0
	1	0	1	0
	0	1	0	1
1E	1	0	0	0
	0	1	0	0
	0	1	0	0
	0	0	1	0
	0	0	0	1
0	0	0	0	0

The simulation diagram of the above circuit is shown below. SPWM is used to generate the PWM signals to the switches of the circuit.

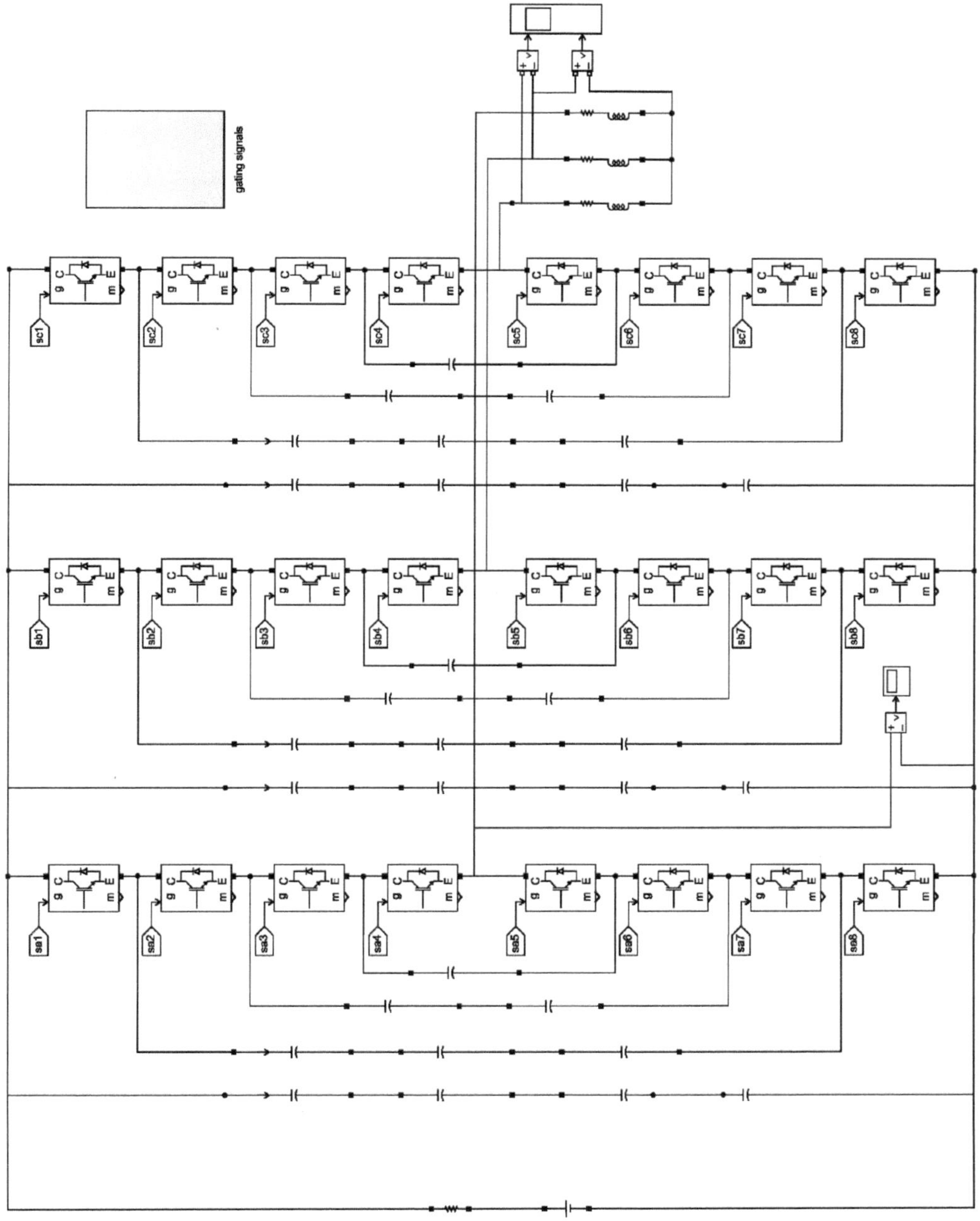

Fig. 4.6.2 (b) Simulink block diagram for three-phase Five-level Flying-capacitor Inverter.

Vab wave form:

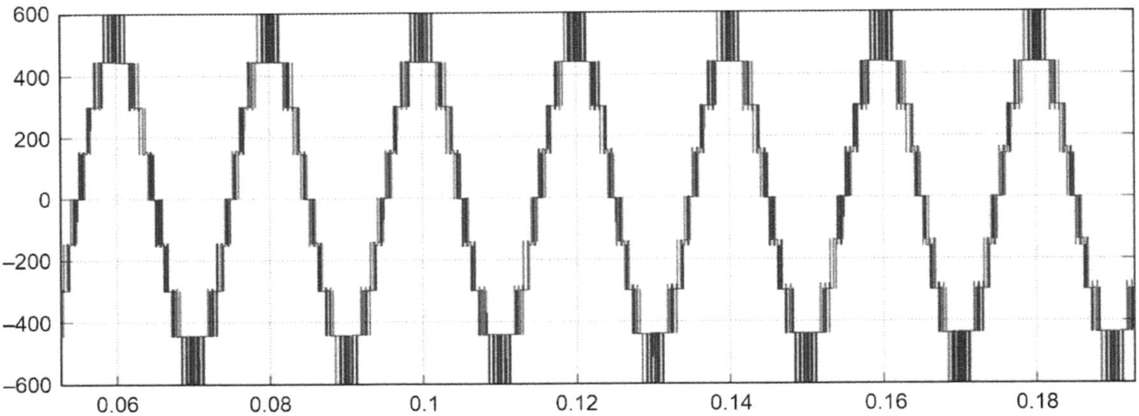

Fig. 4.6.2 (c) Line Voltage across resistive-inductive load, Vab for five-level output.

Van wave form:

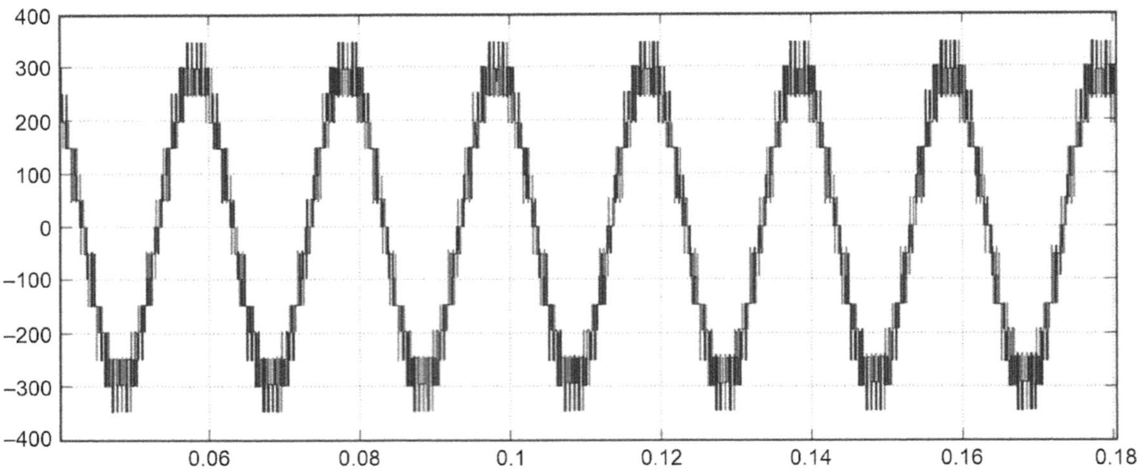

Fig. 4.6.2 (d) Phase Voltage across resistive-inductive load, Van for five-level output.

SPWM Technique:

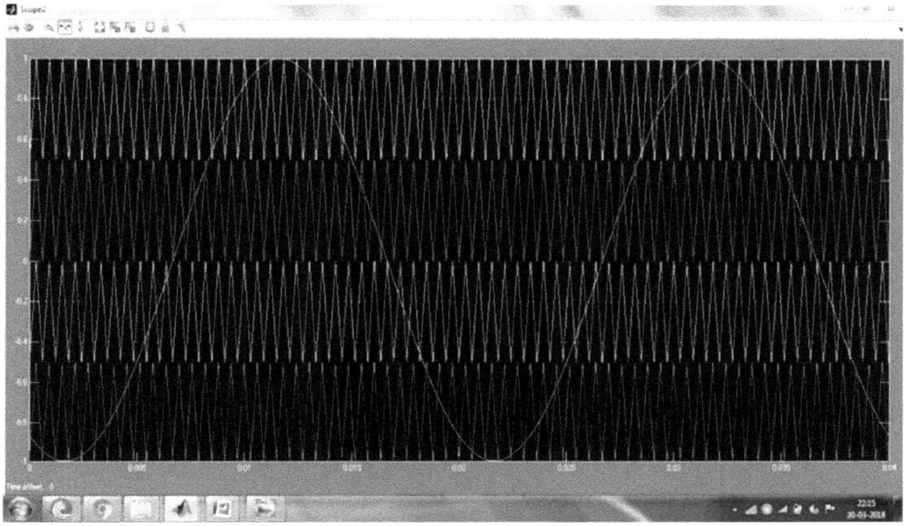

Fig. 4.6.2 (e) Sinusoidal PWM techniques applied to five-level flying-capacitor inverter.

THD of Vab:1.88 %:

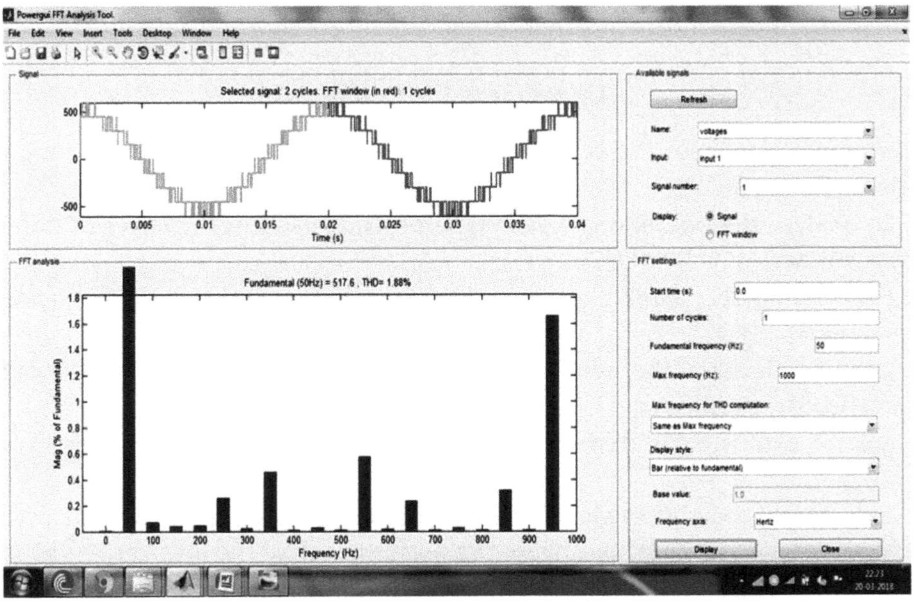

Fig. 4.6.2 (f) Total Harmonic Distortion spectrum for load line voltage.

4.7 Summary

The flying-capacitor inverter is used in many industrial and power systems applications, due to their inherent advantages. The basic construction of flying-capacitor based inverter and its functioning is explained elaborately. They are classified into single-leg configuration and full-bridge configuration. They are also available in single-phase and three-phase inverter category.

The higher levels of output voltages for the flying capacitor inverters depends on the number of dc source capacitors. For producing m-level output voltage from flying-capacitor inverter, (m-1) dc source capacitors are required. Besides, an m-level inverter will require a total of (m - 1) x (m - 2)/2 auxiliary capacitors per phase leg. For example, for m=5, 4 dc source capacitors and 6 auxiliary or flying capacitors are required for the five-level inverter.

The single-phase single-leg and single-phase full-bridge flying-capacitor inverters are presented and discussed their operation by considering various modes and switching patterns for three-level, five-level and seven-level output voltages for resistive-inductive loads. Similarly, the three-phase full-bridge flying-capacitor inverter topologies with their output line voltages are presented for three-level, five-level and seven-level output voltages for star-connected resistive-inductive loads. The load current depends on the nature of the loads.

The m-level configuration is also presented for flying-capacitor inverter. The advantages, disadvantages and applications of all the above discussed multilevel inverters are also given.

The case studies and exercise problems are given in all varieties of flying-capacitor multilevel inverters for practice by modeling and simulation.

Learning Outcomes

- Acquire knowledge about the construction of flying-capacitor inverter of basic three-level operation. (CO1) :: (PO1)
- Ability to analyze the operation of each type of flying-capacitor inverter of different levels of output voltages. (CO4) :: (PO2)
- To be able to apply suitable pulse width modulation techniques for flying-capacitor inverters. (CO3) :: (PO1)
- To be able to develop the simulation model using software tools for flying-capacitor inverters. (CO5) :: (PO5)
- To be able to evaluate the performance of the inverters for various loads by simulation. (CO5) :: (PO3)
- To be able to extend the single-phase flying-capacitor inverter concepts to three-phase flying-capacitor inverter and investigate the performance of the inverter, PWM schemes for three-phase loads. (CO6) :: (PO4)

Questions

4.1 Construct the circuit for a single-phase three-level flying-capacitor bridge Inverter using single-pole triple throw switches for RL loads, and explain the operation of the circuit and draw output voltage waveform?

4.2 Realize the basic circuit of single-phase three-level flying-capacitor by using bidirectional IGBTs and draw the final topology for single-leg arrangement?

4.3 Analyze single-phase single-leg three-level flying-capacitor inverter topology and elaborate the working modes for positive load current with the help of output voltage waveform?

4.4 Analyze single-phase single-leg three-level flying-capacitor inverter topology and give the switching states for obtaining three-level output voltages with positive current?

4.5 Analyze single-phase single-leg three-level flying-capacitor inverter topology for RL load and elaborate the working modes for negative load current with the help of output voltage waveform?

4.6 Discuss the advantages and applications of single-phase three-level flying-capacitor inverter?

4.7 Analyze the topology of single-phase five-level flying-capacitor inverter with single-leg arrangement and bring out detailed operation for five-level output voltages and positive load current?

4.8 Analyze the single-phase five-level flying-capacitor inverter circuit and bring out the switching states for generating five-level output voltages by considering single-leg arrangement?

4.9 Analyze the topology of single-phase five-level flying-capacitor inverter for RL load with single-leg arrangement and bring out detailed operation for five-level output voltages and negative load current?

4.10 Analyze the single-phase five-level flying-capacitor bridge inverter by considering various modes of operations to obtain five-level output voltages for positive load current?

4.11 Analyze the topology of single-phase seven-level flying-capacitor inverter with single-leg arrangement and bring out various modes of operation for positive load current?

4.12 Analyze the single-phase seven-level flying-capacitor inverter circuit and bring out the switching states for generating seven-level output voltages by considering single-leg arrangement and sketch output voltage waveform?

4.13 Analyze the topology of single-phase seven-level flying-capacitor inverter for RL load with single-leg arrangement and bring out various modes of operation for negative load current?

4.14 Analyze the single-phase seven-level flying-capacitor bridge inverter by considering various modes of operations to obtain seven-level output voltages for positive load current?

4.15 Analyze the single-phase seven-level flying-capacitor inverter bridge circuit and bring out the switching states for generating seven-level output voltages and sketch output voltage waveform?

4.16 Analyze the single-phase seven-level flying-capacitor bridge inverter with RL load by considering various modes of operations to obtain seven-level output voltages for negative load current?

4.17 Draw the circuit topology for three-phase three-level flying-capacitor bridge inverter for star-connected load and sketch the three-phase output load voltages?

4.18 Draw the circuit topology for three-phase five-level flying-capacitor bridge inverter for star-connected load and sketch the three-phase output load voltages?

4.19 Draw the circuit topology for three-phase seven-level flying-capacitor bridge inverter for star-connected load and sketch the three-phase output load voltages?

4.20 Draw m-level generalized flying-capacitor inverter with single-leg arrangement? Bring out advantages and disadvantages?

Multiple Choice Questions

4.1 What is the purpose of flying-capacitors in flying-capacitor inverter other than dc link capacitors?

(a) for feed back
(b) for facilitating variable level of voltages
(c) both (a) and (b)
(d) none of the above

4.2 How many dc source capacitors are needed for three-level flying-capacitor inverter?

(a) three
(b) one
(c) two
(d) zero

4.3 How many power semiconductor devices are necessary for three-level Flying-capacitor inverter per leg?

(a) six power semiconductor devices
(b) two power semiconductor devices
(c) three power semiconductor devices
(d) four power semiconductor devices

4.4 What is the purpose of anti-parallel diodes connected across each power semiconductor switching device in multilevel inverter?

(a) source energy transfer to load
(b) to short-circuit the respective conducting PSC devices
(c) load energy transfer to source
(d) none of the above

4.5 How many dc source capacitors are needed for five-level flying-capacitor inverter?

(a) two
(b) four
(c) one
(d) three

4.6 How many power semiconductor devices are necessary for five-level Flying-capacitor inverter per leg?

(a) eight power semiconductor devices
(b) six power semiconductor devices
(c) two power semiconductor devices
(d) four power semiconductor devices

4.7 How many flying-capacitors are required for five-level flying-capacitor Inverter per leg?

(a) six flying-capacitors
(b) four flying-capacitors
(c) eight flying-capacitors
(d) two flying-capacitors

4.8 How many dc source capacitors are needed for seven-level flying-capacitor inverter?

(a) three (b) four
(c) six (d) five

4.9 How many power semiconductor devices are necessary for seven-level flying-capacitor inverter per leg?

(a) ten power semiconductor switching devices
(b) twelve power semiconductor switching devices
(c) six power semiconductor switching devices
(d) eight power semi-conductor switching devices

4.10 How many flying-capacitors are required for seven-level flying-capacitor Inverter per leg?

(a) sixteen flying-capacitors (b) fourteen flying-capacitors
(c) eighteen flying-capacitors (d) fifteen flying-capacitors

4.11 For m-level flying-capacitor inverter, how many switching devices are required per each leg?

(a) (m-1) + (m-2) switching devices
(b) (m-1) x (m-2) switching devices
(c) (m-1) switching devices
(d) 2(m-1) switching devices

4.12 For m-level flying-capacitor inverter, how many auxiliary capacitors are needed per each leg?

(a) (m-1) x (m-2) auxiliary capacitors
(b) (m-1) x (m-2)/2 auxiliary capacitors
(c) (m-1)/2 auxiliary capacitors
(d) (m-1) + (m-2) auxiliary capacitors

4.13 For m-level flying-capacitor inverter, how many dc link source capacitors are required for the circuit?
(a) (m-2) dc link source capacitors
(b) (m-1) x (m-2) dc link source capacitors
(c) (m-1) dc link source capacitors
(d) 2(m-1) dc link source capacitors

4.14 For five-level flying-capacitor inverter (m=5), how many power semiconductors switching devices, dc link source capacitors, and auxiliary capacitors are required per each leg?
(a) 4 switching devices, 8 dc source capacitors, 12 auxiliary capacitors
(b) 8 switching devices, 4 dc source capacitors, 6 auxiliary capacitors
(c) 6 switching devices, 4 dc source capacitors, 6 auxiliary capacitors
(d) 8 switching devices, 6 dc source capa`citors, 4 auxiliary capacitors

4.15 For three-level flying-capacitor inverter (m=3), how many power semiconductors switching devices, dc link source capacitors, and auxiliary capacitors are required per each leg?
(a) 4 switching devices, 2 dc source capacitors, 1 auxiliary capacitor
(b) 6 switching devices, 4 dc source capacitors, 1 auxiliary capacitor
(c) 4 switching devices, 2 dc source capacitors, 2 auxiliary capacitors
(d) 8 switching devices, 6 dc source capacitors, 4 auxiliary capacitors

4.16 For seven-level flying-capacitor inverter (m=7), how many power semiconductors switching devices, dc link source capacitors, and auxiliary capacitors are required per each leg?
(a) 10 switching devices, 8 dc source capacitors, 30 clamping diodes
(b) 8 switching devices, 4 dc source capacitors, 16 clamping diodes
(c) 12 switching devices, 6 dc source capacitors, 15 clamping diodes
(d) 8 switching devices, 6 dc source capacitors, 24 clamping diodes

Exercise Problems

Ex.4.1: Simulate single phase flying-capacitor inverter with single-leg circuit, when dc source voltage V_{dc}=100V, R=20 ohms and L=10 mH and obtain gating signals, output phase voltage and load current by using any software tools available. Use Sinusoidal PWM modulation technique for three-level output.

Ex.4.2: Evaluate the performance of single-phase flying-capacitor inverter with single-leg circuit, when dc source voltage V_{dc}=100V, R=20 ohms and L=10 mH and obtain gating signals, output phase voltage and load current by simulation using any software tools available. Use Carrier-based PWM modulation technique for three-level output.

Ex.4.3: Simulate single phase flying-capacitor inverter with single-leg circuit, when dc source voltage V_{dc}=100V, R=20 ohms and L=10 mH and obtain gating signals, output phase voltage and load current by using any software tools available. Use Space-vector PWM modulation technique for three-level output.

Ex.4.4: Evaluate the performance of single-phase flying-capacitor full-bridge inverter when dc source voltage V_{dc}=100V, R=20 ohms and L=10 mH and obtain gating signals, output line voltage and load current by simulation using any software tools available. Use Sinusoidal PWM modulation technique for three-level output.

Ex.4.5: Evaluate the performance of single-phase flying-capacitor full-bridge inverter when dc source voltage V_{dc}=100V, R=20 ohms and L=10 mH and obtain gating signals, output line voltage and load current by simulation using any software tools available. Use Space-vector PWM modulation technique for three-level output.

Ex.4.6: Simulate single phase flying-capacitor full-bridge inverter when dc source voltage V_{dc}=100V, R=20 ohms and L=10 mH and obtain gating signals, output line voltage and load current by using any software tools available. Use Carrier-based PWM modulation technique for three-level output.

Ex.4.7: Simulate three phase flying-capacitor full-bridge inverter when dc source voltage V_{dc}=100V, R=20 ohms per phase and L=10 mH per phase and obtain output line voltages and load current by using any software tools available. Use Sinusoidal PWM modulation technique for three-level output.

Ex.4.8: Simulate single phase flying-capacitor inverter with single-leg circuit, when dc source voltage V_{dc}=200V, R=20 ohms and L=10 mH and obtain gating signals, output phase voltage and load current by using any software tools available. Use Sinusoidal PWM modulation technique for five-level output.

Ex.4.9: Evaluate the performance of single-phase flying-capacitor inverter with single-leg circuit, when dc source voltage V_{dc}=200V, R=20 ohms and L=10 mH and obtain gating signals, output phase voltage and load current by simulation using any software tools available. Use Carrier-based PWM modulation technique for five-level output.

Ex.4.10: Simulate single phase flying-capacitor inverter with single-leg circuit, when dc source voltage V_{dc}=200V, R=20 ohms and L=10 mH and obtain gating signals, output phase voltage and load current by using any software tools available. Use Space-vector PWM modulation technique for five-level output.

Ex.4.11: Evaluate the performance of single-phase flying-capacitor full-bridge inverter when dc source voltage V_{dc}=200V, R=20 ohms and L=10 mH and obtain gating signals, output line voltage and load current by simulation using any software tools available. Use Sinusoidal PWM modulation technique for five-level output.

Ex.4.12: Evaluate the performance of single-phase flying-capacitor full-bridge inverter when dc source voltage V_{dc}=200V, R=20 ohms and L=10 mH and obtain gating signals, output line voltage and load current by simulation using any software tools available. Use Space-vector PWM modulation technique for five-level output.

Ex.4.13: Simulate single phase flying-capacitor full-bridge inverter when dc source voltage V_{dc}=200V, R=20 ohms and L=10 mH and obtain gating signals, output line voltage and load current by using any software tools available. Use Carrier-based PWM modulation technique for five-level output.

Ex.4.14: Evaluate the performance of three phase flying-capacitor full-bridge inverter when dc source voltage V_{dc}=200V, R=20 ohms per phase and L=10 mH per phase and obtain output line voltages and load current by using any software tools available. Use Sinusoidal PWM modulation technique for five-level output.

Ex.4.15: Simulate single phase flying-capacitor inverter with single-leg circuit, when dc source voltage V_{dc}=200V, R=20 ohms and L=10 mH and obtain gating signals, output phase voltage and load current by using any software tools available. Use Sinusoidal PWM modulation technique for seven-level output.

Ex.4.16: Evaluate the performance of single-phase flying-capacitor inverter with single-leg circuit, when dc source voltage V_{dc}=200V, R=20 ohms and L=10 mH and obtain gating signals, output phase voltage and load current by simulation using any software tools available. Use Carrier-based PWM modulation technique for seven-level output.

Ex.4.17: Simulate single phase flying-capacitor inverter with single-leg circuit, when dc source voltage V_{dc}=200V, R=20 ohms and L=10 mH and obtain gating signals, output phase voltage and load current by using any software tools available. Use Space-vector PWM modulation technique for seven-level output.

Ex.4.18: Evaluate the performance of single-phase flying-capacitor full-bridge inverter when dc source voltage V_{dc}=200V, R=20 ohms and L=10 mH and obtain gating signals, output line voltage and load current by simulation using any software tools available. Use Sinusoidal PWM modulation technique for seven-level output.

Ex.4.19: Evaluate the performance of single-phase flying-capacitor full-bridge inverter when dc source voltage V_{dc}=200V, R=20 ohms and L=10 mH and obtain gating signals, output line voltage and load current by simulation using any software tools available. Use Space-vector PWM modulation technique for seven-level output.

Ex.4.20: Simulate single phase flying-capacitor full-bridge inverter when dc source voltage V_{dc}=200V, R=20 ohms and L=10 mH and obtain gating signals, output line voltage and load current by using any software tools available. Use Carrier-based PWM modulation technique for seven-level output.

Ex.4.21: Evaluate the performance of three phase flying-capacitor full-bridge inverter when dc source voltage V_{dc}=200V, R=20 ohms per phase and L=10 mH per phase and obtain output line voltages and load current by using any software tools available. Use Sinusoidal PWM modulation technique for seven-level output.

Answers to Multiple Choice Questions

4.1	(b)	4.2	(c)	4.3	(d)	4.4	(c)	4.5	(b)
4.6	(a)	4.7	(a)	4.8	(c)	4.9	(b)	4.10	(d)
4.11	(d)	4.12	(b)	4.13	(c)	4.14	(b)	4.15	(a)
4.16	(c)								

CHAPTER

5 Cascaded H-Bridge Inverters

LEARNING OBJECTIVES

- ✓ *To understand and acquire knowledge about construction of cascaded H-bridge multilevel inverters.*
- ✓ *To prepare learners to analyze the operation of three-level, five-level, seven-level, nine-level and m-level cascaded H-bridge inverters.*
- ✓ *To prepare learners to distinguish between symmetrical and asymmetrical cascaded H-bridge multilevel inverters.*
- ✓ *To analyze the transistor-clamped cascaded H-bridge multilevel inverters of higher-order levels.*
- ✓ *To prepare the learners to practice through case studies on cascaded H-bridge multilevel inverters of all varieties.*
- ✓ *To prepare learners to simulate the performance of three-level, five-level, seven-level, and m-level cascaded H-bridge inverters of both symmetrical and asymmetrical configurations with suitable pulse width modulation techniques.*
- ✓ *To prepare learners to simulate the performance of five-level, seven-level, nine-level, twenty-one-level and m-level transistor-clamped cascaded H-bridge inverters of both symmetrical and asymmetrical configurations with suitable pulse width modulation techniques.*
- ✓ *To understand the applications of various levels of cascaded H-bridge inverters.*

5.1 Introduction

The cascaded H-bridge multilevel inverters are made up of from series connected single full bridge inverters, each with independent isolated dc voltage source. The most popular dc sources are batteries, solar photovoltaic cell, fuel cell as independent sources and capacitors and ultracapacitors as dependent sources. Based on the magnitudes of the isolated dc sources CHB inverters are also classified as symmetrical and asymmetrical group. In symmetrical the magnitudes of dc voltage sources are equal and for asymmetrical the dc voltage sources are unequal. For high power and medium voltage applications, a greater number of H-bridge inverters are to be connected in cascade. To build medium load voltages more than one H-bridge need to be connected in cascade and there is a relationship for the requirement of number of H-bridges for required number of voltage levels and independent dc sources. If phase voltage is V_{AN} and m is the number of voltage level, then V_{AN} = m switching takes place. Where as in line voltage waveform V_{AB} = (2m-1) switching states take place. Another relationship is m = (2S+1), where S is the number of dc sources and number of H-bridges and m is the number of levels of output phase voltage. The relationships are tabulated in the table 5.1.

Table 5.1 Summary of m=(2S+1) and V_{AN} = (m) for H-bridge inverters with equal dc source voltages.

Inverter level, m	Number of dc sources, S with equal voltages	Number of H-bridge inverters	Load phase Voltage, V_{AN}
3	1	1	3-levels
5	2	2	5-levels
7	3	3	7-levels
9	4	4	9-levels
11	5	5	11-levels

5.2 Symmetrical Cascaded H-Bridge Inverters

5.2.1 Symmetrical Configuration for Three Level Cascaded H-bridge Inverter with Equal dc Sources

The three-level H-bridge inverter consists of a single H-bridge topology with independent dc voltage source and four switching devices along with their anti-parallel diodes, as shown in Fig.5.1. The corresponding switching pattern is given in Table 5.2 for three-level H-bridge inverter. The circuit can generate three voltage levels as output voltage V_{AN}, namely, $\pm V_{dc}$, and 0. When switches S_1 and S_2 are turned on the output voltage V_{AN} will be $+V_{dc}$, the current flows from dc source V_{dc}, switches S_1, S_2 and the load. When switches S_3 and S_4 are turned on the output voltage will be $-V_{dc}$, the current flows from dc source V_{dc}, switches S_3, S_4 and load. The output voltage V_{AN} will be zero through either switch S_1, diode D_3 or switch S_4, diode D_2 depending up on the direction of the load current.

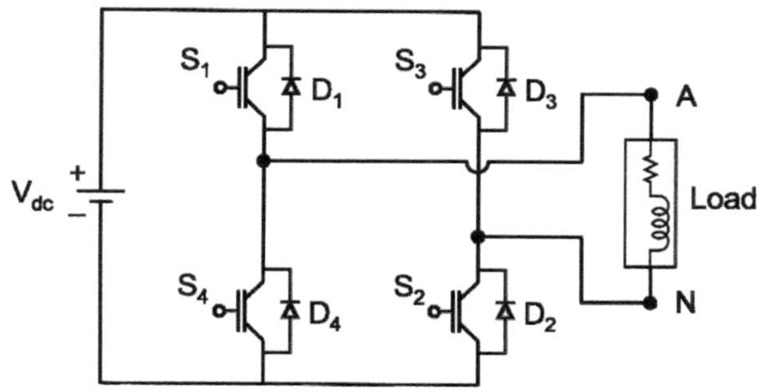

(a)

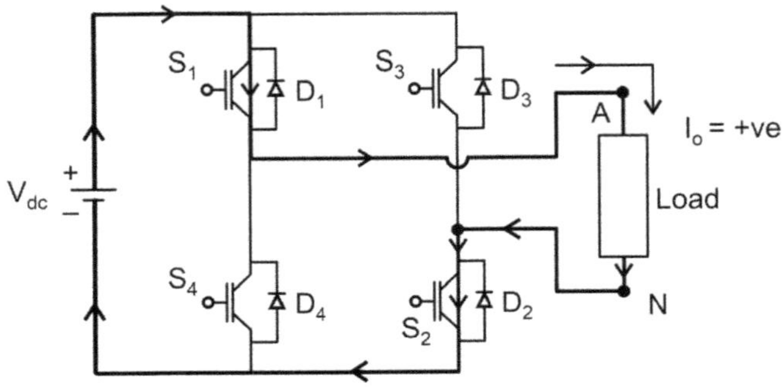

(b) Mode 1 : $V_{AN} = + V_{dc}$

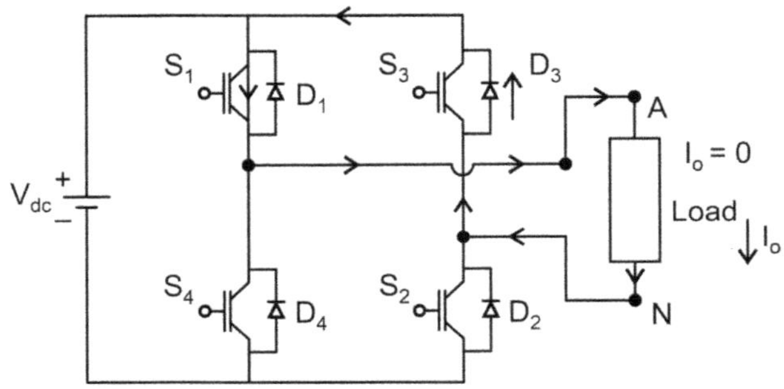

(c) Mode 2 : $V_{AN} = 0$

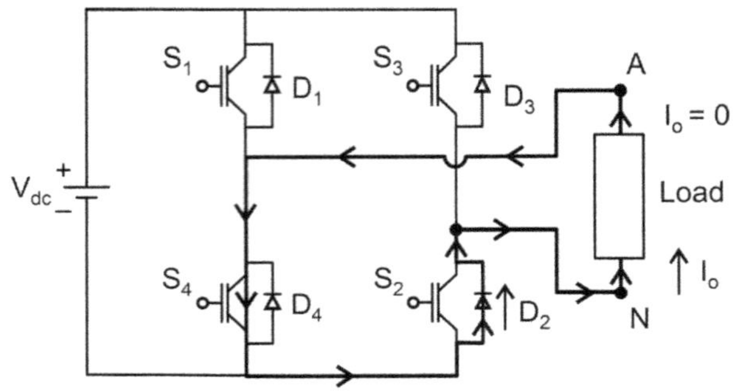

(d) Mode 3 : $V_{AN} = 0$

Fig. 5.1 Symmetrical Configuration for (a) Three-level H-bridge inverter for Single phase load, (b) Mode1: (V_{AN} = + V_{dc}), (c) Mode2 and (d) Mode3: V_{AN} = (0), and (e) Mode4: V_{AN} = (-V_{dc}).

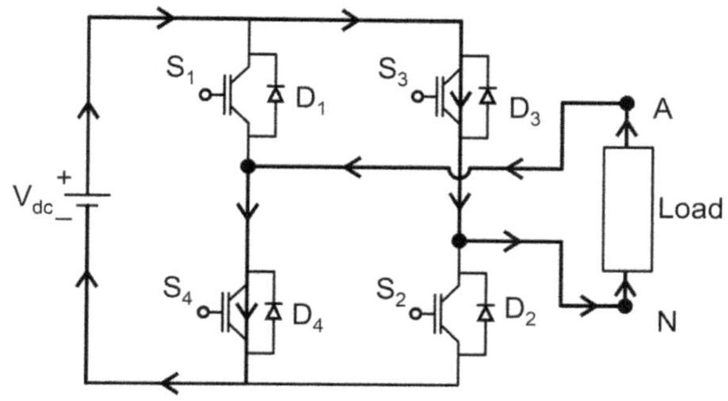

(e) Mode 4 : $V_{AN} = -V_{dc}$

Fig. 5.1 Symmetrical Configuration for (a) Three-level H-bridge inverter for Single phase load, (b) Mode1: (V_{AN} = + V_{dc}), (c) Mode2 and (d) Mode3: V_{AN} = (0), and (e) Mode4: V_{AN} = ($-V_{dc}$).

Table. 5.2 Switching pattern of three-level H-bridge inverter for Symmetrical Configuration.

Modes	Output phase Voltage, V_{AN}	Switching states				Diodes
		S_1	S_2	S_3	S_4	D_n
Mode1	+ V_{dc}	1	1	0	0	0
Mode2	0	1	0	0	0	D_3
Mode3	0	0	0	0	1	D_2
Mode4	$-V_{dc}$	0	0	1	1	0

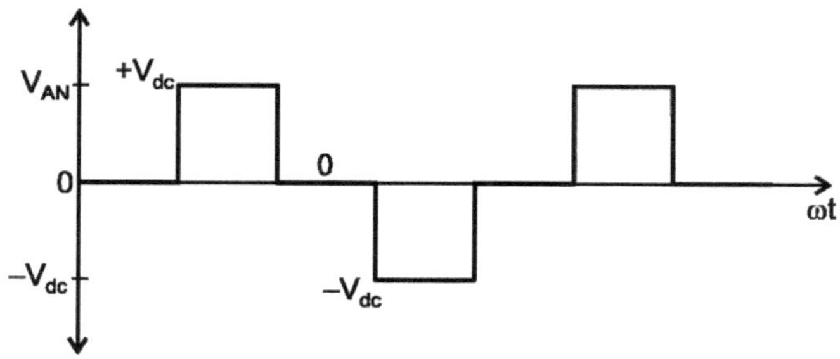

Fig. 5.2 Phase load Voltage (V_{AN}) waveform for three-level H-bridge inverter for Symmetrical Configuration.

For a three-phase system, the output voltages of the three H-bridge inverters can be connected in either Y- or Delta. Three separate single H-bridge inverter modules are arranged to get three-phase H-bridge circuit configuration for three-phase star-connected loads as given in the Fig. 5.3 for three-level operation.

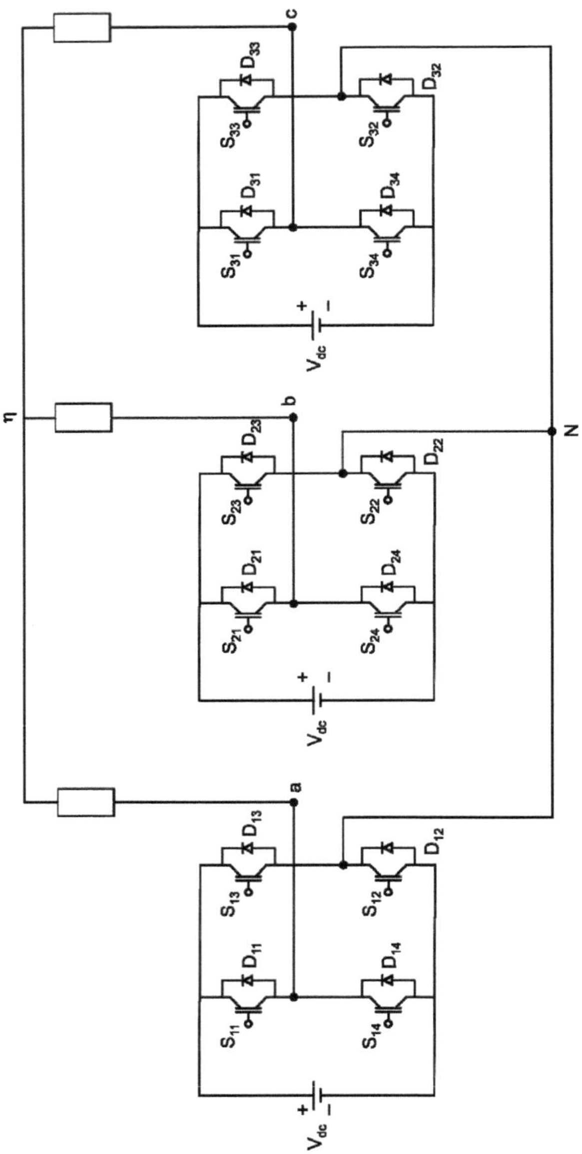

Fig. 5.3 Symmetrical Configuration for Three-level H-bridge inverter for Three-phase loads.

5.2.2 Symmetrical Configuration for Five - Level Cascaded H-bridge Inverter with Equal dc Voltage Sources (with two cells)

The five-level cascaded H-bridge inverter consists of two cells of H-bridge as shown in Fig. 5.4. Each cell has its own individual isolated dc source and four switching devices along with anti-parallel diodes. These two cells are connected in series at the output ac side. At the input side two equal (symmetrical) dc voltage sources are connected as V_{dc} ($V_1=V_2=V_{dc}$). The switching devices of first H-bridge cell are designated as S_1, S_2, S_3 and S_4, and for second H-bridge cell are S_5, S_6, S_7 and S_8. The individual H-bridge cell output voltages are V_1 and V_2 respectively and the series connected cells combined output voltage as V_{AN}. Each H-bridge cell generates three level of voltages, $+V_{dc}$, 0, $-V_{dc}$, whereas the cascaded two-cells H-bridge inverter generates five levels of voltages, $+2V_{dc}$, $+V_{dc}$, 0, $-V_{dc}$, $-2V_{dc}$.

Mode 1: In order to generate $+2V_{dc}$ as net output voltage at V_{AN}, the switching devices S_1, S_2 of first H-bridge cell and S_5, S_6 of second H-bridge cell are to be turned on. The load current passes through first dc source V_1, switching device S_1, output terminal A, load, output terminal N, S_6, second dc source V_2, S_5 and S_2 and back to first dc source V_1, which completes the circuit and $+2V_{dc}$ appears at V_{AN}.

Mode 2: For the output voltage of $+V_{dc}$ at V_{AN}, switching devices S_1, S_2, S_5 are turned on and D_7 into conduction. The load current flows from the first dc voltage source V_1 through switching devices S_1, output terminal A, load, output terminal N, anti-parallel diode D_7, switching devices S_5 and S_2 back to first dc voltage source V_1, thus completes the circuit and $+V_{dc}$ is transferred to V_{AN} (as V_1). Alternately, load current flows from second dc voltage V_2, switching devices S_5, D_3, S_1, output terminal A, load output terminal N, S_6 back to second dc voltage source V_2, thus completes the circuit for $V_{AN} = + V_{dc}$ (as V_2).

Mode 3: To obtain V_{AN} as 0 voltage, switching devices S_2, S_4 and S_6, S_8 are turned on, so that output terminal A is connected to output terminal N through S_4, D_2 and S_8 and D_6 and the load, or output terminal N is connected to output terminal A through S_6, D_8 and S_2, D_4 and load.

Mode 4: For the output voltage of $-V_{dc}$ at V_{AN}, switching devices S_3, S_4, S_8 are turned on and D_6 into conduction. The load current flows from the first dc voltage source V_1, S_3, S_8, D_6, output terminal N, load, output terminal A, S_4, back to source terminal V_1. Alternately, load current flows from second dc voltage source V_2, switching device S_7, output terminal N, load, output terminal A, S_4, D_2 and S_8 back to second dc voltage source V_2, thus completes the circuit for $V_{AN} = -V_{dc}$.

Mode 5: In order to generate $-2V_{dc}$ as net voltage at V_{AN}, switching devices S_3, S_4 and S_7, S_8 are turned on. The load current flows from first dc voltage source V_1, switching devices S_3, S_8, second dc voltage source V_2, S_7, output terminal N, load, output terminal A, S_4 back to first dc voltage source V_1, thus completes the circuit and $-2V_{dc}$ appears at V_{AN}. Thus, circuit shown in Fig. 5.4 consists of two H-bridge inverters connected in series with equal dc sources can generate the three-level stepped voltage as given in Fig. 5.5 as per the switching sequence mentioned in table 5.3.

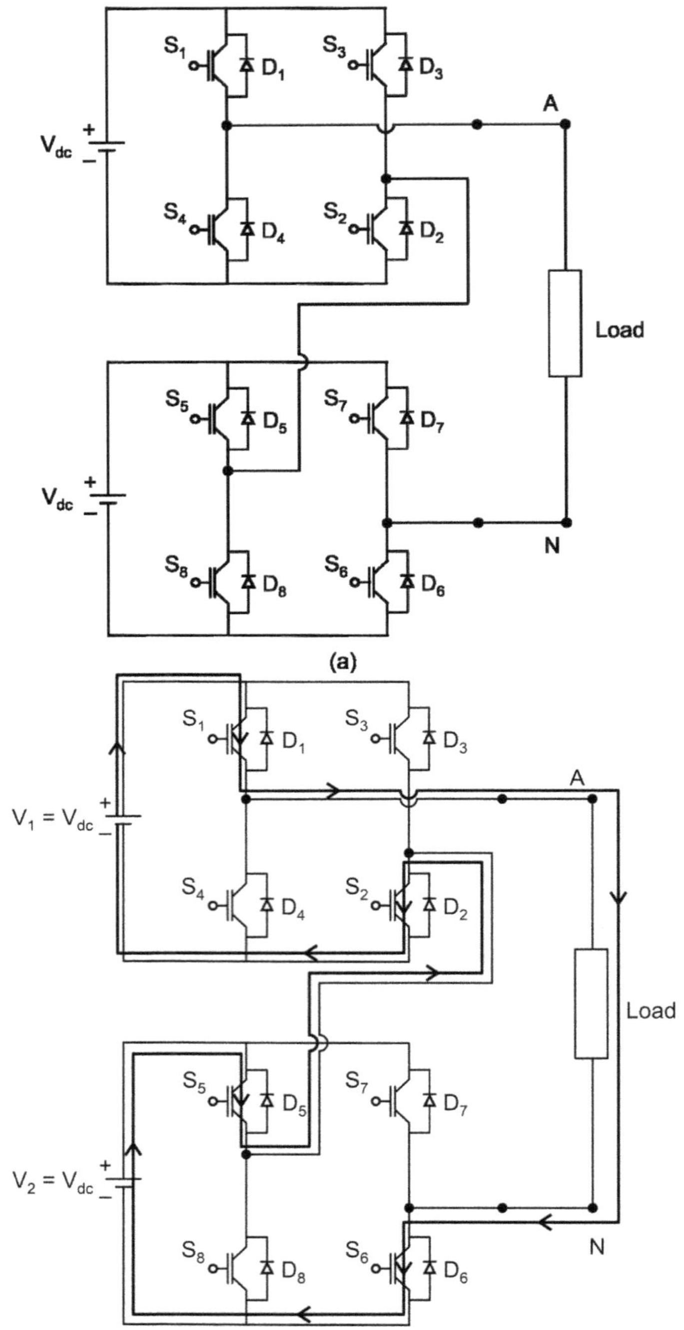

(b) Mode1: $V_{AN} = +2V_{dc}$

Fig. 5.4 Symmetrical configuration for Five-level voltage waveform for Cascaded H-bridge inverter (a) Circuit diagram, (b) Mode1: $V_{AN} = +2V_{dc}$, (c) Mode2: $V_{AN} = +V_{dc}$, (d) Mode3: $V_{AN} = 0$, (e) Mode4: $V_{AN} = -V_{dc}$, and (f) Mode5: $V_{AN} = -2V_{dc}$.

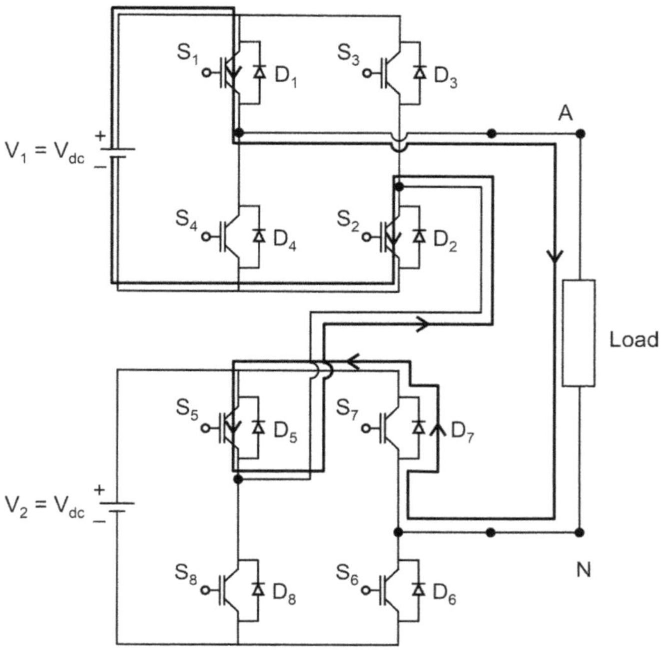

(c) Mode2: $V_{AN} = +V_{dc}$

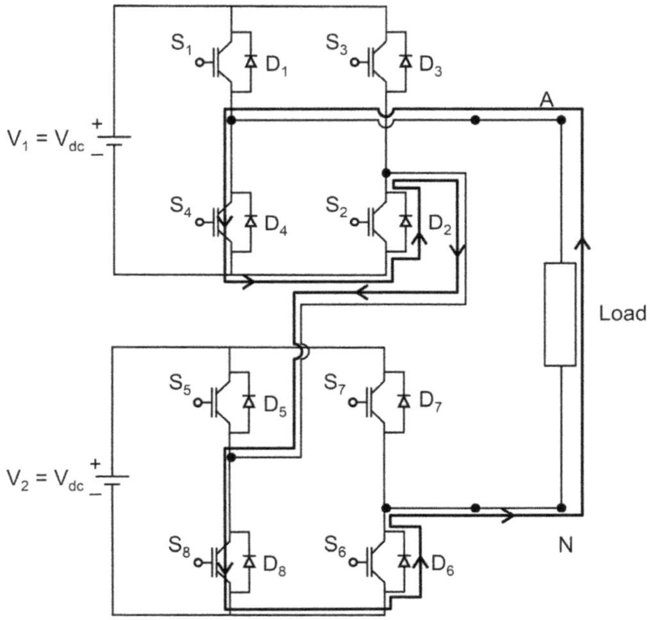

(d) Mode3: $V_{AN} = 0$

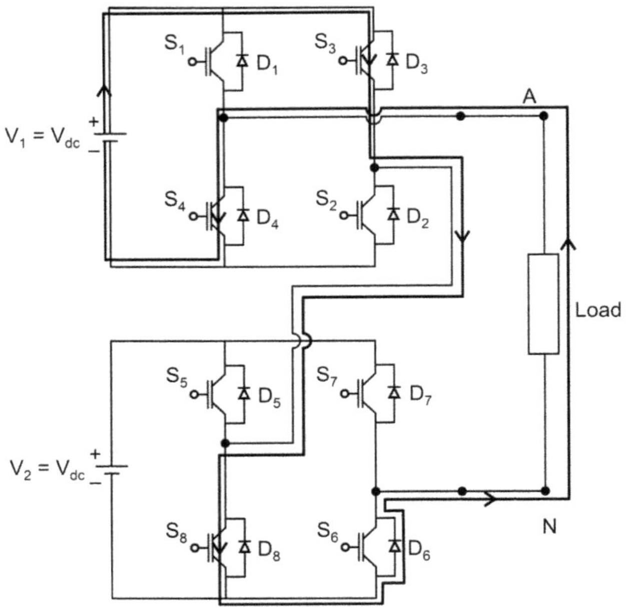

(e) Mode4: $V_{AN} = -V_{dc}$

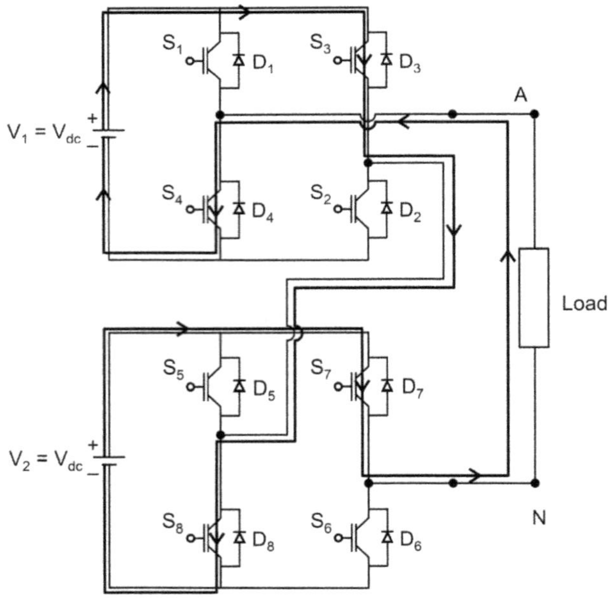

(f) Mode5: $V_{AN} = -2V_{dc}$

Fig. 5.4 Symmetrical configuration for Five-level voltage waveform for Cascaded H-bridge inverter (a) Circuit diagram, (b) Mode1: $V_{AN} = +2V_{dc}$, (c) Mode2: $V_{AN} = +V_{dc}$, (d) Mode3: $V_{AN} = 0$, (e) Mode4: $V_{AN} = -V_{dc}$, and (f) Mode5: $V_{AN} = -2V_{dc}$.

Table 5.3 Switching table for symmetrical configuration for Five-level cascaded H-bridge inverter with equal dc voltage sources.

Modes	Output Phase Voltage, $V_o=V_{AN}$	Switching States of two H-bridge inverter								Anti-parallel Diodes		Voltage sources connected
		S_1	S_2	S_3	S_4	S_5	S_6	S_7	S_8	D of first cell	D of second cell	
Mode1	$+2V_{dc}$	1	1	0	0	1	1	0	0	0	0	V_1+V_2
Mode2	$+V_{dc}$	1	1	0	0	1	0	0	0	0	D_7	V_1
Mode2	$+V_{dc}$	1	0	0	0	1	1	0	0	D_3	0	V_2
Mode3	0	0	0	0	1	0	0	0	1	D_2	D_6	0
Mode3	0	0	1	0	0	0	1	0	0	D_4	D_8	0
Mode4	$-V_{dc}$	0	0	1	1	0	0	0	1	0	D_6	$-V_1$
Mode4	$-V_{dc}$	0	0	0	1	0	0	1	1	D_2	0	$-V_2$
Mode5	$-2V_{dc}$	0	0	1	1	0	0	1	1	0	0	$-V_1-V_2$

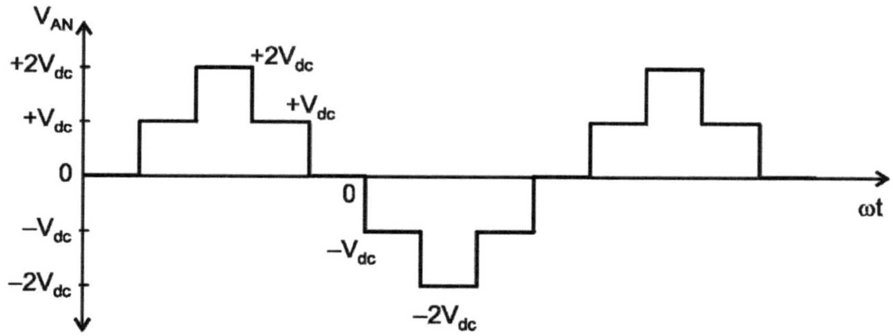

Fig. 5.5 Symmetrical configuration for Five-level voltage waveform for Cascaded H-bridge inverter for load Phase voltage, V_{AN}.

Symmetrical configuration for Five-level Cascaded H-bridge inverter for Three-Phase star connected loads: Similarly, three separate single leg modules are arranged to get three-phase five-level cascaded H-bridge circuit configuration for three-phase star-connected loads as given in the Fig.5.6. The dc source voltages are equal for all the three phases, namely, $V_{A1}=V_{B1}=V_{C1}=V_{A2}=V_{B2}=V_{C2}=V_{dc}$. The three phase load voltages are same as that of single-phase circuit but for a phase shift of 120° between each phase. The phase current waveform depends on the values of RL loads.

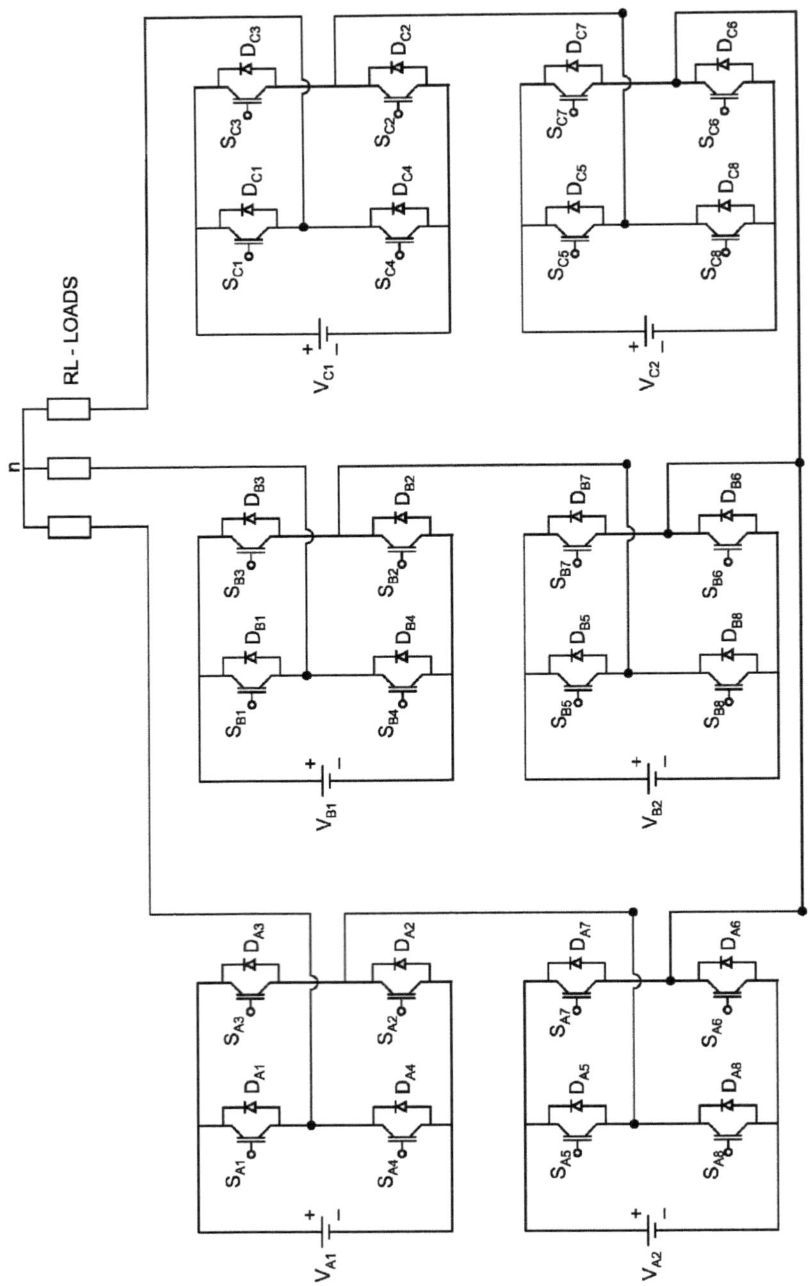

Fig. 5.6 Symmetrical configuration for Five-level Cascaded H-bridge inverter for Three-Phase star connected loads.

5.2.3 Symmetrical Configuration for Seven–Level Cascaded H-bridge Inverter with Equal dc Sources (with three cells)

Three H-bridge inverters with equal dc sources are connected in series (3 cells), namely, H_1, H_2, and H_3 and RL load is connected between one of the output terminals of H_1 and H_3. Each cell has its own individual isolated dc source and four switching devices along with anti-parallel diodes. These three cells are connected in series at the output ac side. At the input side three equal (symmetrical) dc voltage sources are connected as V_{dc} (V_1=V_2=V_3=V_{dc}). The switching devices of first H-bridge cell (H_1) are designated as S_1, S_2, S_3 and S_4, and for second H-bridge cell (H_2) are S_5, S_6, S_7 and S_8, and third H-bridge cell (H_3) are S_9, S_{10}, S_{11} and S_{12}.

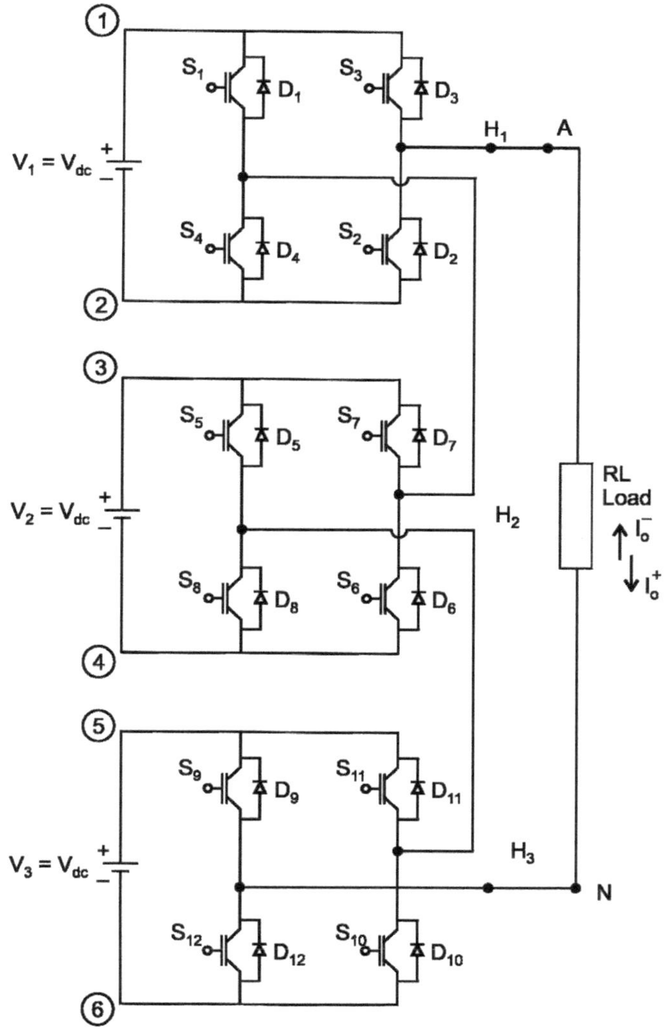

(a)

The individual H-bridge cell output voltages are V_1, V_2, and V_3, respectively and the series connected cells combined output voltage as V_{AN}. Each H-bridge cell generates three level of voltages, $+V_{dc}$, 0, $-V_{dc}$, whereas the cascaded three cells of H-bridge inverter generates seven levels of voltages, $+3V_{dc}$, $+2V_{dc}$, $+V_{dc}$, 0, $-V_{dc}$, $-2V_{dc}$, $-3V_{dc}$. The switching sequence is given in Table 5.4 and seven-level cascaded inverter with equal dc voltage sources is shown in Fig. 5.7 and its operation for seven-levels are explained in detail as follows. In Fig. 5.8 the voltage waveform is shown for seven-levels with equal magnitude of dc sources voltages.

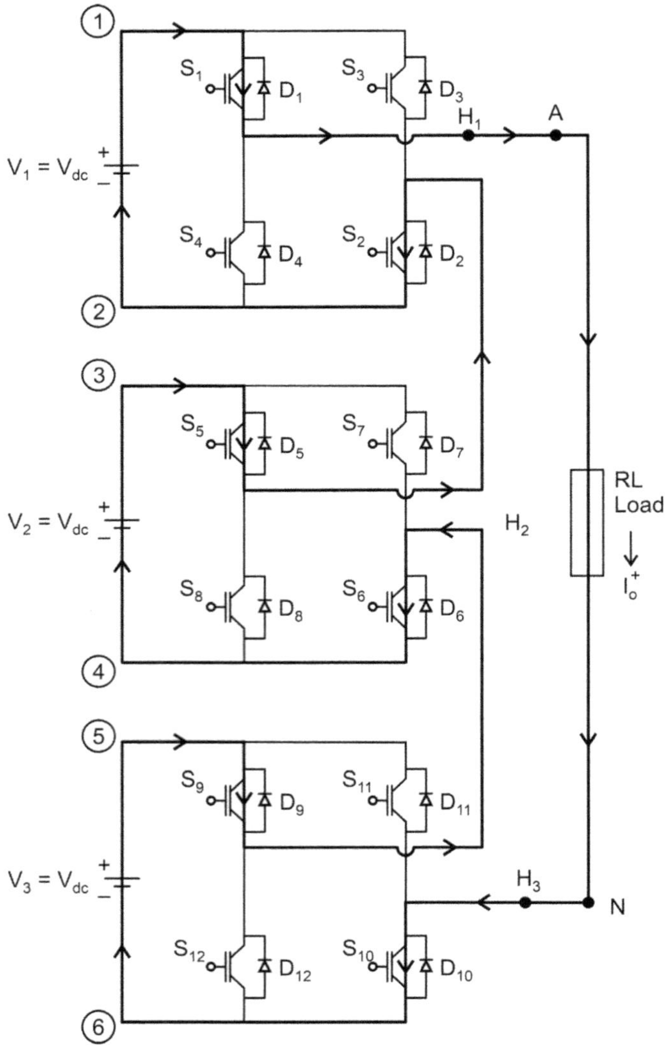

(b) Mode 1: $V_{AN} = +3V_{dc}$

Fig. 5.7 Seven-level cascaded H-bridge inverter (with 3-cells) with equal dc voltage sources for Symmetrical configuration: (a) Main circuit, (b) Mode1: $V_{AN} = +3V_{dc}$, (c) Mode2: $V_{AN} = +2V_{dc}$, (d) Mode3: $V_{AN} = +V_{dc}$, (e) Mode4: $V_{AN} = 0$, (f) Mode5: $V_{AN} = -3V_{dc}$, (g) Mode6: $V_{AN} = -2V_{dc}$, and (h) Mode7: $V_{AN} = -V_{dc}$.

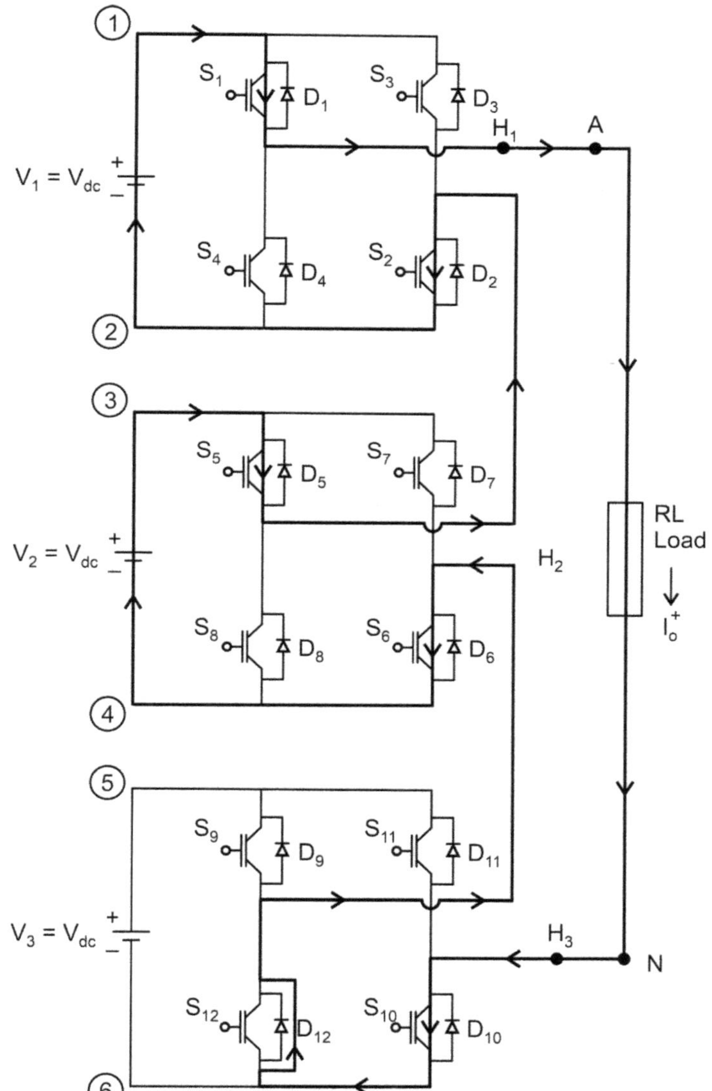

(c) Mode 2: $V_{AN} = +2V_{dc}$

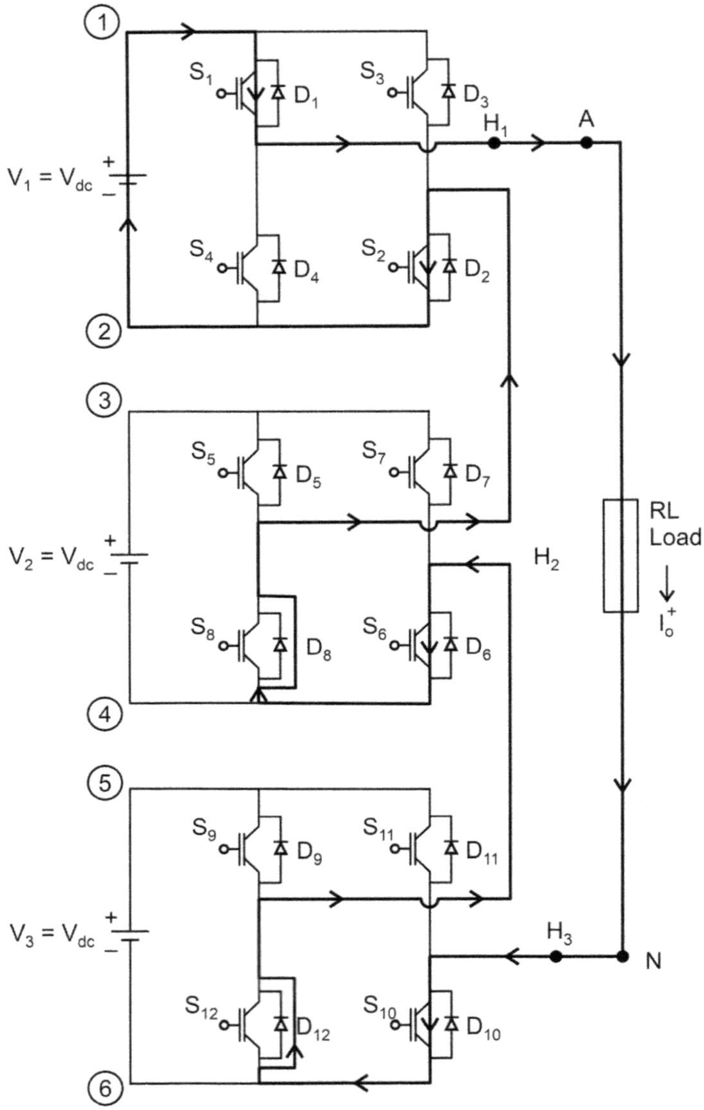

(d) Mode 3: $V_{AN} = +V_{dc}$

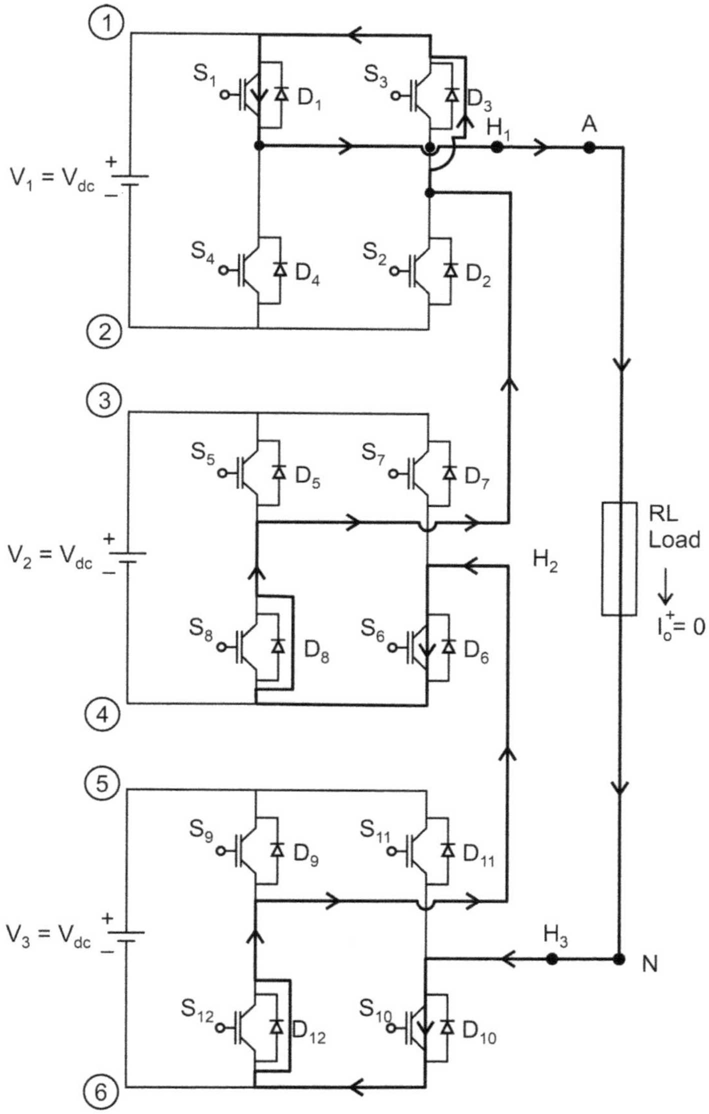

1
2
3
4
5
6
S1
D1
S3
D3
H1
A
V1 = Vdc
S4
D4
S2
D2
S5
D5
S7
D7
V2 = Vdc
RL
Load
H2
S8
D8
S6
D6
Io+ = 0
S9
D9
S11
D11
V3 = Vdc
H3
N
S12
D12
S10
D10

(e) Mode 4: $V_{AN} = 0$

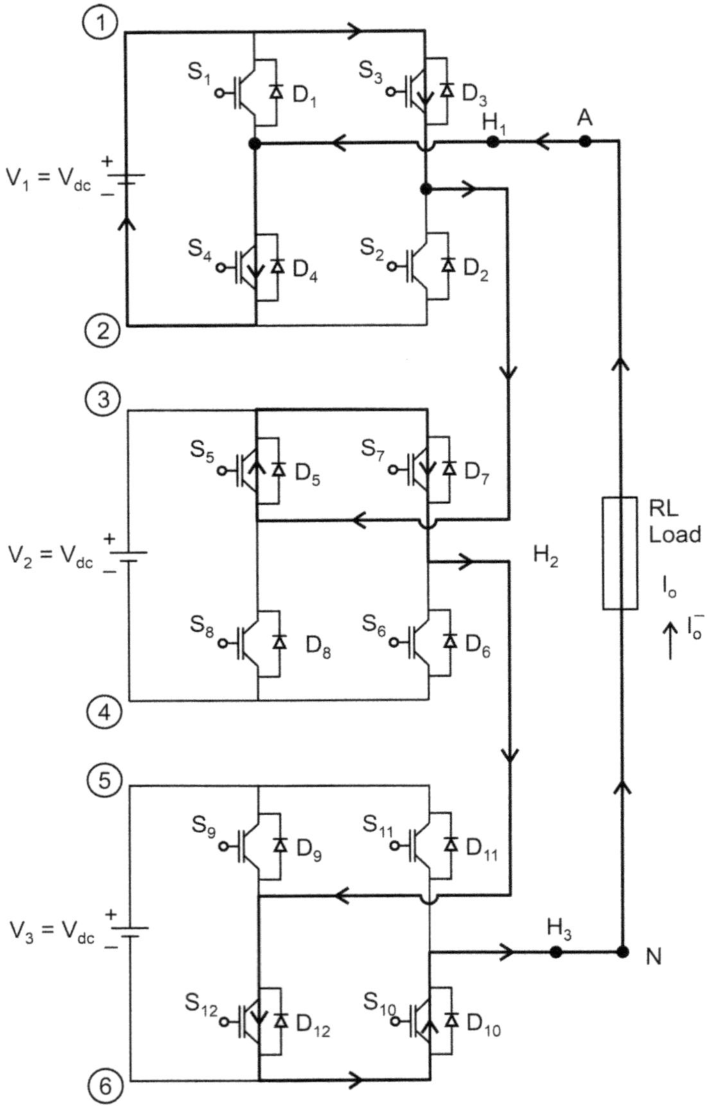

(f) Mode 5: $V_{AN} = -V_{dc}$

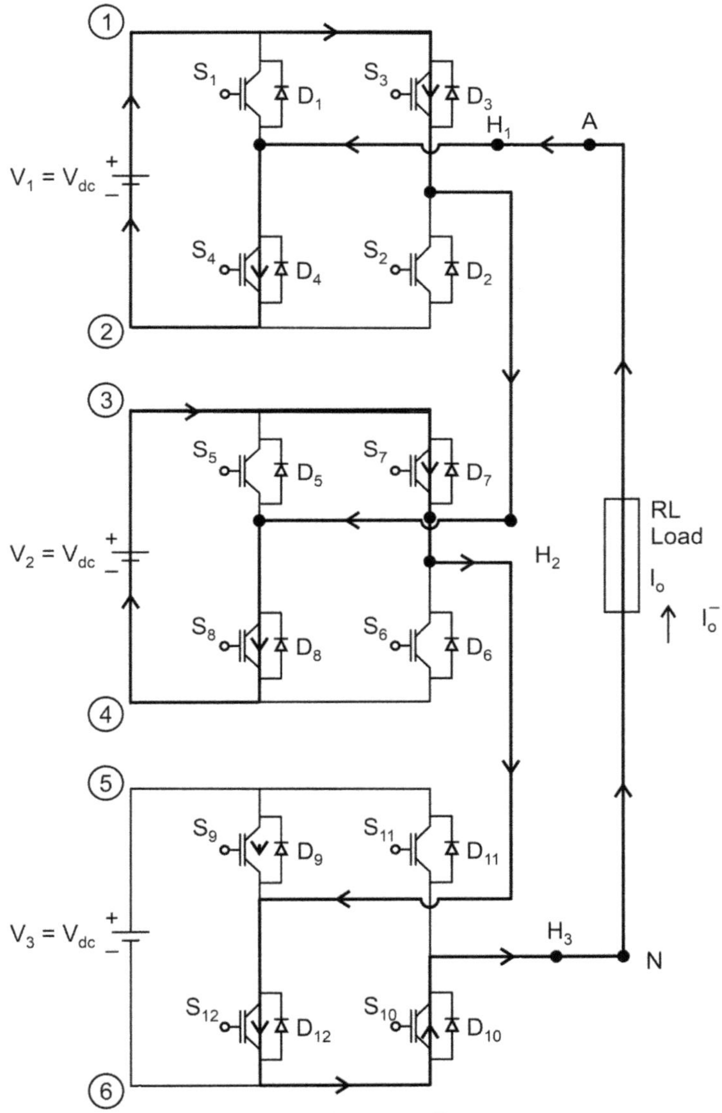

(g) Mode 6: $V_{AN} = -2V_{dc}$

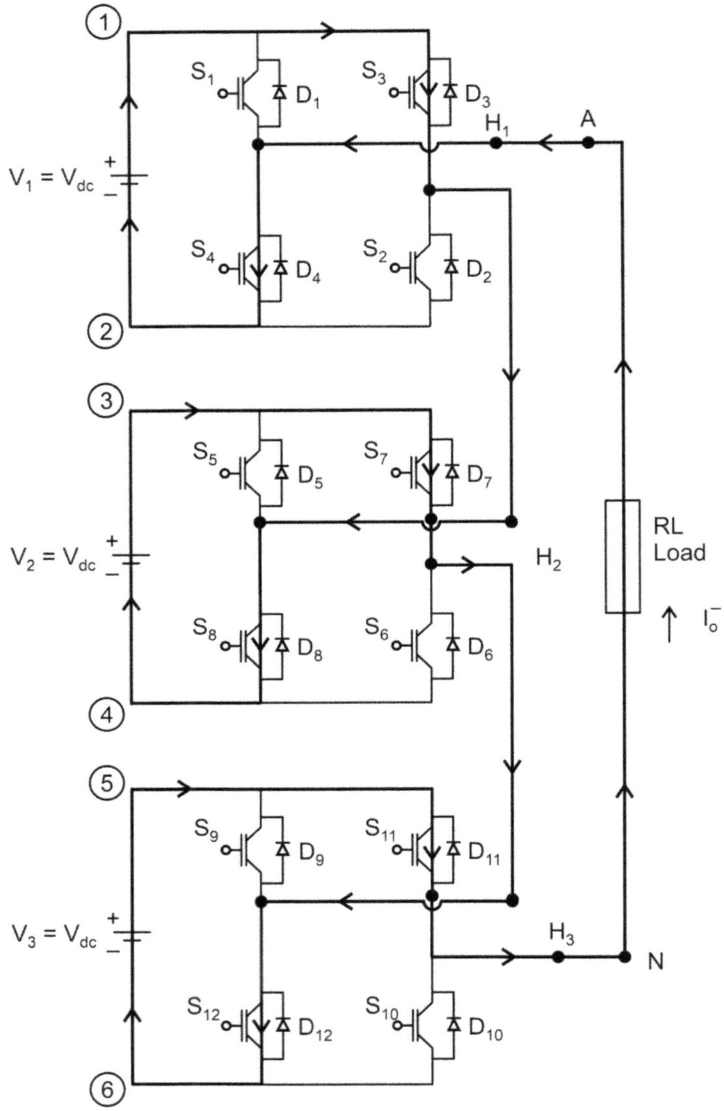

(h) Mode 7: $V_{AN} = -3V_{dc}$

Fig. 5.7 Seven-level cascaded H-bridge inverter (with 3-cells) with equal dc voltage sources for Symmetrical configuration: (a) Main circuit, (b) Mode1: $V_{AN} = +3V_{dc}$, (c) Mode2: $V_{AN} = +2V_{dc}$, (d) Mode3: $V_{AN} = +V_{dc}$, (e) Mode4: $V_{AN} = 0$, (f) Mode5: $V_{AN} = -V_{dc}$, (g) Mode6: $V_{AN} = -2V_{dc}$, and (h) Mode7: $V_{AN} = -3V_{dc}$.

Mode 1: For $V_{AN} = +3V_{dc}$, the conduction takes place through dc source terminal 1 of H_1 bridge, S_1, load A to N, S_{10}, dc source terminal 6 to 5, S_9, S_6, dc source terminal 4 to 3, S_5 of H_2 bridge, S_2, dc source terminal 2 to 1 of H_1 bridge, and load current is positive.

Mode 2: To obtain $V_{AN} = +2V_{dc}$, current flows from dc source terminal 1, S_1, load terminal A to N, S_{10}, D_{12}, S_6, dc source terminal 4 to 3, S_5, S_2, dc source terminal 2 to 1, and load current is positive, (only one possibility is given).

Mode 3: For $V_{AN} = +V_{dc}$, the conduction takes place through dc source terminal 1, S_1, load terminal A to N, S_{10}, D_{12}, S_6, D_8, S_2, dc source terminal 2 to 1, and load current is positive, (only one possibility is shown).

Mode 4: To get $V_{AN} = 0$, current flows from dc source terminal 1, S_1, load terminal A to N, S_{10}, D_{12}, S_6, D_8, D_3, to dc source terminal 1, and load current zero, (only one possibility is shown).

Mode 5: For $V_{AN} = -V_{dc}$, the conduction takes place through dc source terminal 1, S_3, D_5, S_7, S_{12}, D_{10}, load terminal N to A, S_4, dc source terminal 2 to 1, and load current is negative, (only one possibility is shown).

Mode 6: To obtain $V_{AN} = -2V_{dc}$, current flows from dc source terminal 1, S_3, S_8, dc source terminal 4 to 3, S_7, S_{12}, D_{10}, load terminal N to A, S_4, dc source terminal 2 to 1, and load current is negative, (only one possibility is shown).

Mode 7: For $V_{AN} = -3V_{dc}$, the conduction takes place through dc source terminal 1, S_3, S_8, dc source terminal 4 to 3, S_7, S_{12}, dc source terminal 6 to 5, S_{11}, load terminal N to A, S_4, dc source terminal 2 to 1, and load current is negative.

Table. 5.4 Switching table for Seven-level cascaded H-bridge inverter (with 3-cells) with equal dc voltage sources for symmetrical configuration.

Modes	V_{AN}	S_1	S_2	S_3	S_4	S_5	S_6	S_7	S_8	S_9	S_{10}	S_{11}	S_{12}	Diodes
M1	$+3V_{dc}$	1	1	0	0	1	1	0	0	1	1	0	0	----
M2	$+2V_{dc}$	1	1	0	0	1	1	0	0	0	1	0	0	D_{12}
M3	$+V_{dc}$	1	1	0	0	0	1	0	0	0	1	0	0	D_8, D_{12}
M4	0	1	0	0	0	0	1	0	0	0	1	0	0	D_3, D_8, D_{12}
M5	$-V_{dc}$	0	0	1	1	0	0	1	0	0	0	0	1	D_5, D_{10}
M6	$-2V_{dc}$	0	0	1	1	0	0	1	1	0	0	0	1	D_{10}
M7	$-3V_{dc}$	0	0	1	1	0	0	1	1	0	0	1	1	----

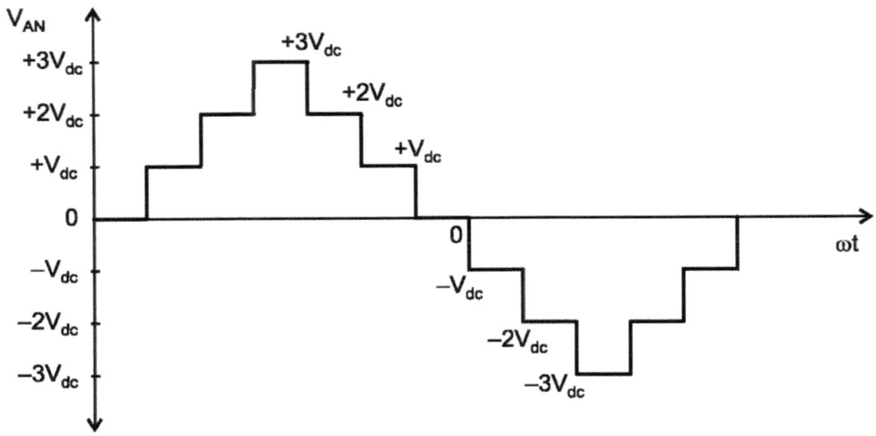

Fig. 5.8 Voltage waveform for seven-level cascaded H-bridge inverter (with 3-cells) with equal dc source voltages for symmetrical configuration.

Similarly, three-independent cascaded (three cells) leg modules are arranged to get three-phase H-bridge circuit configuration for three-phase star-connected loads for seven-level as given in the Fig.5.9.

5.3 Asymmetrical Cascaded H-Bridge Inverters

For understanding Asymmetrical Cascaded H-bridge Inverters, two cases are considered firstly, with two-cascaded H-bridge inverter and second one with three cascaded H-bridge inverters to demonstrate seven-level and thirteen-level output phase voltages, respectively. Asymmetrical configuration will have unequal dc source voltages for each H-bridge connected in series.

5.3.1 Asymmetrical Configuration for Seven Level Cascaded H-bridge Inverter with Unequal dc Voltage Sources (with two-cells)

The two H-bridge inverters with unequal dc voltage sources are connected in series, input dc voltage of first inverter is $V_1=V_{dc}/2$ and input dc voltage of second inverter is $V_2=V_{dc}$. This cascaded H-bridge inverter is operated appropriately switching their switching devices, the inverter can generate seven-level stepped voltages at the ac output V_{AN}. The seven-levels are $+3/2V_{dc}$, $+V_{dc}$, $+V_{dc}/2$, 0, $-V_{dc}/2$, $-V_{dc}$, $-3/2V_{dc}$ when $V_{AN} = V_1+V_2$ ($+3/2V_{dc}$), V_1($+V_{dc}/2$), V_2($+V_{dc}$), $-V_1$($-V_{dc}/2$), $-V_2$($-V_{dc}$) and $-V_1-V_2$ ($-3/2V_{dc}$) and when no source is connected $V_{AN}=0$. The switching sequence is given in Table 5.5 and the seven-level cascaded H-bridge inverter with unequal dc voltage sources is shown in Fig.5.10 and its operation for seven-levels are explained in detail. In Fig.5.11 the seven-level stepped voltage waveform with unequal dc voltage sources is shown.

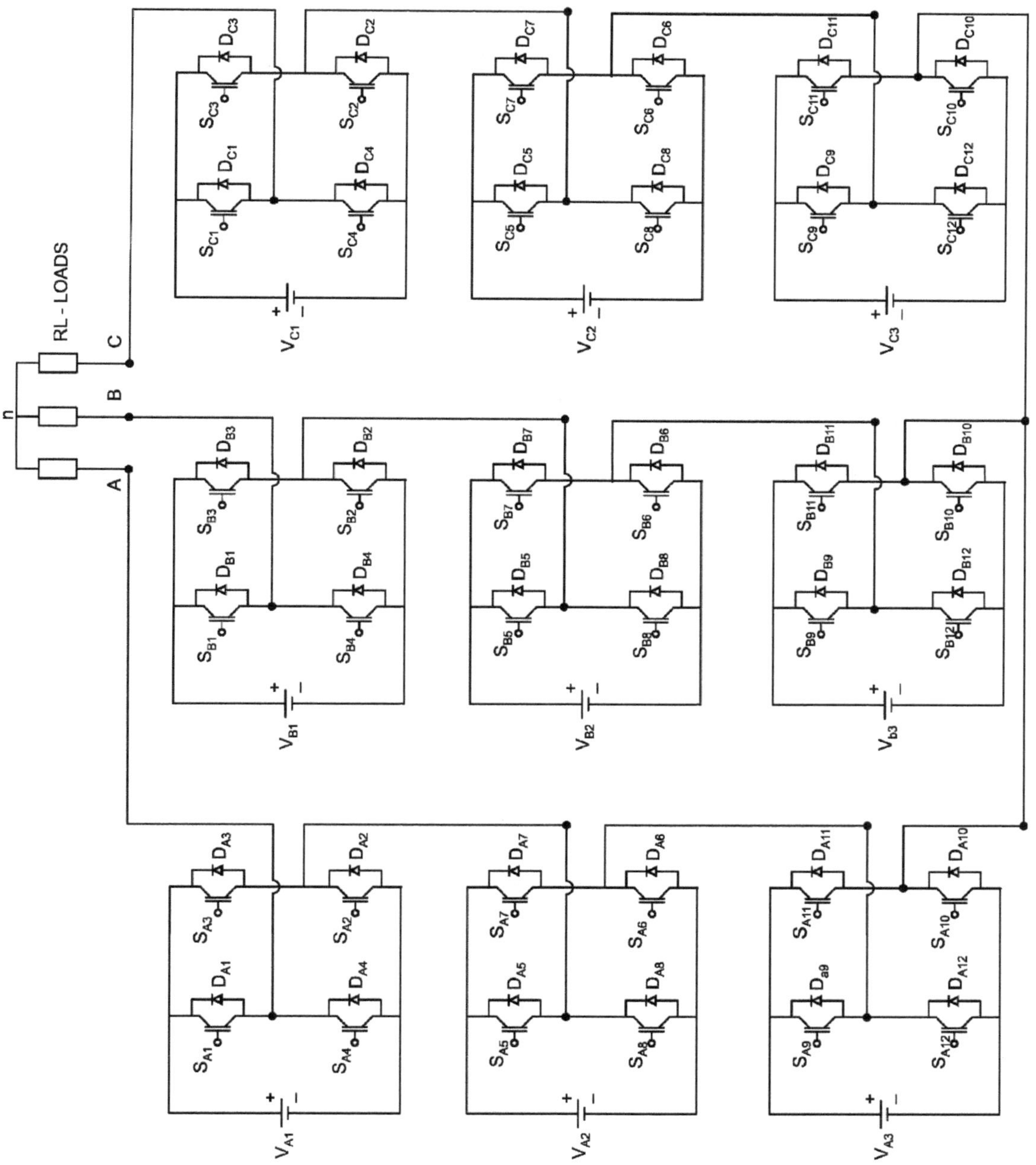

Fig. 5.9 Symmetrical configuration for Seven-level Cascaded H-bridge inverter for Three-Phase star connected loads.

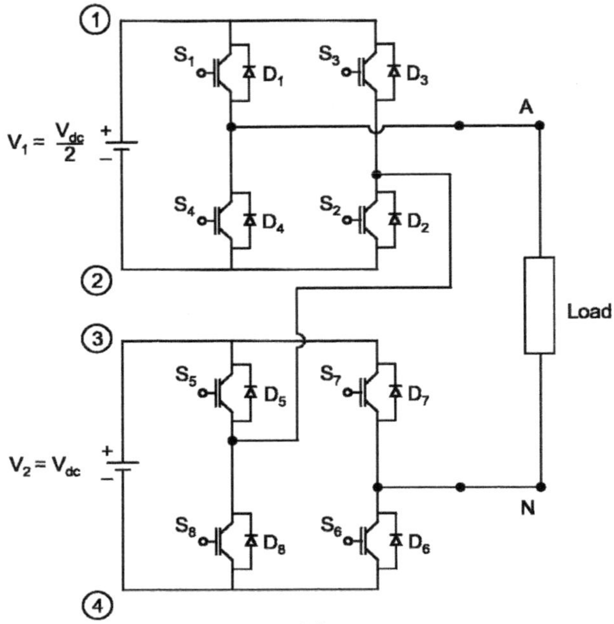

(a)

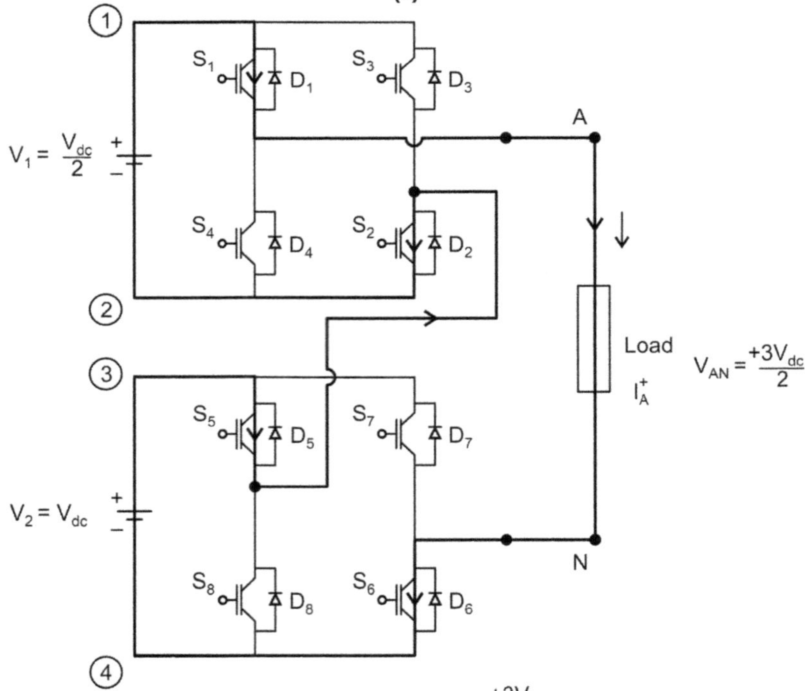

(b) Mode 1: $V_{AN} = \frac{+3V_{dc}}{2}$

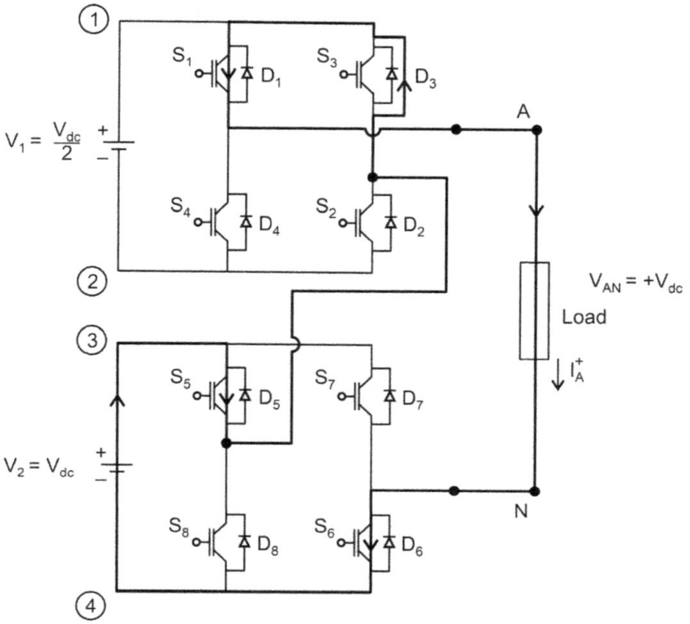

(c) Mode 2: $V_{AN} = +V_{dc}$

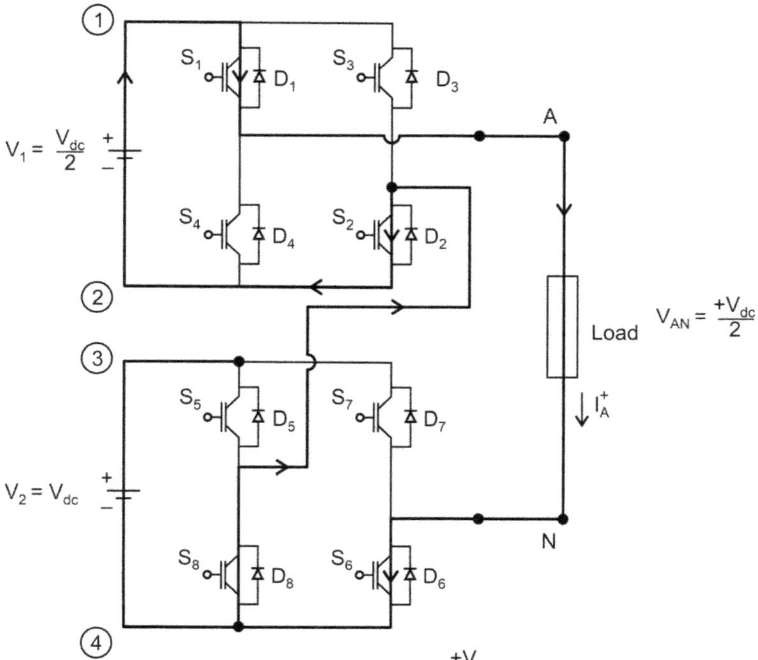

(d) Mode 3 : $V_{AN} = \frac{+V_{dc}}{2}$

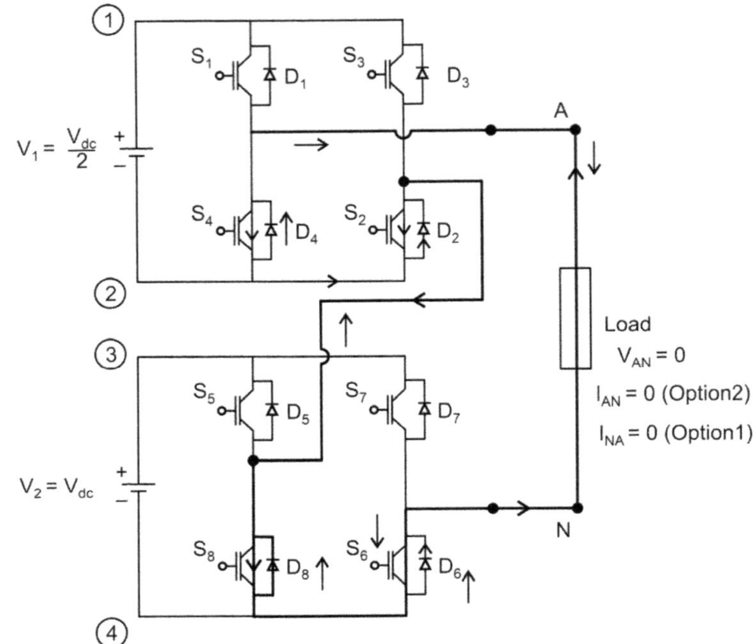

(e) Mode 4: $V_{AN} = 0$

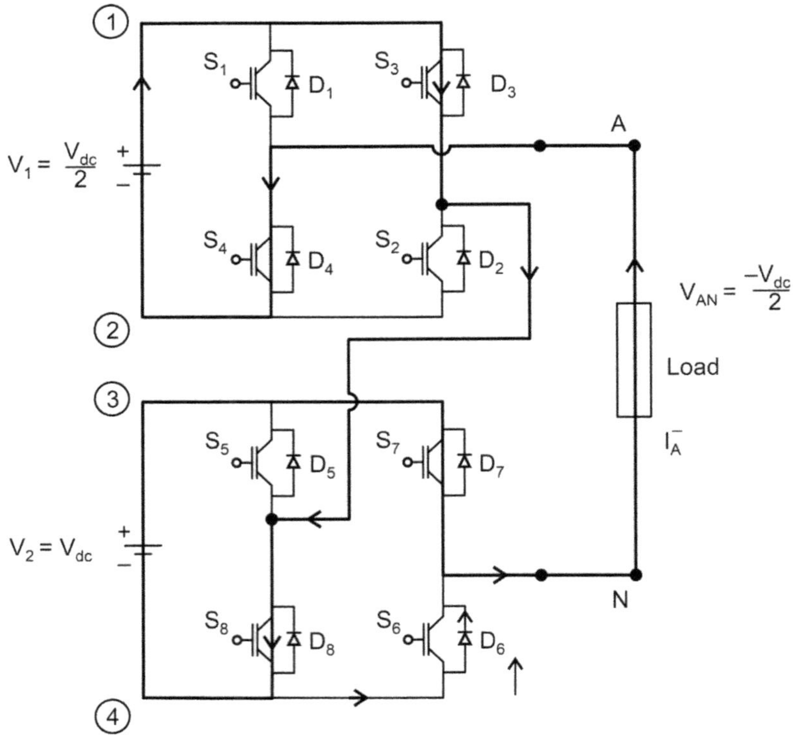

(f) Mode 5 : $V_{AN} = \frac{-V_{dc}}{2}$

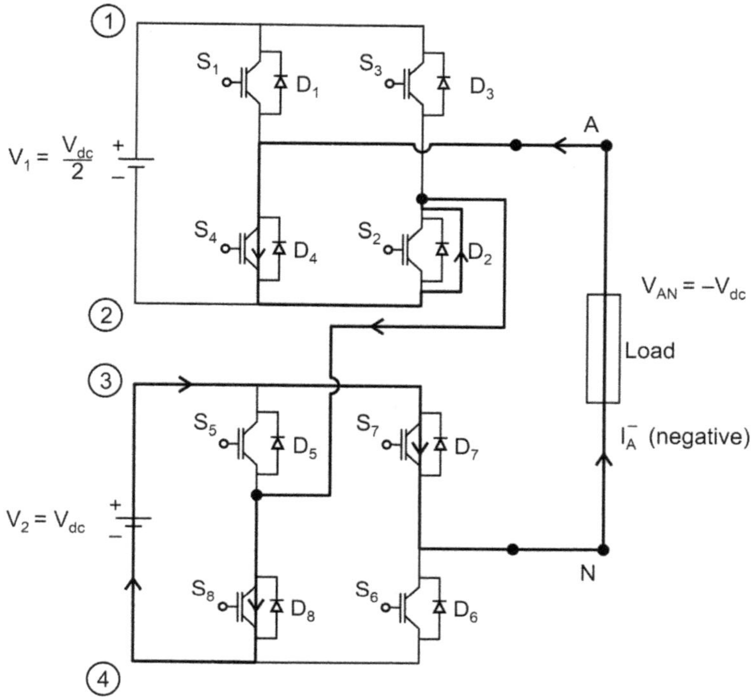

(g) Mode 6 : $V_{AN} = -V_{dc}$

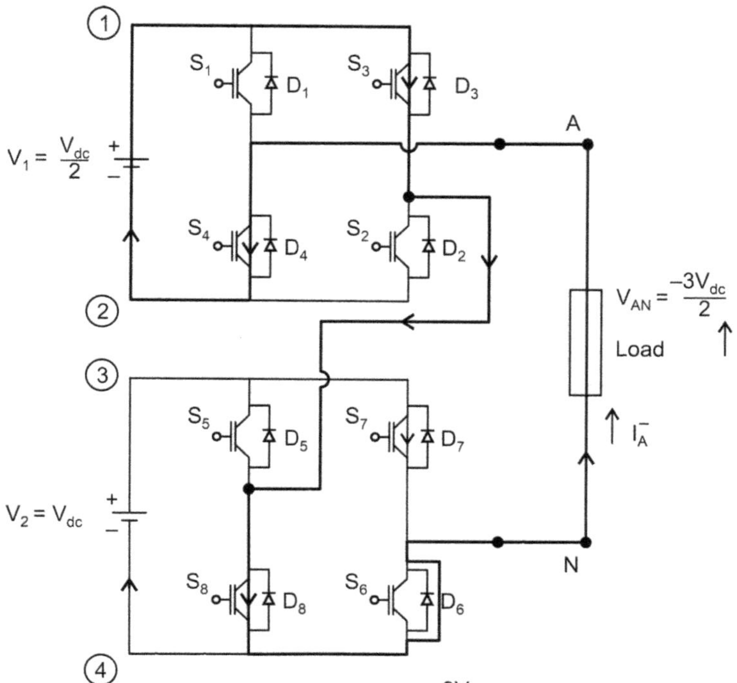

(h) Mode 7 : $V_{AN} = \frac{-3V_{dc}}{2}$

Fig. 5.10 Asymmetrical configuration for Seven-level cascaded H-bridge inverter with unequal dc voltage sources, (a) Main circuit, (b) Mode1: $+3V_{dc}/2$; (c) Mode2: $+V_{dc}$; (d) Mode3: $+V_{dc}/2$; (e) Mode4: 0; (f) Mode5: $-V_{dc}/2$. (g) Mode6: $-V_{dc}$; (h) Mode7: $-3V_{dc}/2$.

Table 5.5 Switching table for Seven-level cascaded H-bridge inverter with unequal dc voltage sources for Asymmetrical configuration.

Modes	Output Voltages, $V_o=V_{AN}$	Switching States of two H-bridge inverter								Anti-parallel Diodes		Voltage sources connected
		S_1	S_2	S_3	S_4	S_5	S_6	S_7	S_8	Diodes of first cell	Diodes of second cell	
Mode1	$+3V_{dc}/2$	1	1	0	0	1	1	0	0	0	0	V_1+V_2
Mode2	$+V_{dc}$	1	0	0	0	1	1	0	0	D_3	0	V_2
Mode3	$+V_{dc}/2$	1	1	0	0	0	1	0	0	0	D_8	V_1
Mode4	0	0	0	0	1	0	0	0	1	D_2	D_6	0
Mode4	0	0	1	0	0	0	1	0	0	D_4	D_8	0
Mode5	$-V_{dc}/2$	0	0	1	1	0	0	0	1	0	D_6	$-V_1$
Mode6	$-V_{dc}$	0	0	0	1	0	0	1	1	D_2	0	$-V_2$
Mode7	$-3V_{dc}/2$	0	0	1	1	0	0	1	1	0	0	$-V_1-V_2$

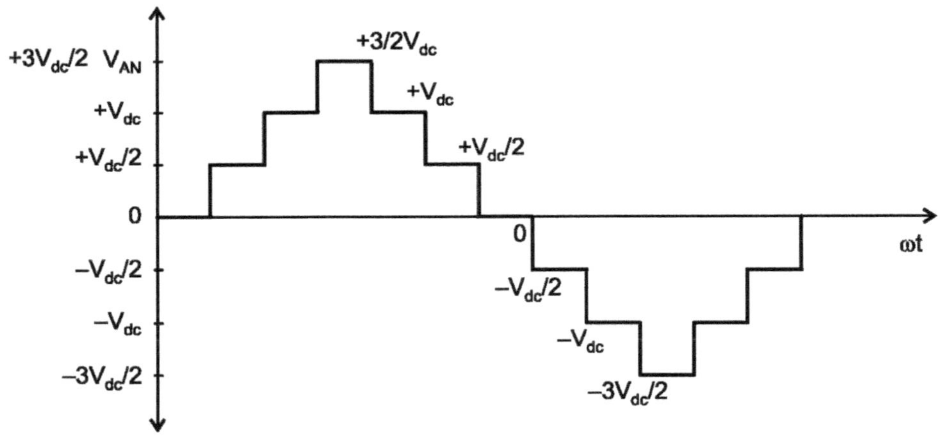

Fig. 5.11 Seven-level stepped voltage waveform for cascaded (2 cells) H-bridge inverter with unequal dc voltage sources for Asymmetrical configuration.

The seven-level Cascaded H-bridge inverter with unequal dc voltage sources as shown in Fig.5.10.

Mode 1: The net output voltage $V_o=V_{AN}$, will be equal to $+3V_{dc}/2$ by switching the devices S_1, S_2 of first H-bridge and S_5, S_6 of second H-bridge of the CHB Inverter, turned-on, the load current flows from First dc source through S_1, output terminal A, load, output terminal N, S_6, second dc source, S_5, S_2 and back to first dc source, thus completes the circuit and V_1+V_2 will be equal to $+3V_{dc}/2$ ($V_{dc}/2 + V_{dc} = 3V_{dc}/2$) appear at V_{AN}.

Mode 2: The $+V_{dc}$ is obtained at the output terminals V_{AN} when switching devices S_1, S_5, S_6 are turned-on and make anti-parallel diode D_3 to conduct, then load current flows from second dc source through S_5, S_1, D_3, output terminal A, load, output terminal N, S_6 and back to second dc source, thus second dc source V_2 is connected as $+V_{dc}$ at V_{AN}.

Mode 3: The CHB inverter can also generate $+V_{dc}/2$ when switching devices S_1, S_2, S_6 are turned-on and make diode D_8 to conduct, then the load current flows from first dc source V_1 through S_1, output terminal A, load, output terminal N, S_6 and D_8, S_2 and back to first dc source, thus circuit completes.

Mode 4: Lastly, V_{AN} will be 0 and there are two ways to obtain zero voltage. This can be achieved by switching one device and one diode in each H-bridge circuit. As per first option, the switching devices S_4 and S_8 are turned-on and diodes D_2 and D_6 are made to conduct. Then the load current flows from output terminal A, through S_4, D_2 and S_8, D_6, output terminal N, load and back to output terminal A, thus circuit completes and generate 0 (zero) voltage at V_{AN}, since no dc source is connected. As per the second option, the switching devices S_2, S_6 are turned-on and diodes D_4, D_8 are made to conduct, then the load current flows from output terminal N, S_6, D_8 and S_2, D_4, output terminal A into the load, thus 0 (zero) voltage is generated at V_{AN}, since no dc source is connected. The seven - level stepped voltage waveform generated by this CHB inverter is shown in Fig. 5.12.

Mode 5: Whereas $-V_{dc}/2$ can be obtained when switching devices S_3, S_4, S_8 are turned-on and make diode D_6 to conduct, then the load current flows from first dc source V_1 through S_3, S_8 and D_6, output terminal N, load, output terminal A, S_4 and back to first dc source V_1, thus V_{AN} is $-V_{dc}/2$, since current flows from terminal N to terminal A through load.

Mode 6: Whereas it can generate $-V_{dc}$ at V_{AN} when switching devices S_4, S_7, S_8 are turned-on and make diode D_2 to conduct, then the load current from the second dc source V_2 through S_7, output terminal A, load, output terminal N, S_4 and D_2, S_8 and back to second dc source. Here the load current flows from output terminal N to output terminal A through load, (Negative Current).

Mode 7: Whereas $-3V_{dc}/2$ can be obtained when switching devices S_3. S_4 of first H-bridge and S_7, S_8 of second H-bridge of CHB inverter are turned-on, then load current will now flow from first dc source through S_3, S_8, second dc source V_2, S_7, output terminal 2, load, output terminal A, S_4, beck to first dc source V_1, which completes the circuit and generated $-3V_{dc}/2$ at V_{AN} as current flows from output terminal N to terminal A through load, (Negative current).

Asymmetrical configuration for Seven-level Cascaded (2 cells) H-bridge inverter for Three-Phase star connected loads: Similarly, three separate single leg modules are arranged to get three-phase seven-level cascaded H-bridge circuit configuration for three-phase star-connected loads as given in the Fig.5.12. The dc source voltages are unequal for all the three phases, namely, $V_{A1}=V_{B1}=V_{C1}= V_{dc}/2$, and $V_{A2}=V_{B2}=V_{C2}=V_{dc}$. The three phase load voltages are same as that of single-phase circuit but for a phase shift of 120^0 between each phase. The phase current waveform depends on the values of RL loads.

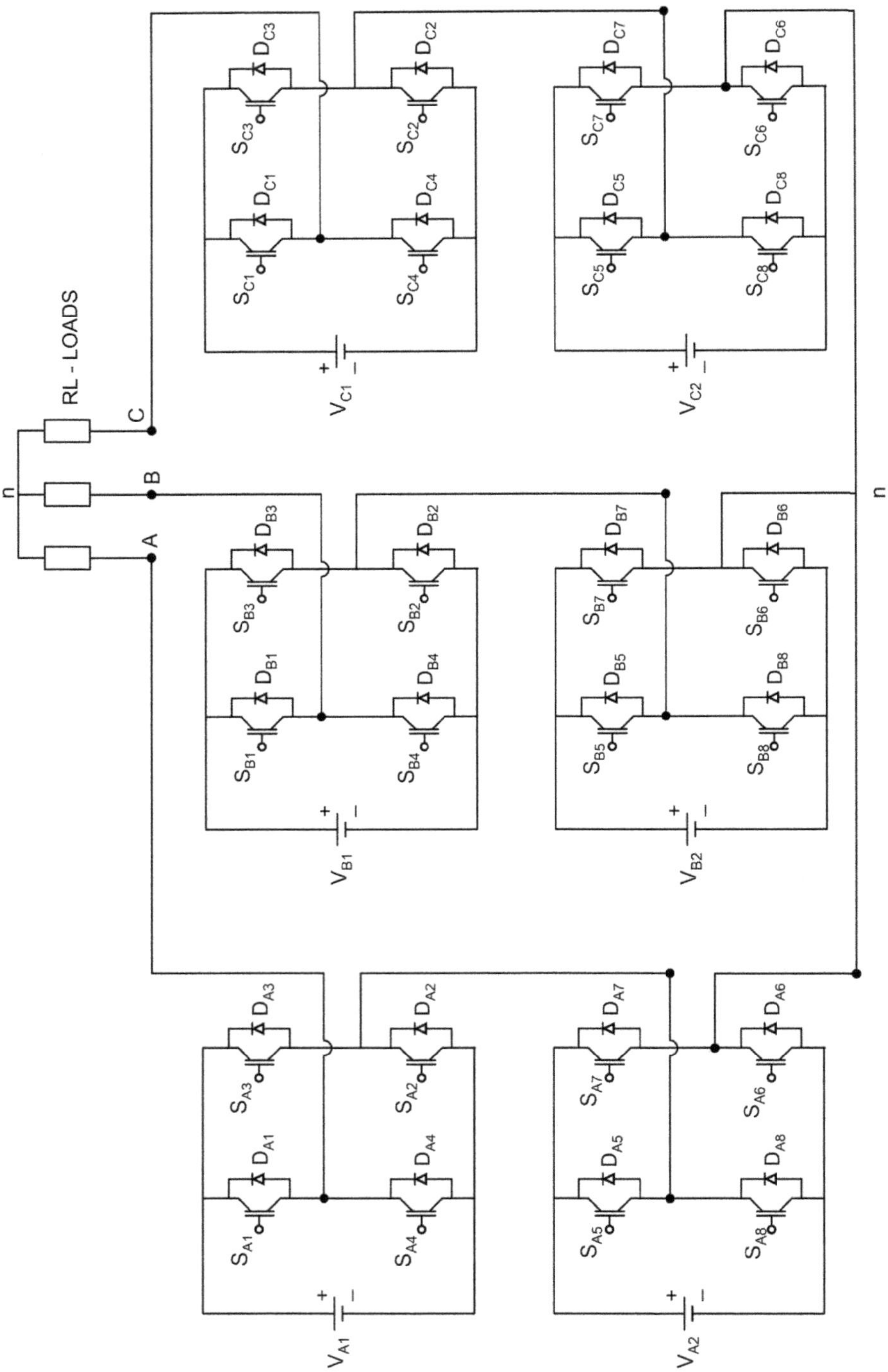

Fig. 5.12 Asymmetrical configuration for Three-phase seven-level cascaded (2 cells) H-bridge inverter for three-phase loads.

5.3.2 Asymmetrical Configuration for Thirteen - Level Cascaded H-bridge Inverter with Unequal dc Voltage Sources (with three-cells)

The load current will be positive if it flows from load terminals A to N, and negative for reverse flow. When three H-bridge inverters are connected in cascade with unequal dc voltage sources for RL load, the net output voltage across load terminals A and N are thirteen-levels, namely $\pm 6V_{dc}$, $\pm 5V_{dc}$, $\pm 4V_{dc}$, $\pm 3V_{dc}$, $\pm 2V_{dc}$, $\pm V_{dc}$, and 0. The operation of thirteen-level cascaded H-bridge inverter with unequal dc voltage sources are explained in detail as follows, with the help of circuit diagram given in Fig.5.13. In Fig. 5.14, the thirteen-level stepped voltage waveform with unequal dc voltage sources is shown.

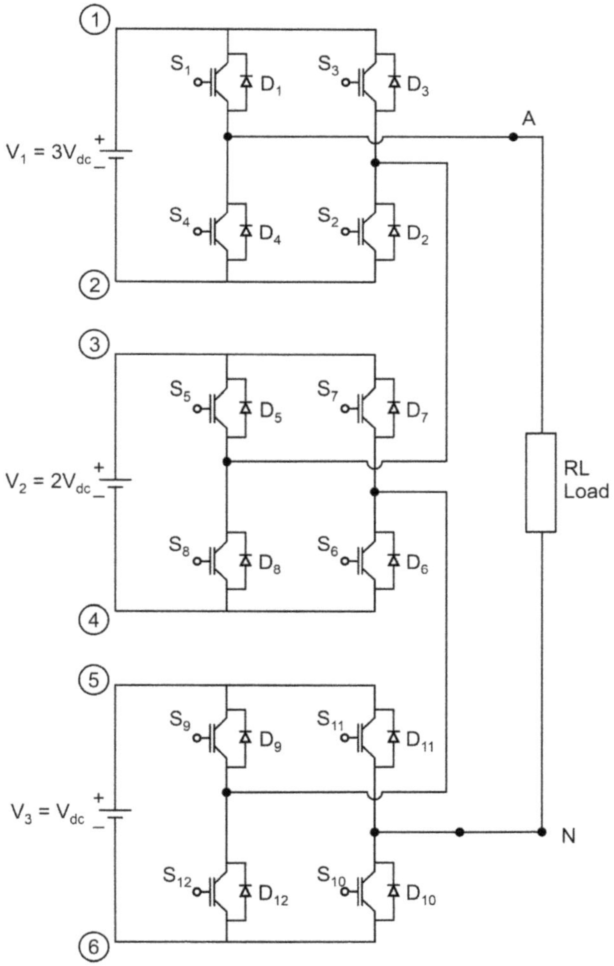

(a)

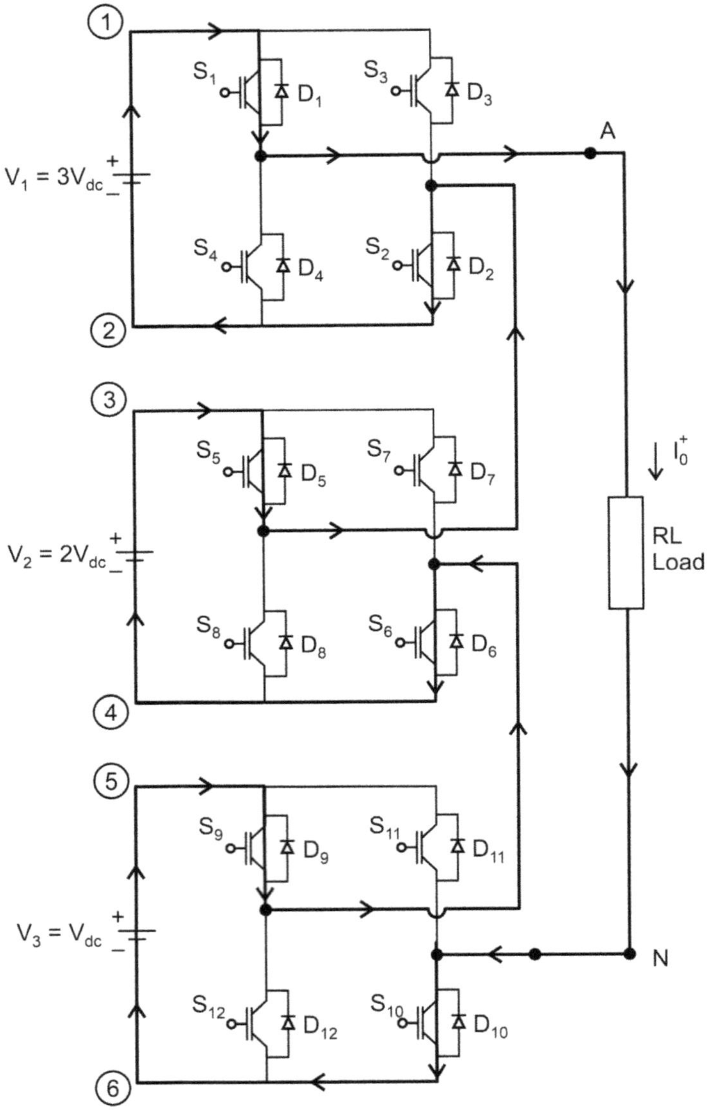

(b) Mode 1 ($V_{AN} = +6V_{dc}$)

Fig. 5.13 Thirteen-level cascaded H-bridge (3 cells) inverter with unequal dc voltage sources circuit for Asymmetrical configuration, (a) Main circuit; (b) Mode1: $V_{AN} = +6V_{dc}$; (c) Mode2: $V_{AN} = +5V_{dc}$; (d) Mode3: $V_{AN} = +4V_{dc}$; (e) Mode4: $V_{AN} = +3V_{dc}$; (f) Mode5: $V_{AN} = +2V_{dc}$; (g) Mode6: $V_{AN} = +V_{dc}$; (h) Mode7: $V_{AN} = 0$; (i) Mode8: $V_{AN} = -V_{dc}$; (i) Mode9: $V_{AN} = -2V_{dc}$; (k) Mode10: $V_{AN} = -3V_{dc}$; (l) Mode11: $V_{AN} = -4V_{dc}$; (m) Mode12: $V_{AN} = -5V_{dc}$; (n) Mode13: $V_{AN} = -6V_{dc}$.

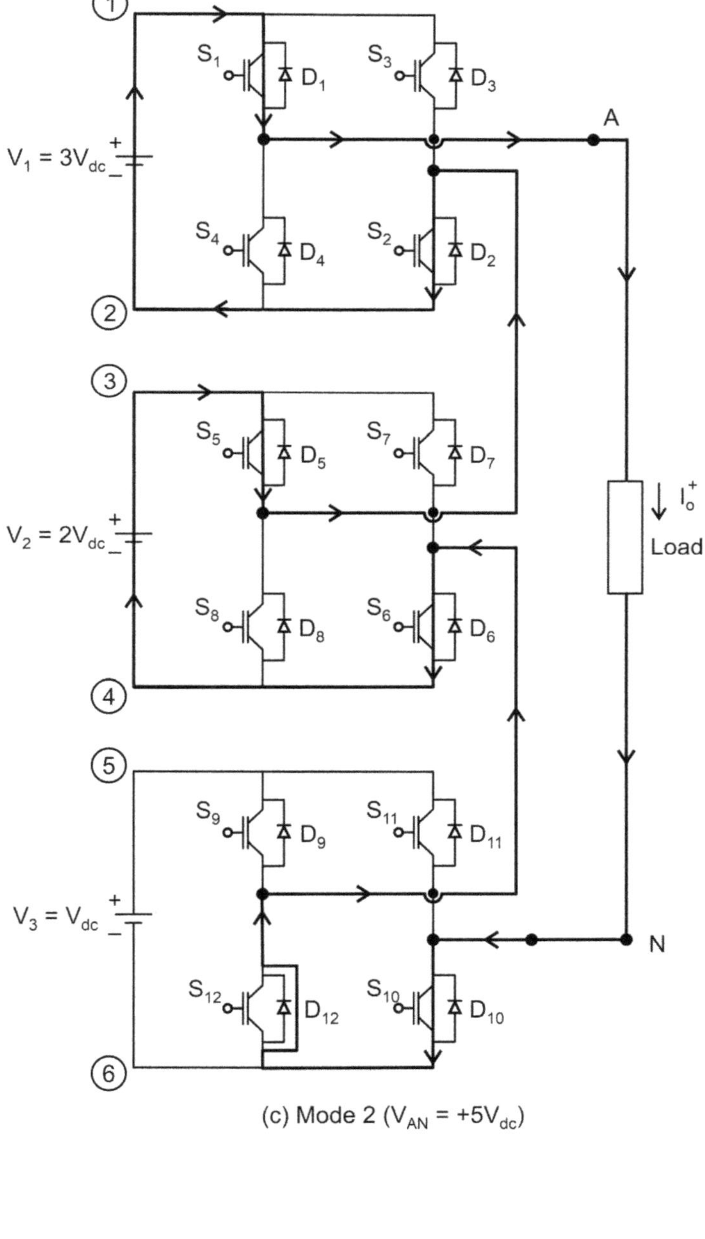

(c) Mode 2 ($V_{AN} = +5V_{dc}$)

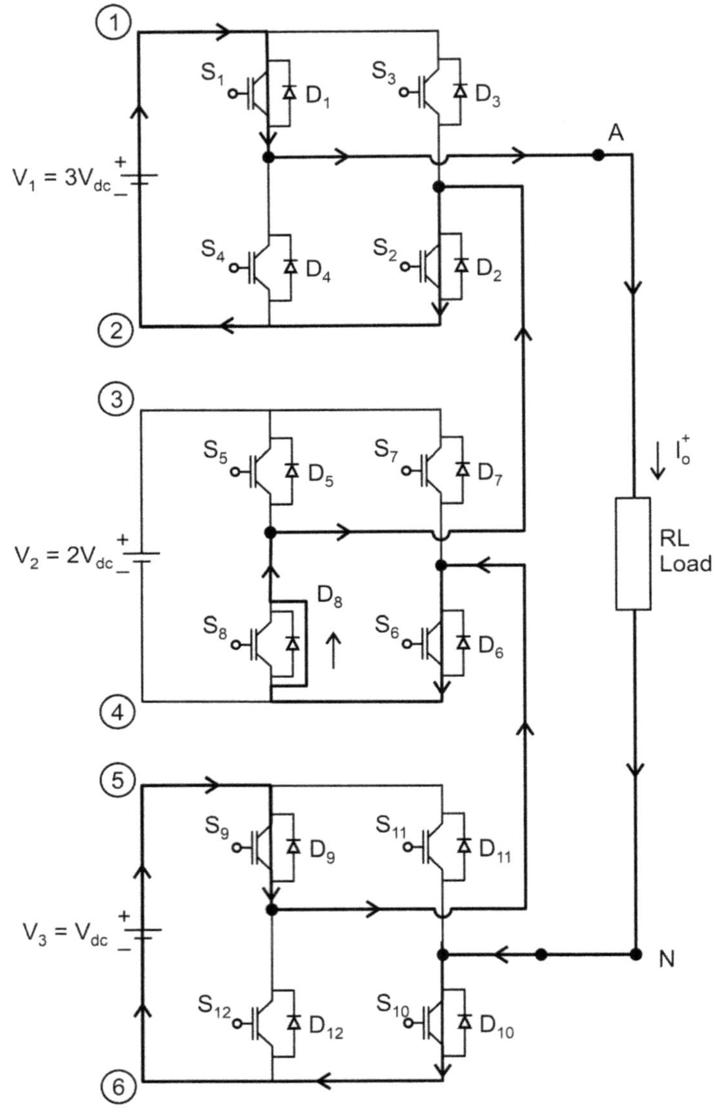

(d) Mode 3 ($V_{AN} = +4V_{dc}$)

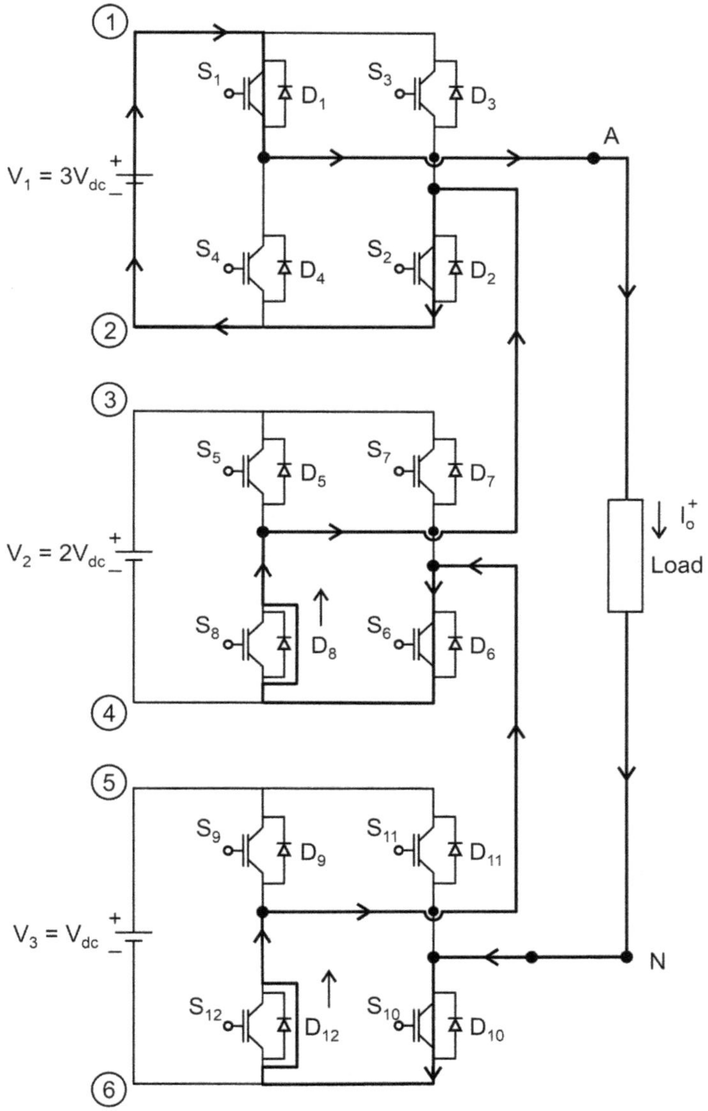

(e) Mode 4 (option(a)) : ($V_{AN} = +3V_{dc}$)

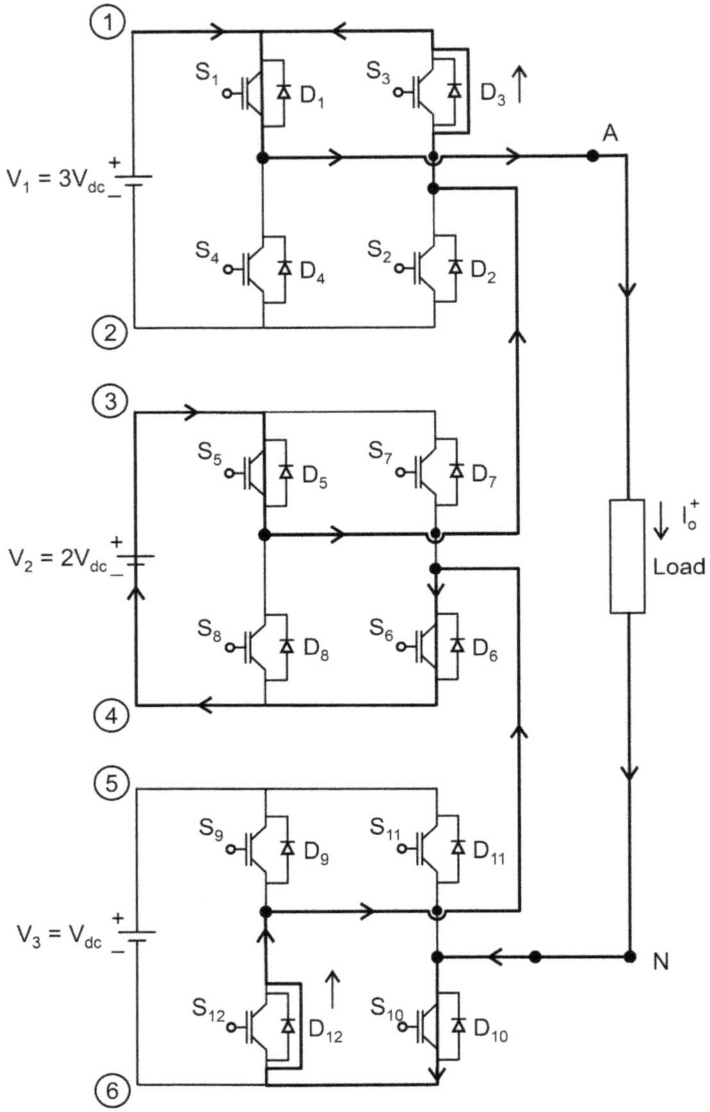

(f) Mode 5 ($V_{AN} = +2V_{dc}$)

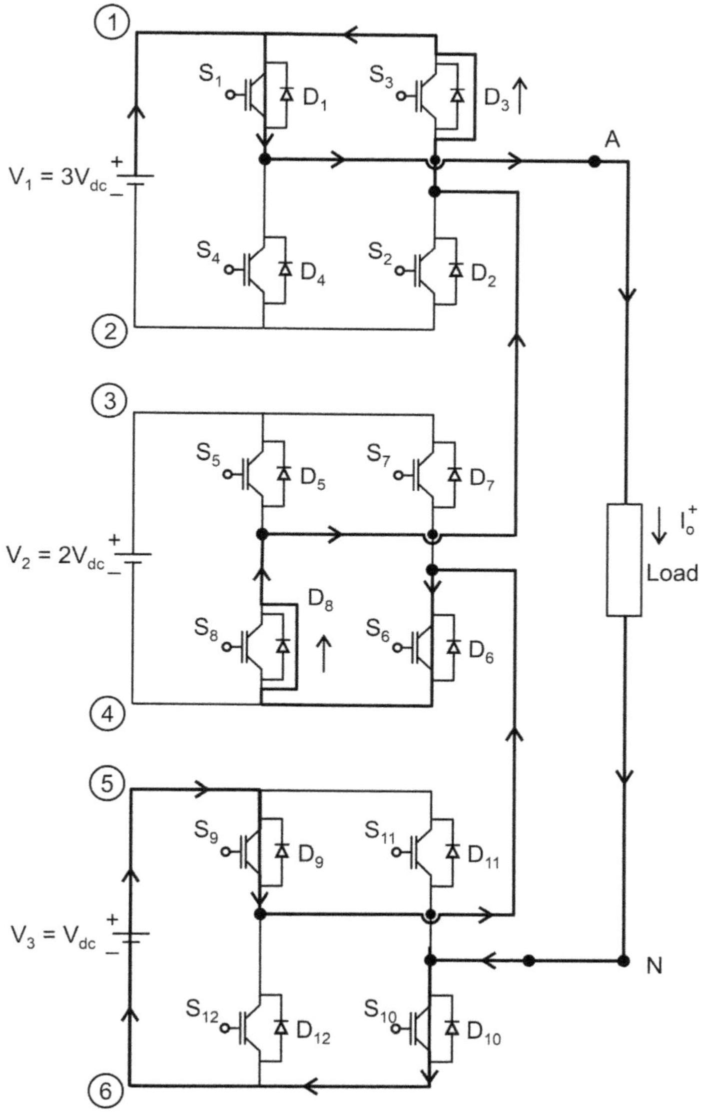

(g) Mode 6 ($V_{AN} = +V_{dc}$)

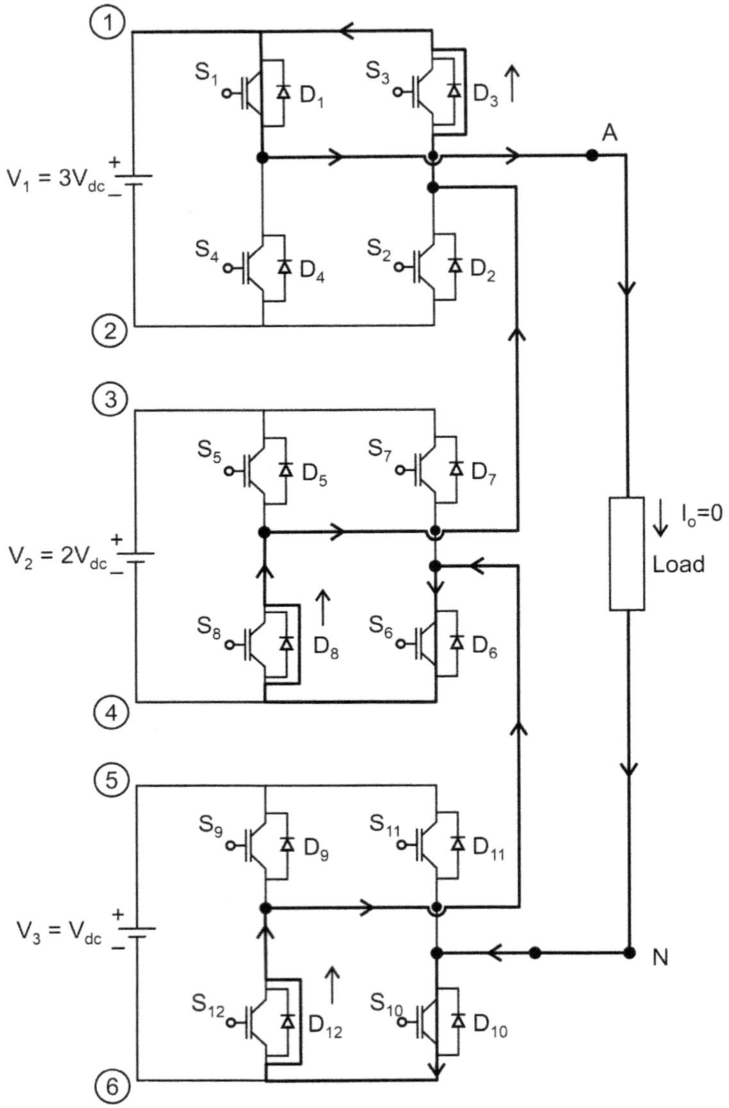

(h) Mode 7(option(a)) : (V_{AN} = 0)

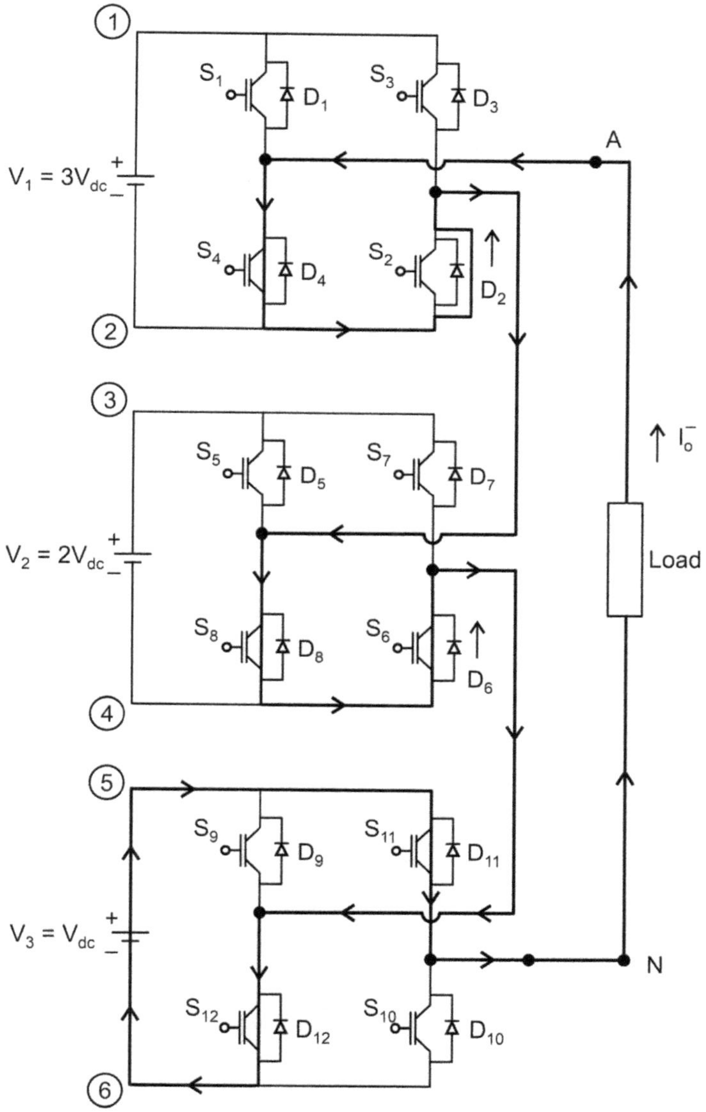

(i) Mode 8 ($V_{AN} = -V_{dc}$)

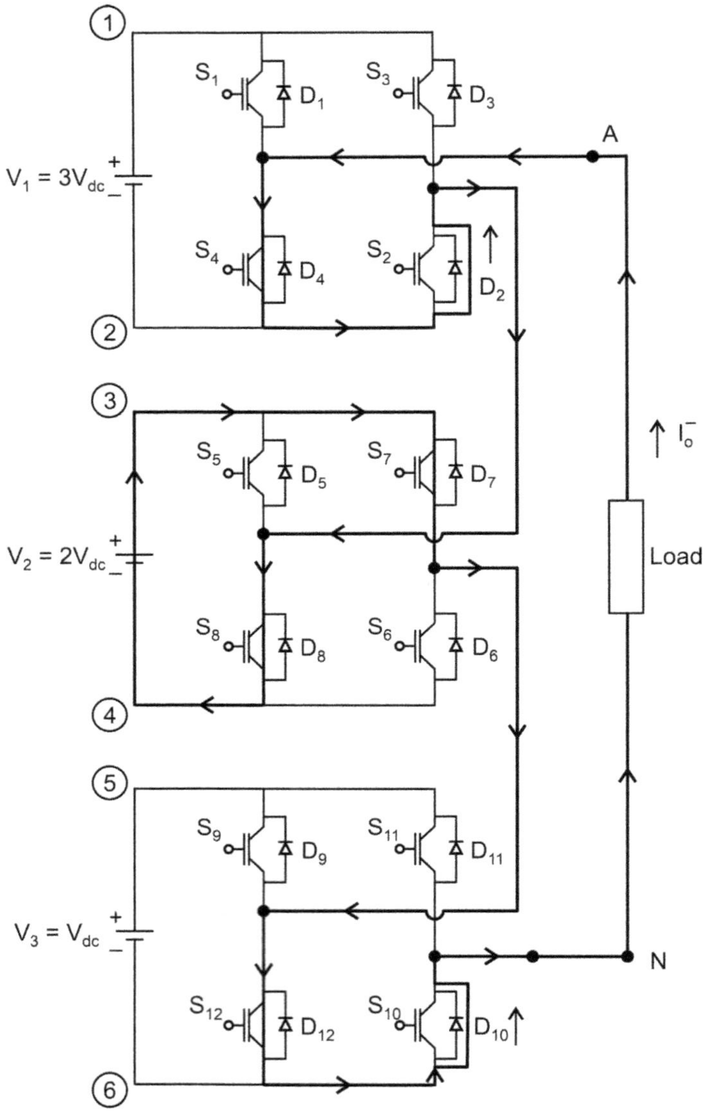

(j) Mode 9 ($V_{AN} = -2V_{dc}$)

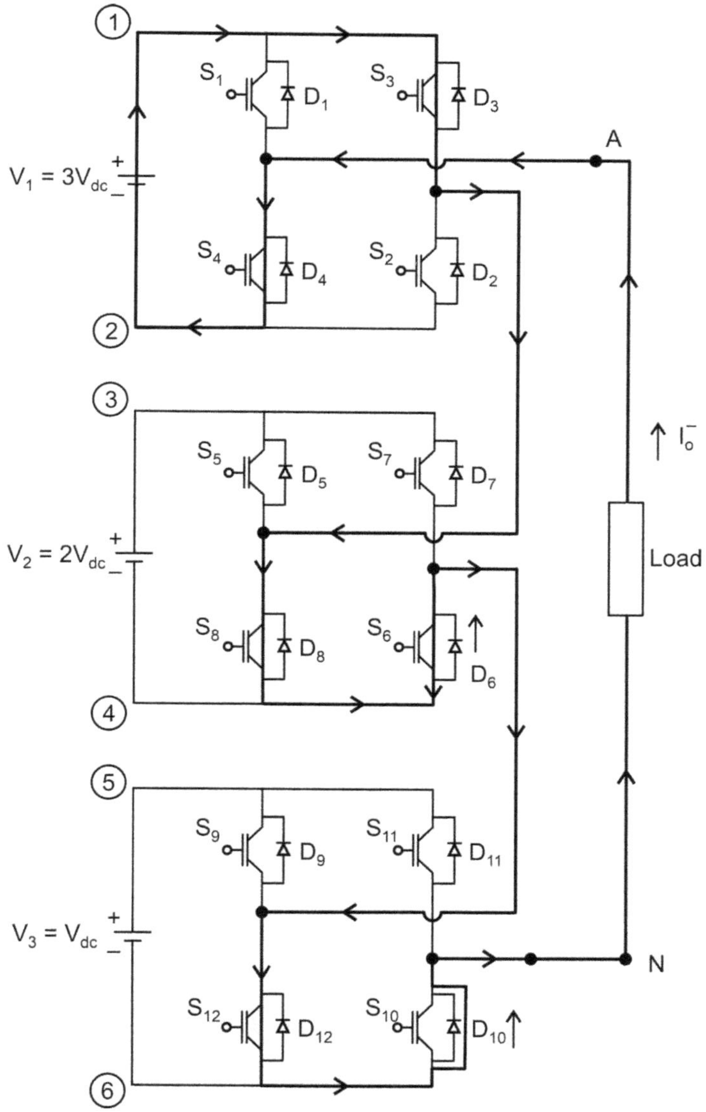

(k) Mode 10(option(a)) : ($V_{AN} = -3V_{dc}$)

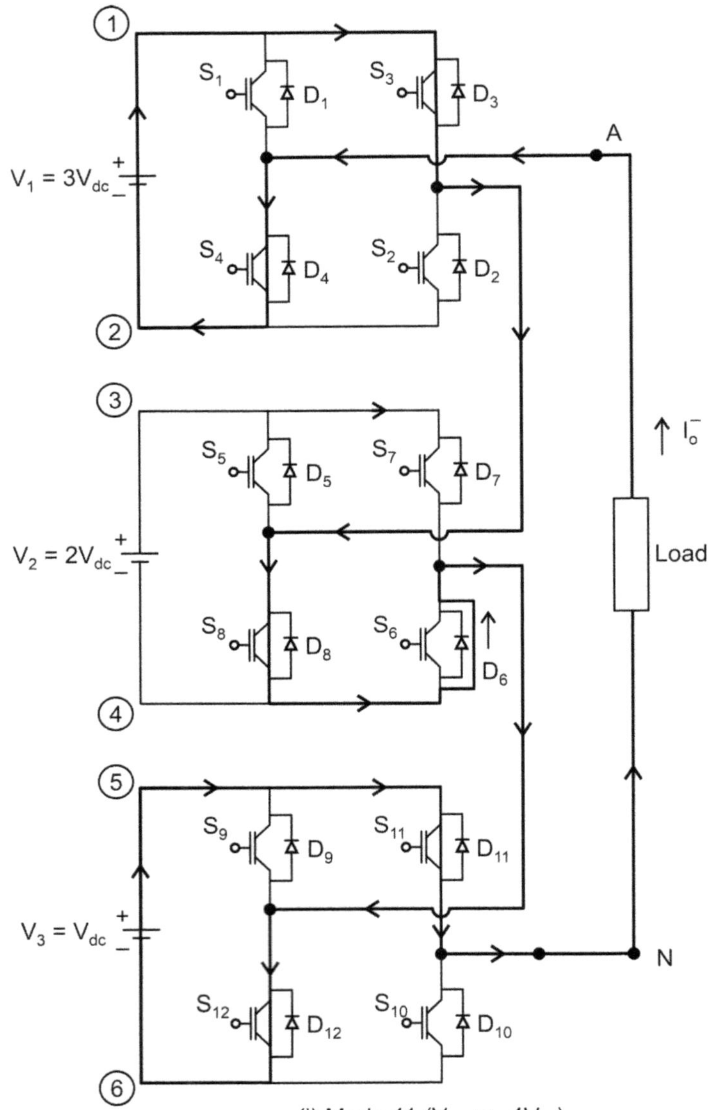

(l) Mode 11 ($V_{AN} = -4V_{dc}$)

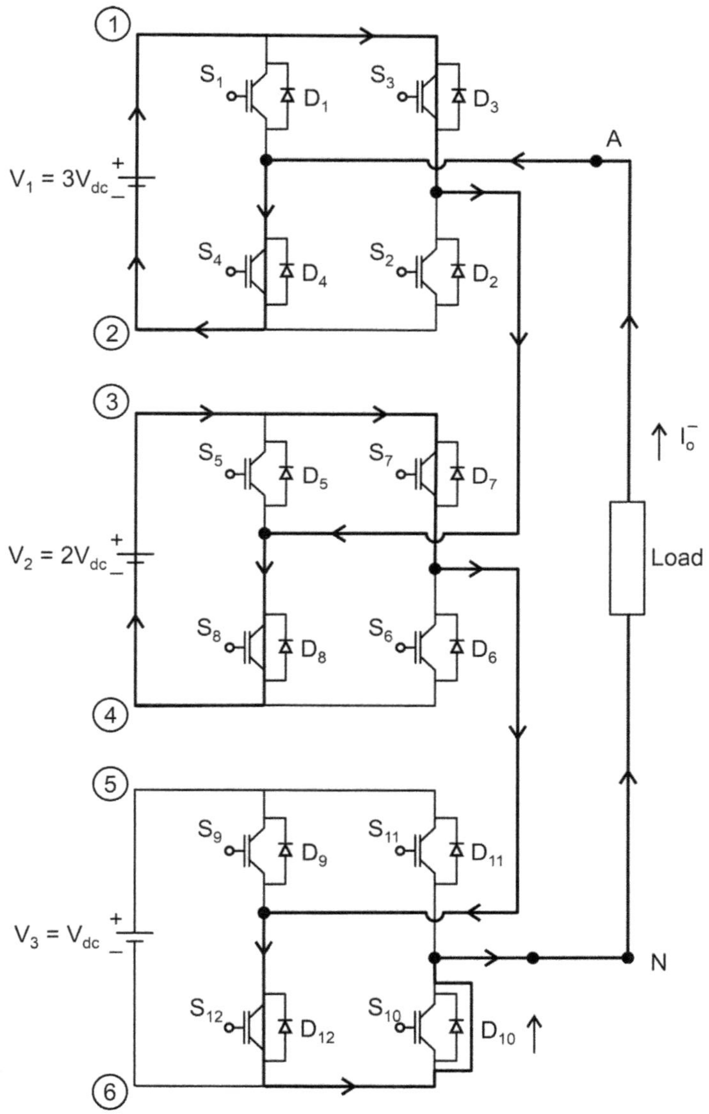

(m) Mode 12 ($V_{AN} = -5V_{dc}$)

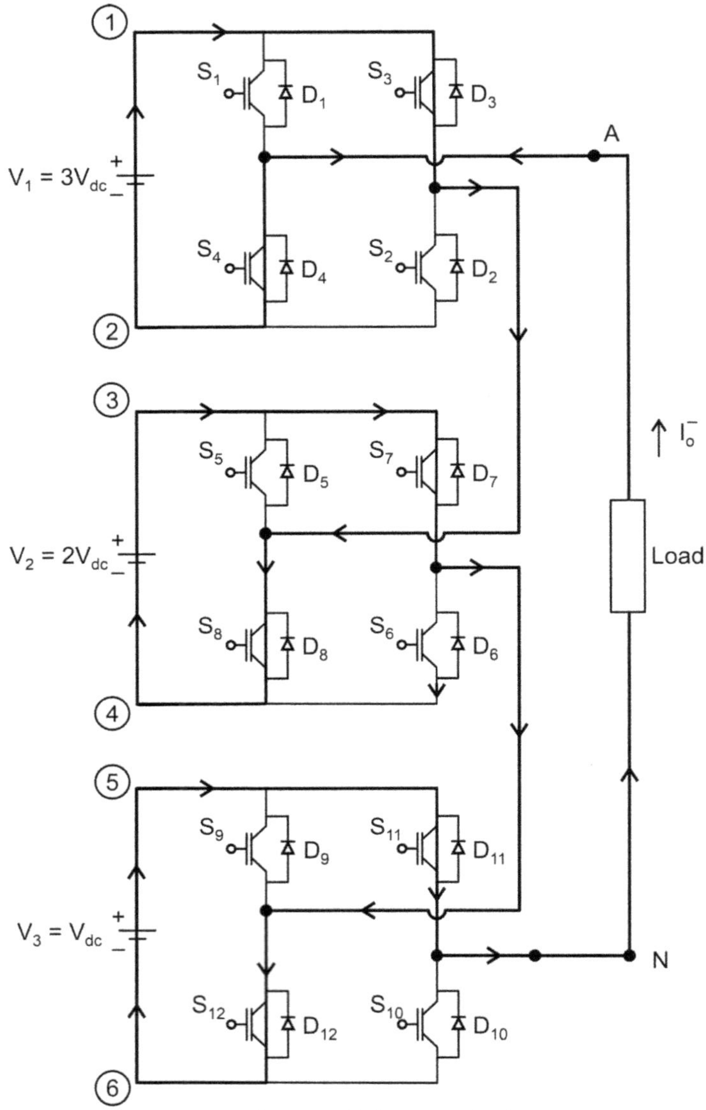

(n) Mode 13 ($V_{AN} = -6V_{dc}$)

Fig. 5.13 Thirteen-level cascaded H-bridge (3 cells) inverter with unequal dc voltage sources circuit for Asymmetrical configuration, (a) Main circuit; (b) Mode1: $V_{AN} = +6V_{dc}$; (c) Mode2: $V_{AN} = +5V_{dc}$; (d) Mode3: $V_{AN} = +4V_{dc}$; (e) Mode4: $V_{AN} = +3V_{dc}$; (f) Mode5: $V_{AN} = +2V_{dc}$; (g) Mode6: $V_{AN} = +V_{dc}$; (h) Mode7: $V_{AN} = 0$; (i) Mode8: $V_{AN} = -V_{dc}$; (j) Mode9: $V_{AN} = -2V_{dc}$; (k) Mode10: $V_{AN} = -3V_{dc}$; (l) Mode11: $V_{AN} = -4V_{dc}$; (m) Mode12: $V_{AN} = -5V_{dc}$; (n) Mode13: $V_{AN} = -6V_{dc}$

Mode 1: To obtain $V_{AN} = +6V_{dc}$, the conduction takes place through source terminal 1, switches S_1, load A to N, S_{10}, source terminals 6 to 5, S_9, S_6, source terminals 4 to 3, S_5, S_2, source terminals 2 to 1, and the load current is positive.

Mode 2: For $V_{AN} = +5V_{dc}$, current flows from source terminal 1, switch S_1, load A to N, switch S_{10}, diode D_{12}, switch S_6, source terminals 4 to 3, and switches S_5, S_2, source terminals 2 to 1, and load current is positive.

Mode 3: To get $V_{AN} = +4V_{dc}$, the conduction takes place from source terminal 1, switch S_1, load A to N, switch S_{10}, source terminals 6 to 5, switches S_9, S_6, diode D_8, switch S_2, source terminals 2 to 1, and load current is positive.

Mode 4: For $V_{AN} = +3V_{dc}$, current flows from either (a) source terminal 1, switch S_1, load A to N, switch S_{10}, diode D_{12}, switch S_6, diode D_8, switch S_2, source terminals 2 to 1, or (b) source terminal 1, switch S_1, load A to N, switch S_{10}, source terminals 6 to 5, switches S_9, S_6, source terminals 4 to 3, switch S_5, diode D_3, and load current is positive.

Mode 5: To obtain $V_{AN} = +2V_{dc}$, conduction takes place through source terminal 1, switch S_1, load A to N, switch S_{10}, diode D_{12}, switch S_6, source terminals 4 to 3, switch S_5, diode D_3, and load current is positive.

Mode 6: For $V_{AN} = +V_{dc}$, current flows from source terminal 1, switch S_1, load A to N, switch S_{10}, source terminals 6 to 5, switches S_9, S_6, diodes D_8, D_3, and load current is positive.

Mode 7: To get $V_{AN} = 0$, conduction takes place through either (a) source terminal 1, switch S_1, load A to N, switch S_{10}, diode D_{12}, switch S_6, diodes D_8, D_3, or (b) switch S_{12}, diode D_{10}, load N to A, switch S_4, diode D_2, switch S_8, diode D_6, and load current is zero.

Mode 8: To obtain $V_{AN} = -V_{dc}$, conduction takes place through source terminal 5, switch S_{11}, load N to A, switch S_4, diode D_2, switch S_8, diode D_6, switch S_{12}, source terminals 6 to 5, and load current is negative.

Mode 9: For $V_{AN} = -2V_{dc}$, current flows from source terminal 3, switches S_7, S_{12}, diode D_{10}, load N to A, switch S_4, diode D_2, switch S_8, source terminals 4 to 3, and load current is negative.

Mode 10: To get $V_{AN} = -3V_{dc}$, conduction takes place either through (a) source terminal 1, switches S_3, S_8, diode D_6, switch S_{12}, diode D_{10}, load N to A, switch S_4, source terminal 2 to 1, or (b) source terminal 1, switches S_3, S_8, source terminal 4 to 3, switches S_7, S_{12}, source terminal 6 to 5, switch S_{11}, load N to A, diode D_1, and load current is negative.

Mode 11: For $V_{AN} = -4V_{dc}$, the current flows from source terminal 1, switches S_3, S_8, diode D_6, switch S_{12}. Source terminal 6 to 5, switch S_{11}, load N to A, switch S_4, source terminal 2 to 1, and load current is negative.

Mode 12: To obtain $V_{AN} = -5V_{dc}$, conduction takes place through source terminal 1, switches S_3, S_8, source terminal 4 to 3, switches S_7, S_{12}, diode D_{10}, load N to A, switch S_4, source terminal 2 to 1, and load current is negative.

Mode 13: For $V_{AN} = -6V_{dc}$, the current flows from switch S_{11}, load N to A, switch S_4, source terminals 2 to 1, switches S_3, S_8, source terminals 4 to 3, switches S_7, S_{12}, source terminals 6 to 5, and load current is negative.

Table 5.6 Switching table for thirteen-level cascaded (3 cells) H-bridge inverter with unequal dc voltage sources for Asymmetrical configuration

Modes	V_o=V_{AN}	S_1	S_2	S_3	S_4	S_5	S_6	S_7	S_8	S_9	S_{10}	S_{11}	S_{12}	Diodes			
M1	$+6V_{dc}$	1	1	0	0	1	1	0	0	1	1	0	0	-	-	-	
M2	$+5V_{dc}$	1	1	0	0	1	1	0	0	0	1	0	0	-	-	D_{12}	-
M3	$+4V_{dc}$	1	1	0	0	0	1	0	0	1	1	0	0	-	D_8	-	-
M4 (a)	$+3V_{dc}$	1	1	0	0	0	1	0	0	0	1	0	0	-	D_8	D_{12}	-
(b)	$+3V_{dc}$	1	0	0	0	1	1	0	0	1	1	0	0	D_3	-	-	-
M5	$+2V_{dc}$	1	0	0	0	1	1	0	0	0	1	0	0	D_3	-	D_{12}	-
M6	$+V_{dc}$	1	0	0	0	0	1	0	0	1	1	0	0	D_3	D_8	-	-
M7 (a)	0	1	0	0	0	0	1	0	0	0	1	0	0	D_3	D_8	D_{12}	-
(b)	0	0	0	0	1	0	0	0	1	0	0	0	1	D_2	D_6	-	D_{10}
M8	$-V_{dc}$	0	0	0	1	0	0	0	1	0	0	1	1	D_2	D_6	-	-
M9	$-2V_{dc}$	0	0	0	1	0	0	1	1	0	0	0	1	D_2	-	-	D_{10}
M10 (a)	$-3V_{dc}$	0	0	1	1	0	0	0	1	0	0	0	1	-	D_6	-	D_{10}
(b)	$-3V_{dc}$	0	0	1	0	0	0	1	1	0	0	1	1	D_1	-	-	-
M11	$-4V_{dc}$	0	0	1	1	0	0	0	1	0	0	1	1	-	D_6	-	-
M12	$-5V_{dc}$	0	0	1	1	0	0	1	1	0	0	0	1	-	-	-	D_{10}
M13	$-6V_{dc}$	0	0	1	1	0	0	1	1	0	0	1	1	-	-	-	-

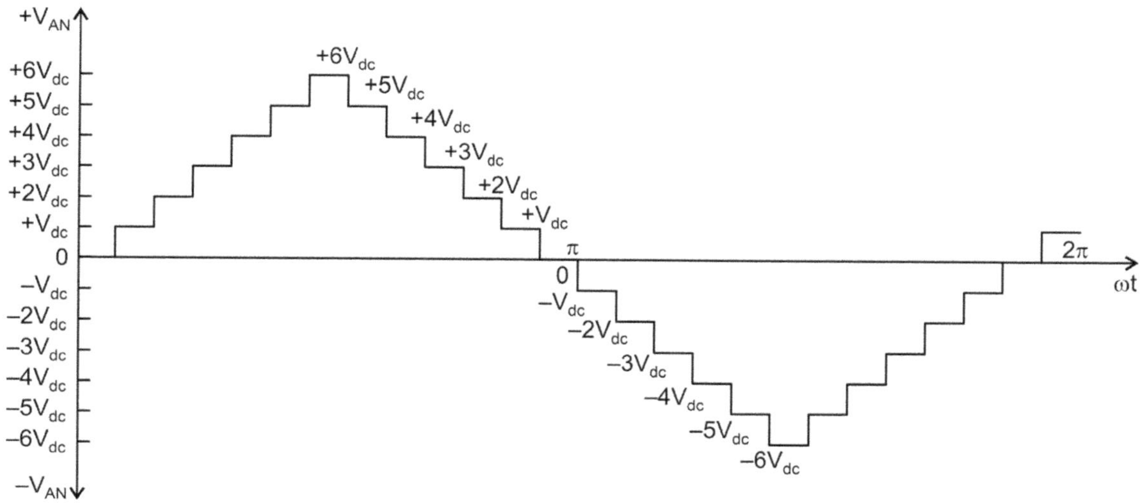

Fig. 5.14 Voltage waveform (V_{AN}) for Thirteen-level cascaded H-bridge inverter with unequal dc voltage sources circuit for Asymmetrical configuration.

Three-phase thirteen-level cascaded H-bridge inverter with unequal dc voltage sources circuit for Asymmetrical configuration: Similarly, three-phase thirteen-level cascaded H-bridge inverter with unequal dc voltage sources circuit for Asymmetrical configuration is shown in Fig.5.15. The dc source voltages are applied as $V_{A1}=V_{B1}=V_{C1}=3V_{dc}$; $V_{A2}=V_{B2}=V_{C2}=2V_{dc}$; and $V_{A3}=V_{B3}=V_{C3}=V_{dc}$. The load phase voltage waveform is same as that of single-phase circuit per phase and for three phase load voltages they are phase shifted by 120^0 between them.

Table 5.7 Summary of various load voltage levels produced by cascaded H-bridge inverter with equal and unequal dc voltages sources for Asymmetrical configuration.

Number of H-bridges	Equal dc voltage sources	Number of load voltage levels	Unequal dc voltage sources	Number of load voltage levels
1	V1	3-level	V1	3-level
2	V1 = V2	5-level	V1 ≠ V2	7-level
3	V1 = V2 = V3	7-level	V1 ≠ V2≠ V3	13-level
4	V1=V2=V3=V4	9-level	V1≠V2≠V3≠V4	21-level

It is concluded that in cascaded H-bridge inverters the voltage levels are also dependent on the type of dc sources voltages, namely equal and unequal dc source voltages. From the table 5.7, it is observed that with two H-bridge inverters connected in cascade produces 5-level load voltages with equal dc source voltages and whereas it produces 7-level load voltages with unequal dc source voltages respectively. Similarly, with three H-bridge inverters connected in cascade produces 7-level load voltages with equal dc source voltages whereas it produces 13-level load voltages with unequal dc source voltages, respectively. And finally, with four H-

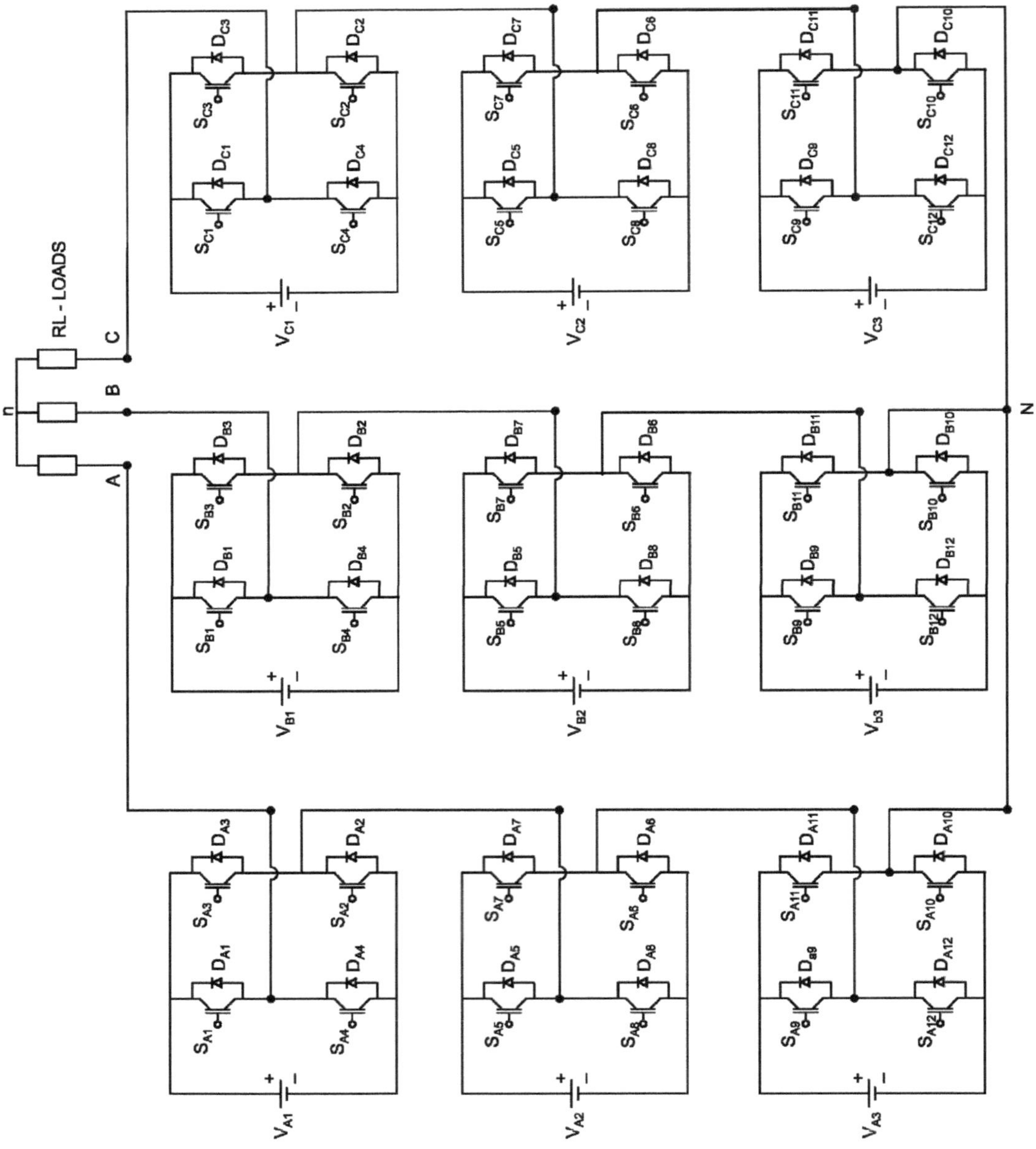

Fig. 5.15 Three-phase Thirteen-level cascaded H-bridge inverter with unequal dc voltage sources circuit for Asymmetrical configuration for three-phase star-connected loads.

bridge inverters connected in cascade produces 9-level load voltages with equal dc source voltages whereas it produces 21-level load voltages with unequal dc source voltages.

The main drawback of Cascaded H-bridge inverters is as voltage levels requirement increases the circuit requires a greater number of H-bridges to be connected in series. Recently, new topology of H-bridge inverter is presented in the literature where with a smaller number of H-bridge inverters, it produces higher-level of load voltages. The same is presented and discussed in the section 5.6.

5.4 Symmetrical Transistor-clamped Cascaded H-Bridge Inverters

New Topology for H-bridge inverters with reduced number of switching devices (or) H-bridges with Symmetrical Configuration is known as Transistor-clamped cascaded H-bridge inverters. The following table 5.8 gives the requirement of number of Cascaded H-Bridge inverters for different voltage levels. The relationship is (n-1)/2 number of H-bridges are required for n-levels of voltages.

As per the table 5.8, for generating 5-level voltages, it requires two H-bridge inverters to be connected in series. To obtain 7-level voltages, three H-bridge inverters are required to be connected in cascade, similarly, to get 9-level voltages, four H-bridge inverters are required to be connected in cascaded with equal dc source voltages.

By using Transistor-clamped H-bridge inverter with reduced number of switching devices, with one H-bridge inverter, can generate 5-level voltages, instead of two H-bridges. And with two new topology H-bridge inverters in series, can generate 9-level voltages against four H-bridges, that means two H-bridges can be reduced. In this section, a 5-level and 9-level voltages are generated by using one H-bridge and two H-bridges in series, respectively, with reduced switching devices are explained.

5.4.1 Symmetrical Configuration of Five-level Transistor-Clamped H-bridge Inverter with Reduced Number of Switching Devices

In order to reduce the requirement of number of switching devices in H-bridge inverters, introduced a transistor-clamped cascaded H-bridge. The main advantage of this topology is that the number of switches required is reduced and also the number of sources. Fig.5.16 shows the new cascaded five level H-bridge multilevel inverter [6]. It has additional one bidirectional switch connected between the first leg of the H-bridge and the capacitor midpoint, enabling five output voltage levels.

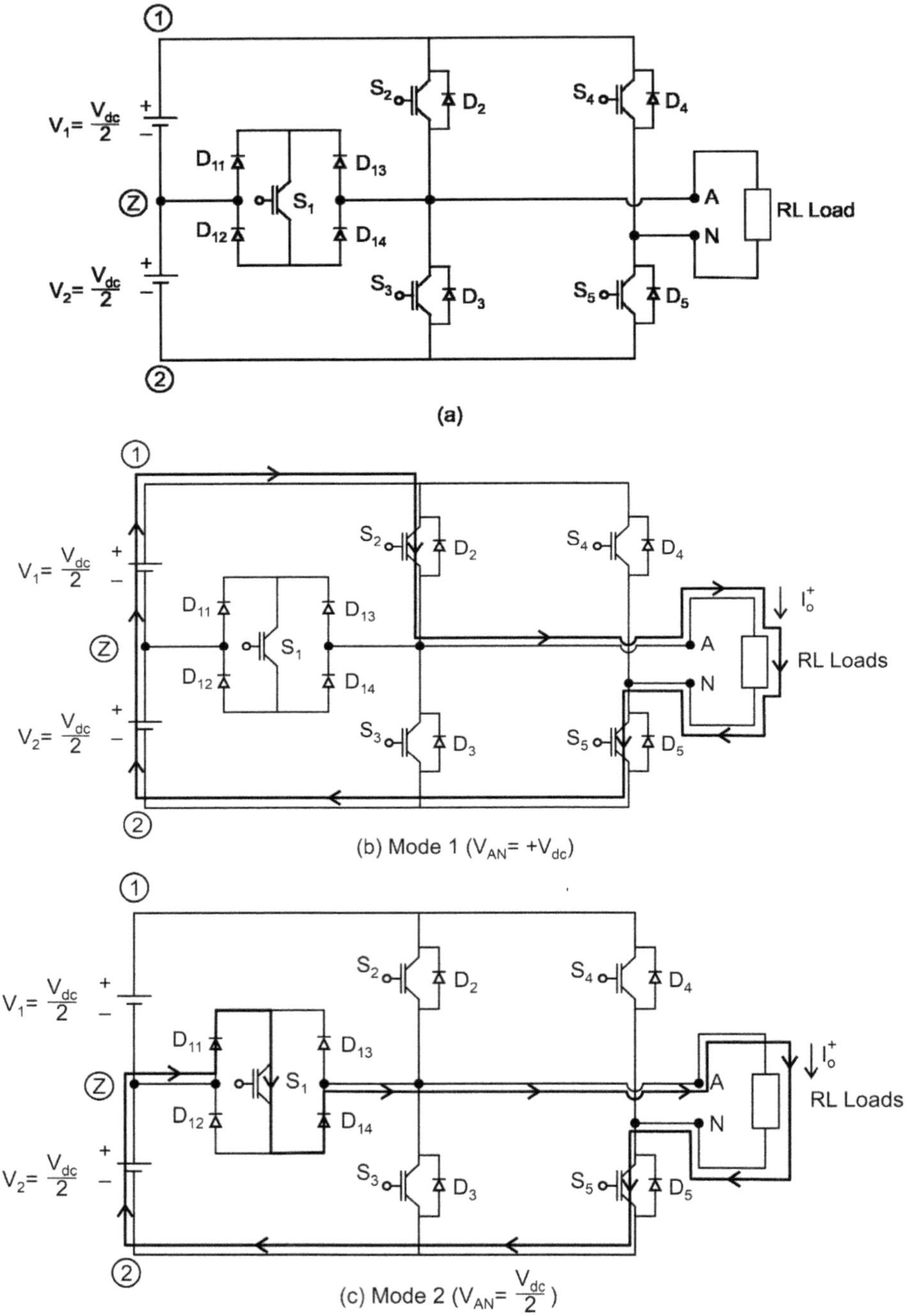

Fig. 5.16 Symmetrical configuration of Transistor-clamped Single H-bridge inverter for five-level output voltage with reduced number of switching devices with equal dc source voltages, (a) Main Circuit, (b) Mode 1: for V_{AN}= +V_{dc}, (c) Mode 2: for V_{AN}= +$V_{dc}/2$, (d) Mode 3: for V_{AN}= 0, (e) Mode 4: for V_{AN}= -$V_{dc}/2$, and (f) Mode 5: for V_{AN}= -V_{dc}.

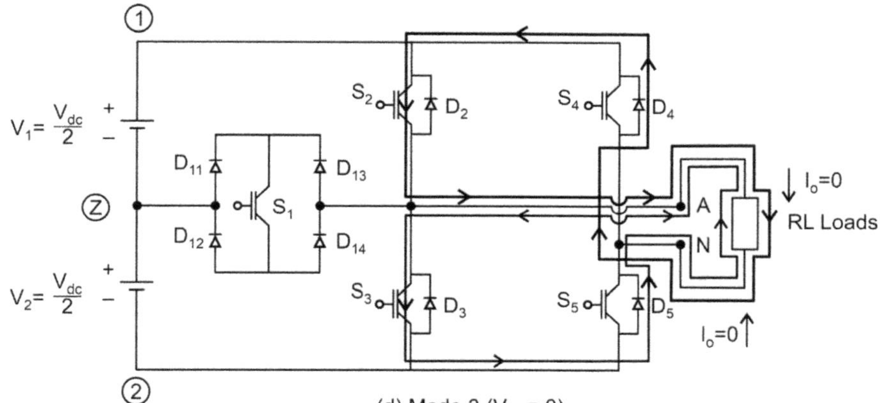

(d) Mode 3 (V_{AN}= 0)

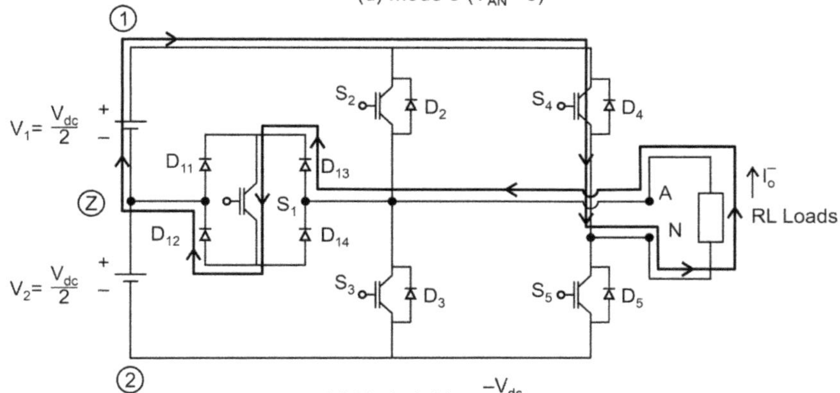

(e) Mode 4 ($V_{AN}= \frac{-V_{dc}}{2}$)

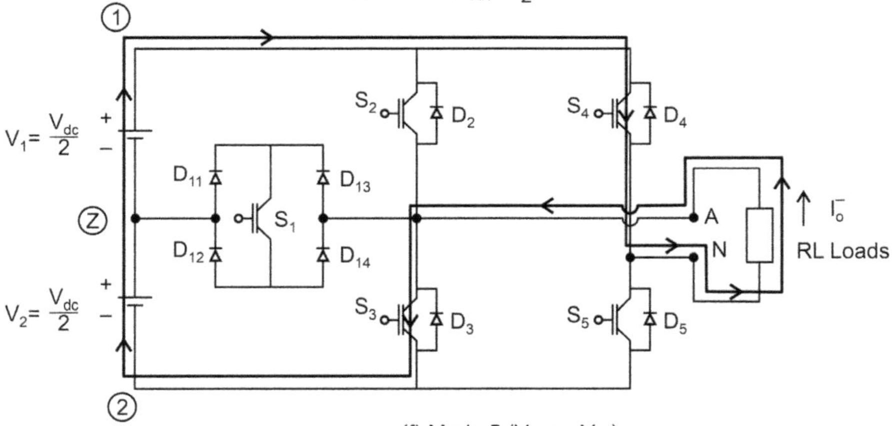

(f) Mode 5 ($V_{AN}= -V_{dc}$)

Fig. 5.16 Symmetrical configuration of Transistor-clamped Single H-bridge inverter for five-level output voltage with reduced number of switching devices with equal dc source voltages, (a) Main Circuit, (b) Mode 1: for $V_{AN}= +V_{dc}$, (c) Mode 2: for $V_{AN}= +V_{dc}/2$, (d) Mode 3: for $V_{AN}= 0$, (e) Mode 4: for $V_{AN}= -V_{dc}/2$, and (f) Mode 5: for $V_{AN}= -V_{dc}$.

In the circuit shown in Fig.5.16 (a)-(f), transistor-clamped symmetrical single H-bridge module is capable of producing five level output voltages. Each inverter module is capable of producing $+V_{dc}$, $+V_{dc}/2$, 0, $-V_{dc}/2$, $-V_{dc}$. It has five output voltage levels i.e. $+V_{dc}$, $+V_{dc}/2$, 0, $-V_{dc}/2$, $-V_{dc}$.

Mode 1: For getting the output voltage $V_{AN} = +V_{dc}$ the switches S_2, S_5 are to be turned on, so that the current flows from source terminal 1, switch S_2, load terminal A to terminal N, switch S_5 to source terminal 2, as positive current.

Mode 2: Similarly, for output voltage $V_{AN} = +V_{dc}/2$ switches S_1, S_5 are to be turned on and S_2 is turned-off, then conduction takes place from source mid-point terminal Z, diode D_{11}, switch S_1, diode D_{14}, load terminal A to N, switch S_5 and back to source terminal 2, with positive current.

Mode 3: For $V_{AN} = 0$ either S_3, D_5 or S_2, D_4 need to be turned on, no current flows;

Mode 4: For $V_{AN} = -V_{dc}/2$ switches S_1, S_4 are to be turned on, then conduction takes place from source terminal 1, switch S_4, load terminal N to A, diode D_{13}, switch S_1, diode D_{12} to source mid-point terminal Z, giving negative current.

Mode 5: For $V_{AN} = -V_{dc}$ switches S_3, S_4 are to be turned on, then conduction takes place from source terminal 1, switch S_4, load terminal N to A, switch S_3 to source terminal 2, giving negative current. The switch combinations are shown in table 5.8.

Table 5.8 Five- level transistor-clamped H-bridge output voltage with reduced number of switches or H-bridges with equal dc sources voltages.

Modes	S_1	S_2	S_3	S_4	S_5	Diodes	Load Voltage, V_{AN}	Load current, I_o
M1	0	1	0	0	1	----	$+V_{dc}$	Positive
M2	1	0	0	0	1	D_{11}, D_{14}	$+V_{dc}/2$	Positive
M3	0	1	1	0	0	D_4, D_5	0	Zero
M4	1	0	0	1	0	D_{12}, D_{13}	$-V_{dc}/2$	Negative
M5	0	0	1	1	0	----	$-V_{dc}$	Negative

(Switch ON = 1, switch OFF = 0)

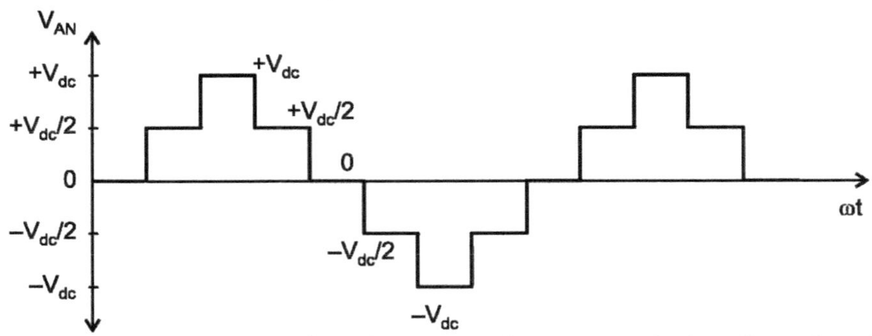

Fig. 5.17 Load Phase Voltage Waveform for new topology symmetrical configuration of single H-bridge five level inverter for reduced number of switching devices.

Three-phase five-level symmetrical configuration of transistor-clamped single H-bridge inverter: Three single H-bridge inverters are connected to form a three-phase circuit for three-phase star-connected RL loads is shown in Fig.5.18. For symmetrical configuration the dc source voltages are equal, namely, $V_{A1}=V_{B1}=V_{C1}=V_{A2}=V_{B2}=V_{C2}=V_{dc}$. The output voltages are obtained for five-levels same waveform as shown for single phase circuit, but for phase shift between phases by 120°.

5.4.2 Symmetrical Configuration of Transistor-clamped Nine-level Cascaded H-bridge Inverter with Reduced Number of Switching Devices or H-bridges

The nine-level output voltage can be produced by using two such bridges in cascade. The circuit for transistor-clamped nine level cascaded H-bridges is shown in Fig. 5.19, the gating signals for the inverter is generated and applied by using multicarrier modulation. This circuit is feeding power to RL load. The total output voltage is sum of the outputs of the inverter modules and the nine voltage levels are $+4V_{dc}$, $+3V_{dc}$, $+2V_{dc}$, $+V_{dc}$, 0, $-V_{dc}$, $-2V_{dc}$, $-3V_{dc}$, $-4V_{dc}$. The advantage of this proposed circuit is having reduced number of switches. The cost and complexity are less in this circuit. To synthesize nine output voltage levels, it employs two independent dc voltage sources of $2V_{dc}$ which are divided into two input sources V_{dc} in order to secure an additional dc voltage source of V_{dc} in each H-bridge. Each inverter module having a bidirectional switch produces 5-levels of output voltage ($+2V_{dc}$, $+V_{dc}$, 0, $-V_{dc}$, $-2V_{dc}$) by controlling the switches. Since every output terminal of the inverter module is connected in series, the output voltage becomes the sum of the terminal voltages of each inverter. The detailed operation is explained as follows:

Mode 1: To obtain $V_{AN} = +4V_{dc}$, the conduction takes place through source terminal 1, switch S_{11}, load A to N, switch S_{24}, source terminals 4 to 3 switches S_{21}, S_{14} and source terminal 2, and the load current is positive.

Mode 2: For $V_{AN} = +3V_{dc}$, the load current flows from source mid-point terminal Z_1, diode D_{15}, switch S_1, diode D_{16}, load A to N, switch S_{24}, source terminals 4 to 3, switches S_{21}, S_{14} and source terminal 2, the load current is positive.

Mode 3: To get $V_{AN} = +2V_{dc}$, the conduction takes place from source mid-point terminal Z_1, diode D_{15}, switch S_1, diode D_{16}, load A to N, switch S_{24}, source terminal 4, source mid-point terminal Z_2, diode D_{25}, switch S_2, diode D_{26}, switch S_{14} and source terminal 2, giving positive load current.

Mode 4: To obtain $V_{AN} = +V_{dc}$, there are two paths, first one is : (a) the current flows from mid-point terminal Z_1, diode D_{15}, switch S_1, diode D_{16}, load A to N, switches S_{24}, D_{22}, S_{14} and source terminal 2, Or the second path is: (b) the current flows from mid-point terminal Z_2, diode D_{25}, switch S_2, diode D_{26}, switches S_{14}, D_{12}, load A to N, switch S_{24} and source terminal 4, the load current is positive.

Mode 5: To get $V_{AN} = 0$, there are two paths for conduction, one is from (a) switches S_{11}, D_{13}, S_{24}, and D_{22}, or another is from (b) switches S_{12}, D_{14}, D_{21}, and S_{23}, where the load current is also zero.

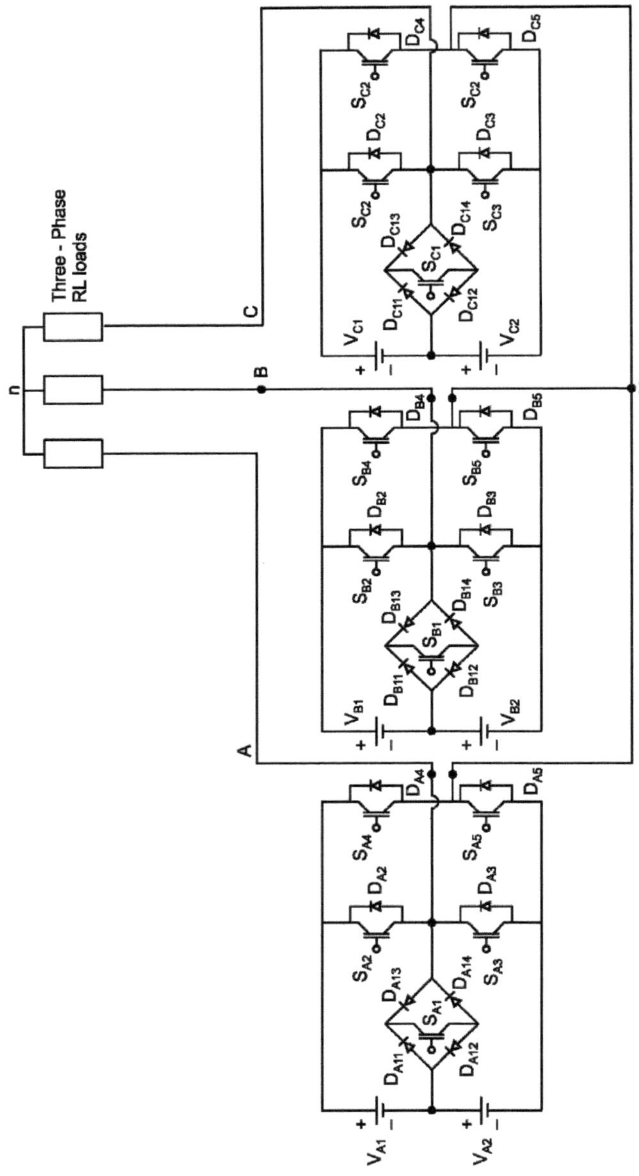

Fig. 5.18 Three-phase circuit for New Topology symmetrical configuration of Single H-bridge inverter for five-level output voltage with reduced number of switching devices with equal dc source voltages.

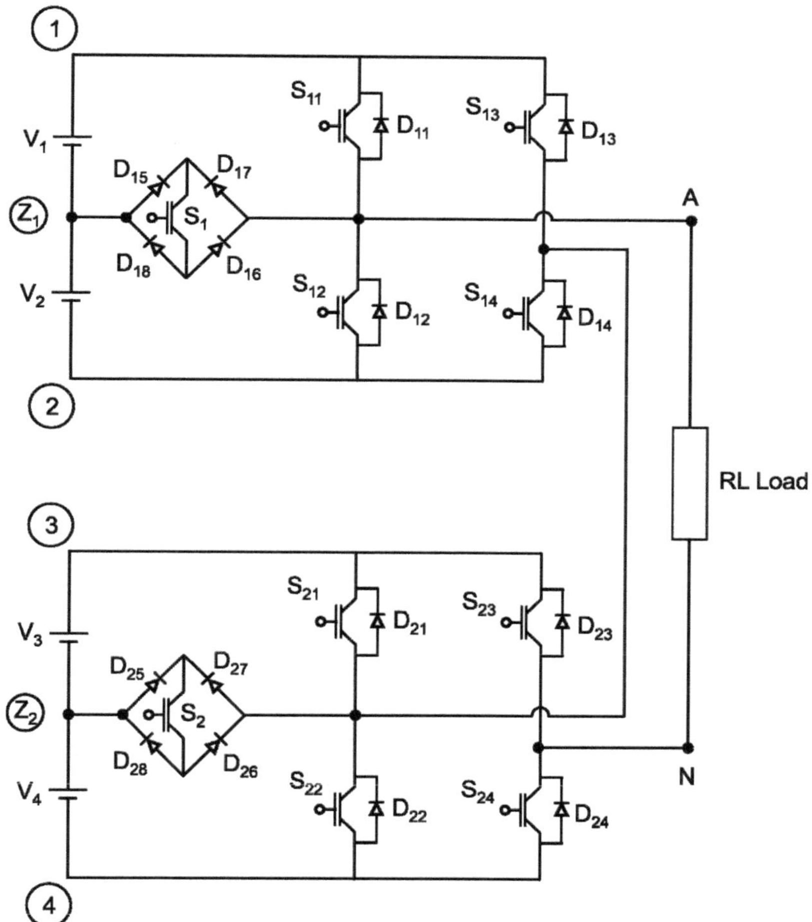

(a)

Fig. 5.19 Transistor-clamped Nine-level Cascaded H-bridge inverter with reduced number of devices with symmetrical configuration, (a) Main Circuit, (b) Mode1: for $V_{AN} = +4V_{dc}$, (c) Mode2: for $V_{AN} = +3V_{dc}$, (d) Mode3: for $V_{AN} = +2V_{dc}$, (e) Mode4: for $V_{AN} = +V_{dc}$, (f) Mode5: $V_{AN} = 0$, (g) Mode6: for $V_{AN}=-V_{dc}$. (h) Mode7: for $V_{AN}=-2V_{dc}$, (i) Mode8: for $V_{AN}=-3V_{dc}$ and (j) Mode9: for $V_{AN}=-4V_{dc}$.

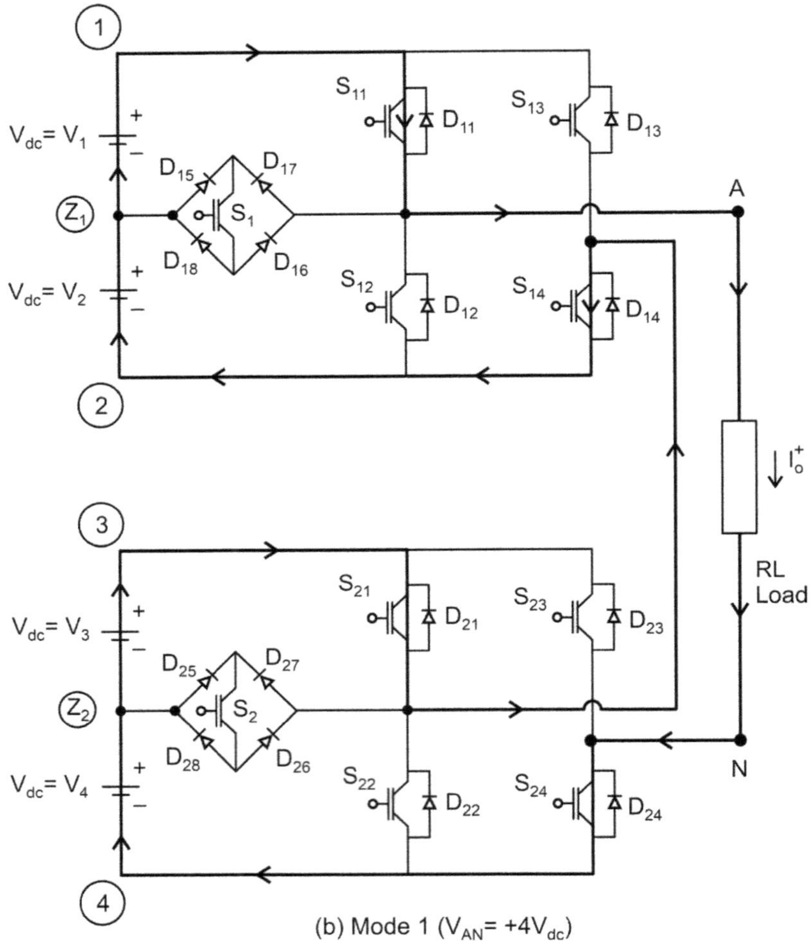

(b) Mode 1 (V_{AN}= $+4V_{dc}$)

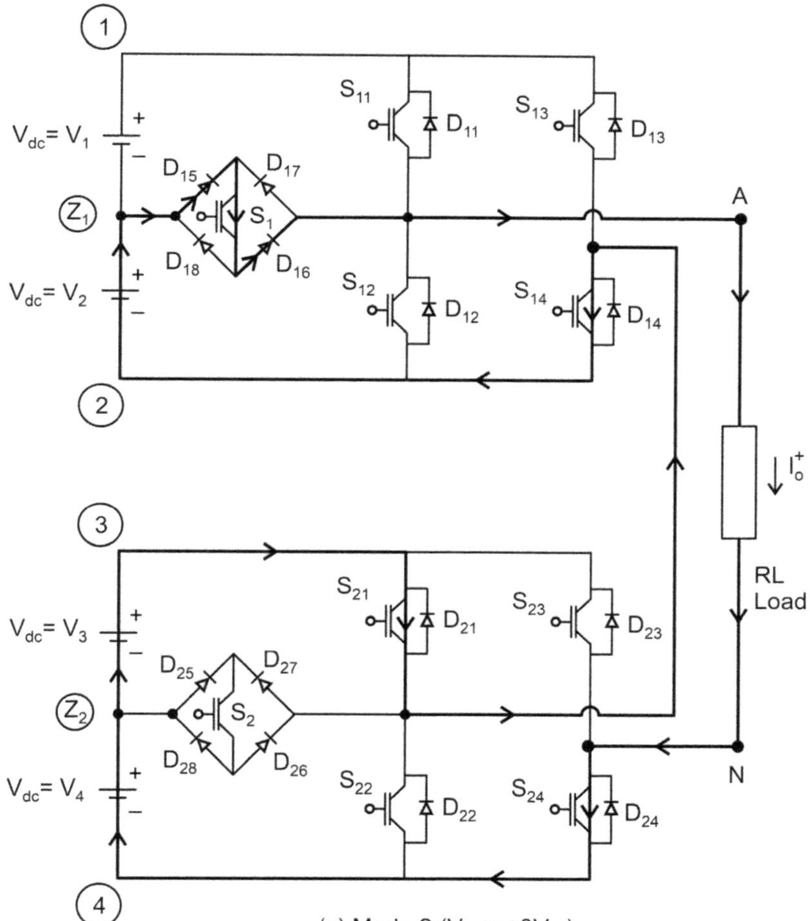

(c) Mode 2 ($V_{AN}= +3V_{dc}$)

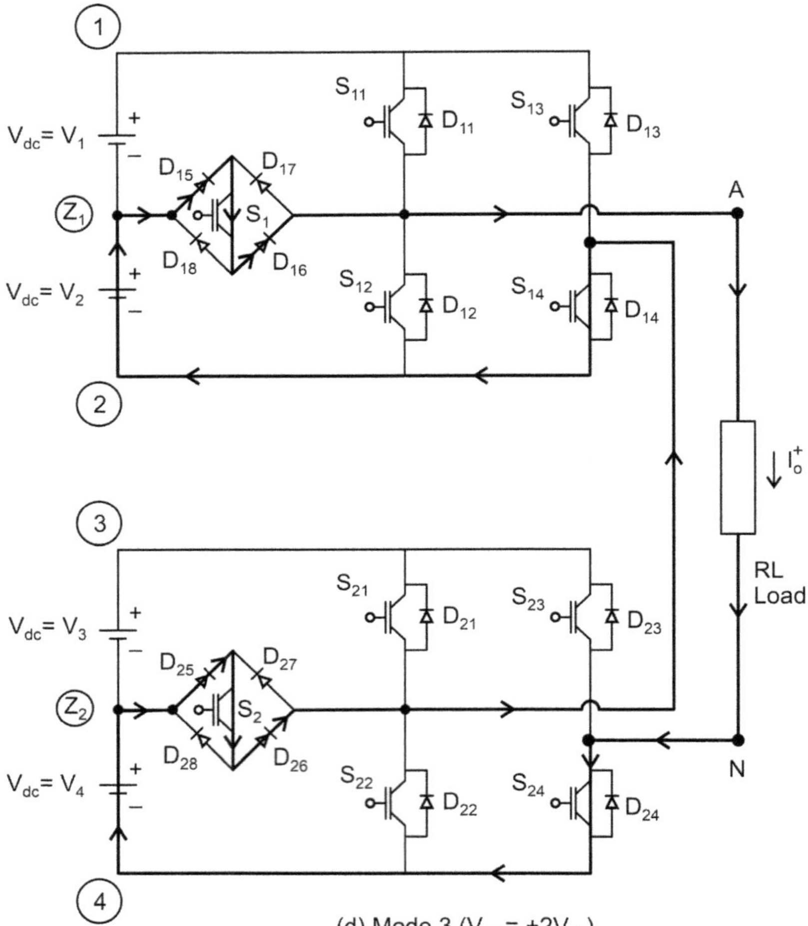

(d) Mode 3 (V_{AN}= +2V_{dc})

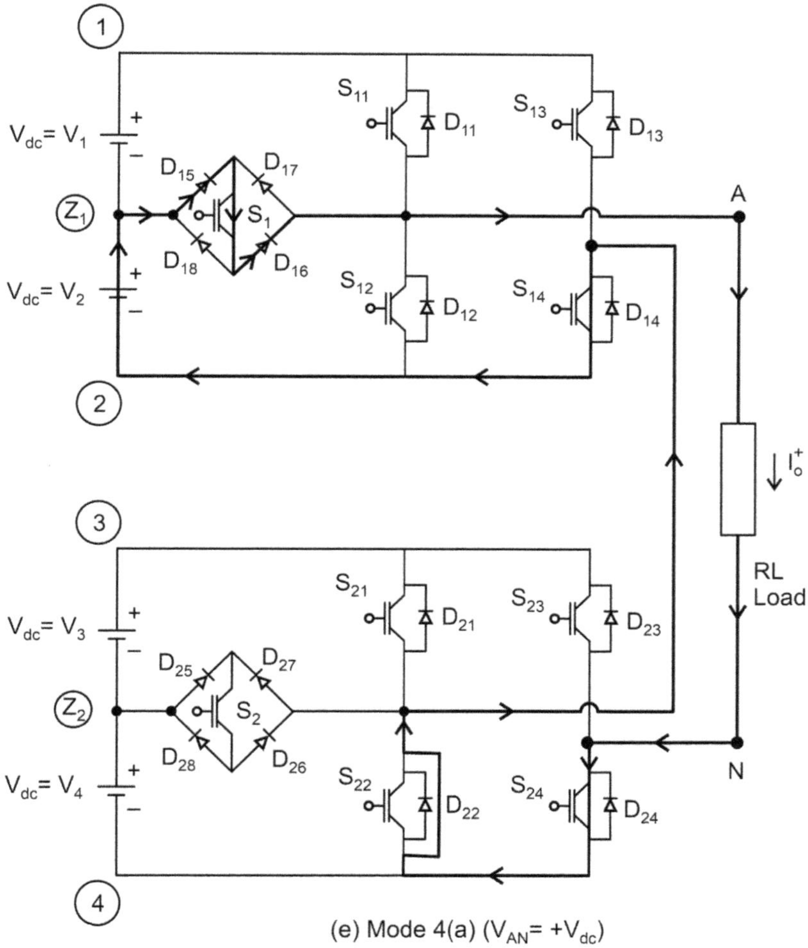

(e) Mode 4(a) ($V_{AN} = +V_{dc}$)

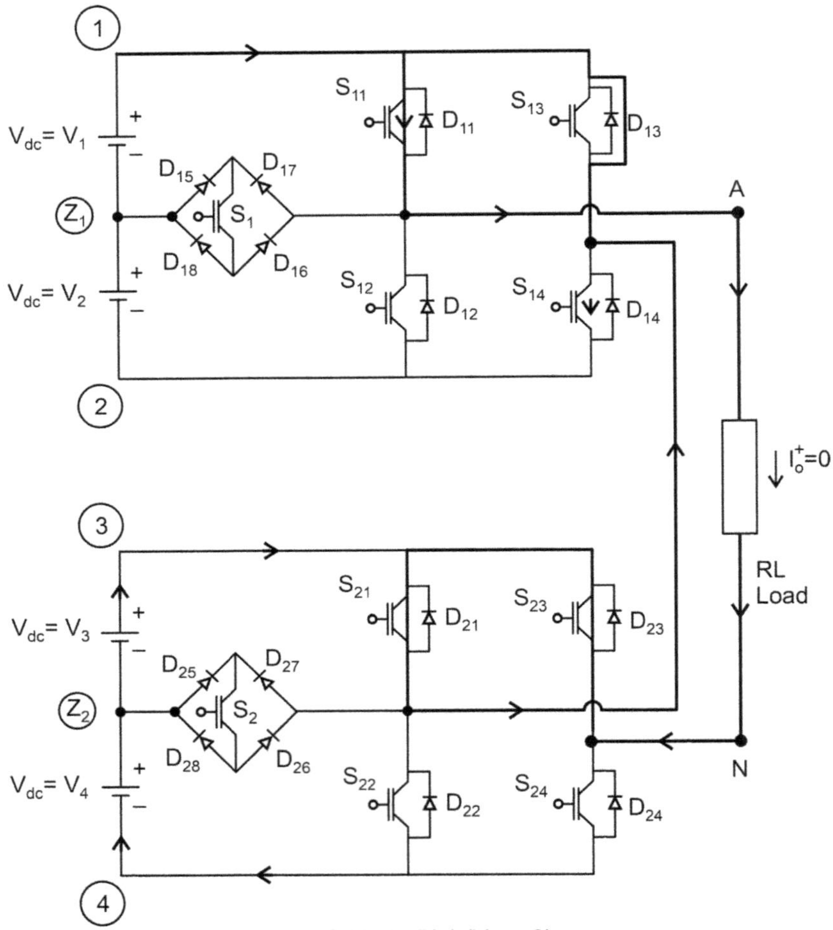

(f) Mode 5(a) (V_{AN}= 0)

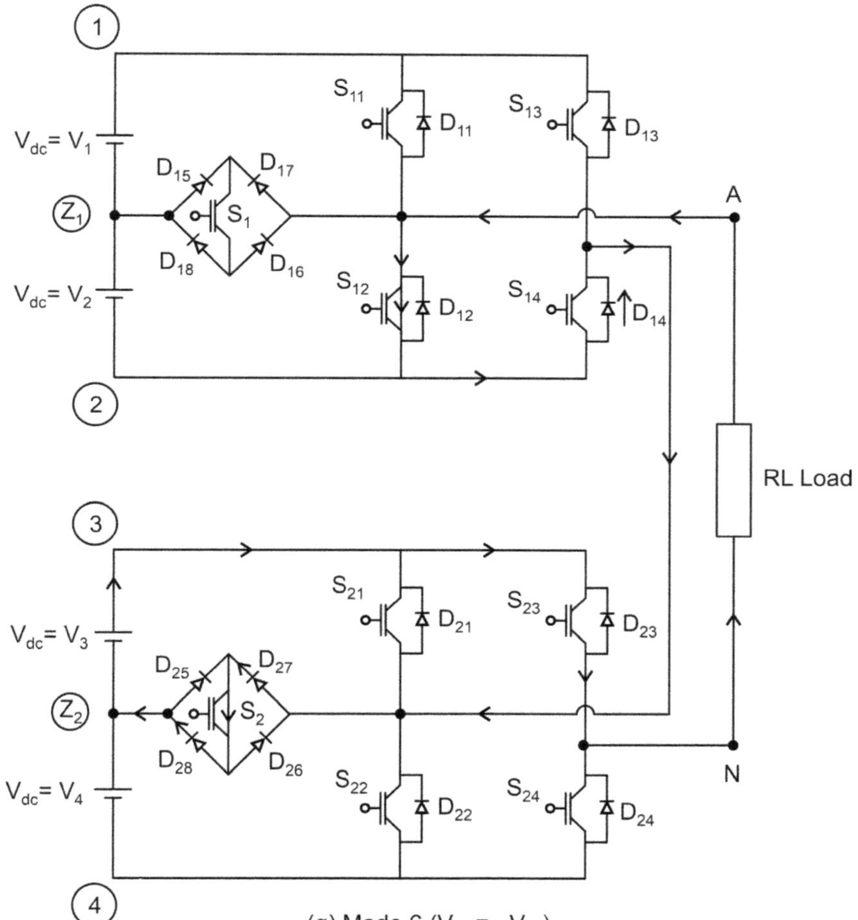

(g) Mode 6 ($V_{AN} = -V_{dc}$)

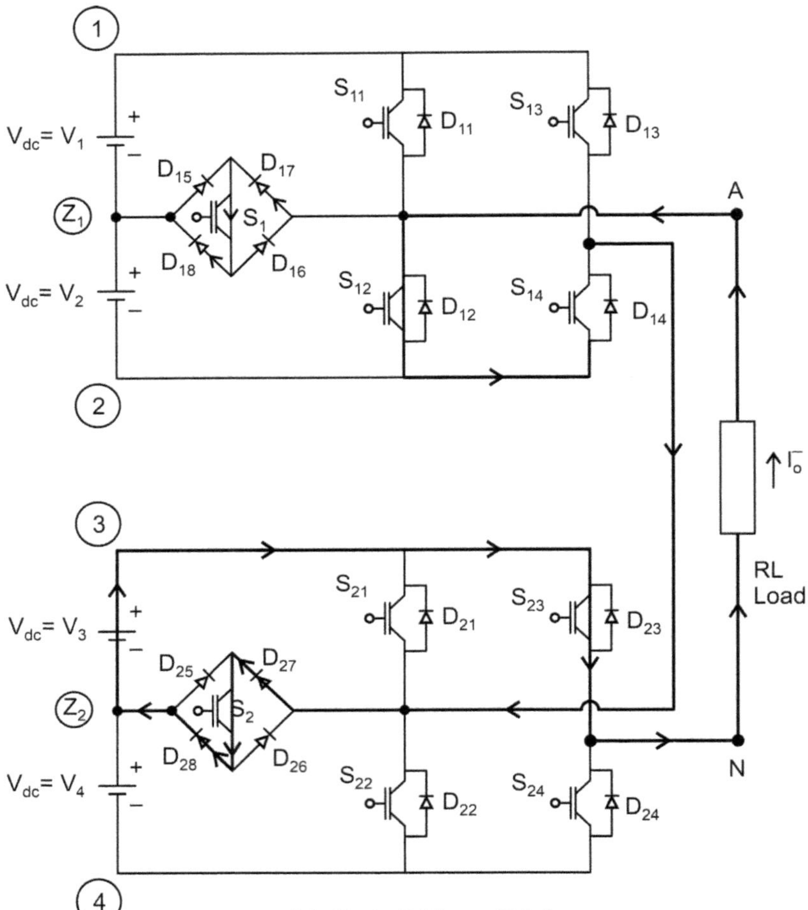

(h) Mode 7 ($V_{AN}= -2V_{dc}$)

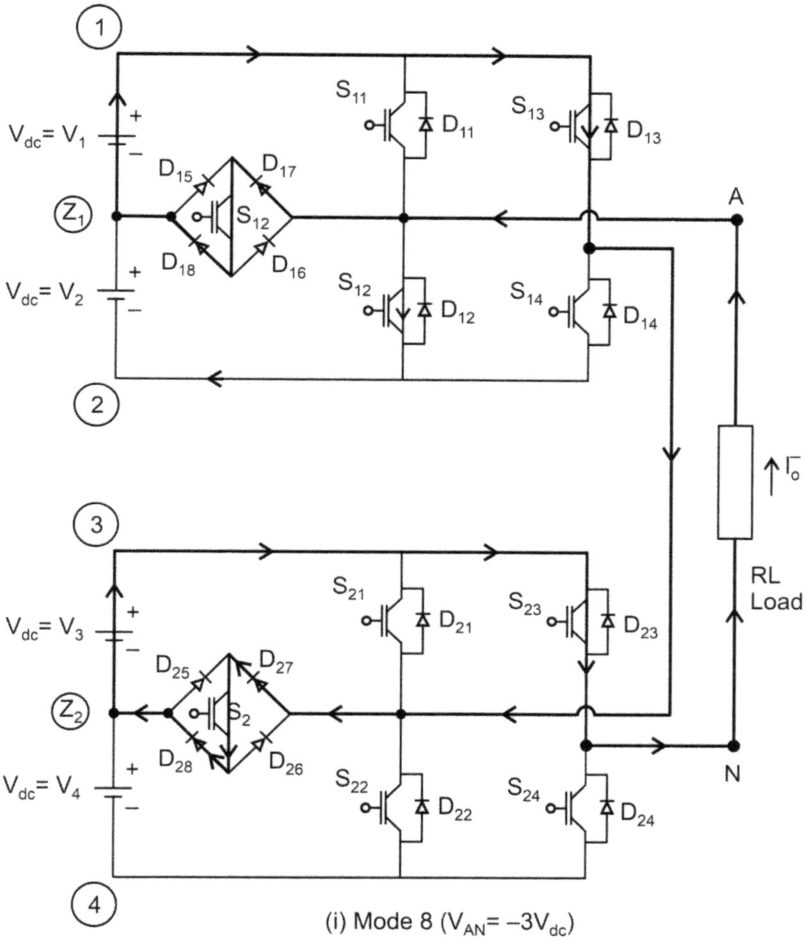

(i) Mode 8 ($V_{AN} = -3V_{dc}$)

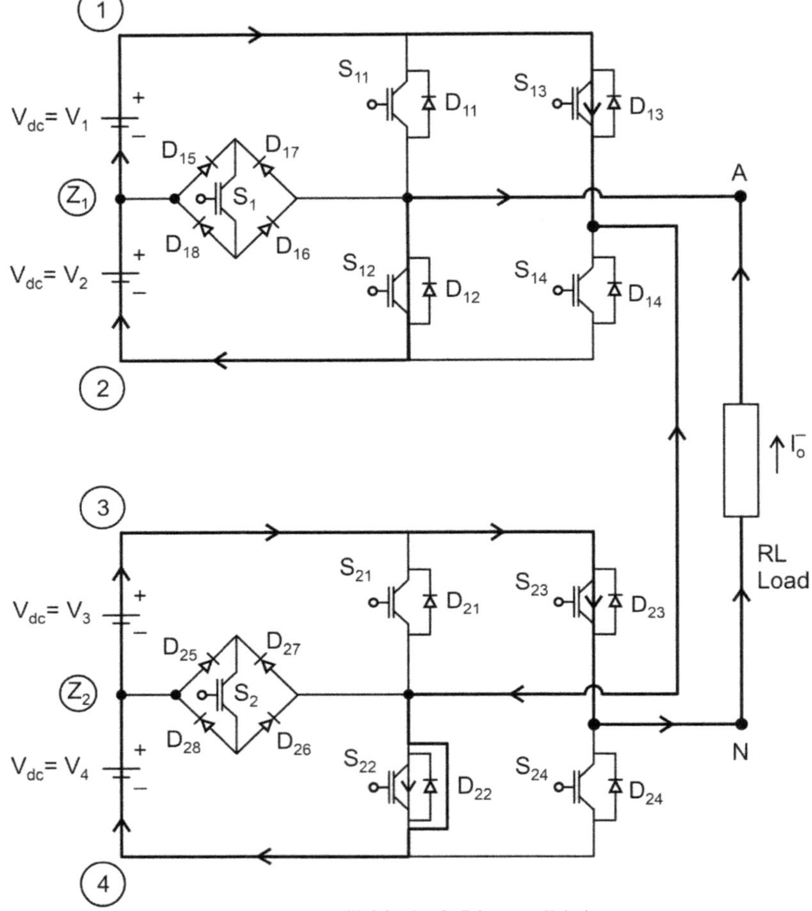

(j) Mode 9 (V_{AN}= $-4V_{dc}$)

Fig. 5.19 Transistor-clamped Nine-level Cascaded H-bridge inverter with reduced number of devices with symmetrical configuration, (a) Main Circuit, (b) Mode1: for V_{AN} = + $4V_{dc}$, (c) Mode2: for V_{AN} = + $3V_{dc}$, (d) Mode3: for V_{AN} = + $2V_{dc}$, (e) Mode4: for V_{AN} = $+V_{dc}$, (f) Mode5: V_{AN} = 0, (g) Mode6: for V_{AN} = $-V_{dc}$. (h) Mode7: for V_{AN} = $-2V_{dc}$, (i) Mode8: for V_{AN} = $-3V_{dc}$ and (j) Mode9: for V_{AN} = $-4V_{dc}$.

Table 5.9 Symmetrical configuration of transistor-clamped Nine- level cascaded H-bridge output voltage with reduced number of switches or H-bridges.

Modes	Load Voltage V_{AN}	S_1	S_{11}	S_{12}	S_{13}	S_{14}	S_2	S_{21}	S_{22}	S_{23}	S_{24}	antiparallel Diodes	Diodes across S_1/S_2
M1	$+4V_{dc}$	0	1	0	0	1	0	1	0	0	1	----	----
M2	$+3V_{dc}$	1	0	0	0	1	0	1	0	0	1	----	D_{15}, D_{16}
M3	$+2V_{dc}$	1	0	0	0	1	1	0	0	0	1	----	D_{15}, D_{16}, D_{25}, D_{26}
M4 (a)	$+V_{dc}$	1	0	0	0	1	0	0	0	0	1	---- D_{22}	D_{15}, D_{16}
M5 (a)	0	0	1	0	0	0	0	0	0	0	1	D_{13}, D_{22}	----
M6	$-V_{dc}$	0	0	1	0	0	1	0	0	1	0	D_{14}	D_{27}, D_{28}
M7	$-2V_{dc}$	1	0	0	1	0	1	0	0	1	0	D_{27}, D_{28}	D_{17}, D_{18}
M8	$-3V_{dc}$	0	0	1	1	0	1	0	0	1	0	----	D_{27}, D_{28}
M9	$-4V_{dc}$	0	0	1	1	0	0	0	1	1	0	----	----

Mode 6: To obtain $V_{AN} = -V_{dc}$, there are two possible conduction paths, one is (a) from source terminal 3, switch S_{23}, load N to A, switches S_{12}, D_{14}, diode D_{27}, switch S_2 and diode D_{28}, and Z_2. or another is (b) from source terminal 1, switches S_{13}, S_{22}, D_{24}, load N to A, diode D_{17}, switch S_1, diode D_{18} and mid-point source terminal Z_1, yielding negative load current.

Mode 7: To get $V_{AN} = -2V_{dc}$, conduction starts from source terminal 1, switch S_{13}, diode D_{27}, switch S_2, diode D_{28}, source terminals Z_2, to 3, switch S_{23}, load N to A, diode D_{17}, switch S_1, diode D_{18}, and mid-point source terminal Z_1, results in negative load current.

Mode 8: To obtain $V_{AN} = -3V_{dc}$, current flows from source terminal 1, switch S_{13}, diode D_{27}, switch S_2, diode D_{28}, mid-point source terminal Z_2, source terminal 3, switch S_{23}, load N to A, switch S_{12} and source terminal 2 for negative load current.

Mode 9: For $V_{AN} = -4V_{dc}$, conduction takes place from source terminal 1, switches S_{13}, S_{22}, source terminals 4 to 3, switch S_{23}, load N to A, switch S_{12}, and source terminal 2 and load current is negative.

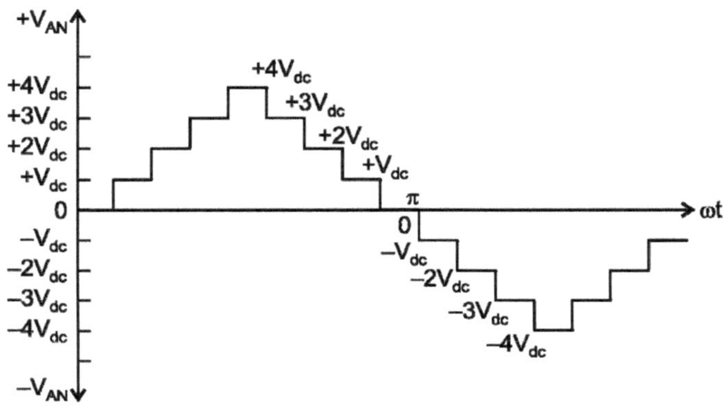

(a)

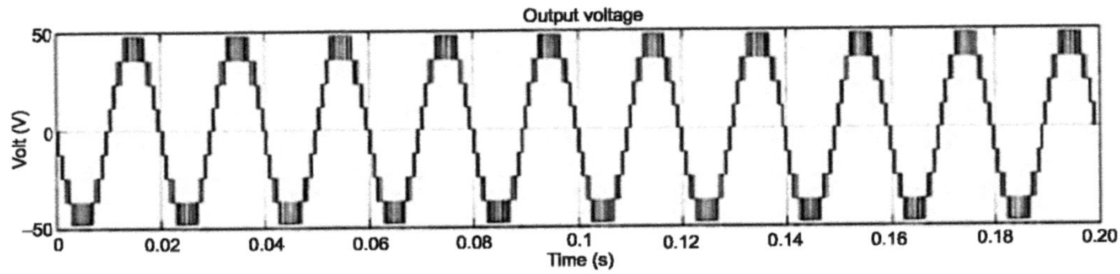

(b)

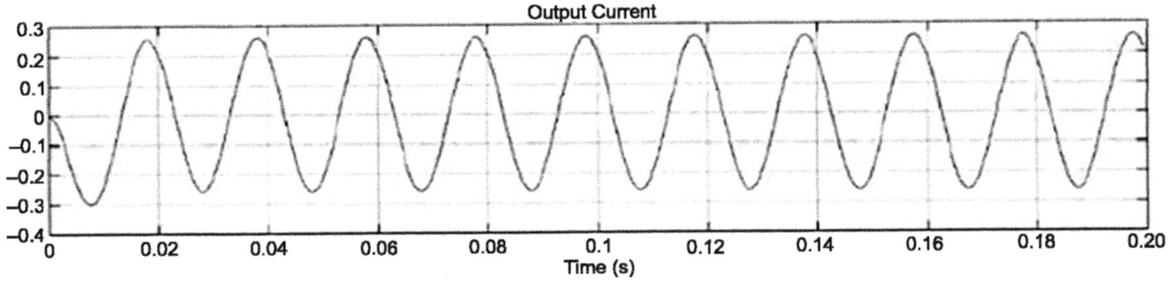

(c)

Fig. 5.20 The output voltage and current waveforms of 9-level with symmetrical configuration of transistor-clamped cascaded H-bridge inverter with reduced number of switches or H-bridges (a) load phase voltage, V_{AN}, (b) typical simulated load phase voltage, V_{AN}, and (c) load current Io, for resistive & inductive load.

The developed nine-level transistor-clamped Cascaded H-bridge Multilevel inverter using two H-bridges has many merits such as reduce number of switches, lower EMI, less harmonic distortion and lower % of THD.

Three-phase nine-level transistor-clamped cascaded H-bridge inverter using two H-bridges per phase with symmetrical configuration: Three-phase nine-level transistor-clamped cascaded H-bridge inverter using two H-bridges per phase with symmetrical configuration, i.e. equal dc source voltages is given in Fig. 5.21, for star-connected load.

The dc source voltages are $V_{A1}=V_{B1}=V_{C1}=V_{A2}=V_{B2}=V_{C2}=V_{A3}=V_{B3}=V_{C3}=V_{A4}=V_{B4}=V_{C4}=V_{dc}$. The operation of each phase is similar to the single-phase circuits and the output phase voltages for nine-levels are similar to the single-phase output voltage but shifted by 120^0 from each other to get the three-phase output voltage waveform.

5.5 Asymmetrical configuration of Transistor-clamped Cascaded H-Bridge Inverters

Asynchronous configuration of transistor-clamped H-bridge inverter with reduced number of switching devices or H-bridges are presented and discussed. In asymmetrical configuration of transistor-clamped H-bridge with single H-bridge with reduced number of switching devices produces seven-level output voltage. And with asymmetrical configuration of transistor-clamped cascaded H-bridge with two-bridges with reduced number of switching devices produces twenty-one-level output voltage. The load current depends on the nature of the load.

5.5.1 Asymmetrical Configuration of Seven-Level Transistor-clamped Single H-bridge Inverter with Reduced Number of Switching Devices or H-bridges

In order to further reduce the requirement of number of switching devices in transistor-clamped single H-bridge inverters, introduced unequal dc source voltages for H-bridge. The main advantage of this topology is that the number of switches required is further reduced and also the number of sources. Fig.5.22 shows the seven-level transistor-clamped cascaded H-bridge multilevel inverter [6]. It has additional one bidirectional switch connected between the first leg of transistor-clamped H-bridge and the capacitor midpoint, enabling seven level output voltage when compared to symmetrical configuration (i.e., with equal dc source voltages) of transistor-clamped single H-Bridge module which produces only five-level output voltage.

In the circuit shown in Fig.5.22 (a) – (h), asymmetrical configuration of transistor-clamped single H-bridge module with unequal dc source voltages is capable of producing seven - level output voltage. It has seven output voltage levels i.e. $+3V_{dc}$, $+2V_{dc}$, $+V_{dc}$, 0, $-V_{dc}$, $-2V_{dc}$, $-3V_{dc}$.

Mode 1: For getting the output voltage $V_{an} = +3V_{dc}$ the switches S_2, S_5 are to be turned on, so that the current flows from source terminal 1, switch S_2, load terminal A to terminal N, switch S_5 to source terminal 2, as positive current.

Mode 2: Similarly, for output voltage $V_{AN} = +2V_{dc}$ switches S_1, S_5 are to be turned on and S_2 is turned-off, then conduction takes place from source mid-point terminal Z, diode D_{11}, switch S_1, diode D_{14}, load terminal A to N, switch S_5 and back to source terminal 2, with positive current.

Mode 3: For $V_{AN} = +V_{dc}$, from dc source terminal 1, switch S_2, load terminal A to N, switch S_5, diodes D_3, D_{13}, switch S_1, diode D_{12}, source terminal Z, with positive current.

Mode 4: For $V_{AN} = 0$ either switches (a) S_3, D_5 or (b) S_2, D_4 need to be turned on, no current flows through the load;

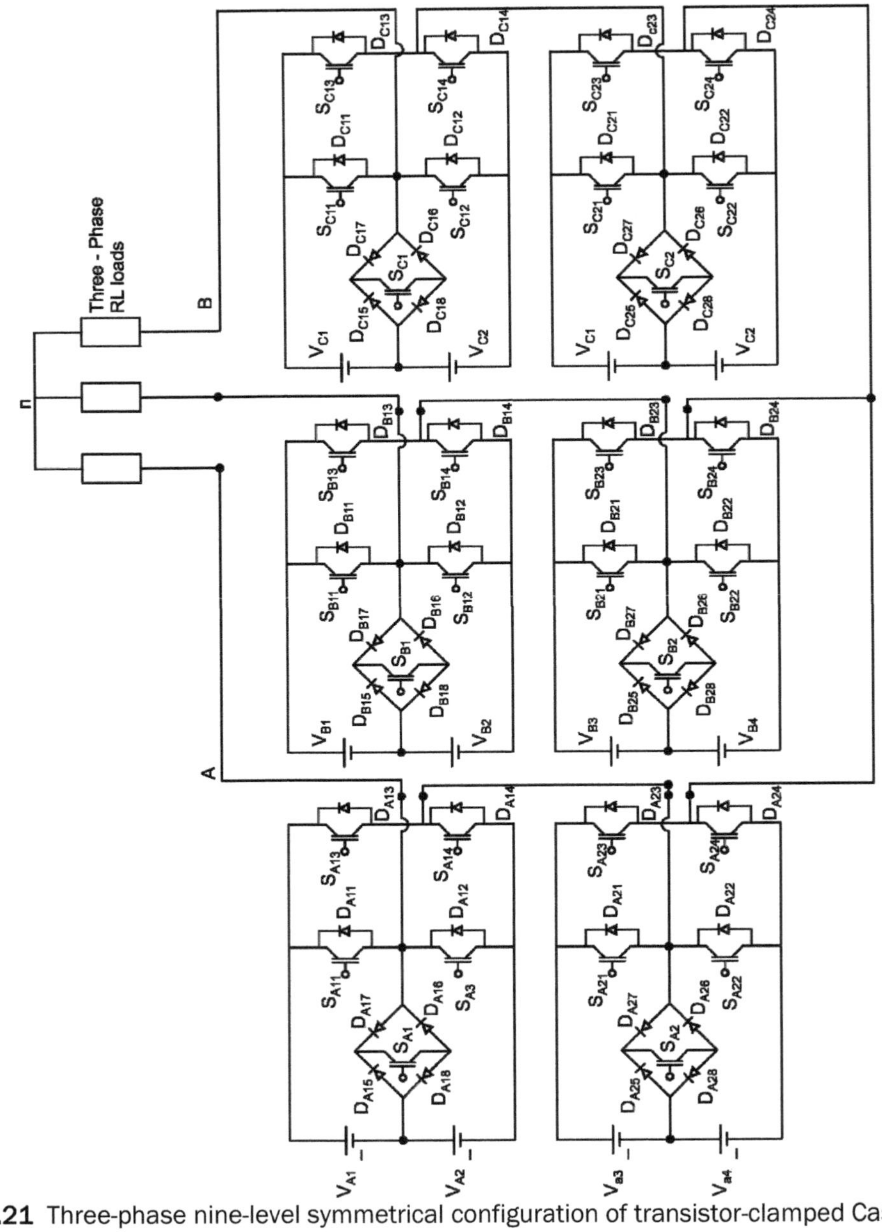

Fig. 5.21 Three-phase nine-level symmetrical configuration of transistor-clamped Cascaded H-bridges (2 cells) inverters with reduced number of switches or H-bridges for three-phase star-connected loads.

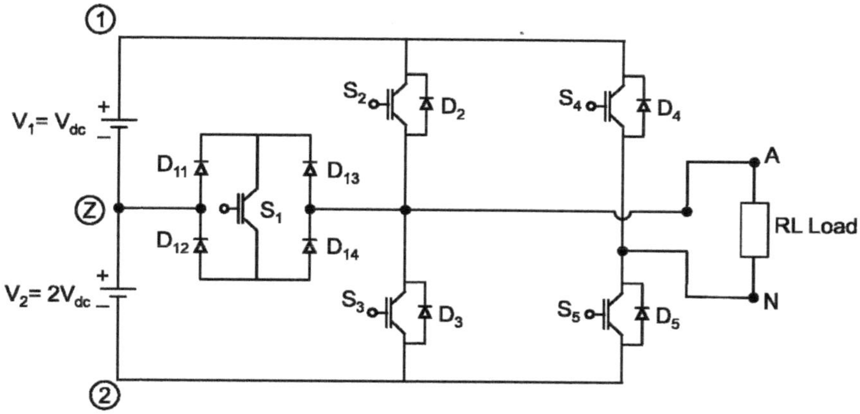

(a)

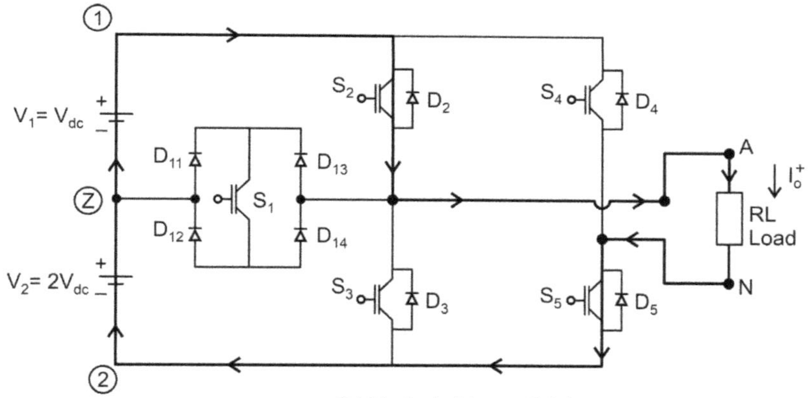

(b) Mode 1: (V_{AN}= +3V_{dc})

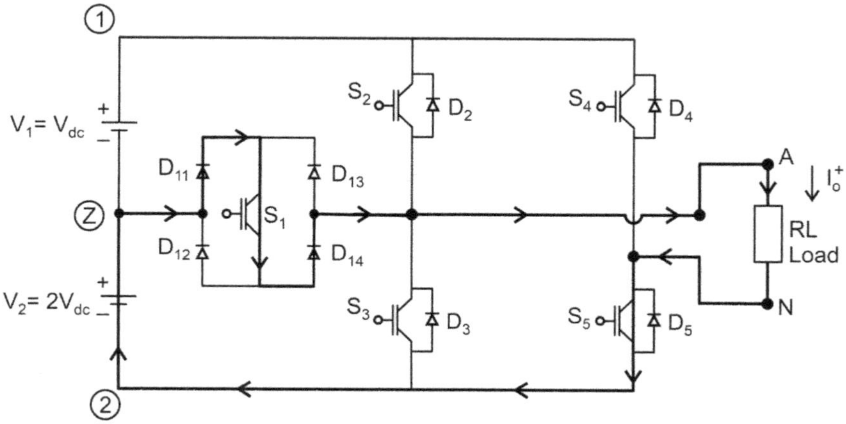

(c) Mode 2 : (V_{AN}= +2V_{dc})

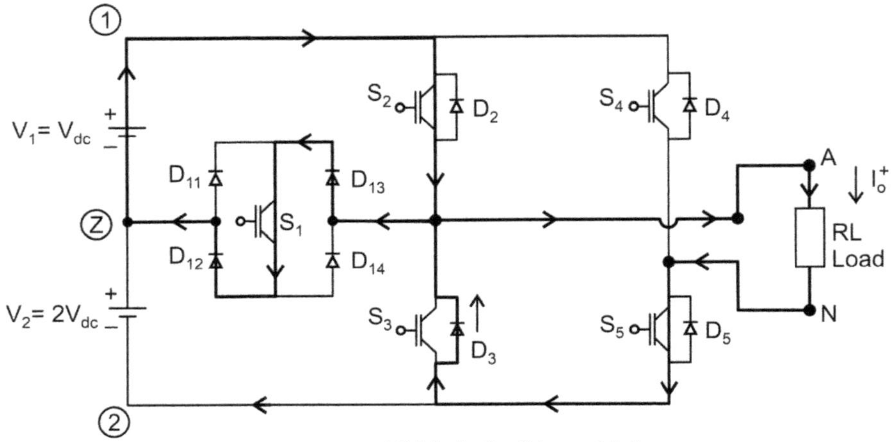

(d) Mode 3 : (V_{AN}= +V_{dc})

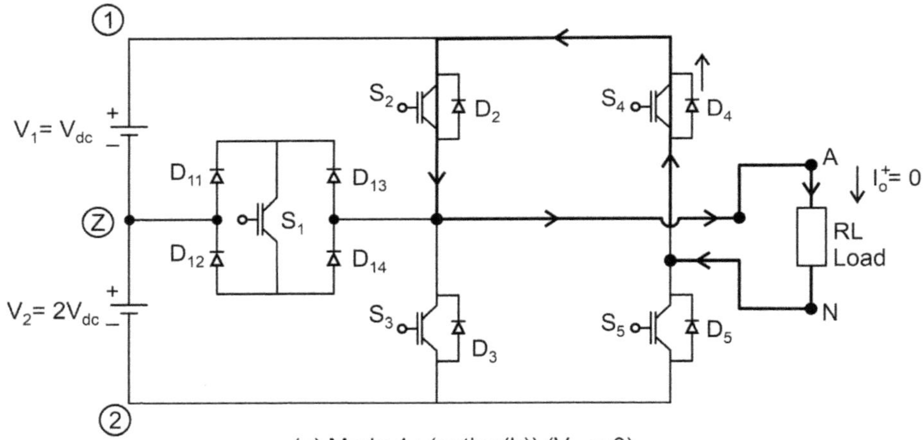

(e) Mode 4 : (option(b)) (V_{AN}= 0)

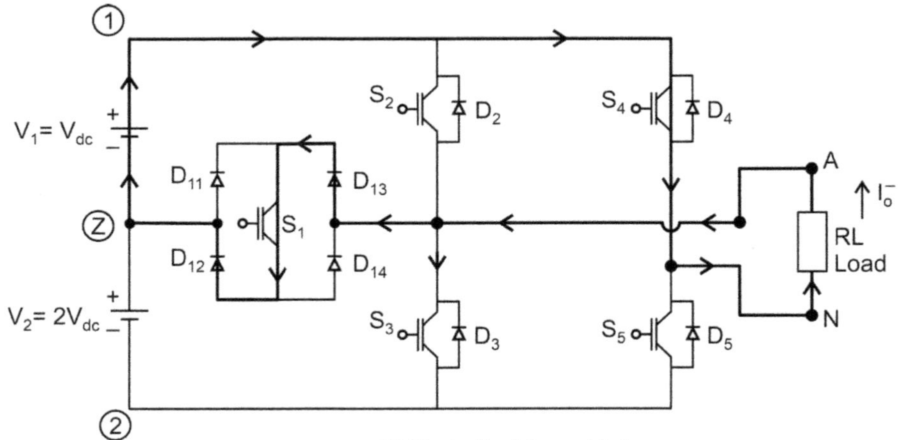

(f) Mode 5 : (V_{AN}= –V_{dc})

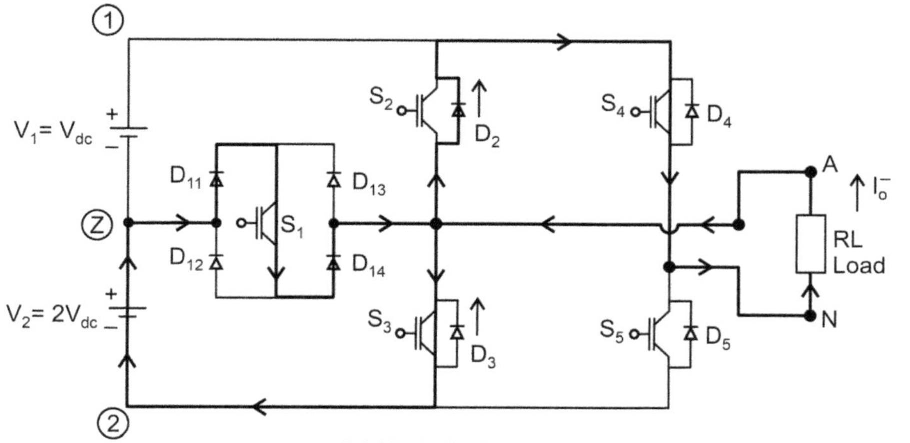

(g) Mode 6 : (V_{AN}= –2V_{dc})

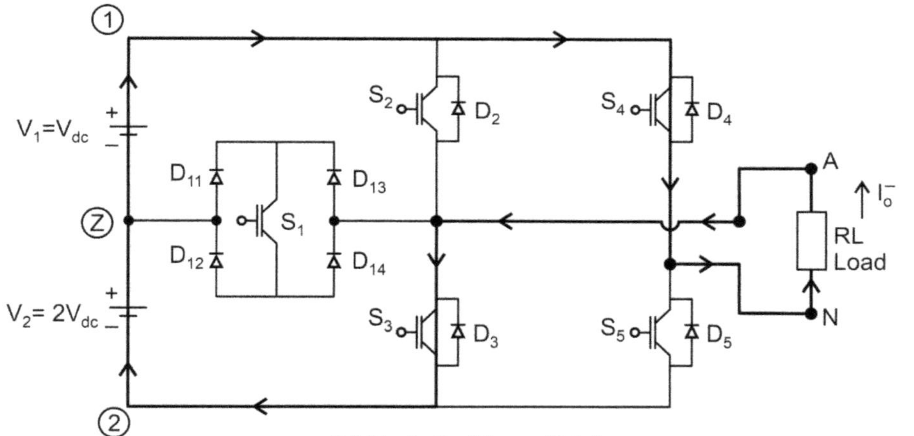

(h) Mode 7 : (V_{AN}= –3V_{dc})

Fig. 5.22 Asymmetrical configuration of transistor-clamped Seven-level with Single H-bridge inverter, (a) Main circuit, (b) Mode1: V_{AN} = +3V_{dc}; (c) Mode2: V_{AN} = +2V_{dc}; (d) Mode3: V_{AN} = +V_{dc}; (e) Mode4: V_{AN} = 0; (f) Mode5: V_{AN} = –3V_{dc}; (g) Mode6: V_{AN} = –2V_{dc}; and (h) Mode7: V_{AN} = –V_{dc}.

Mode 5: For V_{AN} = -V_{dc}, current flows from dc source terminal 1, switch S_4, load terminal N to A, diode D_{13}, switch S_1, diode D_{12}, mid-point source terminal Z to source terminal 1, and with negative load current. The switch combinations are shown in table 5.10.

Mode 6: For V_{AN} =-2V_{dc}, conduction takes place through dc source terminal 1, switch S_4, load terminal N to A, switch S_3, dc source terminal 2, mid-point source terminal Z, diode D_{11}, switch S_1, diodes D_{14}, D_2, switch S_4, and negative load current.

Mode 7: For V_{AN} = -3V_{dc}, from dc source terminal 1, switch S_3, load terminal N to A, switch S_4, dc source terminal 2, with negative current.

Table 5.10 Asymmetrical configuration of Seven- level single H-bridge output voltage with reduced number of switches or H-bridges.

Modes	S_1	S_2	S_3	S_4	S_5	Diodes	Load Voltage, V_{AN}	Load current, I_o
M1	0	1	0	0	1	----	$+3V_{dc}$	Positive
M2	1	0	0	0	1	D_{11}, D_{14}	$+2V_{dc}$	Positive
M3	1	1	0	0	1	D_3, D_{12}, D_{13}	$+V_{dc}$	Positive
M4	0	0 or 1	1 or 0	0	0	D_4 - D_5	0	Zero
M5	1	0	0	1	0	D_{12}, D_{13}	$-V_{dc}$	Negative
M6	1	0	1	1	0	D_2, D_{11}, D_{14}	$-2V_{dc}$	Negative
M7	0	0	1	1	0	----	$-3V_{dc}$	Negative

(Switch ON = 1, switch OFF = 0)

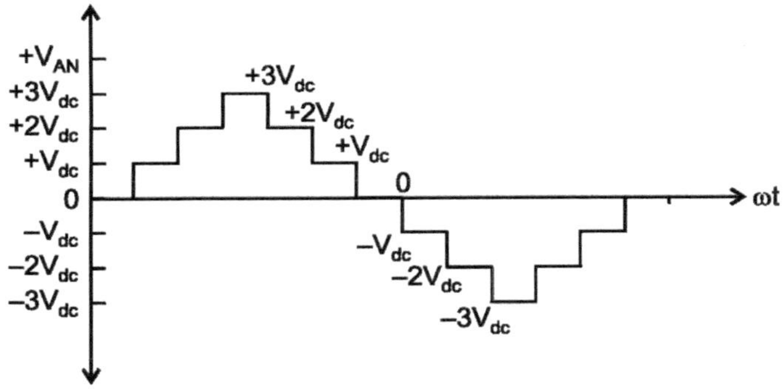

Fig. 5.23 Asymmetrical configuration of seven-level transistor-clamped single H-bridge inverter with reduced number of switches: Load phase voltage waveform (with unequal dc source voltages).

Three-phase asymmetrical configuration of seven-level transistor-clamped single H-bridge inverter: Three single H-bridge inverters are connected to form a three-phase circuit for three-phase star-connected RL loads. For asymmetrical configuration the dc source voltages are unequal, namely, $V_{A1} = V_{B1} = V_{C1} = V_{dc}$ and $V_{A2} = V_{B2} = V_{C2} = 2V_{dc}$. The output voltages are obtained for seven-levels same waveform as shown for single phase circuit, but for phase shift between phases by 120°.

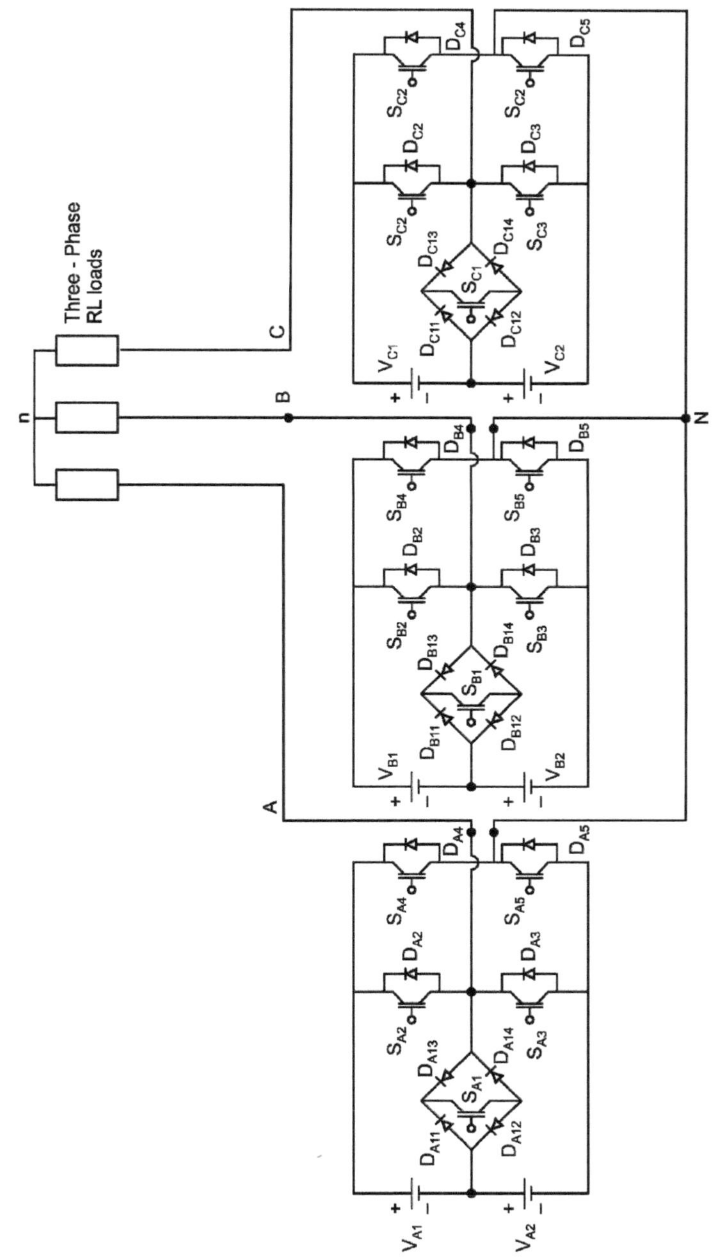

Fig. 5.24 Three-phase asymmetrical configuration of transistor-clamped single H-bridge inverter with reduced number of switches or H-bridges for three-phase star-connected RL loads (with unequal dc source voltages).

5.5.2 Asymmetrical Configuration of Twenty-one-level Transistor-Clamped Cascaded Two-H-bridge Inverter with Reduced Number of Switching Devices or H-bridge

Asymmetrical configuration shown in Fig. 5.25, where two-identical H-bridges are connected in cascade with four different dc source voltages, namely, $V_1 = 4V_{dc}$, $V_2 = 3V_{dc}$, $V_3 = 2V_{dc}$, and $V_4 = V_{dc}$. This configuration has one mid-point source terminal for transistor-clamping for each H-bridge circuit. The transistor-clamped cascaded H-bridge inverter with 2-cells (with unequal dc source voltages) are producing twenty-one-level output phase voltages. The complete circuit employs 8-bidirectional controlled power semiconductor devices with anti-parallel diode (IGBTs) and 2-unidirectional controlled device (IGBTs) associated with 4-diode bridge rectifier arrangement for each H-bridge. The detailed operation of the circuit configuration given in Fig. 5.25 (a)-(v) for 21 modes is explained as follows:

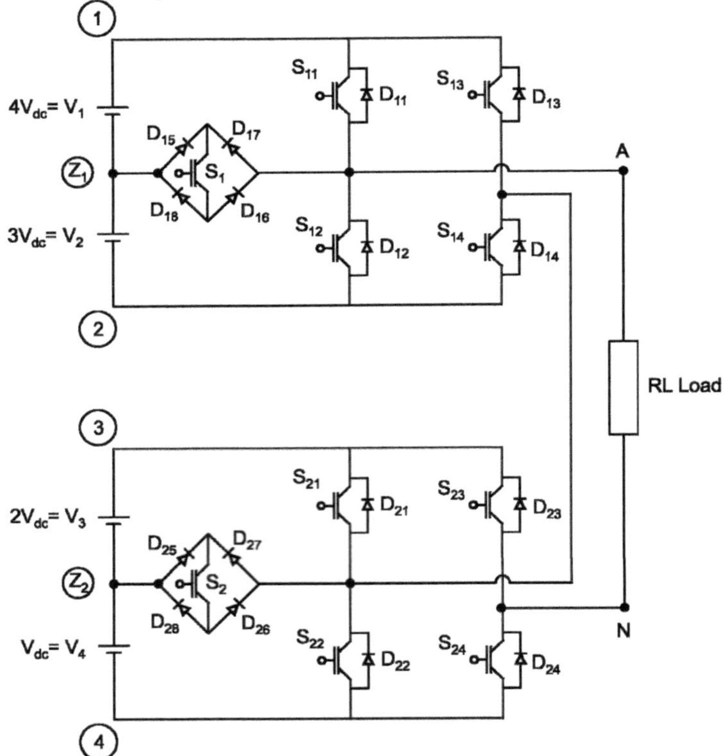

(a) Main circuit

Fig. 5.25 (a) Main circuit of transistor-clamped twenty-one level cascaded H-bridge inverter, (b) Mode1: $V_{AN} = +10V_{dc}$; (c) Mode2: $V_{AN} = +9V_{dc}$; (d) Mode3: $V_{AN} = +8V_{dc}$; (e) Mode4: $V_{AN} = +7V_{dc}$; (f) Mode5: $V_{AN} = +6V_{dc}$; (g) Mode6: $V_{AN} = +5V_{dc}$; (h) Mode7: $V_{AN} = +4V_{dc}$; (i) Mode8: $V_{AN} = +3V_{dc}$; (j) Mode9: $V_{AN} = +2V_{dc}$; (k) Mode10: $V_{AN} = +V_{dc}$; (l) Mode11: $V_{AN} = 0$; (m) Mode12: $V_{AN} = -V_{dc}$. (n) Mode13: $V_{AN} = -2V_{dc}$; (o) Mode14: $V_{AN} = -3V_{dc}$; (p) Mode15: $V_{AN} = -4V_{dc}$; (q) Mode16: $V_{AN} = -5V_{dc}$ (r) Mode 17: $V_{AN} = -6V_{dc}$ (s) Mode18: $V_{AN} = -7V_{dc}$; (t) Mode19: $V_{AN} = -8V_{dc}$; (u) Mode20: $V_{AN} = -9V_{dc}$; (v) Mode21: $V_{AN} = -10V_{dc}$.

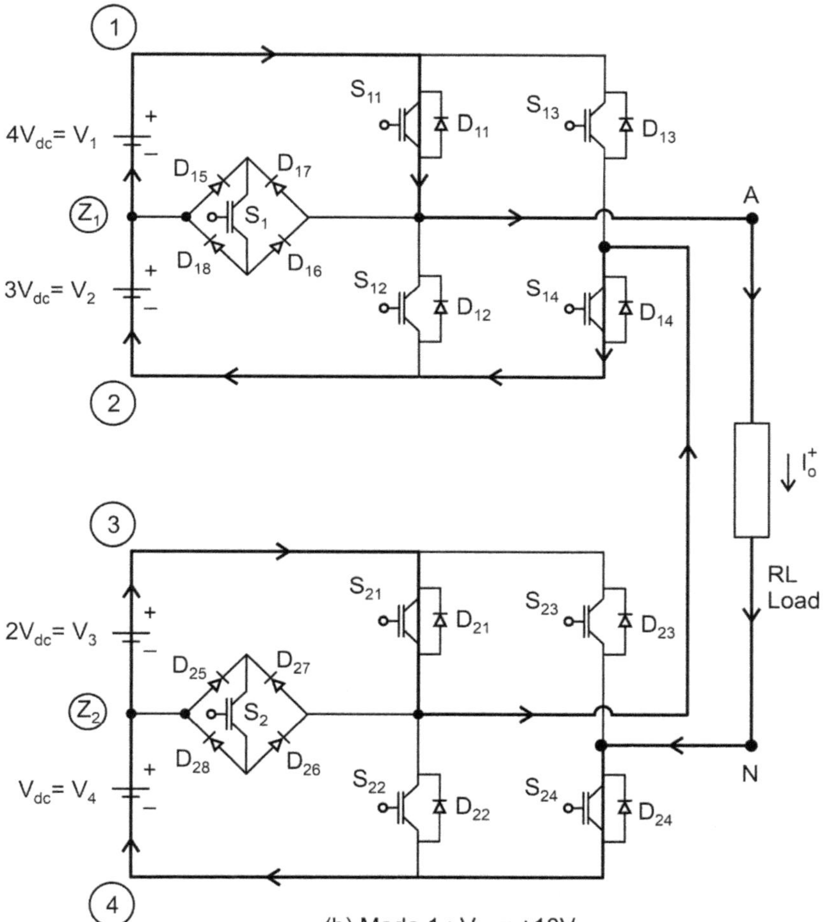

(b) Mode 1 : $V_{AN} = +10V_{dc}$

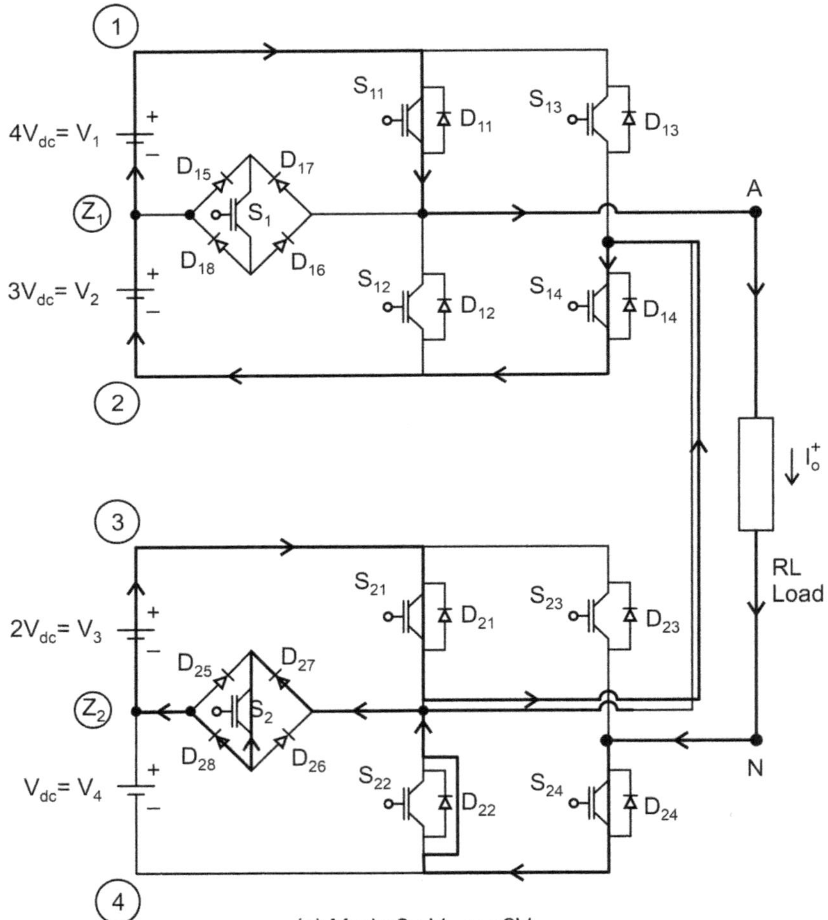

(c) Mode 2 : $V_{AN} = +9V_{dc}$

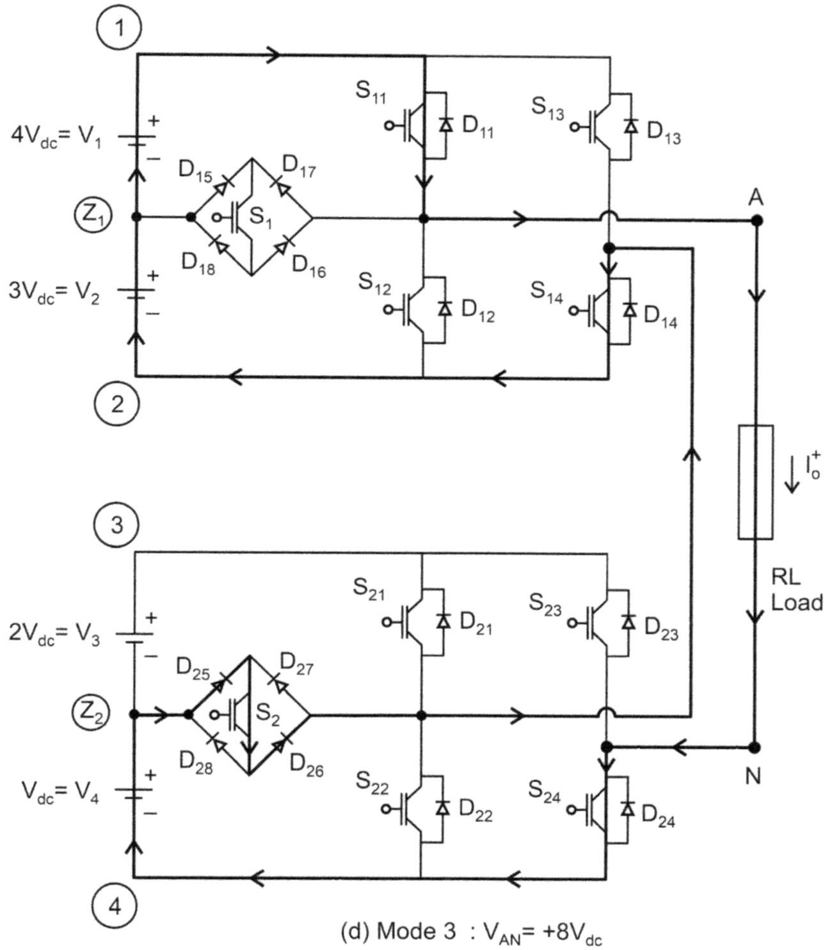

(d) Mode 3 : $V_{AN} = +8V_{dc}$

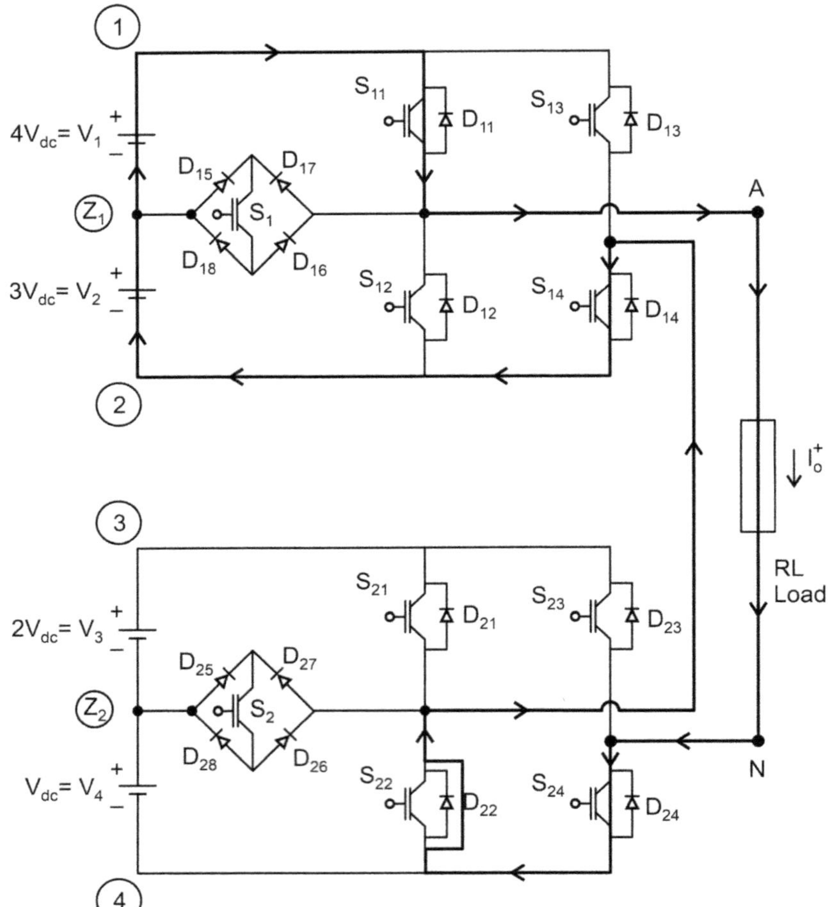

(e) Mode 4(a) : $V_{AN} = +7V_{dc}$

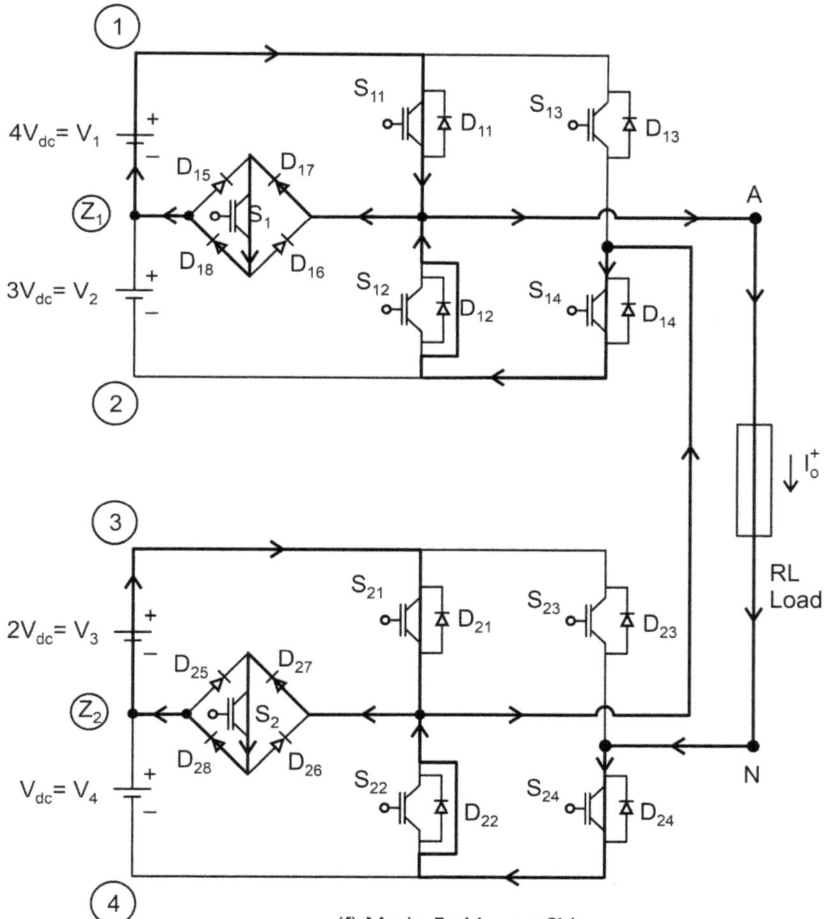

(f) Mode 5 : $V_{AN} = +6V_{dc}$

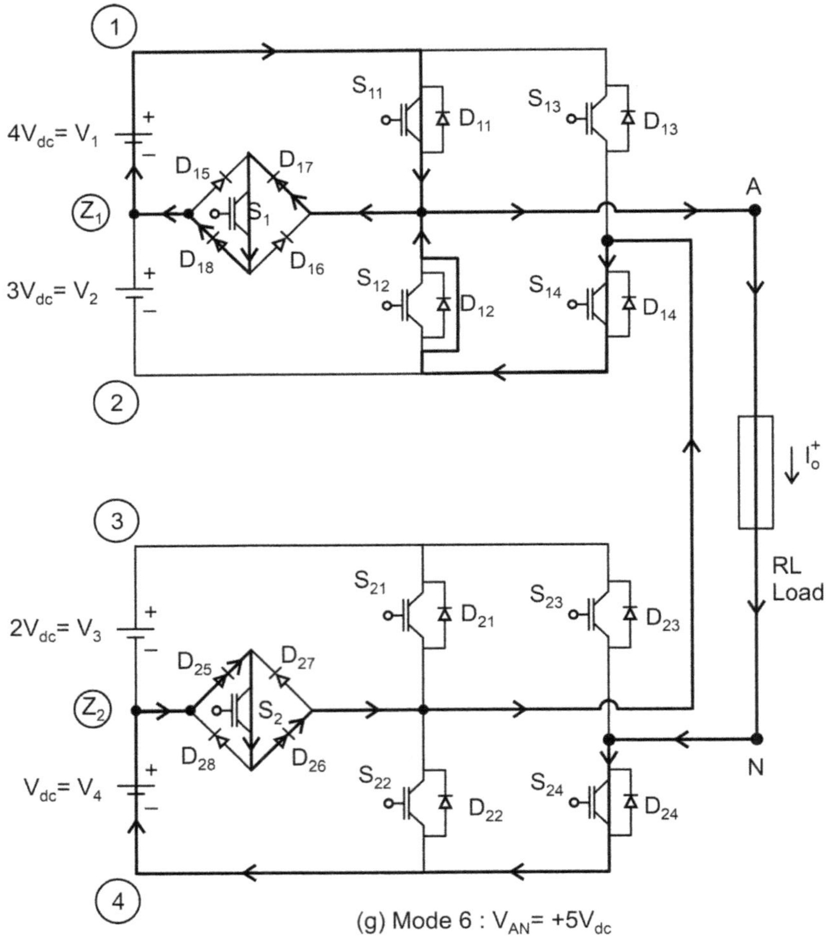

(g) Mode 6 : $V_{AN} = +5V_{dc}$

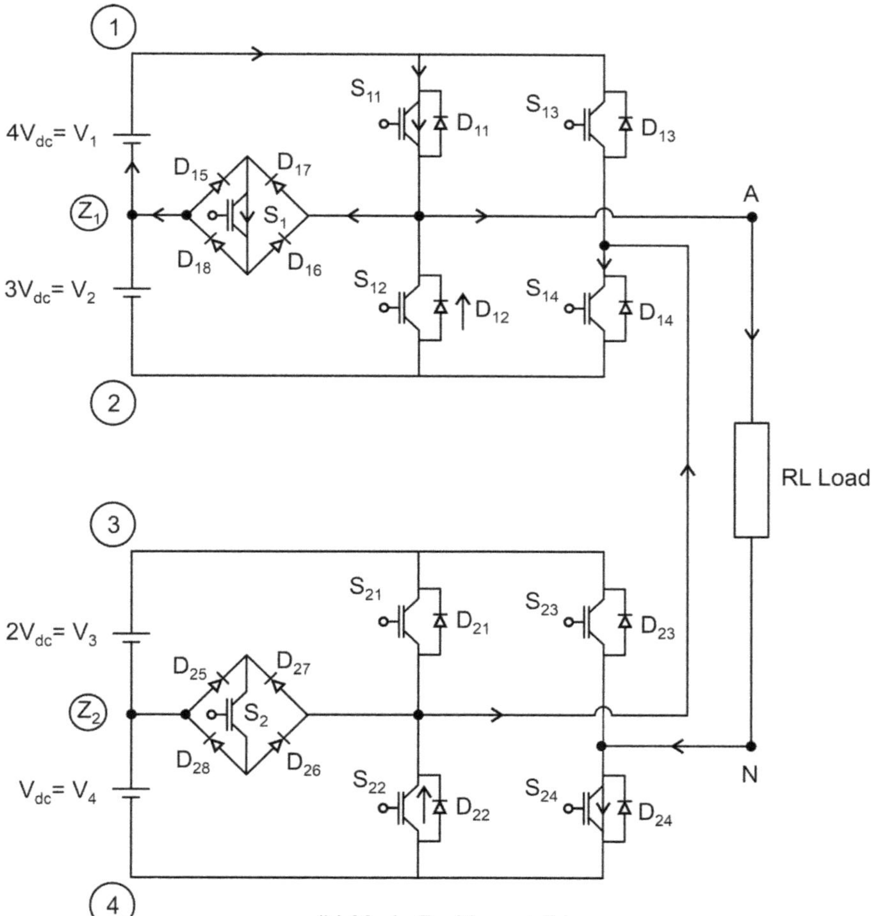

(h) Mode 7 : $V_{AN} = +4V_{dc}$

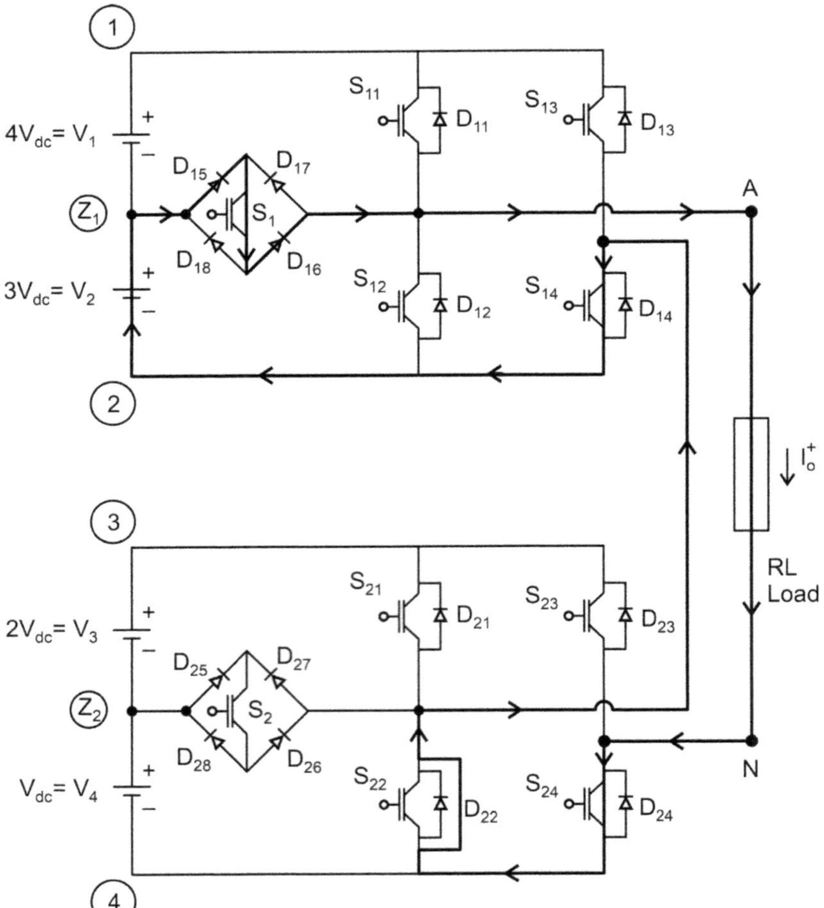

(i) Mode 8 : $V_{AN} = +3V_{dc}$

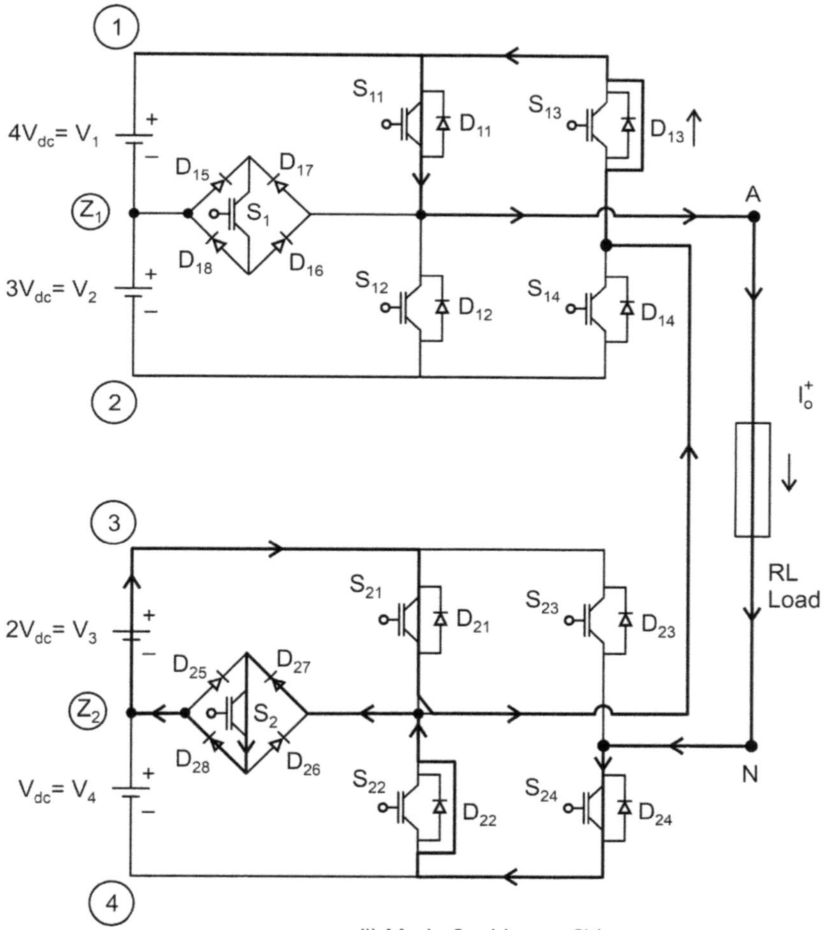

(j) Mode 9 : $V_{AN} = +2V_{dc}$

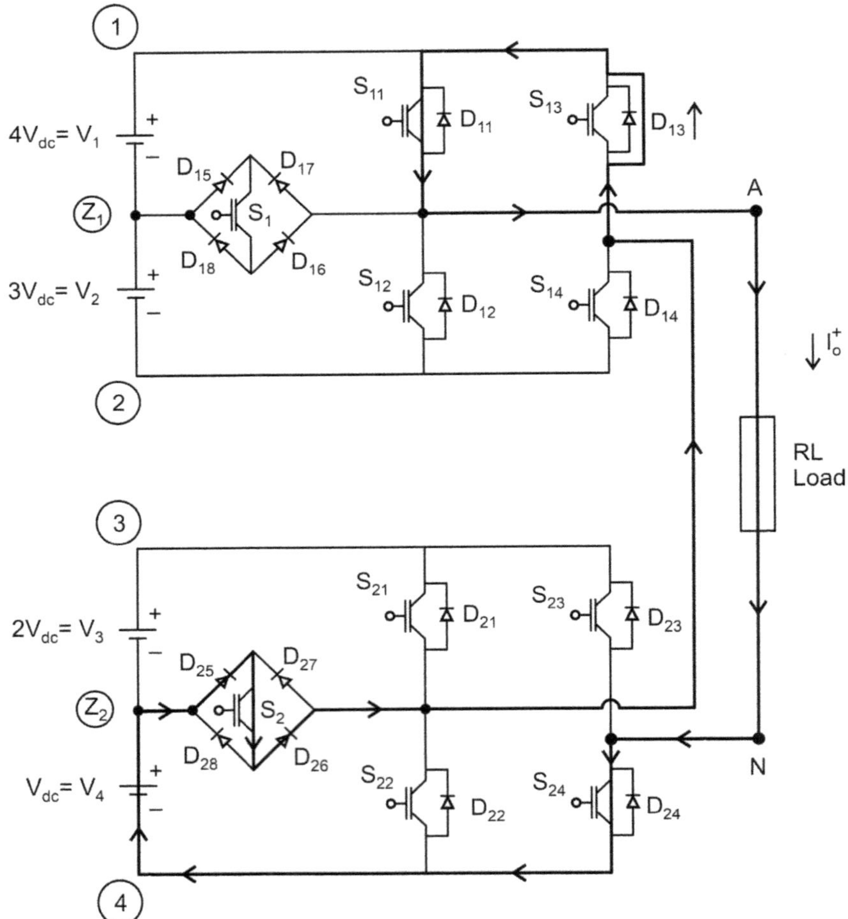

(k) Mode 10 : $V_{AN} = +V_{dc}$

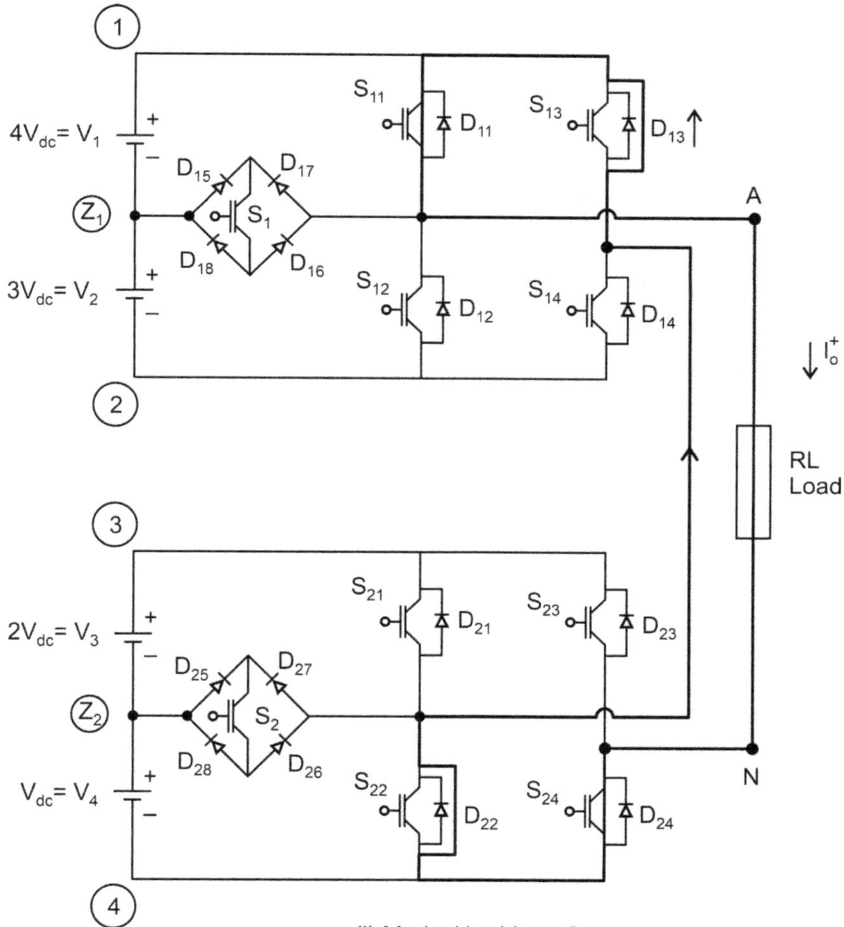

(l) Mode 11 : $V_{AN} = 0$

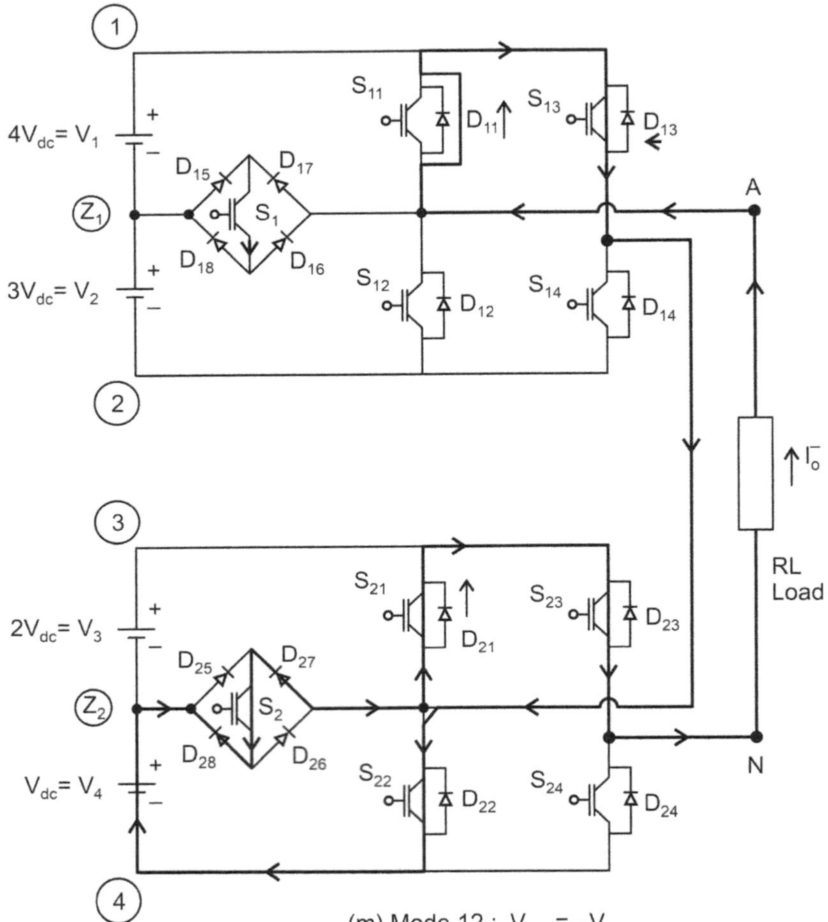

(m) Mode 12 : $V_{AN} = -V_{dc}$

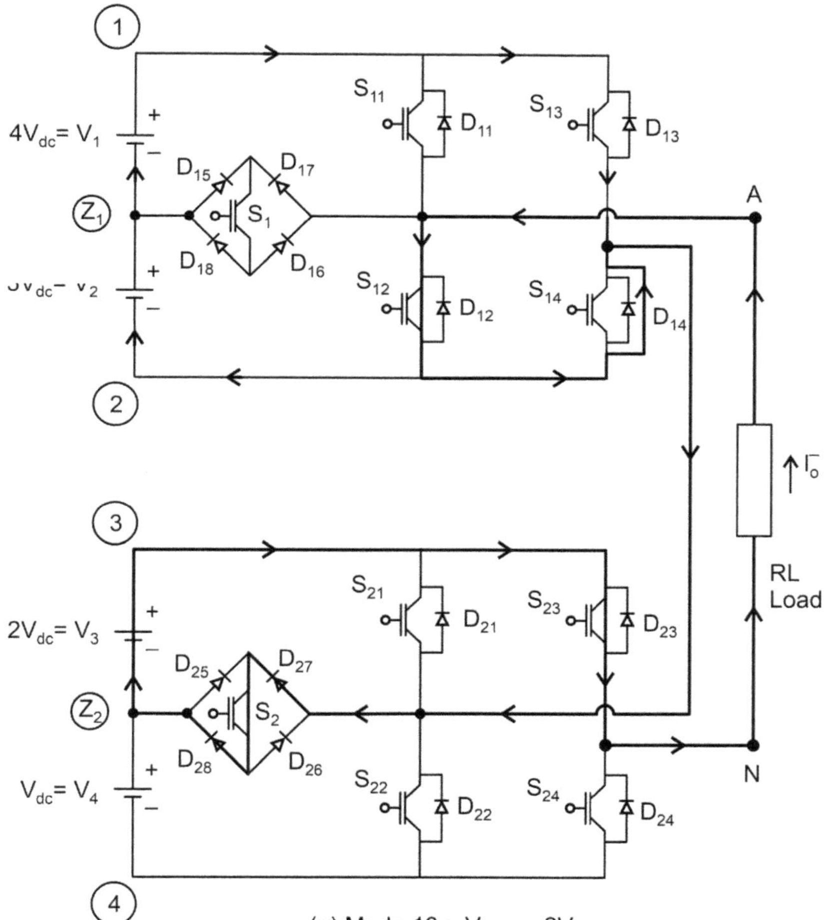

(n) Mode 13 : $V_{AN} = -2V_{dc}$

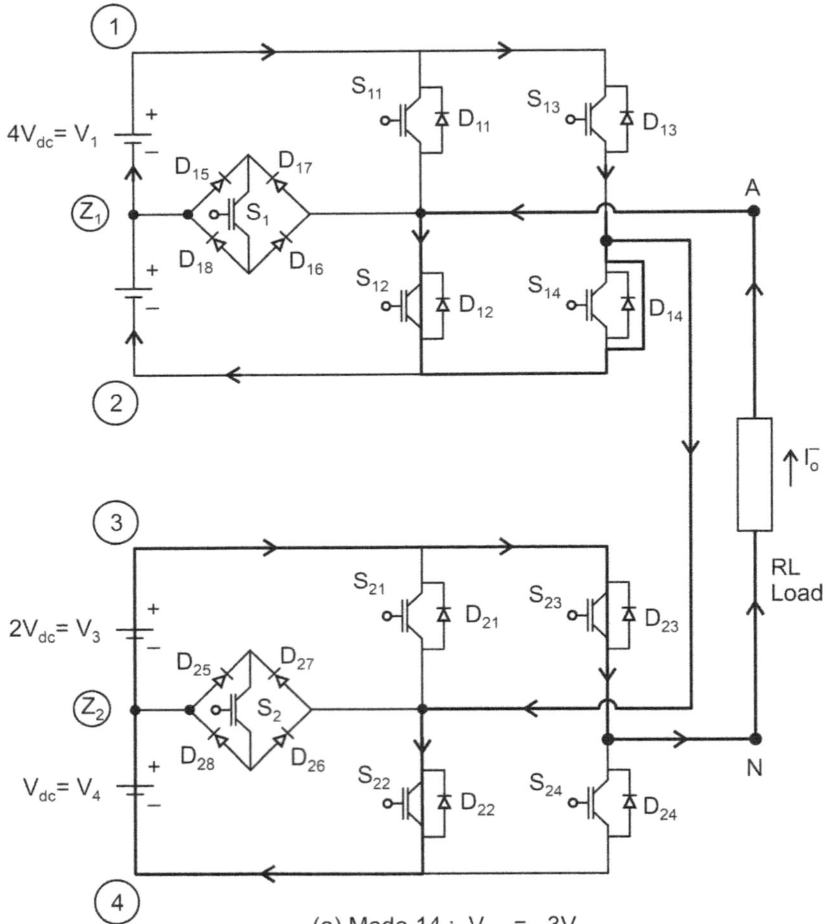

(o) Mode 14 : $V_{AN} = -3V_{dc}$

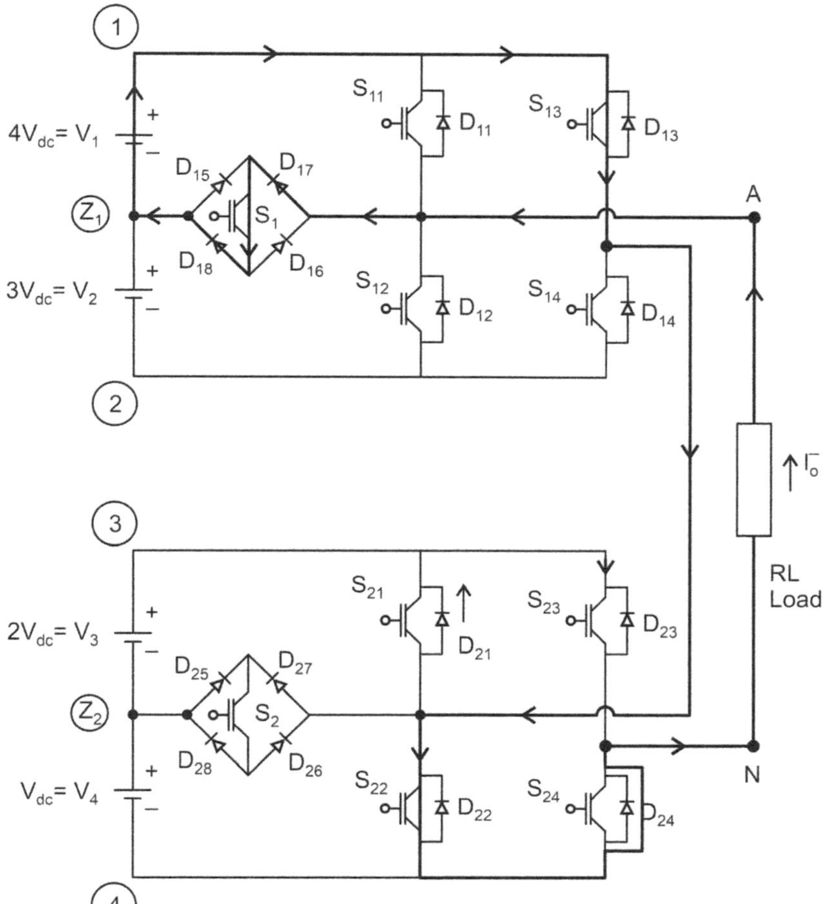

(p) Mode 15 : $V_{AN} = -4V_{dc}$

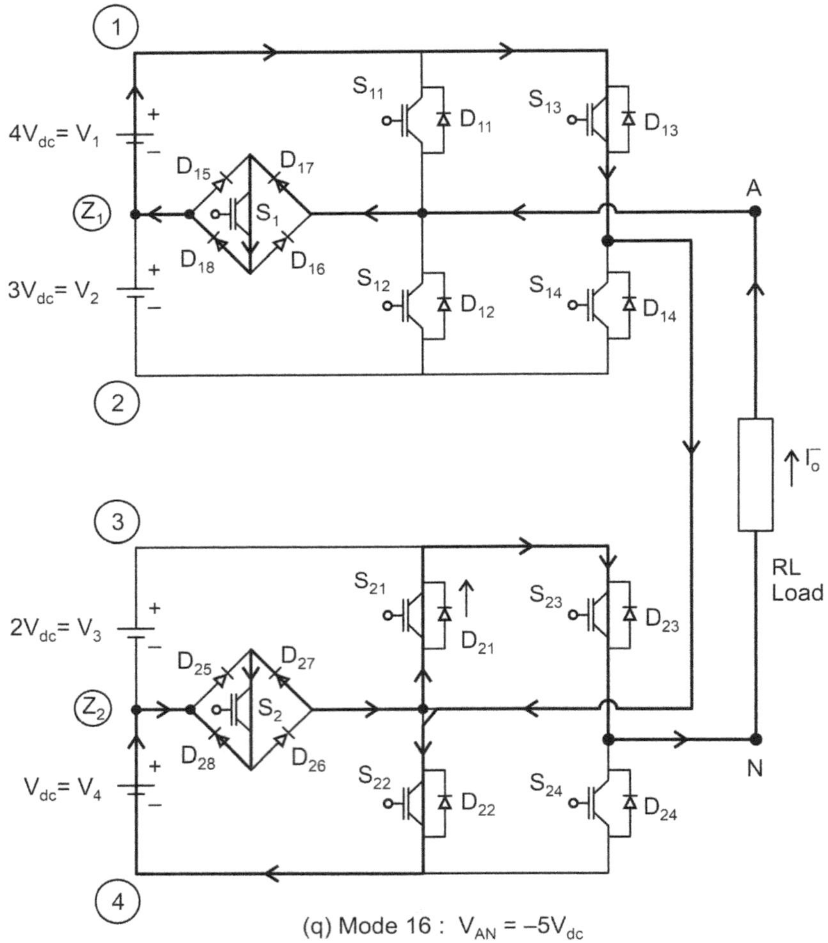

(q) Mode 16 : $V_{AN} = -5V_{dc}$

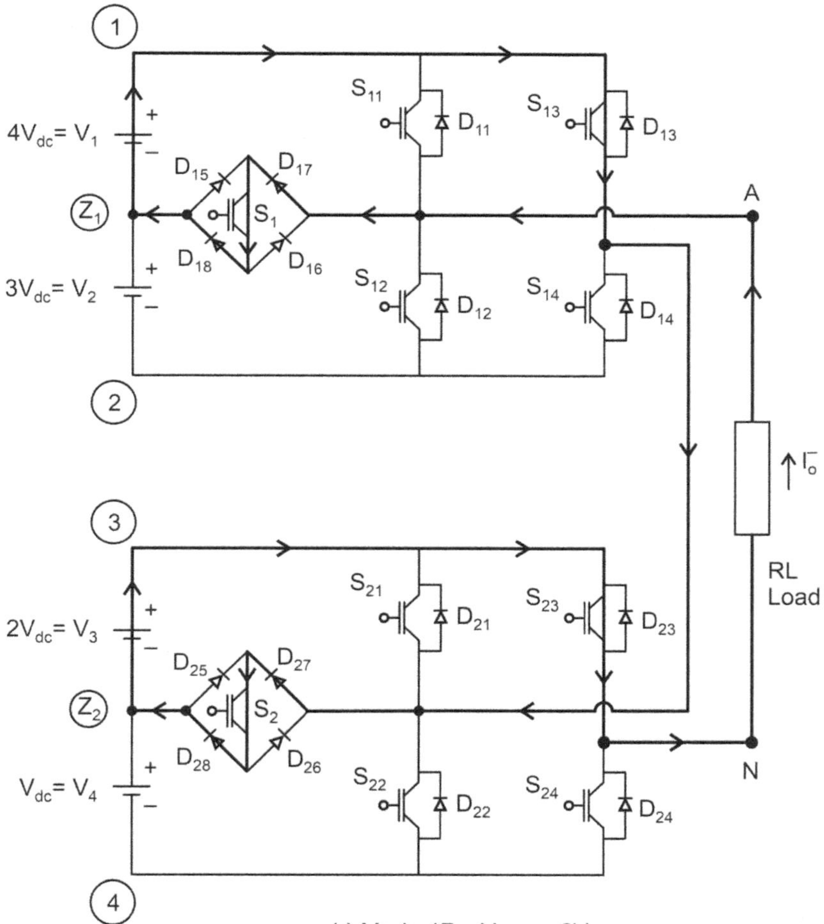

(r) Mode 17 : $V_{AN} = -6V_{dc}$

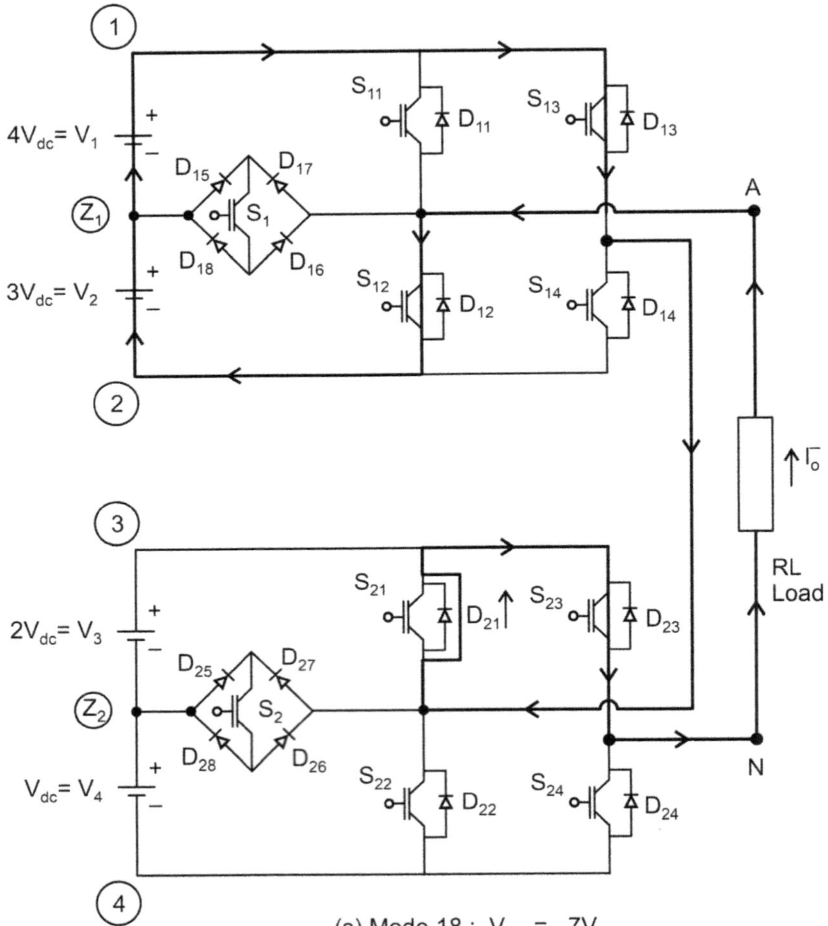

(s) Mode 18 : $V_{AN} = -7V_{dc}$

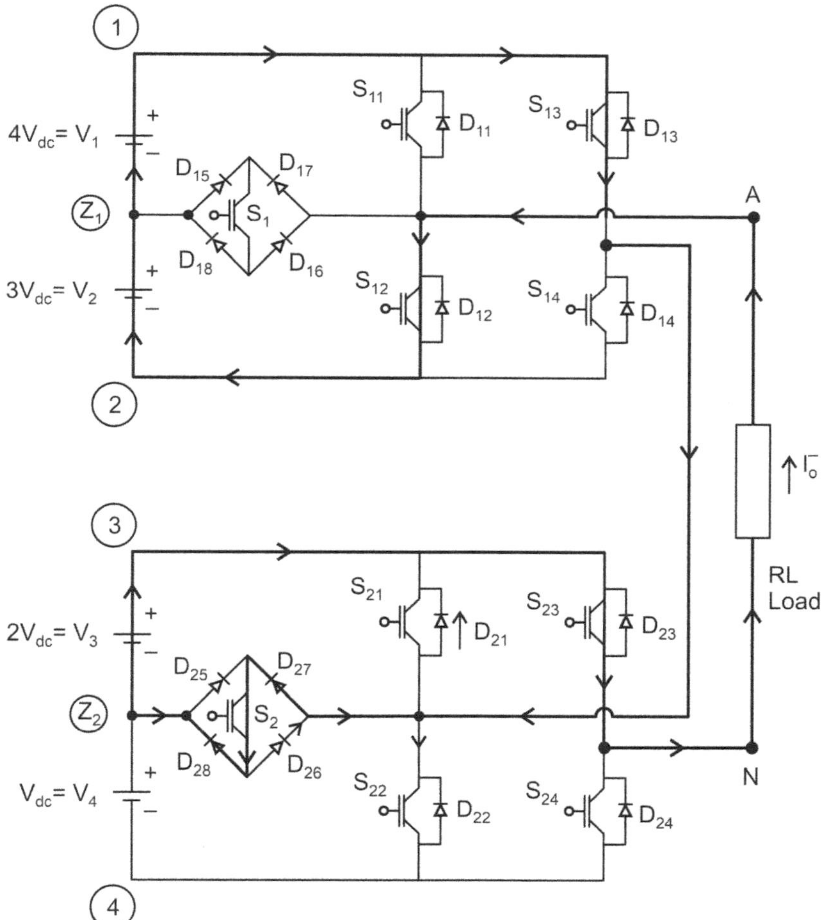

(t) Mode 19 : $V_{AN} = -8V_{dc}$

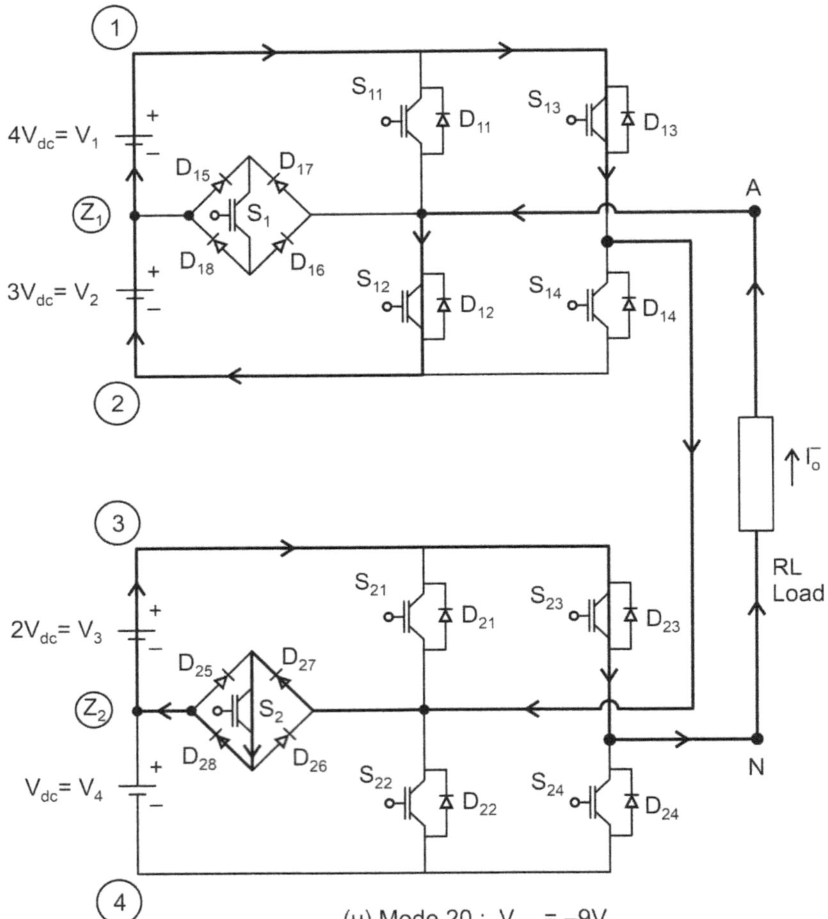

(u) Mode 20 : $V_{AN} = -9V_{dc}$

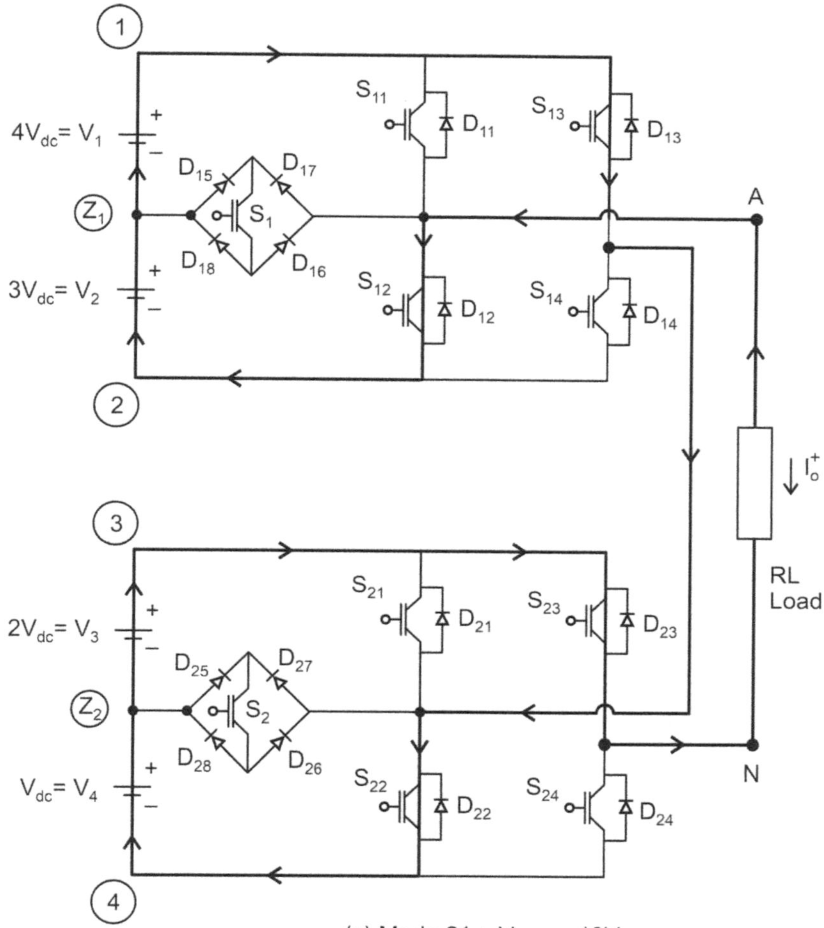

(v) Mode 21 : $V_{AN} = -10V_{dc}$

Fig. 5.25 (a) Main circuit of transistor-clamped twenty-one level cascaded H-bridge inverter, (b) Mode1: $V_{AN} = +10V_{dc}$; (c) Mode2: $V_{AN} = +9V_{dc}$; (d) Mode3: $V_{AN} = +8V_{dc}$; (e) Mode4: $V_{AN} = +7V_{dc}$; (f) Mode5: $V_{AN} = +6V_{dc}$; (g) Mode6: $V_{AN} = +5V_{dc}$; (h) Mode7: $V_{AN} = +4V_{dc}$; (i) Mode8: $V_{AN} = +3V_{dc}$; (j) Mode9: $V_{AN} = +2V_{dc}$; (k) Mode10: $V_{AN} = +V_{dc}$; (l) Mode11: $V_{AN} = 0$; (m) Mode12: $V_{AN} = -V_{dc}$. (n) Mode13: $V_{AN} = -2V_{dc}$; (o) Mode14: $V_{AN} = -3V_{dc}$; (p) Mode15: $V_{AN} = -4V_{dc}$; (q) Mode16: $V_{AN} = -5V_{dc}$ (r) Mode 17: $V_{AN} = -6V_{dc}$ (s) Mode18: $V_{AN} = -7V_{dc}$; (t) Mode19: $V_{AN} = -8V_{dc}$; (u) Mode20: $V_{AN} = -9V_{dc}$; (v) Mode21: $V_{AN} = -10V_{dc}$

Mode 1: For $V_{AN} = +10V_{dc}$, all four dc sources will be contributing for the load phase voltage, the conduction takes place through dc source terminal 1, switch S_{11}, load terminal A to N, switch S_{24}, dc source terminal 4 to 3, switches S_{21}, S_{14}, dc source terminal 2 to 1, and load current is positive.

Mode 2: To obtain V_{AN} = $+9V_{dc}$, the conduction takes place through dc source terminal 1, switch S_{11}, load terminal A to N, switch S_{24}, diodes D_{22}, D_{27}, switch S_2, diode D_{28}, mid-point source terminal Z_2 to 3, switches S_{21}, S_{14}, dc source terminal 2 to 1, and load current is positive.

Mode 3: For V_{AN} = $+8V_{dc}$, current flows from dc source terminal 1, switch S_{11}, load terminal A to N, switch S_{24}, dc source terminal 4 to Z_2, diode D_{25}, switch S_2, diode D_{26}, switch S_{14}, dc source terminal 2 to 1, and load current is positive.

Mode 4: To get V_{AN} = $+7V_{dc}$, the conduction takes place in two pathways (a) one is through dc source terminal 1, switch S_{11}, load terminal A to N, switch S_{24}, diode D_{22}, switch S_{14}, dc source terminal 2 to 1, and (b) another is through dc source terminal 1, switch S_{11}, load terminal A to N, switch S_{24}, dc source terminal 4 to 3, switches S_{21}, S_{14}, diodes D_{12}, D_{17}, switch S_1,diode D_{18}, source mid-point terminal Z_1 to 1, with load current positive.

Mode 5: For V_{AN} = $+6V_{dc}$, current flows from dc source terminal 1, switch S_{11}, load terminal A to N, switch S_{24}, diodes D_{22}, D_{27}, switch S_2, diode D_{28}, source mid-point terminal Z_2 to 3, switches S_{21}, S_{14}, diodes D_{12}, D_{17}, switch S_1, diode D_{18}, source mid-point terminal Z_1 to 1, and load current is positive.

Mode 6: To obtain V_{AN} = $+5V_{dc}$, the conduction takes place through dc source terminal 1, switch S_{11}, load terminal A to N, switch S_{24}, terminal 4 to Z_2, D_{25}, S_2, D_{26}, S_{14}, D_{12}, D_{17}, switch S_1, diode D_{18}, source mid-point terminal Z_1 to1, with positive load current.

Mode 7: For V_{AN} = $+4V_{dc}$, current flows from dc source terminal 1, switch S_{11}, load terminal A to N, switch S_{24}, diode D_{22}, switch S_{14}, diodes D_{12}, D_{17}, switch S_1, diode D_{18}, source mid-point terminal Z_1 to 1, and load current is positive.

Mode 8: To obtain V_{AN} = $+3V_{dc}$, the conduction takes place through source mid-point terminal Z_1, diode D_{15}, switch S_1, diode D_{16}, load terminal A to N, switch S_{24}, diode D_{22}, switch S_{14}, dc source terminal 2 to Z_1, with positive load current.

Mode 9: To get V_{AN} = $+2V_{dc}$, current flows from dc source terminal 1, switch S_{11}, load terminal A to N, switch S_{24}, didoes D_{22}, D_{27}, switch S_2, diode D_{28}, source mid-point terminal Z_2 to 3, switch S_{21}, diode D_{13}, switch S_{11}, and load current is positive.

Mode 10: For V_{AN} = $+V_{dc}$, the conduction takes place through switch S_{11}, load terminal A to N, switch S_{24}, dc source terminal 4 to Z_2, diode D_{25}, switch S_2, diodes D_{26}, D_{13}, and load current is positive.

Mode 11: For V_{AN} = 0, the conduction takes place through switch S_{11}, load terminal A to N, switches S_{24}, S_{22}, diode D_{13}, and load current is zero.

Mode 12: To Obtain V_{AN} = $-V_{dc}$, the conduction takes place through source mid-point terminal Z_2, diode D_{25}, switch S_2, diodes D_{26}, D_{21}, switch S_{23}, load terminal N to A, diode D_{11}, switches S_{13}, S_{22}, dc source terminal 4 to Z_2, and load current is negative.

Mode 13: For V_{AN} = $-2V_{dc}$, current flows from dc source terminal 3, switch S_{23}, load terminal N to A, diode D_{14}, switch S_{12}, diode D_{27}, switch S_2, diode D_{28}, source mid-point terminal Z_2 to 3, with negative load current.

Mode 14: To get V_{AN} = $-3V_{dc}$, the conduction takes place through dc source terminal 3, switch S_{23}, load terminal N to A, switch S_{12}, diode D_{14}, switch S_{22}, dc source terminal 4 to 3, and load current is negative.

Mode 15: To obtain V_{AN} = -4V_{dc}, current flows from dc source terminal 1, switches S_{13}, S_{22}, diode D_{24}, load terminal N to A, diode D_{17}, switch S_1, diode D_{18}, source mid-point terminal Z_1 to 1, and load current is negative.

Mode 16: To obtain V_{AN} = -5V_{dc}, the conduction takes place through dc source terminal 1, switch S_{13}, S_{22}, terminal 4 to Z_2, diode D_{25}, switch S_2, diode D_{26}, switch S_{21}, S_{23}, load terminal N to A, diode D_{17}, switch S_1, diode D_{18}, mid-point terminal Z_1 to 1 and load current is negative.

Mode 17: For V_{AN} = –6V_{dc}, current flows from dc source terminal 1, switch S_{13}, diode D_{27}, switch S_2, diode D_{28}, source mid-point terminal Z_2 to 3, switch S_{23}, load terminal N to A, diode D_{17}, switch S_1, diode D_{18}, source mid-point terminal Z_1 to 1, with negative load current.

Mode 18: To get V_{AN} = –7V_{dc}, the conduction takes place through dc source terminal 1, switch S_{13}, diode D_{21}, switches S_{23}, load terminal N to A, S_{12}, dc source terminal 2 to 1, and load current is negative.

Mode 19: For V_{AN} = –8V_{dc}, current flows from dc source terminal 1, switch S_{13}, S_{22}, terminal 4 to Z_2, diode D_{25}, switch S_2, diode D_{26}, D_{21}, switch S_{23}, load terminal N to A, switch S_{12}, Source terminal 2 to 1, and load current is negative.

Mode 20: To get V_{AN} = –9V_{dc}, the conduction takes place through dc source terminal 1, switch S_{13}, diode D_{27}, switch S_2, diode D_{28}, source mid-point terminal Z_2 to 3, S_{23}, terminal N to A, Switch S_{12}, switches S_{21}, S_{14}, dc source terminal 2 to 1, with negative load current.

Mode 21: To obtain V_{AN} = -10Vdc, current flows from dc source terminal 1, switches S_{13}, S_{22}, dc source terminal 4 to 3, switches S_{23}, S_{12}, dc source terminal 2 to 1, with negative load current.

Table 5.11 Asymmetrical configuration of Twenty-one level transistor-clamped cascaded H-bridge output voltage with reduced number of switches or H-bridges.

Modes	V_{AN}	S_1	S_2	S_{11}	S_{12}	S_{13}	S_{14}	S_{21}	S_{22}	S_{23}	S_{24}	Antiparallel diodes	Diodes across S_1 / S_2
M1	+10V_{dc}	0	0	1	0	0	1	1	0	0	1	----	----
M2	+9V_{dc}	0	1	1	1	0	1	1	1	0	1	D_{22}	D_{27}, D_{28}
M3	+8V_{dc}	0	1	1	0	0	1	0	0	0	1	----	D_{25}, D_{26}
M4	+7V_{dc}	0	0	1	0	0	1	0	0	0	1	D_{22}	D_{17}, D_{18}
M5	+6V_{dc}	1	1	1	0	0	1	1	0	0	1	D_{22}, D_{12},	D_{27}, D_{28}, D_{17}, D_{18}
M6	+5V_{dc}	1	1	1	0	0	1	0	0	0	1	D_{12},	D_{25}, D_{26} D_{17}, D_{18}
M7	+4V_{dc}	1	0	1	0	0	1	0	0	0	1	D_{12}, D_{22}	D_{25}, D_{26} D_{17}, D_{18}
M8	+3V_{dc}	1	0	0	0	0	1	0	0	0	1	D_{22}	D_{15}, D_{16}

Contd...

Modes	V_{AN}	S_1	S_2	S_{11}	S_{12}	S_{13}	S_{14}	S_{21}	S_{22}	S_{23}	S_{24}	Antiparallel diodes	Diodes across S_1 / S_2
M9	$+2V_{dc}$	0	1	1	0	0	0	1	0	0	1	D_{13}, D_{24}, D_{22}	D_{27}, D_{28}
M10	$+V_{dc}$	0	1	1	0	0	0	0	0	0	1	D_{13}	D_{25}, D_{26}
M11	0	0	0	1	0	0	0	0	1	0	1	D_{13}	----
M12	$-V_{dc}$	0	1	0	0	1	0	0	1	0	0	D_{21}, D_{23}, D_{11}	D_{25}, D_{26}
M13	$-2V_{dc}$	0	1	0	1	0	0	0	0	1	0	D_{14}	D_{27}, D_{28}
M14	$-3V_{dc}$	0	0	0	1	0	0	0	1	1	0	D_{14}	----
M15	$-4V_{dc}$	1	0	0	0	1	0	0	1	0	0	D_{24}	D_{17}, D_{18}
M16	$-5V_{dc}$	1	1	1	0	1	0	1	1	1	0	----	D_{17}, D_{18}, D_{25}, D_{26}
M17	$-6V_{dc}$	1	1	0	0	1	0	0	0	1	0	----	D_{17}, D_{18}, D_{27}, D_{28}
M18	$-7V_{dc}$	0	0	0	1	1	0	0	0	0	0	D_{21}, D_{23}	----
M19	$-8V_{dc}$	0	1	0	1	1	0	0	1	1	0	---- D_{21}	D_{25}, D_{26}
M20	$-9V_{dc}$	0	1	0	1	1	0	0	0	1	0	----	D_{27}, D_{28}
M21	$-10V_{dc}$	0	0	0	1	1	0	0	1	1	0	----	----

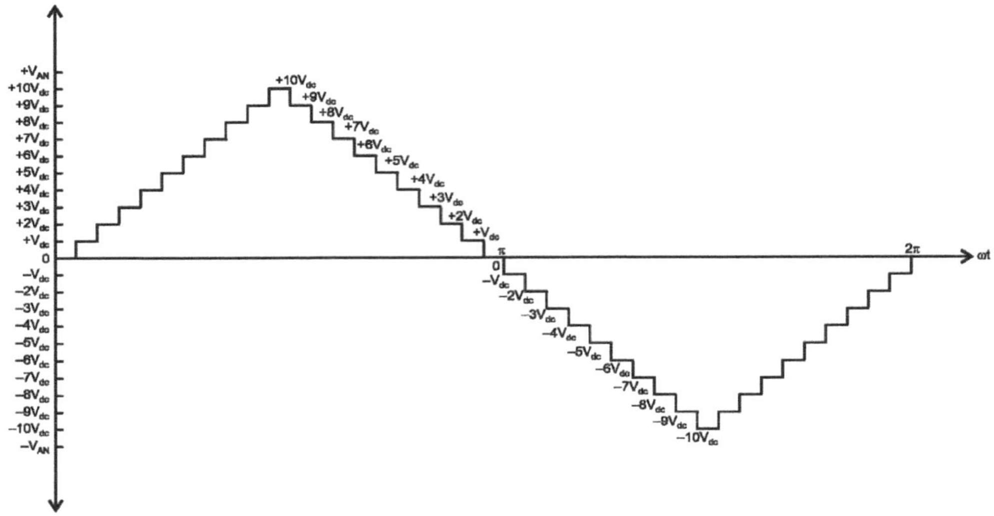

Fig. 5.26 Load phase voltage waveform of Asynchronous configuration of twenty-one level transistor-clamped H-bridge inverter with reduced number of switching devices or H-bridges.

Three-phase Twenty-one level transistor-clamped cascaded two-H-bridge inverter with Asymmetrical configuration: Three-phase Twenty-one level transistor-clamped cascaded H-bridge inverter using two H-bridges per phase with Asymmetrical configuration, i.e. equal dc source voltages.

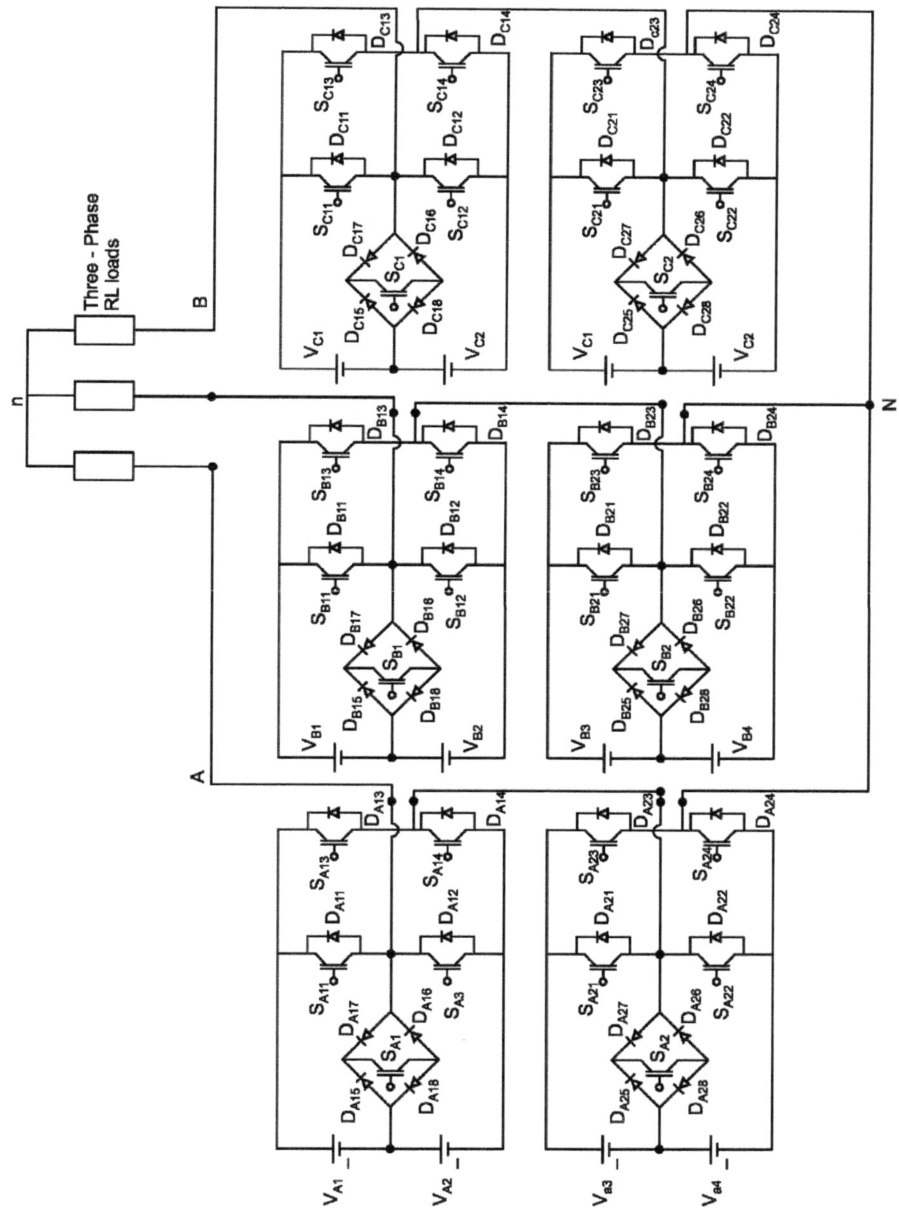

Fig. 5.27 Three-phase twenty-one level asymmetrical configuration of transistor-clamped cascaded H-bridge inverter with reduced number of switches or H-bridges for three-phase star-connected loads.

The dc source voltages are $V_{A1}=V_{B1}=V_{C1}=4V_{dc}$; $V_{A2}=V_{B2}=V_{C2}=3V_{dc}$; $V_{A3}=V_{B3}=V_{C3}=2V_{dc}$; $V_{A4}=V_{B4}=V_{C4}=V_{dc}$. The operation of each phase is similar to the single-phase circuits and the output phase voltages for twenty-one levels are similar to the single-phase output voltage but shifted by 120° from each other to get the three-phase output voltage waveform.

5.6 m-level Cascaded H-bridge Inverter

The generalized structure of a cascaded H-bridge multilevel inverter consists of 'm' number of series connected H-bridge cells, which are fed by independent dc voltage sources. Depending up on the conduction status of four switches in each H-bridge cell, each inverter cell can generate three different voltage levels of $+V_m$, 0, and $-V_m$. The outputs of H-bridge cells are connected in series such that the synthesized voltage waveform is the sum of all of the individual cell outputs.

The output voltage is given by

$$V_{AN} = V_{A1} + V_{A2} + \ldots\ldots\ldots\ldots\ldots\ldots + V_{Am} \qquad \ldots..(5.1)$$

Where the output voltage of the H_1 cell is denoted by V_{A1} and the output of H_m cell is denoted by V_{Am}. The output voltage levels 'm' in a Cascaded H-bridge inverter is defined by m=2S+1, where S is the number of separate dc sources (it implies S number of H-bridges). The generalized structure topology of m-level cascaded H-bridge inverter is given in Fig. 5.28.

The advantages of Cascaded H-bridge inverters are,

- Preferred for producing higher-level output voltages with minimum harmonic content (THD) for high to medium voltage applications.
- Number of possible output voltage levels is more than twice the number of dc voltage sources.
- The series of H-bridges makes for modernized layout and packaging. This will enable the manufacturing process to be done more quickly and cheaply.

The disadvantages of CHB inverters are,

- Separate dc sources are required for each of the H-bridge inverter. This will limit its applications to products that already have multiple separate dc sources readily available.

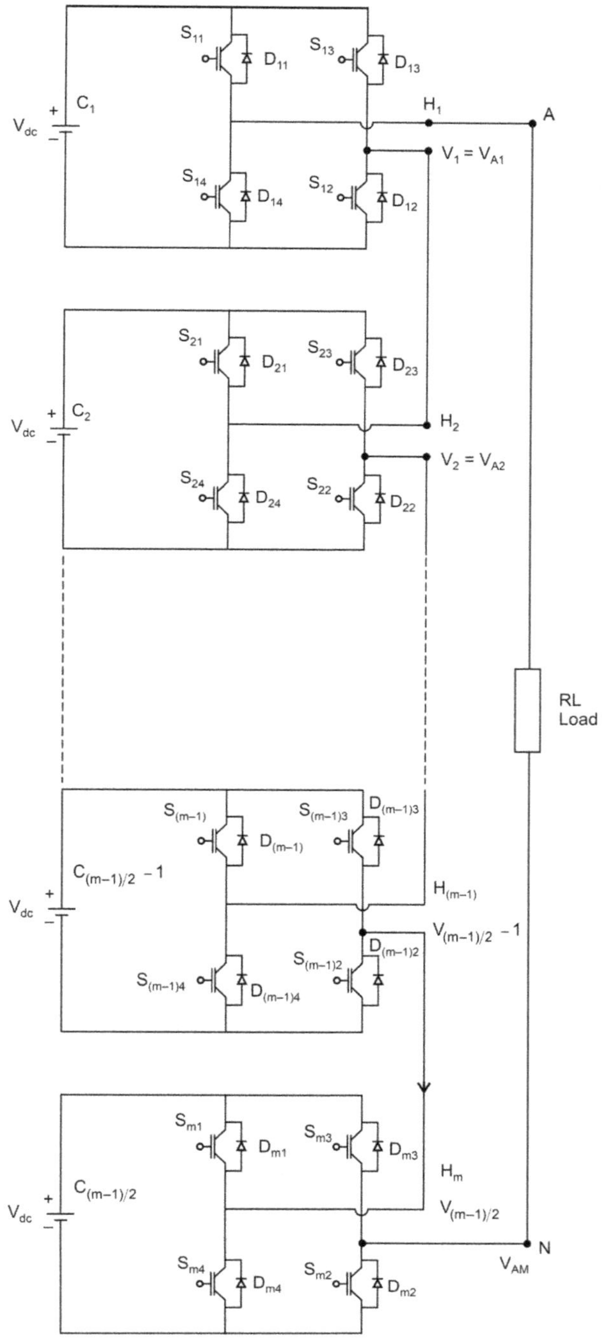

Fig. 5.28 The general structure of m-level cascaded H-bridge multilevel inverter topology with independent dc sources.

Table 5.12 Summary of minimum-level of output voltages obtained from various configurations of cascaded H-bridge inverters.

Serial no.	Number of H-bridge cells connected in series	Type of configuration	Minimum level of output voltage	
1.	Single H-bridge	----	Three-level	3-Level
2.	Two H-bridges	Symmetrical	Five-level	5-level
3.	Three H-bridges	Symmetrical	Seven-level	7-level
4.	Two H-bridges	Asymmetrical	Seven-level	7-level
5.	Three H-bridges	Asymmetrical	Thirteen-level	13-level
6.	Single H-bridge Transistor-clamped	Symmetrical	Five-level	5-level
7.	Two H-bridges Transistor-clamped	Symmetrical	Nine-level	9-level
8.	Single H-bridge Transistor-clamped	Asymmetrical	Seven-level	7-level
9.	Two H-bridges Transistor-clamped	Asymmetrical	Twenty-one-level	21-level

5.7 Case Studies for Cascaded H-bridge Inverters

In this chapter two case studies are considered to demonstrate the working principle of single-phase H-bridge inverter and single-phase cascaded H-bridge inverter for three-level and five-level output load voltages respectively, with resistive load and resistive-inductive loads. The most popularly used multicarrier PWM technique for Cascaded H-bridge inverters are level-shifted technique, since the resultant output load voltage is the summation of individual H-bridge inverter's load voltages. In order to sum the individual H-cells output voltages, each cell output voltage need to maintain the same phase. For this reason, the level-shifted multilevel carrier-based PWM technique is used. In these examples, the level-shifted multicarrier based pulse width modulation techniques are used, namely, (i) In-phase disposition PWM (PD-PWM), (ii) Phase opposition disposition (POD-PWM) and (iii) Alternative Phase opposition disposition (APOD-PWM) techniques, popularly employed for cascaded H—bridge inverter topologies. The simulated results are shown and compared the Total Harmonic Distortion (%THD) of load voltage waveform of each type of H-bridge inverter between the three modulation techniques.

5.7.1 Case Study 1

Simulation of Three-level single-phase H-Bridge (with 1 cell) Inverter with R and RL loads using (i) PD-PWM, (ii) POD-PWM, and (iii) APOD-PWM techniques. The load resistance R=10 ohms and for load resistive-inductive loads, R=10 ohms and L=1mH.

Three Level H-Bridge Inverter:

Consider a single phase with single H-bridge inverter, where four identically rated bidirectional IGBT switches are used with a fixed dc source voltage V_{dc}.

Step 1: Understand the operation of the circuit given in Fig.5.7.1 and write the switching sequence for producing three-level output voltages and tabulate them in Table 5.7.1.

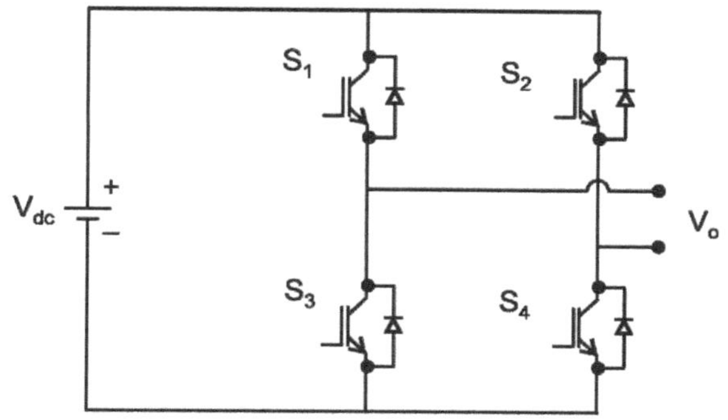

Fig. 5.7.1 (a) Three-level single phase H-bridge inverter.

Table 5.7.1 Switching Pattern for single phase H-bridge inverter for three-level load voltage.

Switches to turn ON	Voltage Level
S_1, S_4	V_{dc}
S_2, S_3	$-V_{dc}$
S_1, S_2 or S_3, S_4	0

Step 2: Based on the switching pattern develop the gating signals for the circuit given, by selecting appropriate carrier signal and reference signal according to the modulation technique applied. For ready reference, the modulation classification chart is given here again, for the application of PD-PWM, POD-PWM, and APOD-PWM techniques.

Modulation Techniques suitable for Multilevel inverters:

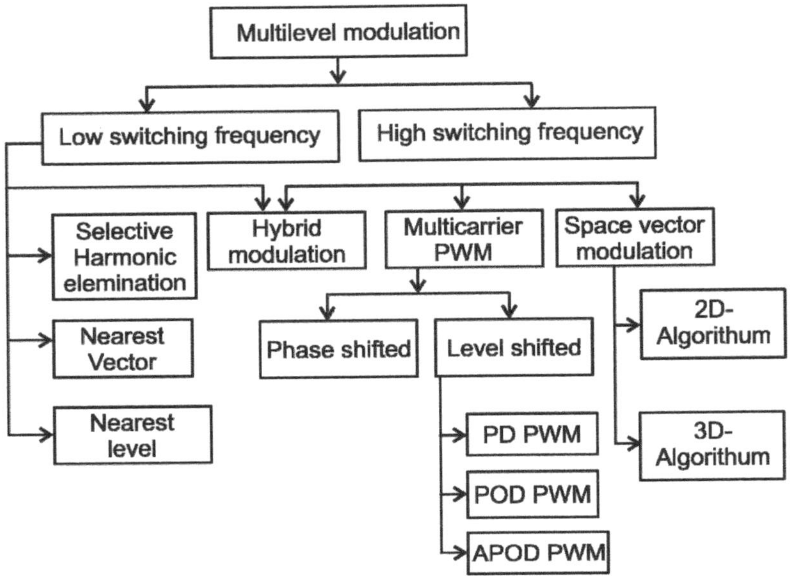

Level-shifted multicarrier based PWM Techniques:

Phase Disposition (PD) PWM Technique: In Phase Disposition PWM Technique all the carrier signals are in same phase. This modulation technique is employed for this three-level H-bridge inverter.

For Three Level H-bridge inverter: The Carrier and Reference signal, and Gating Pulses for first arm and second arm devices are given in the Fig. 5.7.1(b) (i). Two carrier signals considered here which are in-phase disposition with sinusoidal reference signal, the upper signal for generating gating signals for switches S_1 and S_4, the lower signal for generating gating signals for switches S_2 and S_3, respectively. This in turn facilitate the circuit to produce positive and negative half-waves of the output load voltage.

Step 3: Simulate the circuit with proper modulation technique for R and RL loads using any software tools available. (The results are with SIMULINK/MATLAB).

Simulated Results: For single phase three-level H-bridge inverter circuit, when gating signals so developed is applied and simulated separately for case(i) only with R-load of 10 ohms and for case (ii) both R and L loads of 10 ohms and 1 mH, respectively, the resultant output load phase voltage is obtained and the same is presented.

(A) **Phase Disposition (PD) PWM Technique:** Case (i): R-Load with R=10Ω and Case (ii): RL-Load with R=10Ω and L=1mH.

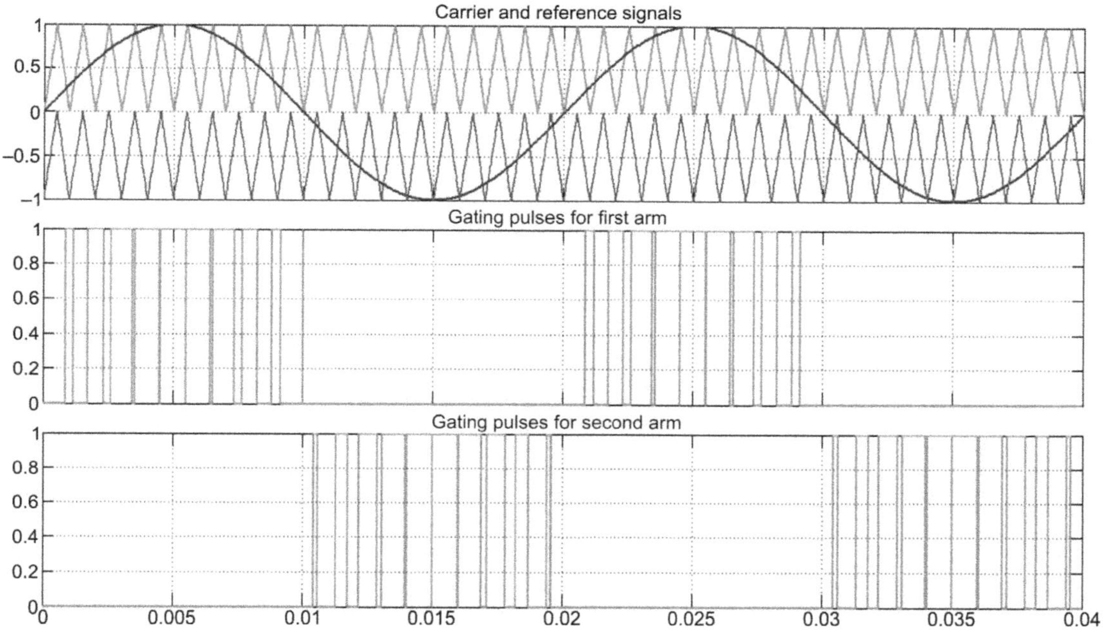

Fig. 5.7.1 (b) Carrier and reference signals, first arm and second arm gating signals for three-level single phase H-bridge (single cell) inverter for PD-PWM technique.

Three - Level Output Voltage with PD technique:

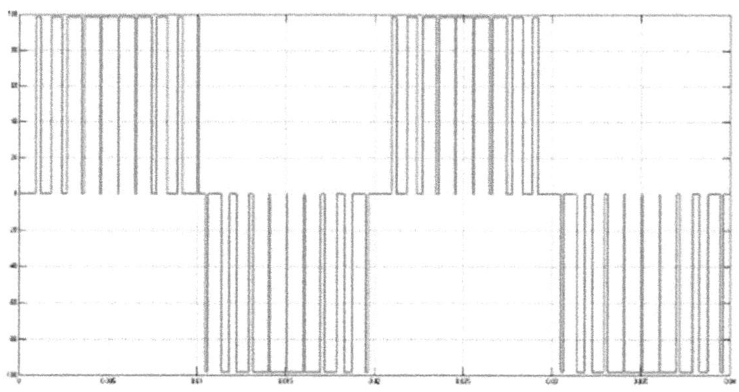

(c)

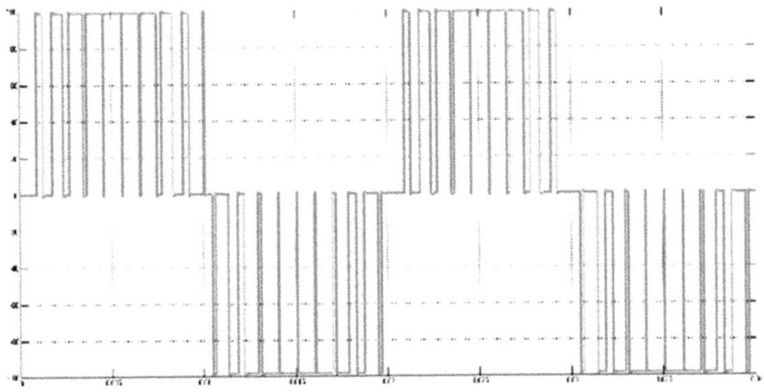

(d)

Fig. 5.7.1 Output phase voltage of three-level single phase H-bridge inverter with PD technique for (c) only R-load and (d) RL – load: (+Vdc, 0, and -Vdc).

(B) Phase Opposition Disposition (POD) PWM Technique: For Phase Opposition Disposition modulation, all the carrier signals above the zero are out of phase with those below the zero by 180° and the reference signal is sinusoidal, compared to generate gating signals for four switches used in the circuit.

For Three Level:

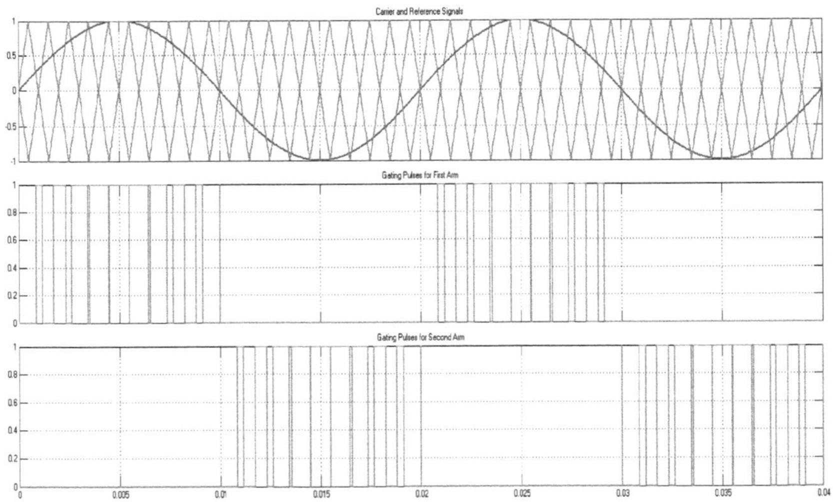

(e)

Fig. 5.7.1 (e) Carrier and reference signals, first arm and second arm gating signals for three-level single phase H-bridge (single cell) inverter for POD-PWM modulation technique.

Three - Level Output Voltage with POD-PWM technique:

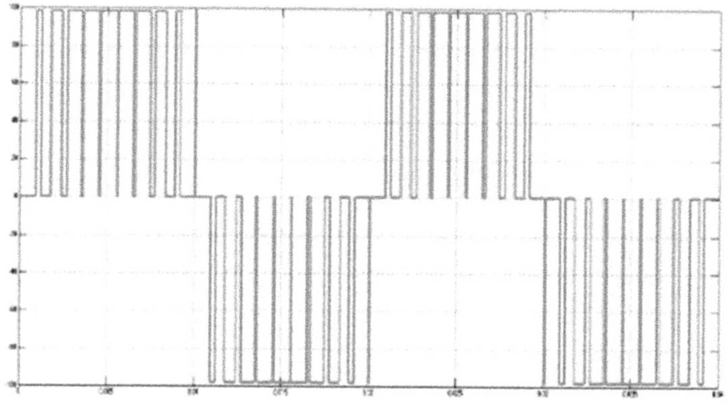

(f)

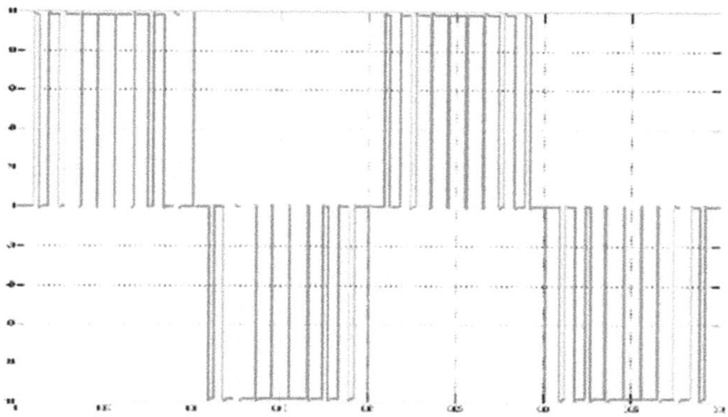

(g)

Fig. 5.7.1 Output phase voltage of three-level single phase H-bridge inverter with POD-PWM technique for (f) only R load and (g) RL -Load: (+Vdc, 0, and -Vdc).

(C) Alternative Phase opposition Disposition (APOD) PWM Technique: For Alternate Phase Opposition Disposition, all the adjacent carrier signals are out of phase by 180°

For Three Level:

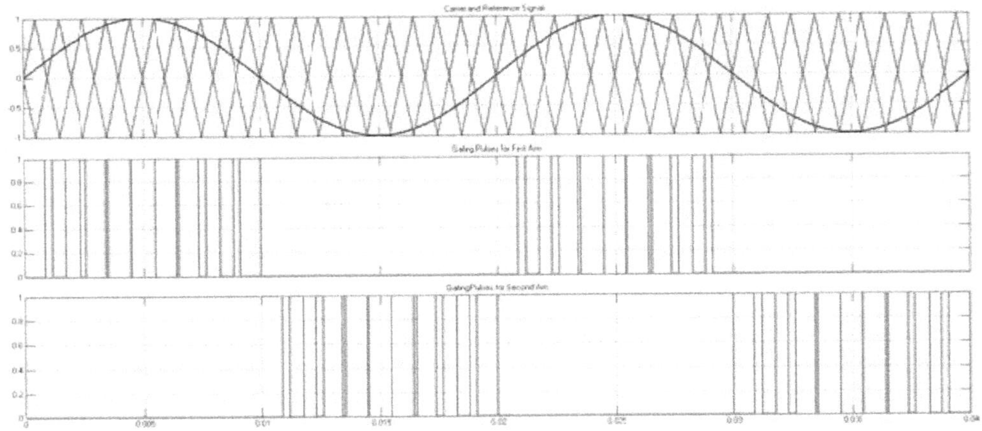

Fig. 5.7.1 (h) Carrier and reference signals, first arm and second arm gating signals for three-level single phase H-bridge (single cell) inverter for APOD-PWM modulation technique.

Simulated Results:

Case(i): Output load phase voltage for R-Load with R=10Ω and Case(ii) for RL load, R=10 Ω, L=1mH.

Three - Level Output Voltage with APOD technique

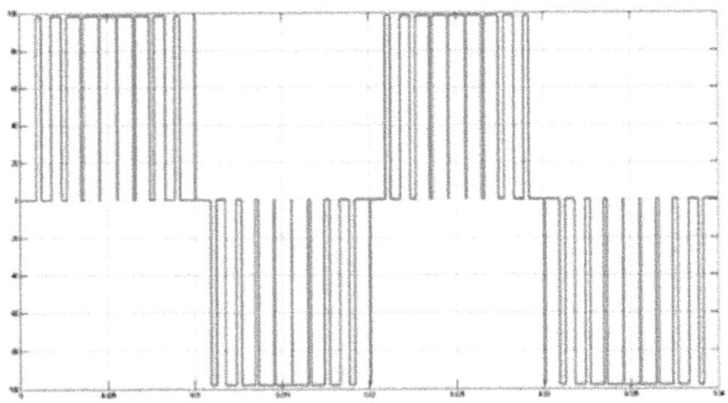

(i)

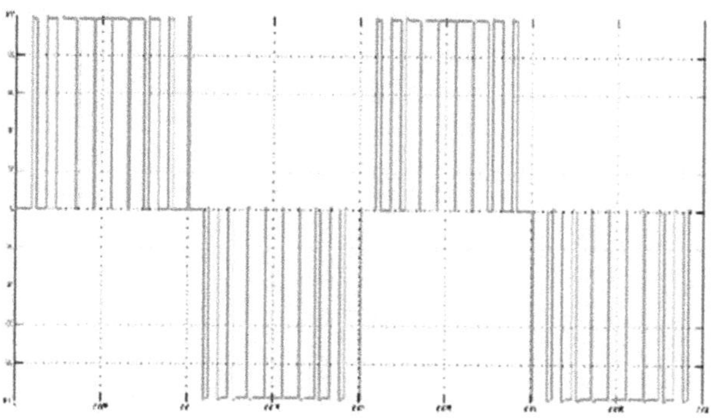

(j)

Fig. 5.7.1 Output phase voltage of three-level single phase H-bridge inverter with APOD-PWM technique for (i) only R load and (j) RL -load: (+Vdc, 0, and -Vdc).

Comparison of THDs of load phase voltage waveforms for different Techniques:

Level	PD	POD	APOD
Three Level - R Load	58.58%	59.61%	59.61%
Three Level – RL Load	58.91%	59.94%	59.94%

5.7.2 Case Study 2

Simulation of Five-level Cascaded H-bridge (with two cells) Inverter with R and RL loads using (i) PD-PWM, (ii) POD-PWM, and (iii) APOD-PWM techniques. The load resistance R=10 ohms and for load resistive-inductive loads, R=10 ohms and L=1mH.

Five Level cascaded H-Bridge Inverter: Consider a single-phase cascaded H-bridge inverter (with 2 cells) for five-level output voltages, where eight identically rated bidirectional IGBT switches are used and for each H-bridge cell a separate fixed magnitude dc source voltage V_{dc} provided.

Step 1: Understand the operation of the circuit given in Fig.5.7.2 and write the switching sequence for producing three-level output voltages and tabulate them in Table 5.7.2.

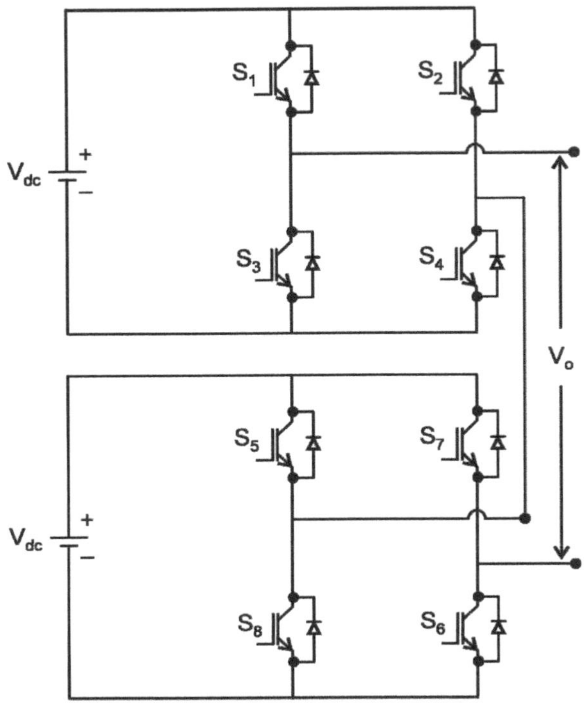

Fig. 5.7.2 (a) Five-level single phase cascaded H-bridge inverter with two-cells.

Table 5.7.2 Switching Pattern for single phase cascaded H-bridge inverter for five-level load voltage.

Switches to turn ON	Voltage Level
S_1, S_4, S_6, S_8	$+V_{dc}$
S_1, S_4, S_5, S_6	$+2V_{dc}$
S_3, S_4, S_6, S_8	0
S_2, S_3, S_6, S_8	$-V_{dc}$
S_2, S_3, S_7, S_8	$-2V_{dc}$

Step 2: Based on the switching pattern develop the gating signals for the circuit given, by selecting appropriate carrier signal and reference signal according to the modulation technique applied, namely, (i) PD-PWM, (ii) POD-PWM, and (iii) APOD-PWM techniques.

(A) Phase Disposition (PD): In Phase Disposition all the carrier signals are in same phase.

For Five Level:

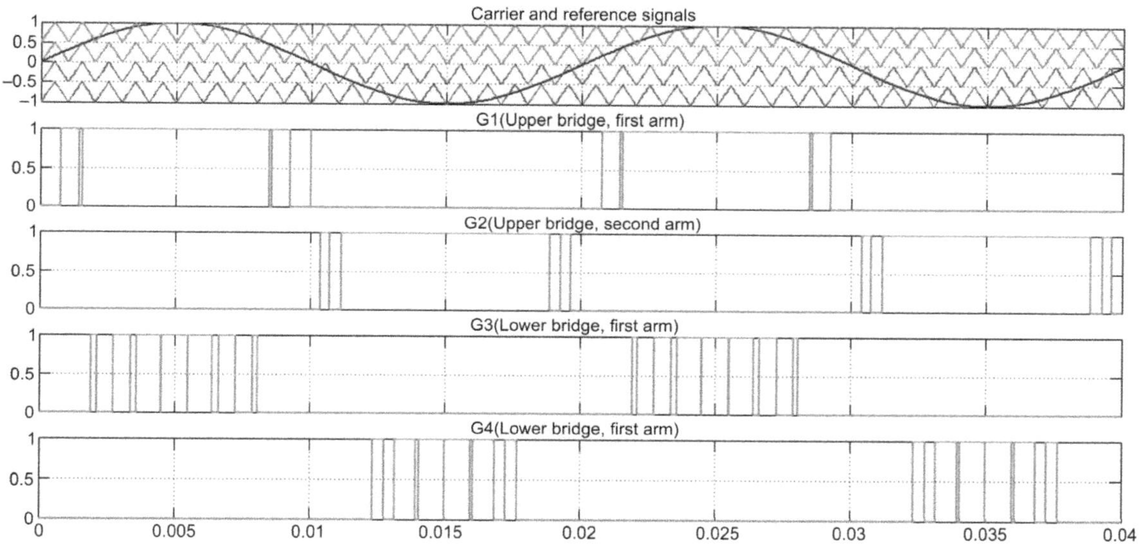

Fig. 5.7.2 (b) Carrier and reference signals, first arm and second arm gating signals for five-level single phase cascaded H-bridge (two cells) inverter for PD-PWM technique.

Step 3: Simulate the circuit with proper modulation technique for R and RL loads using any software tools available. (The results are with SIMULINK/MATLAB).

Case (i): Five - Level Output Voltage with PD technique for R load of 10 ohms, and Case(ii): for RL load, R=10 Ω, L=1mH.

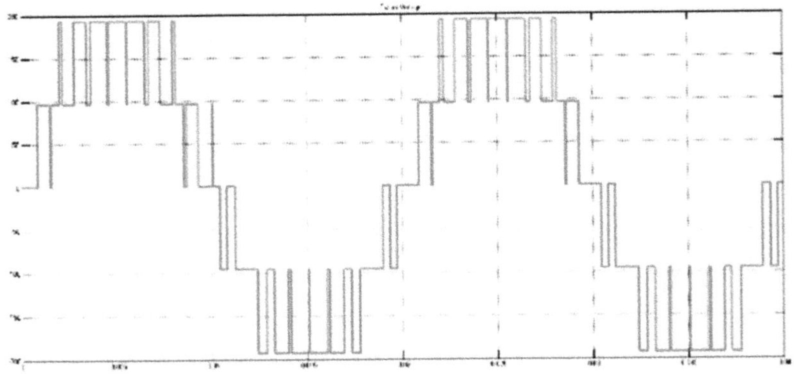

(c)

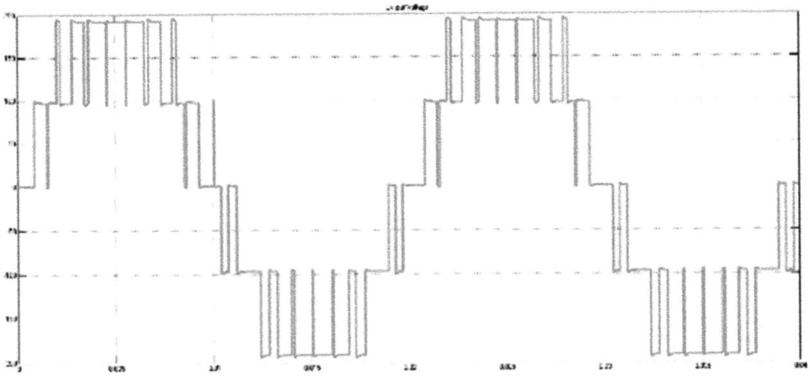

(d)

Fig. 5.7.2 Output phase voltage of five-level single phase cascaded H-bridge inverter with PD-PWM technique (c) for only R-load and (d) for RL -load: (+2Vdc, +Vdc, 0, -Vdc, -2Vdc).

(B) Phase Opposition Disposition (POD): For Phase Opposition Disposition modulation, all the carrier signals above the zero are out of phase with those below the zero by 180° and the reference signal is sinusoidal, compared to generate gating signals for eight switches used in the cascaded circuit. To produce five-level output voltages, four carrier signals are required to generate eight gating signals as per the switching pattern given in the table above.

For Five Level:

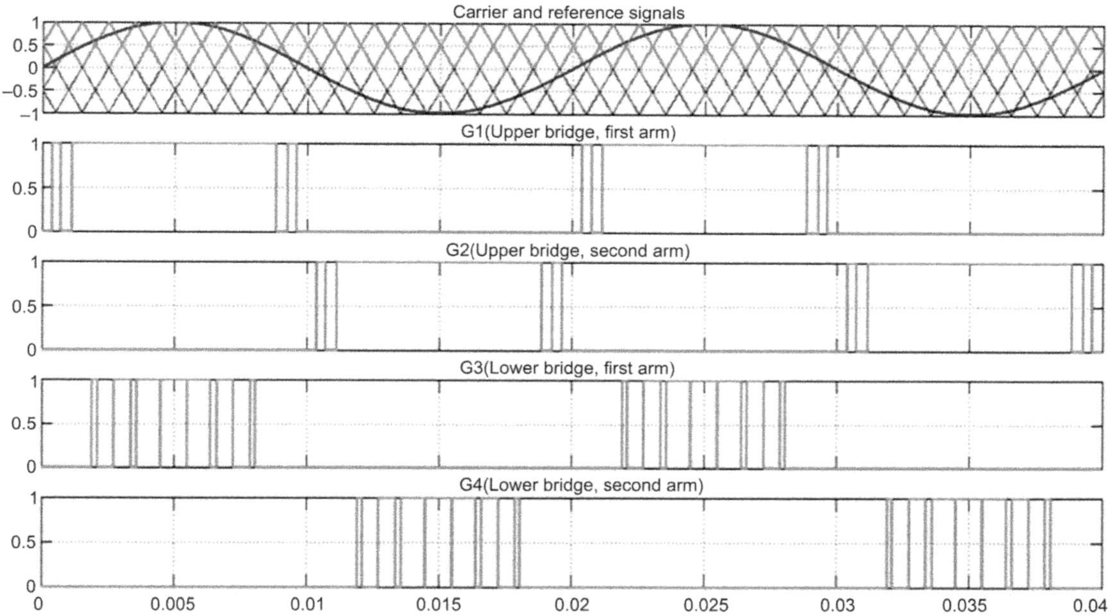

Fig. 5.7.2 (e) Carrier and reference signals, first arm and second arm gating signals for five-level single phase cascaded H-bridge (two cells) inverter for POD-PWM modulation technique.

Five - Level Output Voltage with POD-PWM technique for R load and RL -loads

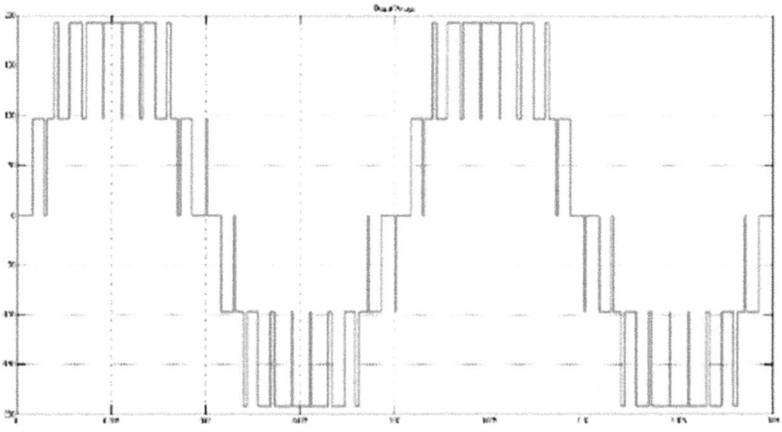

(f)

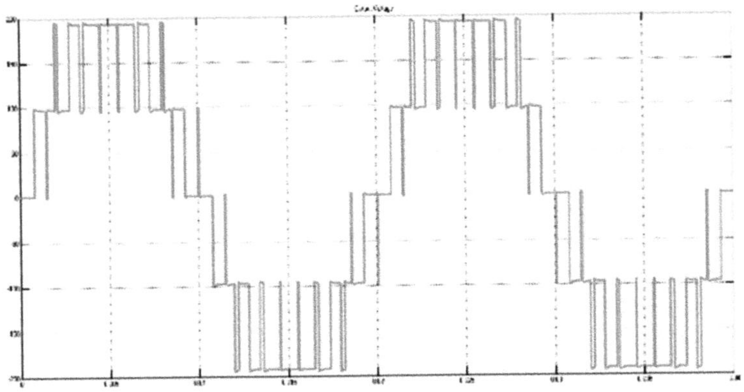

(g)

Fig. 5.7.2 Output phase voltage of five-level single phase cascaded H-bridge inverter with POD-PWM technique (f) for only R load and (g) for RL load: (+2Vdc, +Vdc, 0, -Vdc, -2Vdc).

(C) Alternative Phase Opposition Disposition (APOD-PWM) Technique: For Alternate Phase Opposition Disposition all the adjacent carrier signals are out of phase by 180°.

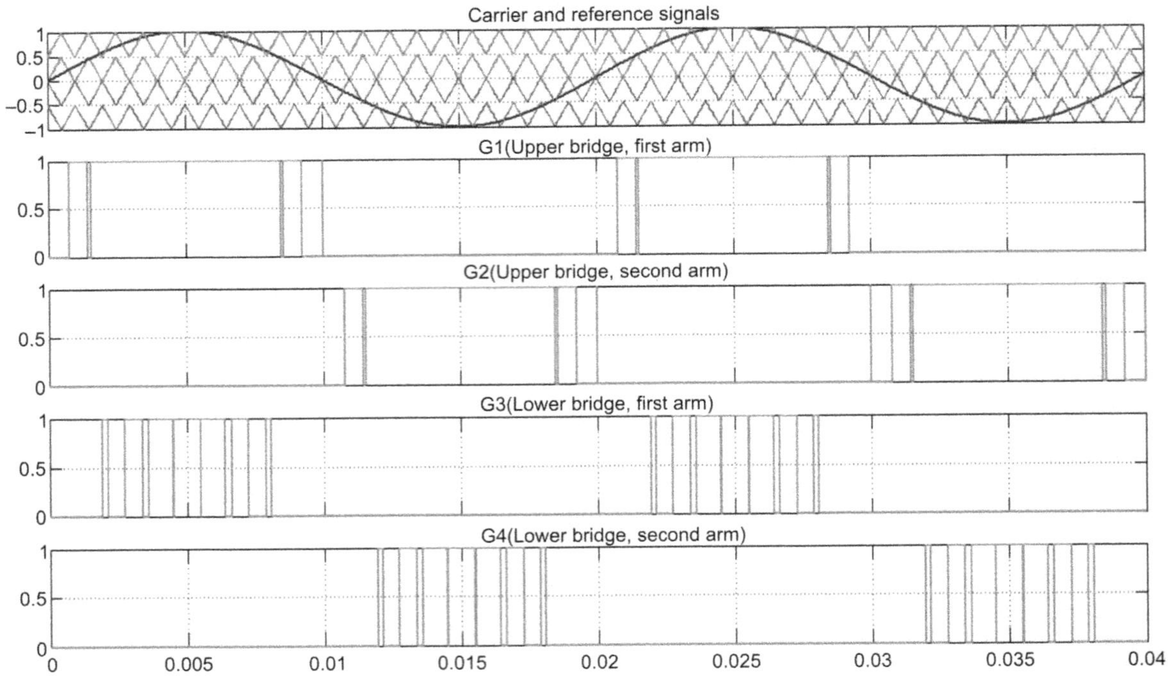

Fig. 5.7.2 (h) Carrier and reference signals, first arm and second arm gating signals for five-level single phase cascaded H-bridge (two cells) inverter for APOD-PWM modulation technique.

Five - Level Output Voltage with APOD-PWM technique for R load and RL- load.

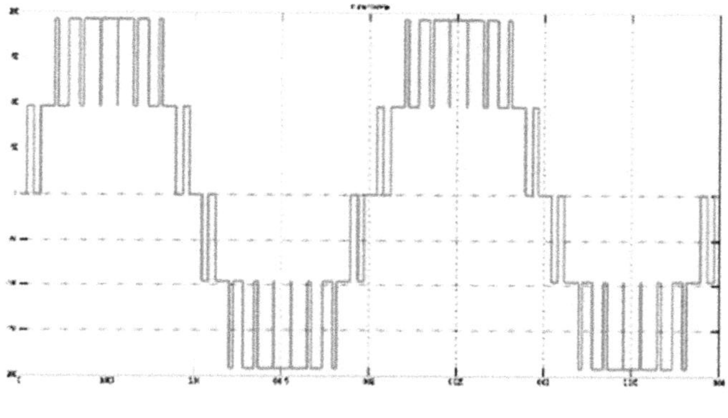

(i)

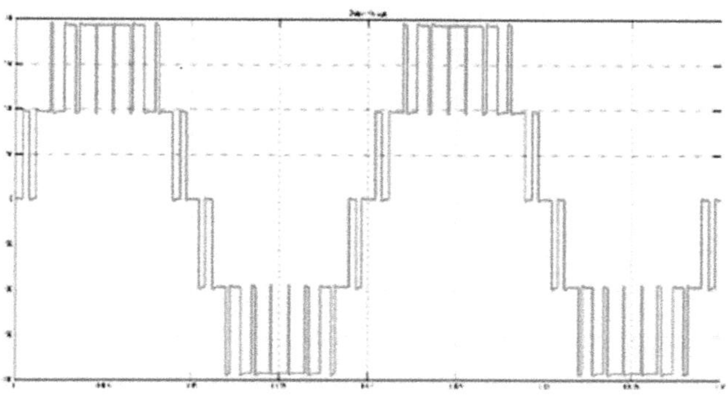

(j)

Fig. 5.7.2 Output phase voltage of five-level single phase cascaded H-bridge inverter with APOD-PWM technique (i) for only R load and (j) for RL load: (+2Vdc, +Vdc, 0, -Vdc, -2Vdc).

Comparison of THDs for different Techniques:

Level	PD	POD	APOD
Five Level – R Load	36.57%	35.48%	36.96%
Five Level – RL Load	36.87%	35.78%	37.28%

5.8 Summary

The multilevel inverters can immediately replace the existing systems that use traditional multi-pulse inverters without the need for transformers. For a Three-phase system, the relationship between the number of levels 'm', and the number of pulses, p, can be formulated by p = (m - 1) x 6.

The cascaded H-bridge inverters are classified as conventional cascaded H-bridge inverter and transistor-clamped cascaded H-bridge inverter, they are further categorized as symmetrical and asymmetrical values of dc source voltages. when dc source voltages of each H-bridge cell are equal, it is called symmetrical configuration and when dc source voltages of each H-bridges are unequal, then it is called asymmetrical configuration. The transistor-clamped cascaded H-bridge inverters have advantages of using reduced number of switching devices when compared to conventional cascaded H-bridge inverters.

The symmetrical configuration conventional cascaded H-bridge inverters are explained with detailed circuits with various modes of operation, switching tables, and output load voltages, for three-level with one-H-bridge, five-level with two-H-bridges and seven-level inverters with three-H-bridges in cascade. The asymmetrical configuration conventional cascaded H-bridge inverters are also presented and explained with detailed circuits with various modes of operation, switching tables, and output load voltages, for seven-level with two-H-bridges and thirteen-level inverters with three-H-bridges.

The transistor-clamped cascaded H-bridge inverters have advantages of using reduced number of switching devices and produces higher-level output load voltages. To produce five-level output load voltage, a transistor-clamped single H-bridge with equal dc source voltage in symmetrical configuration is used and to produce nine-level output load voltage, transistor-clamped two H-bridges with equal dc source voltages in symmetrical configuration is used. They are presented and described with the detailed circuits and various modes of operations, switching tables, and output load voltage waveforms.

Similarly, asymmetrical configuration of transistor-clamped cascaded H-bridge inverters with unequal dc source voltages are explained by using one-H-Bridge it can produce seven-level output load voltage and whereas by using two-H bridges in cascade it can produce twenty-one-level output load voltage. These are explained in detailed by considering circuit diagrams for various modes of operation, switching tables, and output load voltages.

Three-phase circuit configurations are presented for all varieties of multilevel inverters for star-connected RL loads. The load current depends on the nature of the loads. The m-level configurations are also presented. The advantages, disadvantages and applications of all the above discussed multilevel inverters are also given.

The generalized multilevel inverter topology provides a true multilevel structure that can balance each DC voltage level automatically at any number of levels, regardless of active or reactive power conversion, and without any assistance from other circuits. Thus, in principle, it provides a complete multilevel topology that embraces the existing multilevel inverters.

The case studies and exercise problems are given in all varieties of multilevel inverters for practice by modeling and simulation.

Learning Outcomes

- Acquire knowledge about the construction of H-bridge inverter of basic three-level operation (and cascaded H-bridge inverters). (CO1) :: (PO1)
- Ability to analyze the operation of each type of cascaded H-bridge inverter of different levels of output voltages and with both symmetrical and asymmetrical topologies. (CO4) :: (PO2)
- Ability to analyze the operation of each type of cascaded transistor-clamped H-bridge inverter of higher order of levels of output voltages and with both symmetrical and asymmetrical topologies. (CO4) :: (PO2)
- To be able to apply suitable pulse width modulation techniques for cascaded H-bridge inverters. (CO3):(PO1)
- To be able to develop the simulation model using software tools for cascaded H-bridge inverters and transistor-clamped H-bridge inverters. (CO4) :: (PO5)
- To be able to evaluate the performance of the various types of cascaded H-bridge inverters for various loads by simulation. (CO5) :: (PO3)

Questions

5.1 Analyze the Symmetrical Configuration of Three level cascaded H-bridge inverter with equal dc sources which feeds power to RL load and obtain the output voltage?

5.2 Analyze the Symmetrical Configuration of Five - level cascaded H-bridge inverter with equal dc voltage sources when two cells are connected in series with RL load and elaborate the operation with various modes?

5.3 Obtain the switching table for the circuit topology given in question no. 5.2, exhibiting various modes of operation to produce five-level output voltages and draw the output voltage waveform?

5.4 Analyze the Symmetrical configuration for Seven – level cascaded H-bridge inverter with equal dc sources when three cells are connected in series with RL load and elaborate the operation with various modes?

5.5 Obtain the switching table for the circuit topology given in question no. 5.4, exhibiting various modes of operation to produce seven-level output voltages and draw the output voltage waveform?

5.6 Analyze the asymmetrical configuration of Seven-level cascaded H-bridge inverter with unequal dc voltage sources when two-cells are connected in series with RL load and explain various modes of operation with the help of circuit diagrams?

5.7 Obtain the switching table for the asymmetrical circuit topology given in question no. 5.6, Exhibiting various modes of operation to produce seven-level output voltages and draw the output voltage waveform?

5.8 Analyze the asymmetrical configuration of Thirteen - level cascaded H-bridge inverter with unequal dc voltage sources when three-cells are connected in series with RL load and explain various modes of operation with the help of circuit diagrams?

5.9 Obtain the switching table for the asymmetrical circuit topology given in question no. 5.8, exhibiting various modes of operation to produce thirteen-level output voltages and draw the output voltage waveform?

5.10. Draw the three-phase three-level H-bridge circuit configuration for three-phase star-connected loads by employing three separate symmetrical single H-bridge inverters modules are arranged to get the three-phase three-level output voltages for balanced RL loads and sketch the three-phase output voltages?

5.11. Draw the three-phase five-level H-bridge circuit configuration for three phase star-connected RL loads by employing two symmetrical single H-bridges in series for each phase separately and H-bridges modules are arranged to get the three-phase five-level output voltages for balanced RL loads and sketch the three-phase output voltages?

5.12. Draw the three-phase seven-level H-bridge circuit configuration for three phase star-connected loads by employing three symmetrical single H-bridges in series for each phase separately and H-bridges modules are arranged to get the three-phase seven-level output voltages for balanced RL loads and sketch the three-phase output voltages?

5.13. Draw the three-phase seven-level H-bridge circuit configuration for three phase star-connected RL loads by employing two asymmetrical single H-bridges in series for each phase separately and H-bridges modules are arranged to get the three-phase seven-level output voltages for balanced RL loads and sketch the three-phase output voltages?

5.14. Draw the three-phase thirteen-level H-bridge circuit configuration for three phase star-connected RL loads by employing three asymmetrical single H-bridges in series for each phase separately and H-bridges modules are arranged to get the three-phase thirteen-level output voltages for balanced RL loads and sketch the three-phase output voltages?

5.15 Analyze the symmetrical configuration of Five-level transistor-clamped H-bridge inverter with reduced number of switching devices with RL load and elaborate various modes of operation with the help of circuit diagrams?

5.16 Analyze the symmetrical configuration of transistor-clamped Nine-level H-bridge inverter with reduced number of switching devices with RL load employing two single H-bridges in series and elaborate various modes of with the help of circuit diagrams?

5.17 Analyze the asynchronous configuration of transistor-clamped seven - level single H-bridge inverter with reduced number of switching devices with RL load and elaborate various modes of operation for both positive and negative load current with the help of circuit diagrams?

5.18 Analyze the asynchronous configuration of transistor-clamped twenty-one-level cascaded two-H-bridge inverter with reduced number of switching devices with RL load and explain various modes of operation for both positive and negative load current with the help of circuit diagrams?

5.19 Draw the three-phase five-level symmetrical configuration of transistor-clamped H-bridge Inverter by using single H-bridge module for each phase for three phase star-connected RL loads and sketch the three-phase output voltages?

5.20 Draw the three-phase nine-level symmetrical configuration of transistor-clamped H-bridge inverter by using two H-bridge modules for each phase for three phase star-connected RL loads and sketch the three-phase output voltages?

5.21 Draw the three-phase asymmetrical configuration of seven-level transistor- clamped H-bridge inverter by using one H-bridge module in each phase for three-phase star-connected RL loads and Sketch the three-phase output voltages?

5.22 Draw the three-phase asymmetrical configuration of twenty-one-level transistor-clamped H-Bridge inverter by using two H-bridge modules in each phase for three-phase star-connected RL loads and Sketch the three-phase output voltages?

5.23 Draw m-level cascaded H-bridge inverter and give its advantages and disadvantages?

Multiple Choice Questions

5.1 For general m-level cascaded H-bridge inverter, the relationship between number of sources and number of H-bridges (S) and number of levels (m) is,

(a) m = 2(S + 1) (b) m = (2S + 1)

(c) m = (S + 1) (d) m = (S+1)/2

5.2 For deriving three-level output voltage from single phase H-bridge inverter, how many H bridges are needed?

(a) two H-bridges in series (b) three H-bridges in series

(c) one H-bridge only (d) half-bridge only

5.3 For a single phase three-level H-bridge inverter to produce three-level output voltage with positive load current, how many modes of operations are involved?

(a) no modes (b) two modes

(c) four modes (d) three modes

5.4 For a symmetrical single phase five-level cascaded H-bridge inverter to produce a five level output voltage, how many H-bridges are needed?

(a) one H-bridge only (b) two H-bridges in series

(c) three H-bridges in series (d) four H-bridges in series

5.5 For a symmetrical single phase five-level cascaded H-bridge inverter to produce a five-level output voltage, how many dc sources are required and give their relationship?

(a) one dc source, and V_1 (b) two dc sources, and $V_1 \neq V_2$

(c) two dc sources, and $V_1 = V_2$ (d) three dc sources, and $V_1 \neq V_2 \neq V_3$

5.6 What is the net output voltage across the load when a single phase five-level cascaded H-bridge inverter is connected to the load?

(a) $V_o = V_1$ (b) $V_o = V_1 + V_2$

(c) $V_o = V_1 - V_2$ (d) $V_o = V_2$

5.7 How many single-phase H-bridges are used in total, for a three-phase five-level cascaded H-bridge inverter to produce three-phase five-level output voltages for star-connected loads?

(a) three H-bridges (b) five H-bridges

(c) six H-bridges (d) nine H-bridges

5.8 For a symmetrical single phase seven-level cascaded H-bridge inverter to produce a seven-level output load voltage, how many H-bridges are needed?

(a) one H-bridge only (b) two H-bridges in series

(c) three H-bridges in series (d) four H-bridges in series

5.9 For a symmetrical single phase seven-level cascaded H-bridge inverter to produce a seven-level output voltage, how many dc sources are required and give their relationship?

(a) two dc source, and $V_1 = V_2$ (b) three dc sources, and $V_1 = V_2 = V_3$

(c) two dc sources, and $V_1 \neq V_2$ (d) three dc sources, and $V_1 \neq V_2 \neq V_3$

5.10 What is the net output voltage across the load when a symmetrical single phase seven-level cascaded H-bridge inverter is connected to the load?

(a) $V_o = V_1 + V_2 + V_3$ (b) $V_o = V_1 - V_2 - V_3$

(c) $V_o = V_1 - V_2$ (d) $V_o = V_2$

5.11 How many single-phase H-bridges are used in total, for a symmetrical three-phase seven-level cascaded H-bridge inverter to produce three-phase seven-level output voltages for star-connected loads?

(a) three H-bridges (b) five H-bridges

(c) six H-bridges (d) nine H-bridges

5.12 For asymmetrical seven-level cascaded H-bridge inverter to produce a seven-level output load voltage, how many single phase H-bridges are needed?

(a) one H-bridge only (b) two H-bridges in series

(c) three H-bridges in series (d) four H-bridges in series

5.13 For asymmetrical seven-level cascaded H-bridge inverter to produce a seven-level output voltage, how many dc sources are required and give them relationship?

(a) two dc source, and $V_1 \neq V_2$

(b) three dc sources, and $V_1 = V_2 = V_3$

(c) two dc sources, and $V_1 = V_2$

(d) three dc sources, and $V_1 \neq V_2 \neq V_3$

5.14 What is the net output voltage across the load when asymmetrical single phase seven-level cascaded H-bridge inverter is connected to the load?

(a) $V_o = V_1 + V_2 + V_3$ (b) $V_o = V_1 - V_2 - V_3$

(c) $V_o = V_1 + V_2$ (d) $V_o = V_1 - V_2$

5.15 How many single phase H-bridges are used in total, for asymmetrical three-phase seven level cascaded H-bridge inverter to produce three-phase seven-level output voltages for star-connected loads?

(a) three H-bridges (b) six H-bridges

(c) five H-bridges (d) nine H-bridges

5.16 For asymmetrical single phase thirteen-level cascaded H-bridge inverter to produce a thirteen-level output load voltage, how many H-bridges are needed?

(a) one H-bridge only (b) two H-bridges in series

(c) three H-bridges in series (d) four H-bridges in series

5.17 For asymmetrical single phase thirteen-level cascaded H-bridge inverter to produce a thirteen-level output voltage, how many dc sources are required and give the relationship?

(a) two dc source, and $V_1 \neq V_2$ (b) three dc sources, and $V_1 = V_2 = V_3$

(c) two dc sources, and $V_1 = V_2$ (d) three dc sources, and $V_1 \neq V_2 \neq V_3$

5.18 What is the net output voltage across the load when asymmetrical single phase thirteen-level cascaded H-bridge inverter is connected to the load?

(a) $V_o = V_1 + V_2 + V_3$ (b) $V_o = V_1 - V_2 - V_3$

(c) $V_o = V_1 + V_2$ (d) $V_o = V_1 - V_2$

5.19 How many single-phase H-bridges are used in total, for asymmetrical three-phase thirteen-level cascaded H-bridge inverter to produce three-phase thirteen-level output voltages for star-connected loads?

(a) three H-bridges (b) six H-bridges

(c) five H-bridges (d) nine H-bridges

5.20 How many dc sources are required for a symmetrical configuration of Five-level transistor-clamped H-bridge inverter with reduced number of switching devices?

(a) one dc source only

(b) two dc sources of equal magnitude

(c) two dc sources of unequal magnitude
(d) three dc sources of equal magnitude

5.21 How many bidirectional transistors are used in a symmetrical configuration of single phase Five-level transistor-clamped H-bridge inverter with reduced number of switching devices?

(a) two (b) three
(c) four (d) five

5.22 How many transistors are used in total in a symmetrical configuration of single phase Five-level transistor-clamped H-bridge inverter with reduced number of switching devices?

(a) two (b) three
(c) four (d) five

5.23 How many H-bridges are used in a symmetrical configuration of single phase Five-level transistor-clamped H-bridge inverter with reduced number of switching devices?

(a) one (b) two
(c) three (d) four

5.24 How many dc sources are required for a symmetrical configuration of nine-level transistor-clamped H-bridge inverter with reduced number of switching devices?

(a) one dc source only
(b) two dc sources of equal magnitude
(c) three dc sources of unequal magnitude
(d) four dc sources of equal magnitude

5.25 How many bidirectional transistors are used in a symmetrical configuration of single phase nine-level transistor-clamped H-bridge inverter with reduced number of switching devices?

(a) two (b) four
(c) eight (d) nine

5.26 How many unidirectional transistors are used in total in a symmetrical configuration of single phase nine-level transistor-clamped H-bridge inverter with reduced number of switching devices?

(a) two (b) three
(c) four (d) five

5.27 How many H-bridges are used in a symmetrical configuration of three phase nine-level transistor-clamped cascaded H-bridge inverter with reduced number of switching devices to supply three-phase star-connected load?

(a) sixteen (b) six
(c) three (d) nine

5.28 How many H-bridges are used in asymmetrical configuration of single phase twenty-one-level transistor-clamped cascaded H-bridge inverter with reduced number of switching devices?

(a) two (b) four
(c) three (d) six

5.29 What is the minimum level of output voltage produced by using single H-bridge inverter for RL load?

(a) three-level (b) five-level
(c) four-level (d) six-level

5.30 What is the minimum level of output voltage produced by using symmetrical cascaded two H-bridge type inverters for RL load?

(a) Three-level (b) Five-level
(c) Seven-level (d) Nine-level

5.31 What is the minimum level of output voltage produced by using asymmetrical cascaded two H-bridge type inverters for RL load?

(a) Three-level (b) Five-level
(c) Seven-level (d) Nine-level

5.32 What is the minimum level of output voltage produced by using symmetrical cascaded three H-bridge type inverters for RL load?

(a) Three-level (b) Five-level
(c) Seven-level (d) Nine-level

5.33 What is the minimum level of output voltage produced by using asymmetrical cascaded three H-bridge type inverters for RL load?

(a) Nine-level (b) Eleven-level
(c) Twelve-level (d) Thirteen-level

5.34 What is the minimum level of output voltage produced by using symmetrical transistor-clamped single H-bridge type inverter for RL load?

(a) Three-level (b) Five-level
(c) Seven-level (d) Nine-level

5.35 What is the minimum level of output voltage produced by using asymmetrical transistor-clamped single H-bridge type inverter for RL load?

(a) Three-level (b) Five-level
(c) Seven-level (d) Nine-level

5.36 What is the minimum level of output voltage produced by using symmetrical transistor clamped two H-bridge type inverters for RL load?

(a) Three-level (b) Five-level
(c) Seven-level (d) Nine-level

5.37 What is the minimum level of output voltage produced by using asymmetrical transistor-clamped two H-bridge type inverters for RL load?
(a) Nine-level (b) Eleven-level
(c) Thirteen-level (d) Twenty-one-level

Exercise Problems

Ex.5.1: Simulate single phase H-bridge inverter (with single cell) when dc source voltage V_{dc} = 100V, R = 20 ohms and L = 10 mH and obtain gating signals, output phase voltage and load current by using any software tools available. Use Sinusoidal PWM modulation technique for three-level output.

Ex.5.2: Evaluate the performance of single-phase H-bridge inverter (with single cell) when dc source voltage Vdc = 100V, R = 20 ohms and L = 10 mH and obtain gating signals, output phase voltage and load current by using any software tools available. For three-level voltage, use carrier-based PWM modulation technique (a) PD-PWM, (b) POD-PWM and (c) APOD-PWM.

Ex.5.3: Obtain the output phase voltage of single-phase H-bridge inverter when dc source voltage V_{dc} = 100V, R = 20 ohms and L=10 mH and obtain gating signals, output phase voltage and load current by using any software tools available. For three-level voltage employ Space Vector PWM modulation technique.

Ex.5.4: Simulate symmetrical single-phase cascaded H-bridge inverter using two-cells and obtain output phase voltage and load current when dc source voltage V_{dc1} = V_{dc2} = 100V, R = 20 ohms and L = 10 mH by using any software tools available. For five-level voltage, use carrier-based PWM modulation technique (a) PD-PWM, (b) POD-PWM and (c) APOD-PWM.

Ex.5.5: Evaluate the performance of symmetrical single-phase cascaded H-bridge inverter using two-cells and obtain output phase voltage and load current when dc source voltage V_{dc1} = V_{dc2} = 100V, R = 20 ohms and L=10 mH by using any software tools available. For five-level voltage, use Space Vector PWM modulation technique.

Ex.5.6: Obtain the output phase voltage, load current for Asymmetrical single-phase cascaded H-bridge inverter using two-cells when dc source voltage V_{dc1} = 100V, V_{dc2} = 150 V, R = 20 ohms and L=10 mH by using any software tools available. For seven-level voltage, use Space Vector PWM modulation technique.

Ex.5.7: Simulate Asymmetrical single-phase cascaded H-bridge inverter using two-cells when dc source voltage V_{dc1} = 100V, V_{dc2} = 150V, R = 20 ohms and L=10 mH by using any software tools available. For seven-level voltage, obtain output phase voltage and load current waveforms by using carrier-based PWM modulation technique (a) PD-PWM, (b) POD-PWM and (c) APOD-PWM.

Ex.5.8: Obtain the output phase voltage and load current for a symmetrical single-phase cascaded H-bridge inverter using three-cells whose dc source voltages are $V_{dc1} = V_{dc2} = V_{dc3} =$ 100V, and VR = 20 ohms, L = 10mH by using any suitable software tool. For seven-level voltage, use Space Vector PWM modulation Technique.

Ex.5.9: Simulate symmetrical single-phase cascaded H-bridge inverter using three-cells whose dc source voltages are $V_{dc1} = V_{dc2} = V_{dc3} =$ 100V, and R = 20 ohms, L = 10mH by using any suitable software tool. For seven-level voltage, use carrier-based PWM modulation technique (a) PD-PWM, (b) POD-PWM and (c) APOD-PWM.

Ex.5.10: Obtain the output phase voltage and load current for Asymmetrical single-phase cascaded H-bridge inverter using three-cells whose dc source voltages are V_{dc1}=100V, V_{dc2}=150V, V_{dc3}=200V, and R=20 ohms, L=10mH by using any suitable software tool. For seven-level voltage, use Space Vector PWM modulation Technique.

Ex.5.11: Simulate symmetrical single-phase cascaded H-bridge inverter using three-cells whose dc source voltages are V_{dc1} = 100V, V_{dc2} = 150V, V_{dc3} = 200V and R = 20 ohms, L = 10mH by using any suitable software tool. For seven-level voltage, use carrier-based PWM modulation technique (a) PD-PWM, (b) POD-PWM and (c) APOD-PWM.

Ex.5.12: Evaluate the performance of symmetrical single-phase transistor-clamped H-bridge inverter using single cell and obtain output phase voltage and load current when dc source voltage $V_{dc1} = V_{dc2} =$ 100V, R = 20 ohms and L = 10 mH by using any software tools available. For five-level voltage, use Space Vector PWM modulation technique.

Ex.5.13: Evaluate the performance of Asymmetrical single-phase transistor-clamped H-bridge inverter using single cell and obtain output phase voltage and load current when dc source voltage V_{dc1} = 100V, V_{dc2} = 150V, R = 20 ohms and L = 10 mH by using any software tools available. For seven-level voltage, use Space Vector PWM modulation technique.

Ex.5.14: Simulate symmetrical single-phase transistor-clamped H-bridge inverter using two-cells and obtain output phase voltage and load current when dc source voltage $V_{dc1} = V_{dc2} =$ 100V of $cell_1$, and $V_{dc3} = V_{dc4} =$ 100V of $cell_2$, R = 20 ohms and L = 10 mH by using any software tools available. For nine-level voltage, use Space Vector PWM modulation technique.

Ex.5.15: Simulate symmetrical three-phase H-bridge inverter with single cell per phase whose dc source voltages are V_{A1} =100V with 0° phase shift, V_{B1} =100V with 120° phase shift, V_{C1} = 100V with 240° phase shift, and R = 20 ohms per phase and L = 10mH per phase. Obtain output phase voltages and load phase current for star-connected load. Use Sinusoidal PWM technique for three-level voltages.

Ex.5.16: Evaluate the performance of symmetrical three-phase cascaded H-bridge inverter with two-cells per phase whose dc source voltages are $V_{A1} = V_{A2} = 100V$ with 0° phase shift each, $V_{B1} = V_{B2} = 100V$ with 120° phase shift each, $V_{C1} = V_{C2} = 100V$ with 240° phase shift each, and R = 20 ohms per phase and L = 10mH per phase. Obtain output phase voltages and load phase current for star-connected load. Use Sinusoidal PWM technique for five-level voltages.

Ex.5.17: Obtain the performance of symmetrical three-phase cascaded H-bridge inverter with three-cells per phase whose dc source voltages are $V_{A1} = V_{A2} = V_{A3} = 100V$ with 0° phase shift each, $V_{B1} = V_{B2} = V_{B3} = 100V$ with 120° phase shift each, $V_{C1} = V_{C2} = V_{C3} = 100V$ with 240° phase shift each, and R = 20 ohms per phase and L = 10mH per phase. Obtain output phase voltages and load phase current for star-connected load. Use Sinusoidal PWM technique for seven-level voltages.

Answers to Multiple Choice Questions

5.1	(b)	5.2	(c)	5.3	(d)	5.4	(b)	5.5	(c)
5.6	(b)	5.7	(c)	5.8	(c)	5.9	(b)	5.10	(a)
5.11	(d)	5.12	(b)	5.13	(a)	5.14	(c)	5.15	(b)
5.16	(c)	5.17	(d)	5.18	(a)	5.19	(d)	5.20	(b)
5.21	(c)	5.22	(d)	5.23	(a)	5.24	(d)	5.25	(c)
5.26	(a)	5.27	(b)	5.28	(a)	5.29	(a)	5.30	(b)
5.31	(c)	5.32	(c)	5.33	(d)	5.34	(b)	5.35	(c)
5.36	(d)	5.37	(d)						

CHAPTER

6 Introduction to Hybrid Multilevel Inverters

LEARNING OBJECTIVES

- ✓ *To understand and acquire knowledge about various Hybrid multilevel inverters.*
- ✓ *To prepare learners to analyze the operation of various types of Hybrid multilevel inverters of various levels including higher-order levels.*
- ✓ *To prepare learners to simulate the performance of various Hybrid multilevel configurations with suitable pulse width modulation techniques.*
- ✓ *To understand the applications of various levels of cascaded H-bridge inverters.*

6.1 Introduction

Multilevel inverters have been a research hotspot in high-voltage and high-power applications in recent years. Multilevel converters offer significant advantages in medium and high-power applications such as reduction of the switching frequency and losses, reduced electromagnetic interference (EMI) and reduced requirements for series connection of semiconductors. They also provide high quality output waveforms with improved harmonic distortion compared to the two-level converter configurations.

The main multilevel topologies include the neutral point clamped (NPC), the flying capacitor (FC) and the cascaded H-bridges (CHB) converters. These converter topologies have been extensively studied over the last two decades and used in industrial applications. The Diode-clamped (NPC), flying-capacitor, and cascaded multilevel converters are three basic multilevel topologies. To synthesize multilevel output, voltage clamping is one of the most important issues. The meaning of clamping is to limit the switch's terminal voltage in a suitable range by using clamping devices. In the three-existing multilevel-inverter topologies, it can be seen through their names that voltages are clamped by diodes in the diode-clamped multilevel inverter, by capacitors in the flying capacitor multilevel inverter, and by separated voltage sources in the cascaded multilevel topology. However, the three topologies have some disadvantages in certain applications. The diode-clamped multilevel inverter needs separate dc sources or complex control methods when a real power is delivered. Although the flying-capacitor inverter can be used in both real and reactive power conversions, it is difficult to balance the capacitor voltages when it is used for reactive power compensation. The cascaded multilevel inverter needs separate dc voltage sources; therefore, they are more suitable for renewable energy sources such as fuel cell, photovoltaic, biomass, etc.

To overcome the above disadvantages, there is a need to explore new multilevel-inverter topologies with higher performance by changing clamping devices. All the three existing multilevel inverters use only one type of clamping devices in one topology. If two or more types of clamping devices are used in one topology, a series of new multilevel-inverter topologies can be derived. This is also the concept of the hybrid-clamped method.

The combination of the main multilevel converter topologies results in hybrid multilevel converter configurations with the main objective to utilize the advantages that each of the converter offers. Other topologies include the hybrid ANPC (Active Neutral Point Clamped) family of converters and the stacked multicell (SMC) [12] converter. Few basic and popularly used Hybrid multilevel inverters are introduced in this section.

6.2 Hybrid Multilevel Inverters based on Combination of any Two Types of Multilevel Inverters

6.2.1 Hybrid Multilevel Inverter with Half-bridge Modules

Cascaded half-bridge inverters employ half-bridge modules connected in series instead of the full H-Bridges. These inverters are one of the alternatives to the conventional cascaded full H-Bridge inverters.

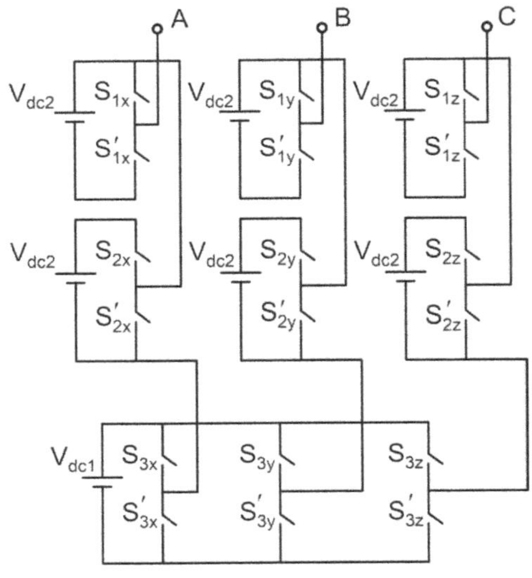

(a)

The modular multilevel inverters (MMI) employ series connections of pairs of half-bridge modules. These modules are connected in delta forming a three-phase system and the capacitors of DC links do not need isolated DC supplies, since the voltage across each half-bridge module DC link capacitor can be actively controlled. The hybrid cascaded half-bridge inverter uses an alternative connection of half-bridge modules to eliminate the output DC level. The three - phase system is reached through a star- connection. The modules are also connected in pairs and the inverter is able to provide just odd levels in the output phase voltages. This type of inverter requires a higher number of insulated DC sources for the same number of levels of a Cascaded H-Bridge. However, lower active power levels are processed in the DC sources.

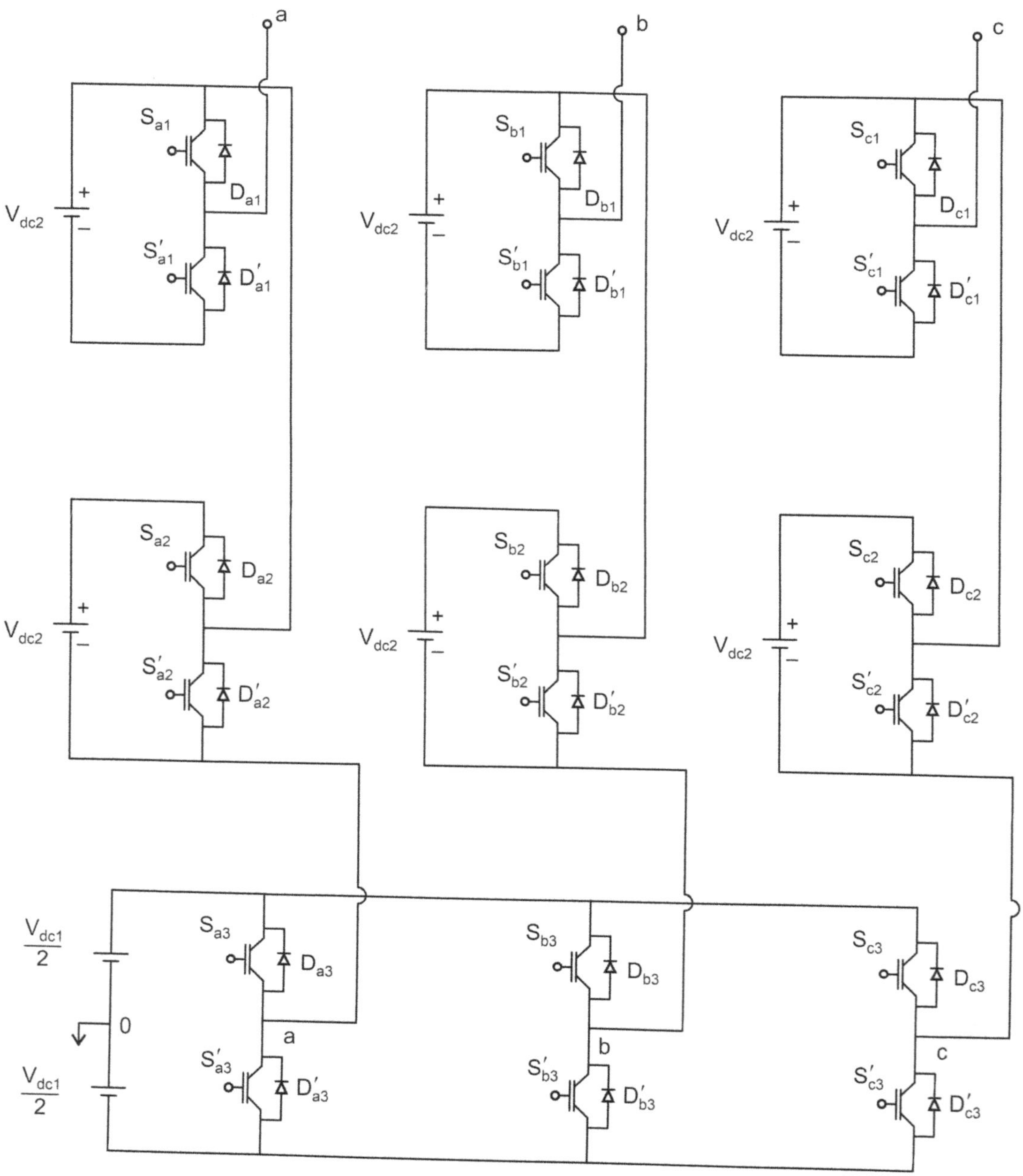

(b)

Fig. 6.1 (a) Basic circuit of Hybrid multilevel inverter based on half-bridge modules (b) Circuit topology realization with IGBTs.

This hybrid cascaded half-bridge inverter makes use of a three-phase inverter shown as a Voltage Source Inverter in Fig. 6.1 where each output is series connected to a pair or multiple pairs (cascade) of half-bridge inverters connected with inverse polarity as shown in Fig. 6.1. Here the special connection of the half-bridge modules guarantees that no DC level is observed at the output voltages. In the symmetrical version the modularity is preserved. Even though the number of insulated sources is increased for the same number of voltage levels as for the CHB or HCHB this inverter lowers the ratings of these devices since the average current that is drawn from each 6-pulse rectifier feeding a half-bridge module is lower when compared to an H-bridge based inverter. Thus, higher power levels can be achieved for a given transformer/rectifier circuit.

The assumptions made for the analysis of circuit are:

(i) the switching devices are ideal

(ii) the DC sources are constant positive voltages

(iii) parasites are neglected

(iv) the virtual center point of the VSI′s DC-link drawn in Fig. 6.1 is assumed as reference for the voltages.

Considering a phase-leg composed of a pair of half-bridge modules and a phase-leg of the VSI, the possible operation stages of this inverter are shown in Fig.6.2 (a) for phase '*a*' with negative value of load current and (b) for phase '*a*' with positive value of load current. It is observed that the output voltage V_a can assume six different values, which are given for Vo, with o = *a, b, c,* in Table 6.1. These output voltage levels depend on the DC sources voltages V_{dc2} and V_{dc1} and on the states of switches Sjo and Sjo', with j = 1, 2, 3. Based on these results, the hybrid cascaded half-bridge inverter can be operated with a number of levels N-level varying from four to six given that

$$\text{N-level} = 4;\ \text{if } V_{dc2} = V_{dc1}$$

$$\text{N-level} = 5;\ \text{if } V_{dc2} = V_{dc1}/2$$

$$\text{N-level} = 6;\ \text{if } V_{dc1} \neq V_{dc2} \neq V_{dc1}/2 \qquad \ldots..(6.1)$$

The possible operation stages for '*a*' phase-leg is given in Table 6.1

The circuit operation is divided into eight modes and explained in detail as per the table 6.1 and as shown in Fig. 6.2 (a) and (b).

(a) Output Voltage V_o, with negative load current (I_a-) for the phase '*a*' circuit:

Mode 1: For $V_o = -V_{dc2} - V_{dc1}/2$, with negative load current I_a, the current flows from load phase '*a*' through diodes D_{a1}, D_{a2}, dc source voltage V_{dc2} of second cell, diode D_{a3}, and dc source oltage $V_{dc1}/2$ to zero terminal.

ode 2: For $V_o = -V_{dc}/2$, the current flows from load phase '*a*' through switch S_{a1}', dc source tage V_{dc2} of first cell, diode D_{a2}, dc source voltage V_{dc2} of second cell, diode D_{a3}, dc source ge $V_{dc1}/2$ to zero terminal.

3: For $V_o = -V_{dc}/2$, the current flows from load phase '*a*' through diode D_{a1}, switch S_{a2}' D_{a3}, dc source voltage $V_{dc1}/2$ to zero terminal.

Mode 4: For $V_o = +V_{dc2} - V_{dc1}/2$, the current flows from load phase '*a*' through switch S_{a1}', dc source voltage V_{dc2} of first cell, switch S_{a2}', Diode D_{a3}, dc source voltage $V_{dc1}/2$ to zero terminal.

(b) Output Voltage V_o, with positive load current (I_a+) for the phase '*a*' circuit:

Mode 5: For $V_o = -V_{dc2} + V_{dc1}/2$, with positive load current I_a, the current flows from dc source voltage $V_{dc1}/2$, through switch S_{a3}, diode D_{a2}', dc source voltage V_{dc2} of first cell, diode D_{a1}', load phase '*a*' to the load.

Mode 6: For $V_o = +V_{dc1}/2$, the current flows from dc source voltage $V_{dc1}/2$, through switch S_{a3}, diode D_{a2}', switch S_{a1}, load phase '*a*' to the load.

Mode 7: For $V_o = +V_{dc1}/2$, the current flows from dc source voltage $V_{dc1}/2$, through switch S_{a3}, dc source voltage $V_{dc}/2$ of second cell, switch S_{a2}, dc source voltage V_{dc2} of first cell, diode D_{a1}', load phase '*a*' to the load.

Mode 8: For $V_o = +V_{dc2} + V_{dc1}/2$, the current flows from dc source voltage $V_{dc1}/2$, through switch S_{a3}, dc source voltage $V_{dc}/2$ of second cell, switch S_{a2}, switch S_{a1}, load phase '*a*' to the load.

Table 6.1 Resulting output phase voltage.

Resulting Output Phase Voltage (V_o, with o = a, b, c) as a Function of the Switching States and of the DC Sources Values V_{dc2} and V_{dc1}, where Sjo (j = 1, 2, 3) corresponds to the State of Switch S_{jo} and $S_{jo}' = S_{jo}$*. (Switch On: 1, Switch Off: 0.)

Modes	S_{1o} /S_{1o}'	S_{2o} /S_{2o}'	S_{3o} /S_{3o}'	Vo, Output Voltage	Case 1 (4-level): $V_{dc2}=V_{dc1}=V_{dc}$	Case 2 (5-level): $V_{dc2}=V_{dc1}/2=V_{dc}$	Case 3 (6-level): $V_{dc2}=V_{dc1}/3=V_{dc}$	Diodes
Mode1	0/0	0/0	0/0	$-V_{dc2} - V_{dc1}/2$	$-3V_{dc}/2$	$-2V_{dc}$	$-5V_{dc}/2$	D_{a1}, D_{a2}, D_{a3}
Mode2	0/1	0/0	0/0	$-V_{dc1}/2$	$-V_{dc}/2$	$-V_{dc}$	$-3V_{dc}/2$	D_{a2}, D_{a3}
Mode3	0/0	0/1	0/0	$-V_{dc1}/2$	$-V_{dc}/2$	$-V_{dc}$	$-3V_{dc}/2$	D_{a1}, D_{a3}
Mode4	0/1	0/1	0/0	$+V_{dc2} - V_{dc1}/2$	$+V_{dc}/2$	0	$-V_{dc}/2$	D_{a3}
Mode5	0/0	0/0	1/0	$-V_{dc2} + V_{dc1}/2$	$-V_{dc}/2$	0	$+V_{dc}/2$	D_{a1}', D_{a2}'
Mode6	1/0	0/0	1/0	$+V_{dc1}/2$	$+V_{dc}/2$	$+V_{dc}$	$+3V_{dc}/2$	D_{a2}'
Mode7	0/0	1/0	1/0	$+V_{dc1}/2$	$+V_{dc}/2$	$+V_{dc}$	$+3V_{dc}/2$	D_{a1}'
Mode8	1/0	1/0	1/0	$+V_{dc2} + V_{dc1}/2$	$+3V_{dc}/2$	$+2V_{dc}$	$+5V_{dc}/2$	-

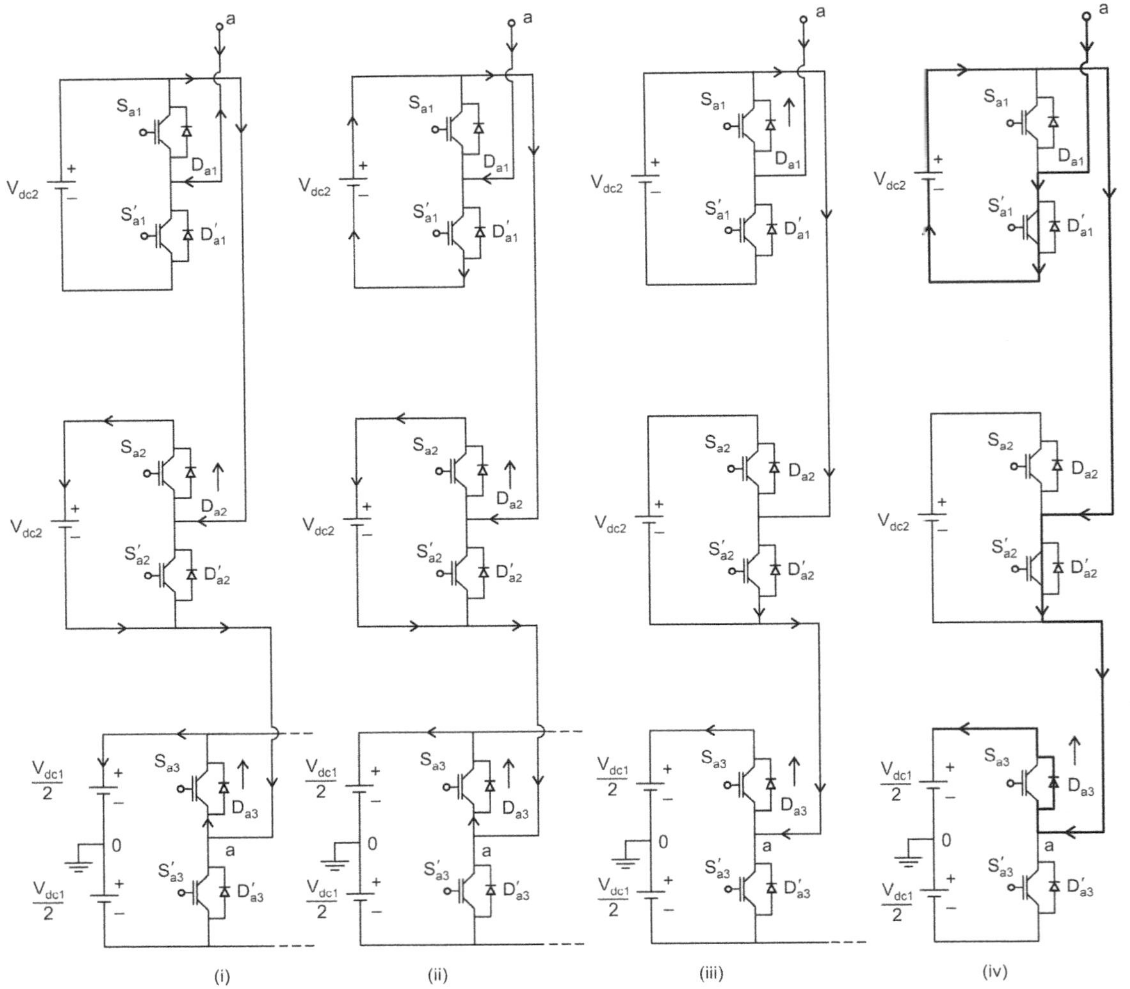

Fig. 6.2 (a) Circuit Operation with Negative Load Current (I_a -), (i) Mode1, (ii) Mode2, (iii) Mode3, and (iv) Mode4.

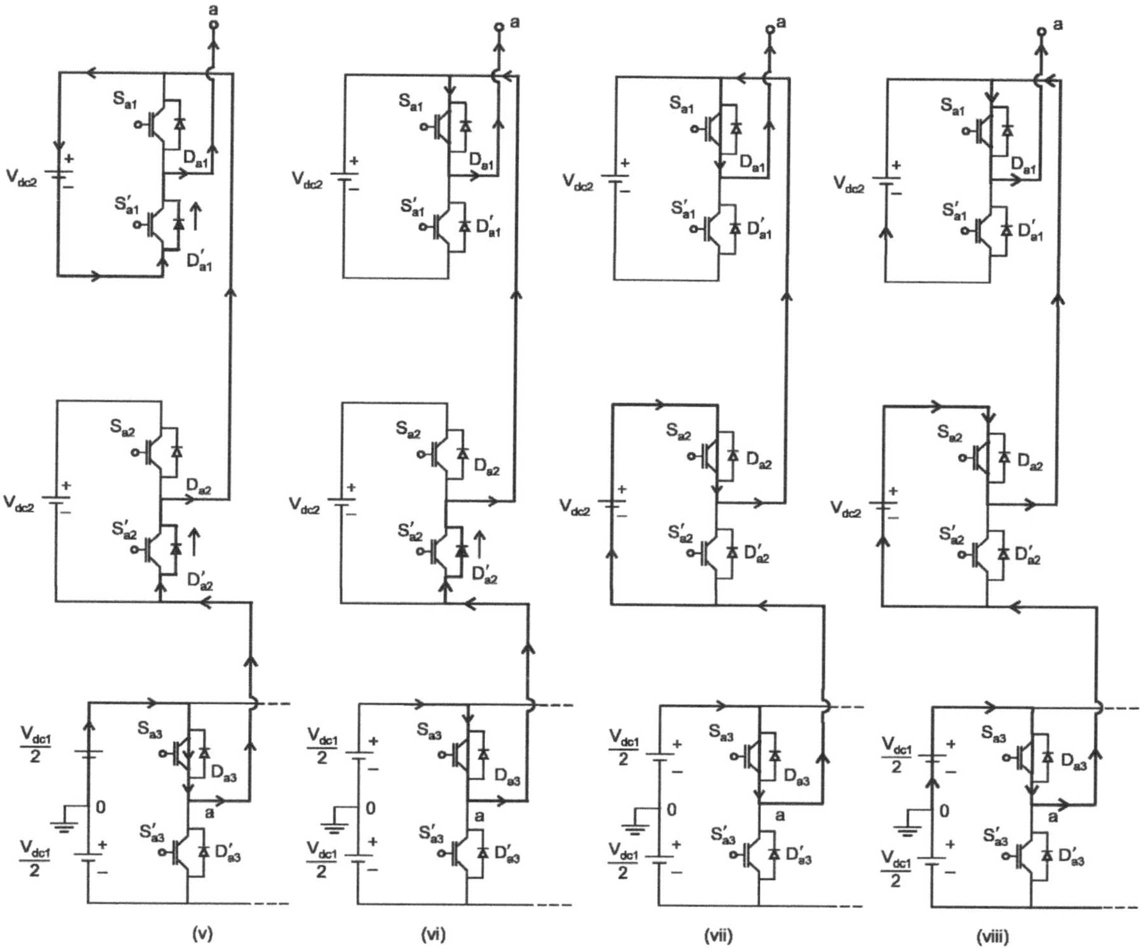

Fig. 6.2 (b) Circuit Operation with Positive Load Current (Ia+), (v) Mode 5, (vi) Mode 6, (vii) Mode 7, and (viii) Mode 8.

6.2.2 New Symmetrical Hybrid Multilevel Inverter

With a proper driving pattern for switches S_1 – S_4, and for the switches of the H-Bridge, it is possible to obtain a voltage waveform between the points a-b as shown in Fig. 6.3. Then the circuit of Fig. 6.3 behaves as a five-level output voltage single-phase inverter. The single leg switches block voltages of value ′ V_{dc}′ and with proper modulation strategy they operate at high frequency (a few kHz). On the other hand, the H-bridge switches S_5, S_6, S_7 and S_8 must block a higher voltage level of '2V_{dc}'. However, these switches operate only in a semi cycle of the output voltage. Thus, they operate at low frequency commutating at zero voltage. Thus, this inverter can also be classified among the hybrid multilevel inverters group.

The multilevel inverters based on the H-bridge symmetrical cascade have a number of levels in its output voltage given by 2N+1. For circuit shown in Fig. 6.3 output voltage levels are also obtained through the same expression 2N+1 level where N is the number of DC sources.

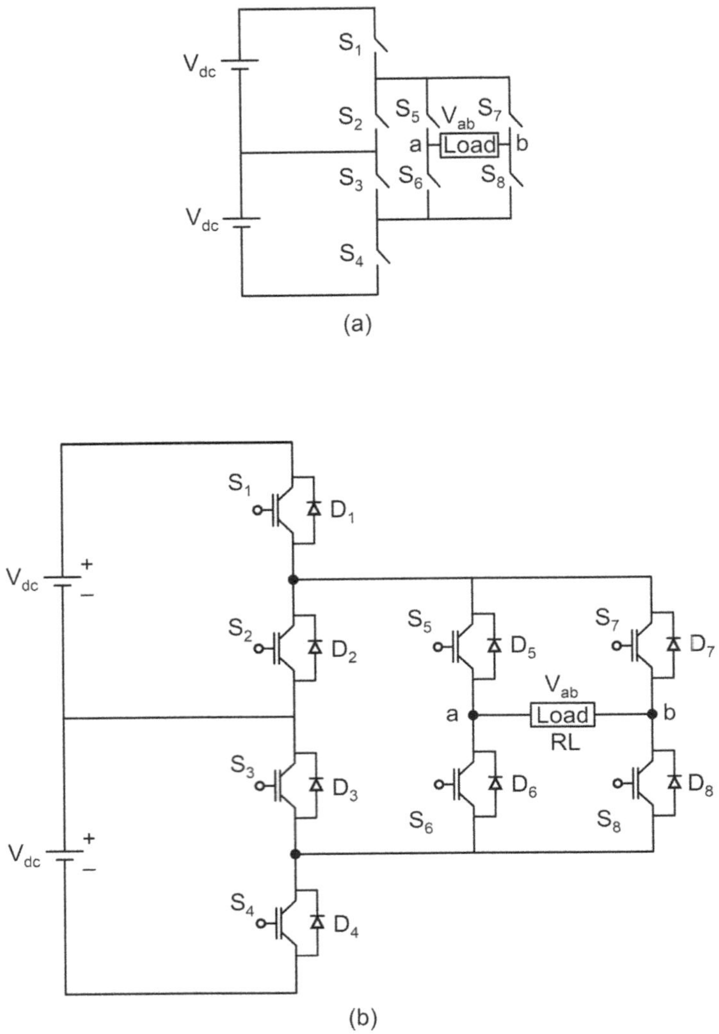

Fig. 6.3 New Symmetrical Hybrid Multilevel Inverter, (a) Basic circuit and (b) Realization of circuit with IGBTs.

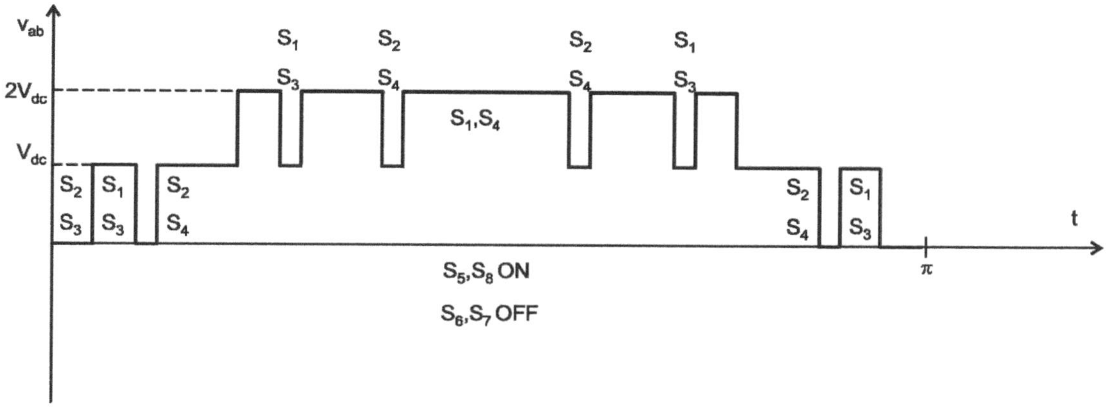

Fig. 6.4 Output voltage waveform for New symmetrical Hybrid Multilevel Inverter.

This can be implemented with the configuration shown in Fig. 6.3 where switches S_5 to S_8 are connected as a full-bridge inverter that is responsible for switching the load terminals according to the gate's signals. Fig. 6.4 shows the possible load voltage V_{ab} for the specified switching conditions. It is seen that the pairs S_5 & S_8 and S_6 & S_7 are turned on complementarily in order to generate, respectively, positive and negative voltages. The three-level DC–DC inverter switches S_1 to S_4 are switched according to a proper modulation pattern in order to generate a desired load voltage. Therefore, the inverter shown in Fig. 6.3 is a five-level single-phase inverter where switches S_1 to S_4 operate at high frequency and are rated for half of the DC-link voltage V_{dc}. Switches S_5 to S_8 are rated for the full DC-link voltage $2V_{dc}$. On the other hand, switches S_5 to S_8 can be implemented with low-frequency devices such as GTOs, integrated gate commutated thyristors (IGCT), and others, since they switch a single time per load-voltage period under zero voltage. Based on this strategy, the proposed inverter is a symmetric (equal DC sources) hybrid (multiple carrier frequencies) multilevel inverter. Furthermore, the number of levels can be increased by cascading multiple single-phase inverters. This can be achieved with other topologies as well. The total number of levels across the load terminals V_{ab} for the mentioned topology is given by $V_{ab} = 2N + 1$ where N is the total number of DC sources.

6.2.3 Hybrid-Clamped Five-level Inverter

All the three existing multilevel inverters namely, Diode-clamped inverter, Flying-capacitor inverter and Transistor-clamped cascaded H-bridge inverter, use only one type of clamping devices in each topology. If two or more types of clamping devices are used in one topology, a series of new multilevel-inverter topologies can be derived. This is also the concept of the hybrid-clamped method.

A topology clamped by diodes and capacitors was proposed in, and one leg of a five-level topology is shown in Fig. 6.5. The purpose of the topology is to solve the problems of dc-link source capacitor voltages unbalancing and higher blocked voltages of the inner switches in traditional diode-clamped multilevel inverters. Compared with the diode-clamped multilevel-inverter topology, the functions of the added clamping capacitors are as follows:

1. providing blocked voltages for the inner switches at turn-off;
2. providing bidirectional current paths; and
3. realizing the dc-link capacitor voltage balancing.

This topology is similar to the flying-capacitor topology in which the clamping capacitors contribute to the voltage synthesis. Each of the two topologies use redundant switching states of the middle voltage levels to balance the capacitor voltages. For the same reason as the flying-capacitor multilevel converter, this hybrid-clamped multilevel converter topology cannot keep the capacitor voltages balancing in pure reactive conversions and is not suitable for reactive power compensation.

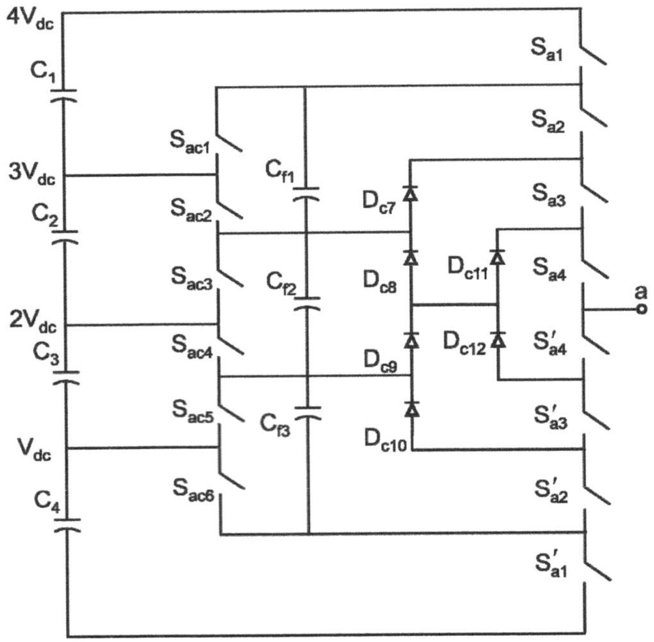

(a)

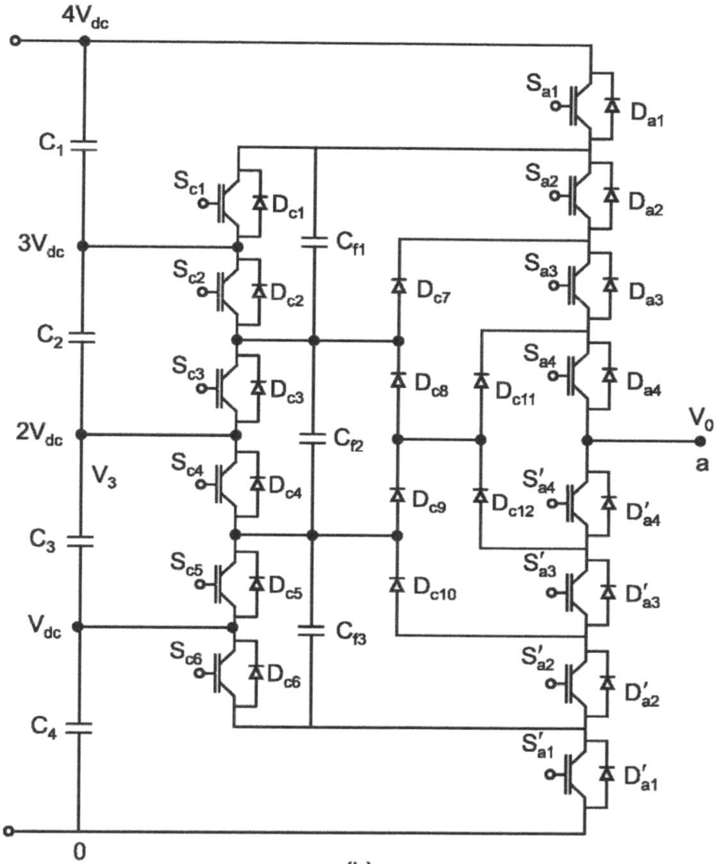

(b)

Fig. 6.5 Hybrid Clamped Five-level Inverter (a) Basic Circuit, (b) realization of circuit with IGBTs.

Fig. 6.5 shows one leg of the new hybrid-clamped five-level inverter. In this topology, switching devices S_{a1}–S_{a4} and S_{a1}' –S_{a4}' are the main switching devices used to produce the desired output voltage. S_{c1}–S_{c6} are the clamping switching devices, and D_{c1}–D_{c12} are the clamping diodes. The clamping switching devices S_{c1}–S_{c6} and the flying capacitors C_{f1}–C_{f3} maintain the dc-link capacitor voltages in balance. It can be seen from Fig. 3 that only the clamping devices nearest to the dc source side are active switches with anti-paralleled diodes, while others are only clamping diodes. This topology can easily be extended to any voltage levels. For one leg of this m-level topology, $(m - 1)$ dc-link source capacitors, $2(m - 1)$ main switching devices, $2(m - 2)$ clamping switching devices, $(m - 1) * (m - 2)$ clamping diodes, and $(m - 2)$ flying capacitors are required to complete the operation. All the capacitors sustain the same voltage, and all the switching devices and diodes have the same voltage stresses.

Table 6.2(a) Switching pattern and its output voltages for positive current.

Modes	Switching States								Current path through Capacitors	Output Voltage V_o	Diodes
	S_{a1}	S_{a2}	S_{a3}	S_{a4}	S_{a4}'	S_{a3}'	S_{a2}'	S_{a1}'			
Mode 1	1	1	1	1	0	0	0	0	$C_1 + C_2 + C_3 + C_4$	$+4\ V_{dc}$	-
Mode 2	0	1	1	1	0	0	0	0	$C_{f1} + C_{f2} + C_{f3}$	$+3\ V_{dc}$	D_{a1}'
	1	0	1	1	1	0	0	0	$C_1+C_2+C_3+C_4-C_{f1}$		-
Mode3	0	0	1	1	0	0	0	0	$C_{f2} + C_{f3}$	$+2\ V_{dc}$	D_{a1}', D_{c7}
	1	0	0	1	1	1	0	0	$C_1+C_2+C_3+C_4-C_{f1}-C_{f2}$		
Mode4	0	0	0	1	0	0	0	0	C_{f3}	$+\ V_{dc}$	D_{a1}', D_{c9}, D_{c11}
	1	0	0	0	1	1	1	0	$C_1+C_2+C_3+C_4-C_{f1}-C_{f2}-C_{f3}$		-
Mode5	0	0	0	0	0	0	0	0	----	0	D_{a1}', D_{a2}', D_{a3}', D_{a4}'

The main and clamping switches connecting to various capacitors are given as follows.

(i) S_{a1} and S_{c1} are in parallel with dc link capacitor C_1

(ii) S_{c1} and S_{c2} are in parallel with flying capacitor C_{f1}

(iii) S_{c2} and S_{c3} are in parallel with dc link capacitor C_2

(iv) S_{c3} and S_{c4} are in parallel with flying capacitor C_{f2}

(v) S_{c4} and S_{c5} are in parallel with dc link capacitor C_3

(vi) S_{c5} and S_{c6} are in parallel with flying capacitor C_{f3} and

(vii) S_{c6} and S_{a1}' are in parallel with dc link capacitor C_4

These switches should be complementary switching pairs, i.e., when one switch is on, the other is off, and vice versa. The switching pattern and its output voltages Vo is shown in Table 6.2(a) for positive load current and (b) for negative load current.

Various modes of operation: The Hybrid-clamped multilevel synthesis principle is explained by presenting a five-level circuit as an example. The dc bus negative point is taken for reference. Five voltage levels are produced by the following five modes as per the switching states as shown in Fig. 6.6 (a)-(e), for bidirectional current paths corresponding to each voltage level.

Table 6.2(b) Switching pattern and its output voltages for negative current.

Modes	Switching States								Current path through Capacitors	Conducting Diodes	Output Voltage V_o
	S_{a1}	S_{a2}	S_{a3}	S_{a4}	S_{a4}'	S_{a3}'	S_{a2}'	S_{a1}'			
Mode 1	0	0	0	0	0	0	0	0	$C_1 + C_2 + C_3 + C_4$	D_{a4}, D_{a3}, D_{a2}, D_{a1}	$-4\ V_{dc}$
Mode 2	0	0	0	0	0	0	0	1	$C_{f1} + C_{f2} + C_{f3}$	D_{a4}, D_{a3}, D_{a2}	$-3\ V_{dc}$
Mode3	0	0	0	0	1	0	0	1	$C_{f2} + C_{f3}$	D_{c8}, D_{c12}	$-2\ V_{dc}$
Mode4	0	0	0	0	1	1	0	1	C_{f3}	D_{c10}	$-\ V_{dc}$
Mode 5	0	0	0	0	1	1	1	1	-	-	0

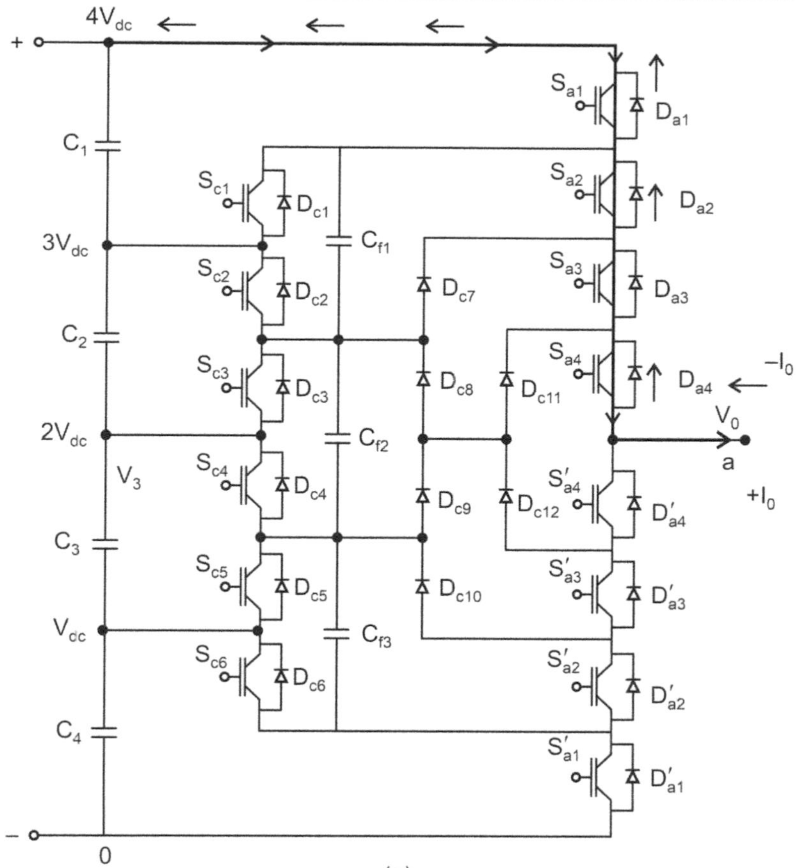

(a)

Fig. 6.6 (a) Mode 1: $V_o = +\ 4\ V_{dc}$ with $+\ I_o$: thick line and $V_o = -4V_{dc}$ with $-\ I_o$: broken arrows.

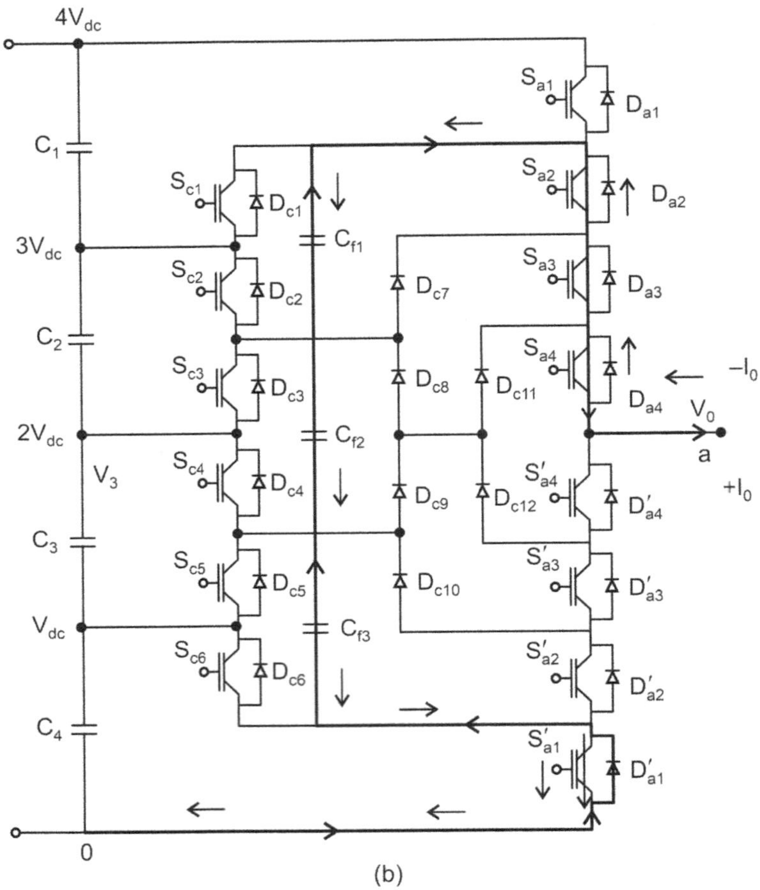

Fig. 6.6 (b) Mode 2: $V_O = +3V_{dc}$ with $+I_o$: thick line; and $V_o = -3V_{dc}$ with $-I_o$: broken arrows.

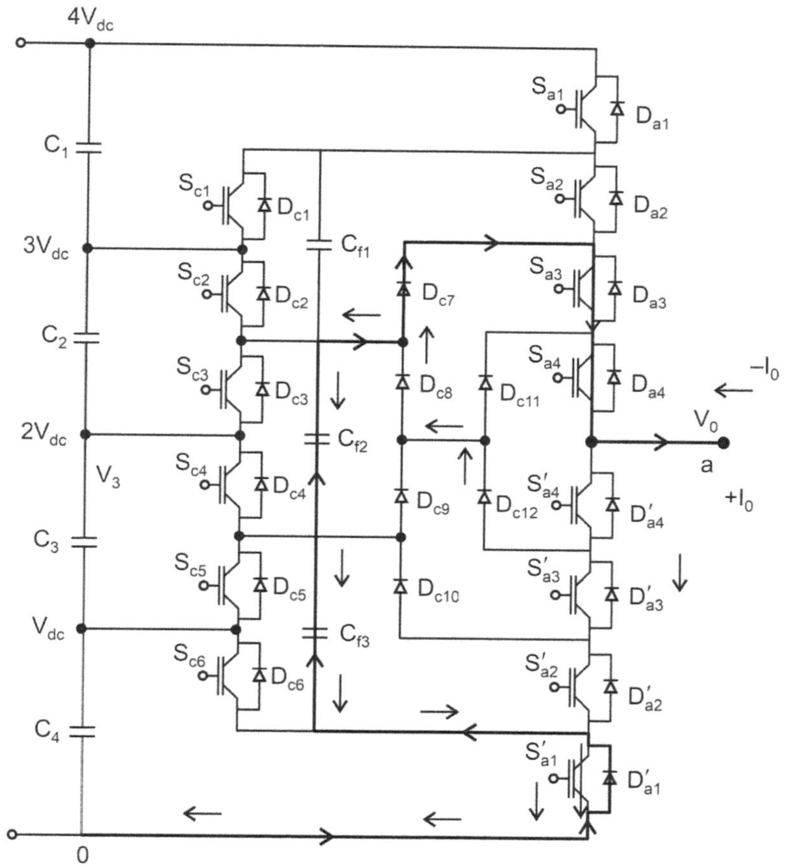

(c)

Fig. 6.6 (c) Mode 3: $V_o = +2V_{dc}$ with $+I_o$: thick line; and $V_o = -2V_{dc}$ with $-I_o$: broken arrows.

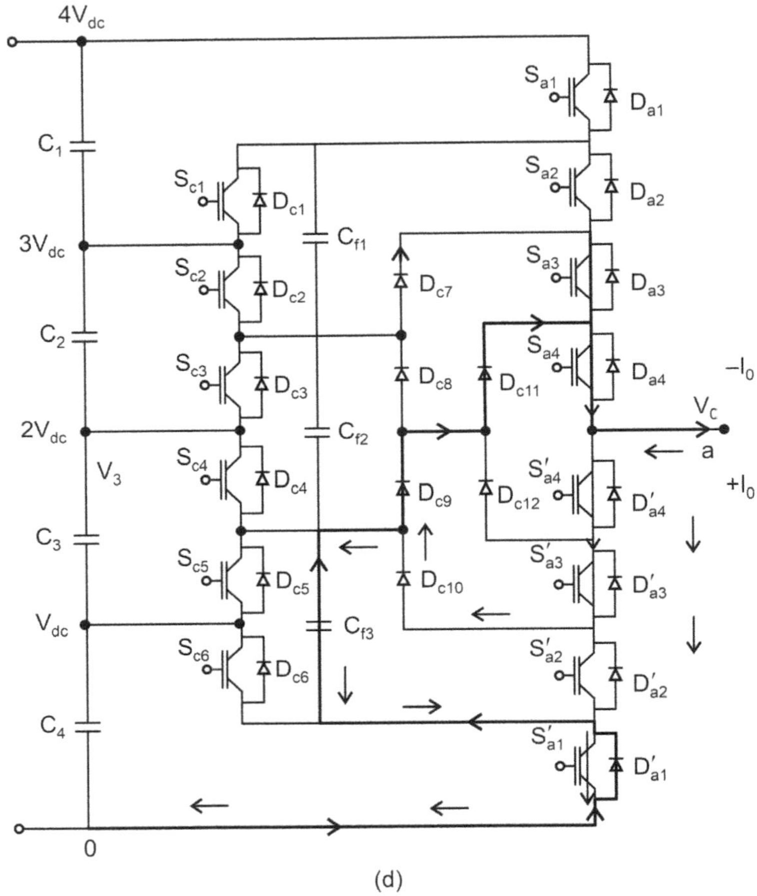

(d)

Fig. 6.6 (d) Mode 4: $V_o = +V_{dc}$ with ^+I_o: thick line; and $V_o = -V_{dc}$ with $-I_o$: broken arrows.

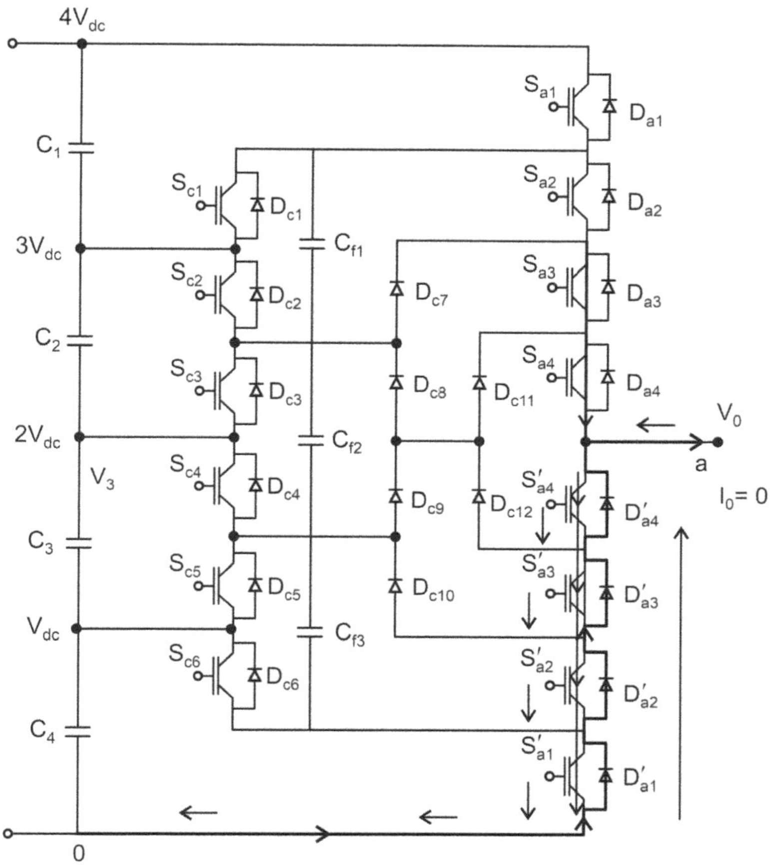

(e)

Fig. 6.6 (e) Mode 5: $V_o = 0$ with + I_o: thick line; and $V_o = 0$ with – I_o: broken arrows.

Fig. 6.6 Hybrid-clamped five-level inverter operating modes for both +Io, and -Io, (a) Mode 1: Vo=+4Vdc, (b) Mode 2: Vo=+3Vdc, (c) Mode 3: Vo=+2Vdc, (d) Mode 4: Vo=+Vdc, and (e) Mode 5: Vo=0, volts.

Mode 1: For voltage level $V_o = +4V_{dc}$, turn on four main switches S_{a1}, S_{a2}, S_{a3}, S_{a4} and three clamping switches S_{c2}, S_{c4}, S_{c6}.

(i) For positive load current (I_o) when current flows into load terminal '*a*', the conduction takes place from top terminal of dc link source capacitor C_1, through main switching devices S_{a1}, S_{a2}, S_{a3}, S_{a4}, to the load terminal '*a*'. This conduction path is shown in Fig.6.6 (a) with thick line.

(ii) For negative load current (I_o) when current flows into the circuit from load terminal '*a*', the conduction takes place from load terminal '*a*', through anti-parallel diodes D_{a4}, D_{a3}, D_{a2}, D_{a1}, to the top terminal of dc link source capacitor C_1. For which the load voltage V_o

= $-4V_{dc}$. This conduction path is shown in Fig.6.6 (a) with reverse arrows marked separately (broken arrows).

Mode 2: For voltage level $V_o = +3V_{dc}$, there are two possible switching states, (a) Turn on four main switches S_{a2}, S_{a3}, S_{a4}, S_{a1}' and three clamping switches S_{c1}, S_{c3}, S_{c5} and (b) Turn on four main switches S_{a1}, S_{a3}, S_{a4}, S_{a4}' and three clamping switches S_{c2}, S_{c4}, S_{c6}. Here one possibility mentioned in (a) is considered for both positive and negative load current flow.

(i) For positive load current (I_o) when current flows into load terminal '*a*', the conduction takes place from the bottom terminal of dc link source capacitor C_4 (zero reference terminal), through diode D_{a1}', flying capacitors C_{f3}, C_{f2}, C_{f1}, main switching devices S_{a2}, S_{a3}, S_{a4}, to the load terminal '*a*'. This conduction path is shown in Fig.6.6 (b) with thick line.

(ii) For negative load current (I_o) when current flows into the circuit from load terminal '*a*', the conduction takes place from load terminal '*a*', through anti-parallel diodes D_{a4}, D_{a3}, D_{a2}, to flying capacitors C_{f1}, C_{f2}, C_{f3}, main switching device S_{a1}', to the zero reference. For which the load voltage $V_o = -3V_{dc}$. This conduction path is shown in Fig.6.6 (b) with reverse arrows marked separately (broken arrows).

Mode 3: For voltage level V_o=$+2V_{dc}$, there are two possible switching states, (a) Turn on four main switches S_{a3}, S_{a4}, S_{a4}', S_{a1}' and three clamping switches S_{c1}, S_{c3}, S_{c5}, and (b) Turn on four main switches S_{a1}, S_{a4}, S_{a4}', S_{a3}' and three clamping switches S_{c2}, S_{c4}, S_{c6}. Here one possibility mentioned in (a) is considered for both positive and negative load current flow.

(i) For positive load current (I_o) when current flows into load terminal '*a*', the conduction takes place from the bottom terminal of dc link source capacitor C_4 (zero reference terminal), through anti-parallel diode D_{a1}', flying capacitors C_{f3}, C_{f2}, clamping diode D_{c7}, main switching devices S_{a3}, S_{a4}, to the load terminal '*a*'. This conduction path is shown in Fig.6.6 (c) with thick line.

(ii) For negative load current (I_o) when current flows into the circuit from load terminal '*a*', the conduction takes place from load terminal '*a*', through main switching device S_{a4}', clamping diode D_{c12}, D_{c8}, to flying capacitors C_{f2}, C_{f3}, main switching device S_{a1}', to the zero-reference terminal. For which the load voltage $V_o = -2V_{dc}$. This conduction path is shown in Fig.6.6 (c) with reverse arrows marked separately (broken arrows).

Mode 4: For voltage level $V_o = +V_{dc}$, there are two possible switching states, (a) Turn on four main switches S_{a4}, S_{a4}', S_{a3}', S_{a1}' and three clamping switches S_{c1}, S_{c3}, S_{c5}, and (b) Turn on four main switches S_{a1}, S_{a4}', S_{a3}', S_{a2}' and three clamping switches S_{c2}, S_{c4}, S_{c6}. Here one possibility mentioned in (a) is considered for both positive and negative load current flow.

(i) For positive load current (I_o) when current flows into load terminal '*a*', the conduction takes place from the bottom terminal of dc link source capacitor C_4 (zero reference terminal), through anti-parallel diode D_{a1}', flying capacitor C_{f3}, clamping diodes D_{c9}, D_{c11}, main switching device S_{a4}, to the load terminal '*a*'. This conduction path is shown in Fig.6.6 (d) with thick line.

(ii) For negative load current (I_o) when current flows into the circuit from load terminal '*a*', the conduction takes place from load terminal '*a*', through main switching devices S_{a4}', S_{a3}', clamping diode D_{c10}, to flying capacitors C_{f3}, main switching device S_{a1}', to the zero-reference terminal. For which the load voltage V_o = -V_{dc}. This conduction path is shown in Fig.6.6 (d) with reverse arrows marked separately (broken arrows).

Mode 5: For voltage level V_o = 0, turn on four main switches S_{a1}, S_{a2}, S_{a3}, S_{a4} and three clamping switches S_{c1}, S_{c3}, S_{c5}.

(i) For positive load current (I_o) when current flows into load terminal '*a*', the conduction takes place from bottom terminal of dc link source capacitor C_4, (zero-reference terminal) through anti-parallel diodes D_{a1}', D_{a2}', D_{a3}', D_{a4}', to the load terminal '*a*'. This conduction path is shown in Fig.6.6 (e) with thick line.

(ii) For negative load current (I_o) when current flows into the circuit from load terminal '*a*', the conduction takes place from load terminal '*a*', through main switching devices S_{a4}', S_{a3}', S_{a2}', S_{a1}', to the bottom terminal of dc link source capacitor C_4 (zero-reference terminal). For which the load voltage V_o = 0. This conduction path is shown in Fig.6.6 (e) with reverse arrows marked separately (broken arrows).

6.2.4 New Hybrid Asymmetrical Multilevel H-Bridge Inverter

Following the principle of increasing the number of voltage levels in output waveform with less switching devices, inverter topology is shown in Fig. 6.7. The new H-bridge topology with an auxiliary bidirectional switch can output maximum five level voltage waveform (+2V_{dc}, +V_{dc}, 0, −V_{dc}, −2V_{dc}).

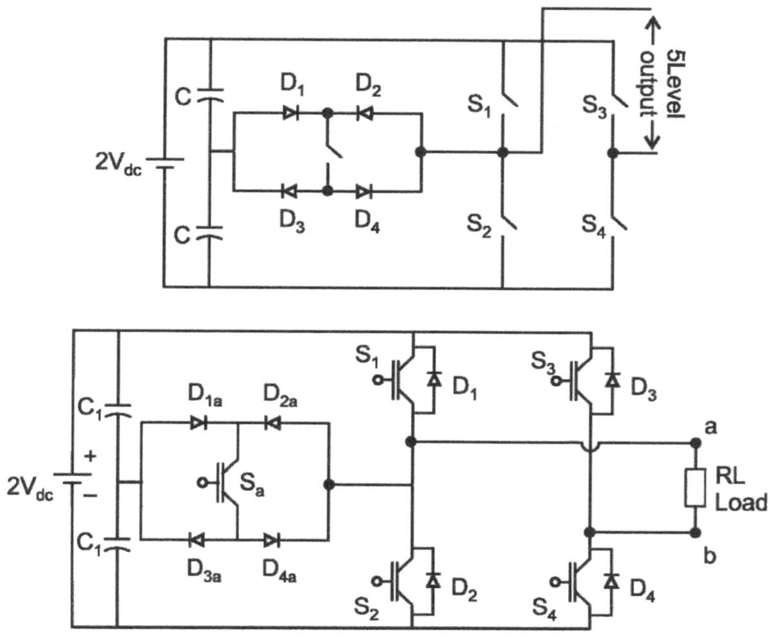

Fig. 6.7 New Hybrid Asymmetrical Multilevel H-Bridge Inverter.

The switching combinations required to generate the five-level output waveform is as given in Table 6.3. In this configuration the two capacitors in the capacitive voltage divider are connected directly across the DC bus and since all switching combinations are activated in an output cycle the dynamic voltage balance between the two capacitors is automatically restored.

Table 6.3 Switching Combinations for five-level output voltage.

Modes	Switch S_a	Switch S_1	Switch S_2	Switch S_3	Switch S_4	Output Voltage, V_{ab}	Diodes
Mode 1	0	1	0	0	1	$+2V_{dc}$	-
Mode 2	1	0	0	0	1	$+V_{dc}$	D_{1a}, D_{4a}
Mode 3	0	0	0	0	1	0	D_2
Mode 4	1	0	0	1	0	$-V_{dc}$	D_{2a}, D_{3a}
Mode 5	0	0	1	1	0	$-2V_{dc}$	-

Mode 1: For $V_{ab} = +2V_{dc}$, with positive load current Ia, the conduction takes place from the dc source voltage $2V_{dc}$, switch S_1, load terminals '*a*' to '*b*', switch S_4, and back to dc source terminal.

Mode 2: For $V_{ab} = +V_{dc}$, with positive load current Ia, the conduction takes place from the dc source lower capacitor C_1, (mid point z) through diode D_{1a}, switch S_a, D_{4a}, load terminals '*a*' to '*b*' switch S_4, and back to dc source terminal.

Mode 3: For $V_{ab} = 0$ volts, with zero load current Ia, the conduction takes place from load terminals '*a*' to '*b*', switch S_4, switch D_2 to the load terminal '*a*'.

Mode 4: For $V_{ab} = -V_{dc}$, with negative load current Ia, the conduction takes place from dc source upper capacitor C_1, though switch S_3, load terminals '*b*' to '*a*', diode D_{2a}, switch S_a, diode D_{3a}, back to dc source upper capacitor C_1, through mid point Z.

Mode 5: For $V_{ab} = -2V_{dc}$, with negative load current Ia, the conduction takes place from the dc source voltage $2V_{dc}$, switch S_3, load terminals '*b*' to '*a*', switch S_2, back to the dc source voltage terminal.

Thus, a five-level output voltage is produced from the circuit given in the above figure.

6.2.5 Hybrid H-bridge type diode clamped Multilevel Inverter (H-bridge based NPC)

A five level H-Bridge based NPC inverter consists of two voltage dividing capacitors and eight semiconductor switches which is equally shared by two legs. The two upper switches act in opposite manner to two bottom switches on each leg. Also, two series connected diodes are connected on each leg and are shorted at the neutral point z (mid-point).

Figure 6.8 shows a schematic diagram of single phase five level H-bridge type diode clamped Multilevel Inverter. Here too the similar output voltage states exist: $+V_{dc}$, $+V_{dc}/2$, 0, $-V_{dc}/2$ and $-V_{dc}$. At any moment, some four switches are in active state and subsequent complementary switches are inactive for each stage of output voltage, dc source V_{dc} is divided equally by two capacitors $C_1 = C_2 = V_{dc}/2$. Table 6.4 explicates the switching arrangement of five level H Bridge type Neutral Point Clamped five-level inverter.

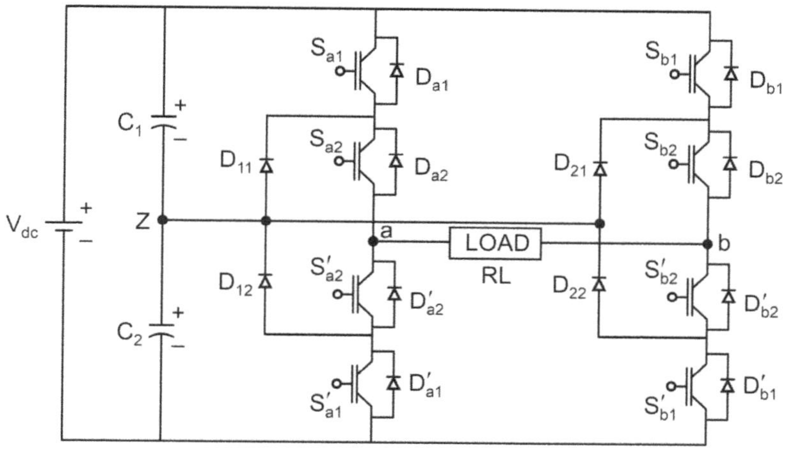

Fig. 6.8 Schematic diagram of H-Bridge based five level NPC MLI.

Table 6.4 Switching arrangement for five levels H-Bridge type diode-clamped multilevel inverter.

Modes	Voltage levels	S_{a1}	S_{a2}	S_{a1}'	S_{a2}'	S_{b1}	S_{b2}	S_{b1}'	S_{b2}'	Clamped Diodes
Mode 1	$+V_{dc}$	1	1	0	0	0	0	1	1	---
Mode 2	$+V_{dc}/2$	1	1	0	0	0	0	0	1	D_{22}
Mode 3	0	1	1	0	0	0	0	0	0	D_{b2}, D_{b1}
Mode 4	$-V_{dc}/2$	0	0	1	1	0	1	0	0	D_{21}
Mode 5	$-V_{dc}$	0	0	1	1	1	1	0	0	---

Table 6.5 Comparison of constructional differences among the MLI topologies.

Type of MLI	Conventional DCMLI (For 5 level)	Cascaded H Bridge MLI (For 5 level)	H Bridge type DCMLI (For 5 level)
Main power devices	8	8	8
Main diodes	8	8	8
Clamping diodes	6	0	4
DC bus capacitors	4	0	2
Number of legs	1	4	2
Number of DC sources	1	2	1
Magnitude of DC sources to get V_{dc} output	$2V_{dc}$	$V_{dc}/2 + V_{dc}/2$	V_{dc}

Mode 1: For $V_{ab} = +V_{dc}$, with positive load current I_a, the conduction takes place from upper terminal of C_1, switches S_{a1}, S_{a2}, load terminals a to b and switches S_{b2}', S_{b1}' to the bottom terminal of C_2.

Mode 2: From $V_{ab} = + V_{dc}/2$, with positive load current I_a, the conduction takes place from upper terminal of C_1, through switches S_{a1}, S_{a2}, load terminals a to b, switch S_{b2}', D_{22} and mid point terminal Z.

Mode 3: For $V_{ab} = 0$, conduction takes place from upper terminal of C_1, through switches Sa_1, Sa_2, load terminals a to b, and through diodes D_{b2}, D_{b1} leading to short-circuit for zero output voltage.

Mode 4: For $V_{ab} = -V_{dc}/2$, with negative load current, I_a, the conduction takes place from mid-point terminal Z, through diode D_{21}, switch S_{b2}, load terminals b to a, switches S_{a2}', S_{a1}' to bottom terminal of C_2.

Mode 5: For $V_{ab} = -V_{dc}$ with negative load current I_a, the conduction takes place from upper terminal of C_1, through switches S_{b1}, S_{b2}, load terminals b to a, switches S_{a2}', S_{a1}' to the bottom terminal of C_2.

Simulation results for Hybrid H-bridge based NPC Inverter:

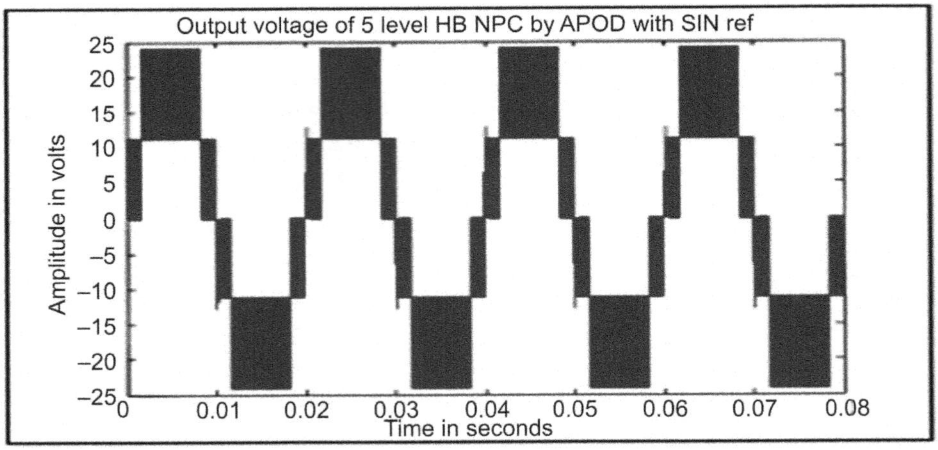

Fig. 6.9 Five level Voltage waveform and its Harmonic profile of HB NPC with SPWM.

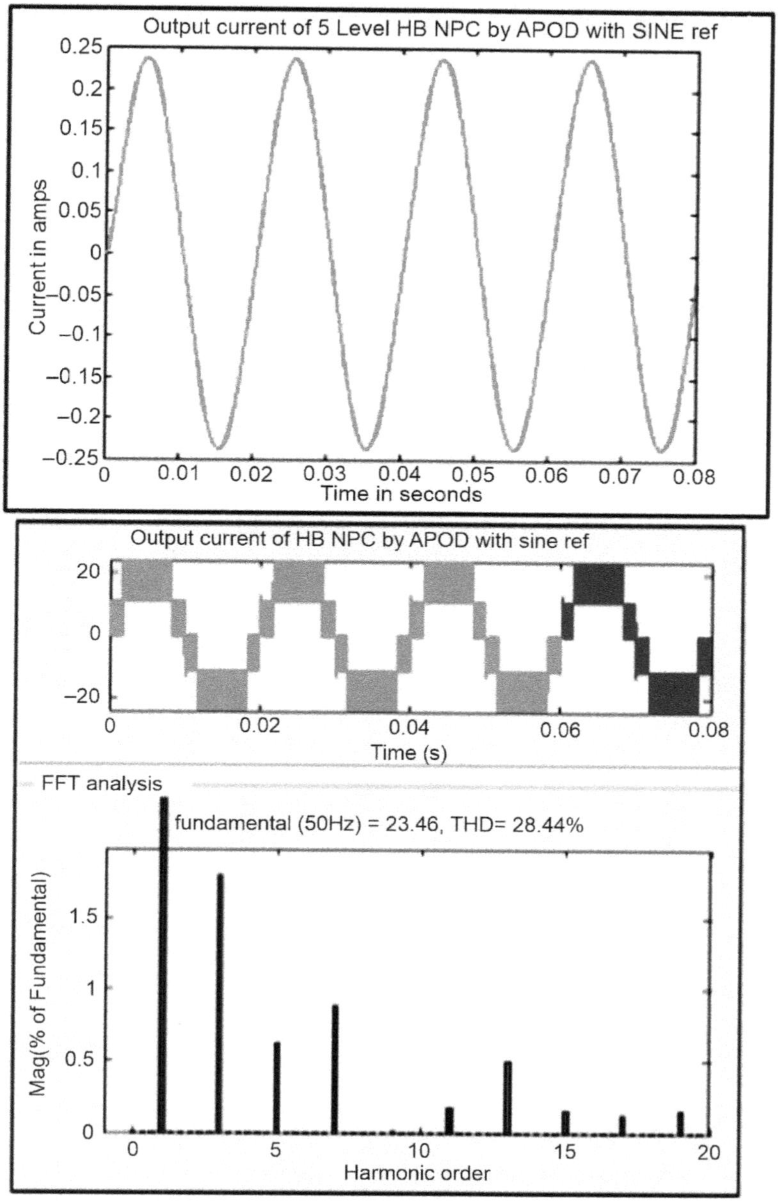

Fig. 6.10 Five level Current waveform and its Harmonic profile of HB NPC with SPWM.

Consequently, for single sourced applications like fuel cell or PV connected inverter, the H-bridge type diode clamped MLI can be recommended by the way to achieve better AC voltage with minimum %THD and number of components.

6.2.6 Hybrid Flying Capacitor and Active NPC Inverter

The seven-level Flying Capacitor based ANPC inverter consists of a three-level ANPC (Active Neutral Point) converter and two Flying-Capacitor cells connected as shown in Fig. 6.11. In order to simplify the analysis, we consider the capacitor of the last Flying-Capacitor cell is maintained at a voltage equal to V_{dc}. The seven levels at the output of the inverter are then derived by maintaining the flying capacitor of the second cell to a voltage of $2V_{dc}$ and each of the two DC-link capacitors to $3V_{dc}$, with total DC-link voltage of $6V_{dc}$. The voltage stress across the outer (S_1) switches of the topology is higher compared to the stress across the Flying Capacitor cells (S_{c1}–S_{c3}) and the converter requires series connection of switching devices or different semiconductor technologies to be used in the configuration, for the converter to reach its full rating. In order to minimize the stress across the outer switches, their switching frequency is selected equal to the fundamental frequency at the output.

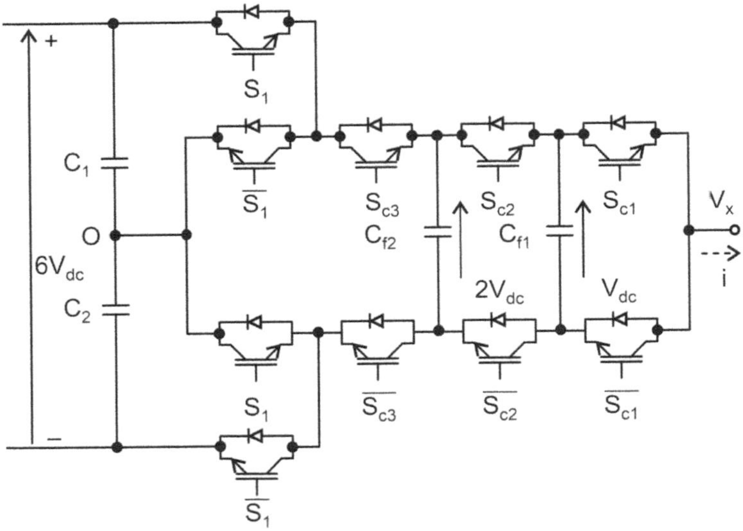

Fig. 6.11 Configuration of the seven-level flying capacitor based active NPC inverter

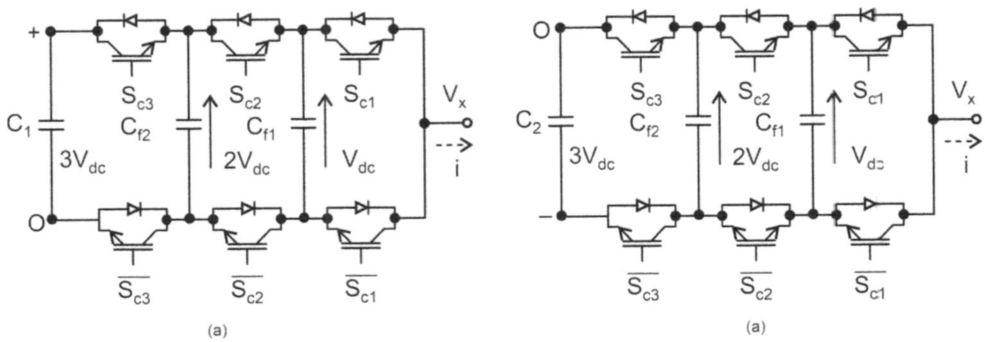

Fig. 6.12 Equivalent circuits during half-periods of operation (a) Positive half-period and (b) Negative half-period.

The inverter operates in two distinct half-periods during a fundamental period, one at which the upper DC-link capacitor (C_1) is connected to a positive phase and a negative phase where the lower DC-link capacitor (C_2) is used. During these half-periods, the topology simplifies to a four-level Flying-Capacitor inverter and for which the Flying Capacitor balancing requirements can be identified. The equivalent circuits for one phase of the converter during the two half-periods of operation are given in Fig. 6.12. The limitation of fundamental frequency in the outer S_1 switches of the topology limits the available states of the three-level ANPC to four. This also limits the number of states that generate the zero-voltage level to one for each of the half-periods in the waveform. The state that produces the zero-voltage level during the positive half period (V_9) is denoted as (+) 0 while the state (V_8) that generates the zero-voltage level during the negative half-period is denoted as (−) 0. The limitation in the switching states of the three level ANPC, in combination with the two states of each Flying Capacitor cell, offer a total of sixteen switching states for the seven level ANPC topology and are given in Table 6.6.

Table 6.6 Switching States of the Seven-Level Flying-Capacitor based ANPC Inverter.

Switching States	S_1	S_{c3}	S_{c2}	S_{c1}	Output Voltage
V_1	0	0	0	0	$-3V_{dc}$
V_2	0	0	0	1	$-2V_{dc}$
V_3	0	0	1	0	$-2V_{dc}$
V_4	0	1	0	0	$-2V_{dc}$
V_5	0	0	1	1	$-V_{dc}$
V_6	0	1	0	1	$-V_{dc}$
V_7	0	1	1	0	$-V_{dc}$
V_8	0	1	1	1	(-) 0
V_9	1	0	0	0	(+) 0
V_{10}	1	0	0	1	$+V_{dc}$
V_{11}	1	0	1	0	$+V_{dc}$
V_{12}	1	1	0	0	$+V_{dc}$
V_{13}	1	0	1	1	$+2V_{dc}$
V_{14}	1	1	0	1	$+2V_{dc}$
V_{15}	1	1	1	0	$+2V_{dc}$
V_{16}	1	1	1	1	$+3V_{dc}$

The voltage levels of $\pm V_{dc}$ and $\pm 2V_{dc}$ can be acquired with three different switching combinations each. The three different switching states have disparate effect on both the neutral point voltage and the voltages across each of the two flying capacitor cells. Utilization of the available redundancies provides the means for regulating the Flying capacitor voltages to the required levels, ensuring the proper operation of the converter.

A three-phase seven-level ANPC converter is considered, in order to illustrate the operation and the performance of the topology. The total DC-link voltage of $6V_{dc}$ is maintained constant and split to two identical flying capacitors, (C_{f1} and C_{f2}). The reference value for the voltage of the two flying capacitors is then equal to V_{dc} and $2V_{dc}$ for cell 1 and cell 2 respectively.

The sinusoidal PWM modulation is used to illustrate the operation. Fig. 6.13 shows the output phase voltage for the seven-level converter under steady state operation. The multilevel voltage waveform results in high quality currents as shown in Fig. 6.14.

The hybrid converter is based on the combination of a three-level ANPC converter with two flying capacitor cells. The sinusoidal PWM modulation is applied independently to each phase of the converter and is suitable for single or multi-phase applications. The SPWM modulator is capable of providing high quality waveforms while regulating the voltages of the Flying Capacitor cells to the reference value, balancing the DC-link capacitor voltages and maintaining the switching frequency of the individual switches low. The modulator is also capable of regulating the voltages of the capacitors under dynamic and transient changes in the operation of the converter.

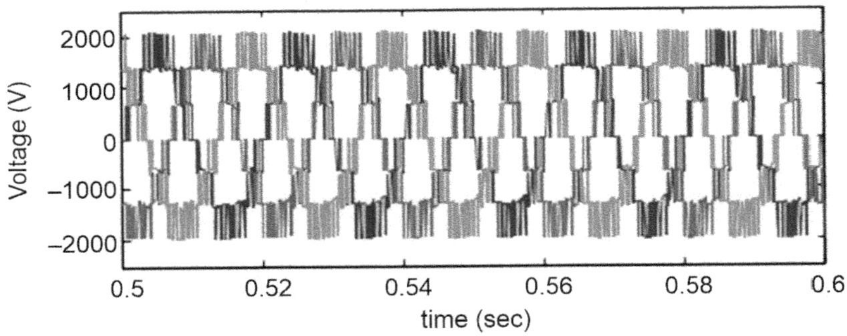

Fig. 6.13 Phase voltages of the seven-level ANPC Inverter with FC.

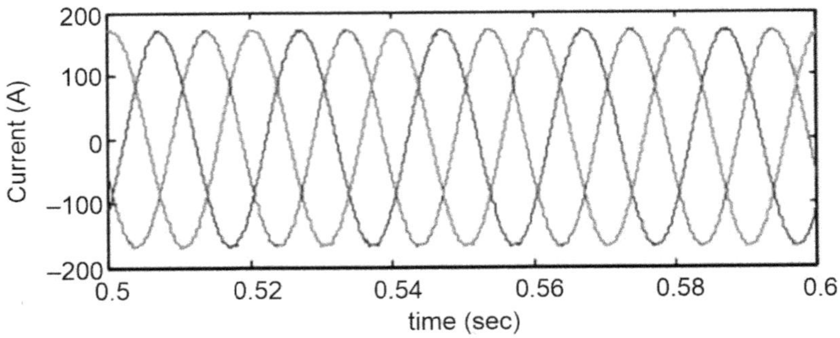

Fig. 6.14 Load currents of the seven-level ANPC Inverter with FC.

6.3 Hybrid Inverter with Reduced Power Semiconductor Switches

6.3.1 Various Configurations of Hybrid Inverters

In this section, in order to increase the number of output voltage levels and reduce the number of power switches, driver circuits, and the total cost of the inverter, a new topology of cascaded multilevel inverters is proposed. It is important to note that in the new topology, the unidirectional power switches are used. Then, to determine the magnitude of the dc voltage sources, a new algorithm is proposed. Moreover, the new topology is compared with other topologies from different points of view such as the number of IGBTs, number of dc voltage sources, the variety of the values of the dc voltage sources, and the value of the blocking voltages per switch. Finally, the performance of the proposed topology in generating all voltage levels through a seven -level and 31-level inverters.

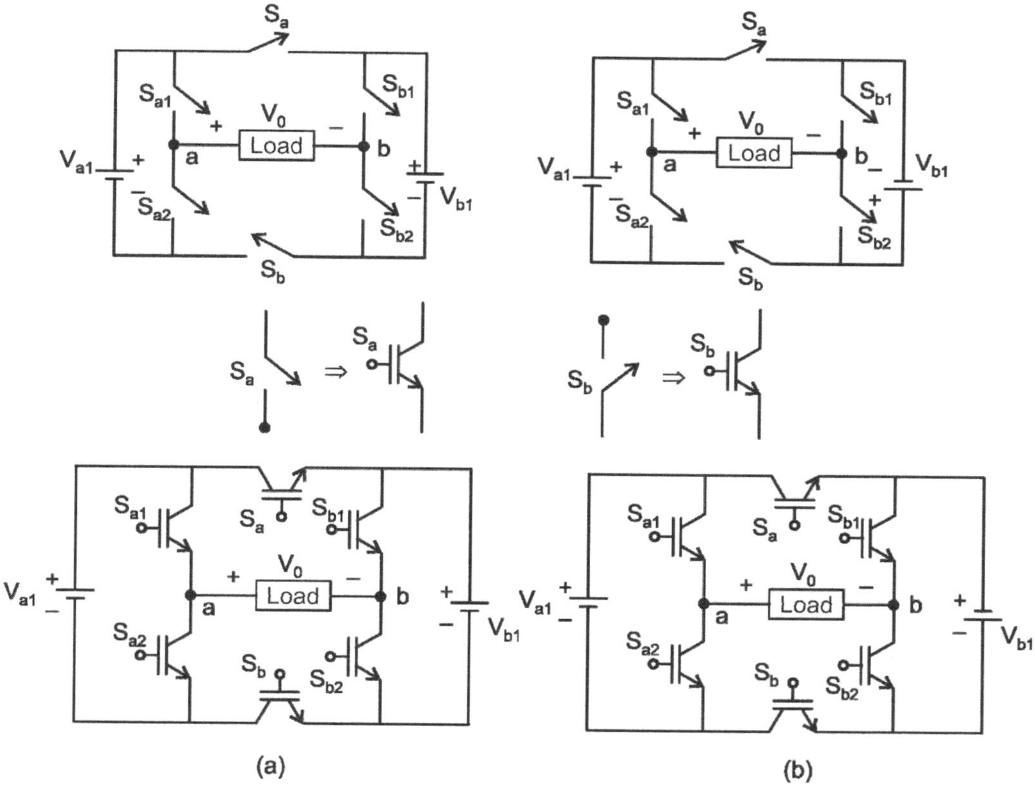

Fig. 6.15 Proposed seven-level inverters. (a) First proposed topology. (b) Second proposed topology.

In Fig. 6.15, two new topologies are proposed for a seven-level inverter. As shown in Fig. 6.15 (a) and (b) the proposed topologies are obtained by adding two unidirectional power switches and one dc voltage source to the H-bridge inverter structure. In other words, the proposed inverters are comprised of six unidirectional power switches (S_a, S_b, S_{a1}, S_{a2}, S_{b1}, S_{b2}) and two dc voltage sources (V_{a1}, and V_{b1}). In this section, these topologies are called developed H-bridge. As shown in Fig. 6.15, the switches are realized by IGBTs, the simultaneous turn-on of S_{a1} and S_{a2} (or S_{b1} and S_{b2}) causes the voltage sources to short-circuit. Therefore, the simultaneous turn-on of the mentioned switches must be avoided. In addition, S_a and S_b should not turn on, simultaneously. The difference in the topologies illustrated in Fig. 6.15 (a) and (b) is in the connection of the dc voltage sources polarity, in (a) Both V_{a1} and V_{b1} are with positive polarity, and in (b) V_{a1} is with positive polarity and V_{b1} is with negative polarity.

Table 6.7 Shows the output voltages of the proposed inverters for different states of the switches. In this table, 1 and 0 indicate the ON – and OFF – states of the switches, respectively.

Modes	S_{a1}	S_{a2}	S_{b1}	S_{b2}	S_a	S_b	V_o (Fig.6.15(a))	V_o (Fig.6.15(b))
Mode 1	1	0	0	1	0	1	V_{a1}	V_{a1}
Mode 2	1	0	0	1	1	0	V_{b1}	$-V_{b1}$
Mode 3	1	0	1	0	0	1	$V_{a1} - V_{b1}$	$V_{a1} + V_{b1}$
Mode 4	1	0	1	0	1	0	0	0
	0	1	0	1	0	1		
Mode 5	0	1	1	0	1	0	$-V_{a1}$	$-V_{a1}$
Mode 6	0	1	1	0	0	1	$-V_{b1}$	$+V_{b1}$
Mode 7	0	1	0	1	1	0	$-(V_{a1} - V_{b1})$	$-(V_{a1} + V_{b1})$

The proposed seven-level inverter circuit is comprising of six unidirectional switches (S_a, S_b, S_{a1}, S_{a2}, S_{b1}, S_{b2}) and two dc voltage sources (V_{a1}, and V_{b1}). The seven - Level Cascaded Multilevel Inverter has seven modes of operation for the topology given in Fig. 6.15 (a) and they are as follows:

Mode 1: In this mode, the output voltage is (V_{a1}) to generate this output voltage level, switches S_{a1}, S_{b2} and S_b are turned on. Switches S_{a2}, S_{b1} and S_a are turned off. The conduction takes place from dc source V_{a1}, switch S_{a1}, load terminals '*a*' to '*b*' switch S_{b2}, central bottom switch S_b and back to dc source V_{a1}. The load current I_o is positive.

Mode 2: In this mode, the output voltage is (V_{b1}) to generate this output voltage level, switches S_{a1}, S_{b2} and S_a are turned on. Switches S_{a2}, S_{b1} and S_b are turned off. The conduction takes place from dc source V_{b1}, switch S_{b2}, load terminals '*b*' to '*a*' switch S_{a1}, central top switch S_a and back to dc source V_{b1}. The load current I_o is negative.

Mode 3: In this mode, the output voltage is ($V_{a1} - V_{b1}$) to generate this output voltage level, switches S_{a1}, S_{b1} and S_b are turned on. Switches S_{a2}, S_{b2} and S_a are turned off. The conduction takes place from dc source V_{a1}, switch S_{a1}, load terminals '*a*' to '*b*' switch S_{b1}, dc source V_{b1} and central bottom switch S_b. The load current I_o is positive.

Mode 4: In this mode, the output voltage is ($V_o = V_{ab} = 0$) to generate this output voltage level, switches S_{a2}, S_{b2} and S_b are turned on. Switches S_{a1}, S_{b1} and S_a are turned off. The conduction takes place from switch S_{b2}, central bottom switch S_b, switch S_{a2}, load terminals '*a*' to '*b*'. The load current I_o is zero.

Mode 5: In this mode, the output voltage is ($-V_{a1}$) to generate this output voltage level, switches S_{a2}, S_{b1} and S_a are turned on. Switches S_{a1}, S_{b2} and S_b are turned off. The conduction takes place from dc source V_{a1}, central top switch S_a, switch S_{b1}, load terminals '*b*' to '*a*' switch S_{a2} and back to dc source V_{a1}. The load current I_o is negative.

Mode 6: In this mode, the output voltage is ($-V_{b1}$) to generate this output voltage level, switches S_{a2}, S_{b1} and S_b are turned on. Switches S_{a1}, S_{b2} and S_a are turned off. The conduction takes place from dc source V_{b1}, switch S_{b1}, load terminals '*b*' to '*a*', switch S_{a2}, central bottom switch S_b and back to dc source V_{b1}. The load current I_o is negative.

Mode 7: In this mode, the output voltage is $-(V_{a1} - V_{b1})$ to generate this output voltage level, switches S_{a2}, S_{b2} and S_a are turned on. Switches S_{a1}, S_{b1} and S_b are turned off. The conduction takes place from dc source V_{a1}, central top switch Sa, dc source V_{b1}, switch S_{b2}, load terminals '*b*' to '*a*' switch S_{a2} and back to dc source V_{a1}. The load current I_o is negative.

As it is obvious from Table 6.8, if the values of the dc voltage sources are equal, the number of voltage levels decreases to three. Therefore, the values of dc voltage sources should be different to generate more voltage levels without increasing the number of switches and dc voltage sources. Considering Table 6.8, to generate all voltage levels (odd and even) in the proposed topology shown in Fig. 6.15 (a), the magnitudes of V_{a1} and V_{b1} should be considered 3pu and 1pu, respectively. Similarly, for the topology shown in Fig. 6.15 (a), the magnitudes of V_{a1} and V_{b1} should be considered 2pu and 1pu, respectively. Considering the aforementioned explanations, the total cost of the proposed topology in Fig. 6.15 (b) is low because dc voltage sources with low magnitudes are needed. By developing the seven-level inverter shown in Fig. 6.15 (b), the 31-level inverter shown in Fig. 6.16 can be proposed by adding two more switches in series and one dc source per phase leg. This topology consists of ten unidirectional power switches and four dc voltage sources.

According to Fig. 6.16, if the power switches of (S_{a1}, S_{a2}), (S_{a3}, S_{a4}), (S_{b1}, S_{b2}), and (S_{b3}, S_{b4}) turn on simultaneously, the dc voltage sources of V_{a1}, V_{a2}, V_{b1}, and V_{b2} will be short-circuited, respectively. Therefore, the simultaneous turn-on of these switches should be avoided. In addition, S_a and S_b should not turn on simultaneously. It is important to note that the 31-level topology can be provided through the structure presented in Fig. 6.16 where the only difference will be in the polarity of the applied dc voltage sources, both V_{a1} and V_{a2} are with positive polarity and both V_{b1} and V_{b2} are with negative polarity. By developing the proposed 31-level inverter, a 127 -level inverter can be proposed by extending two more switches in series and one additional dc source per phase leg as shown in Fig. 6.17. This topology consists of 14 unidirectional power switches and 6 dc voltage sources where V_{a1}, V_{a2}, V_{a3} are with positive polarity and V_{b1}, V_{b2}, V_{b3} are with negative polarity. Similarly, by developing the proposed basic topologies, a generalized topology, as shown in Fig. 6.18, can be proposed.

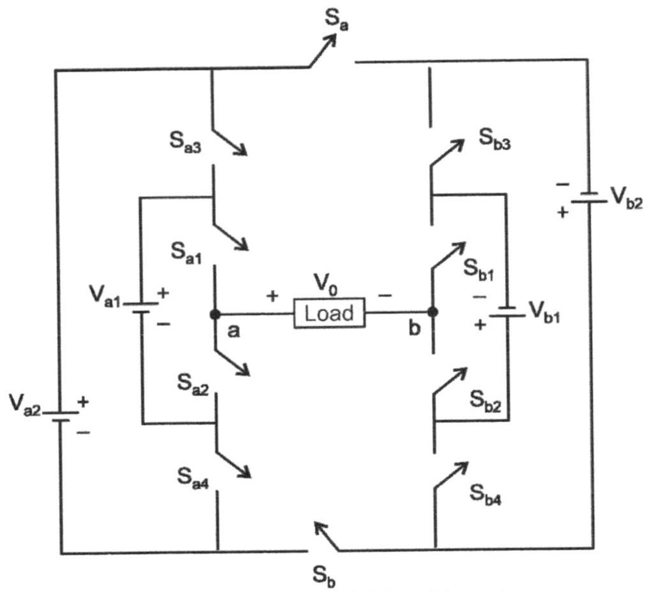

Fig. 6.16 Proposed 31-level Inverter.

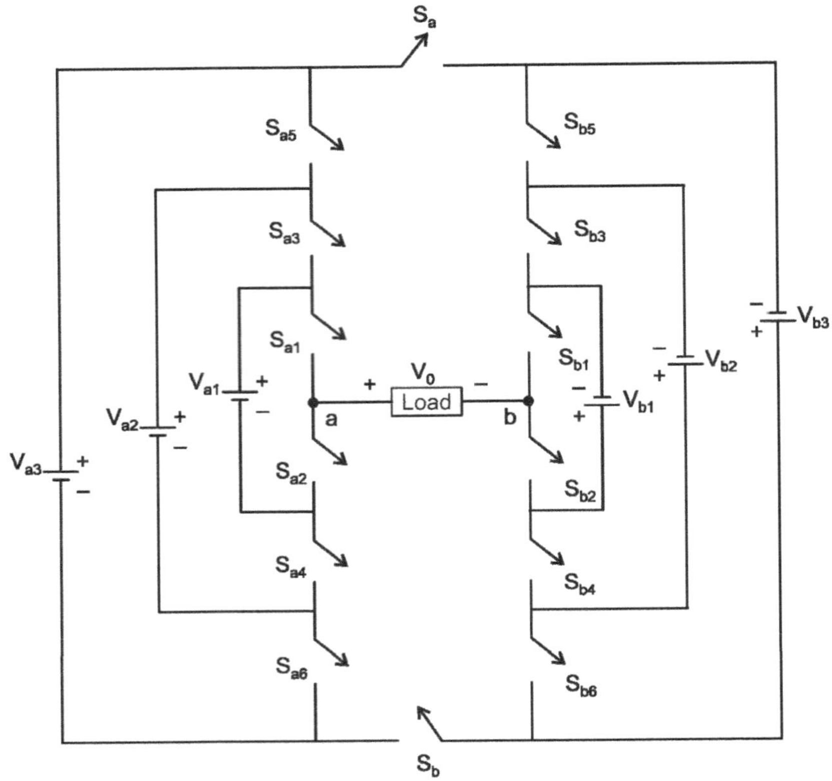

Fig. 6.17 Proposed 127-level Inverter.

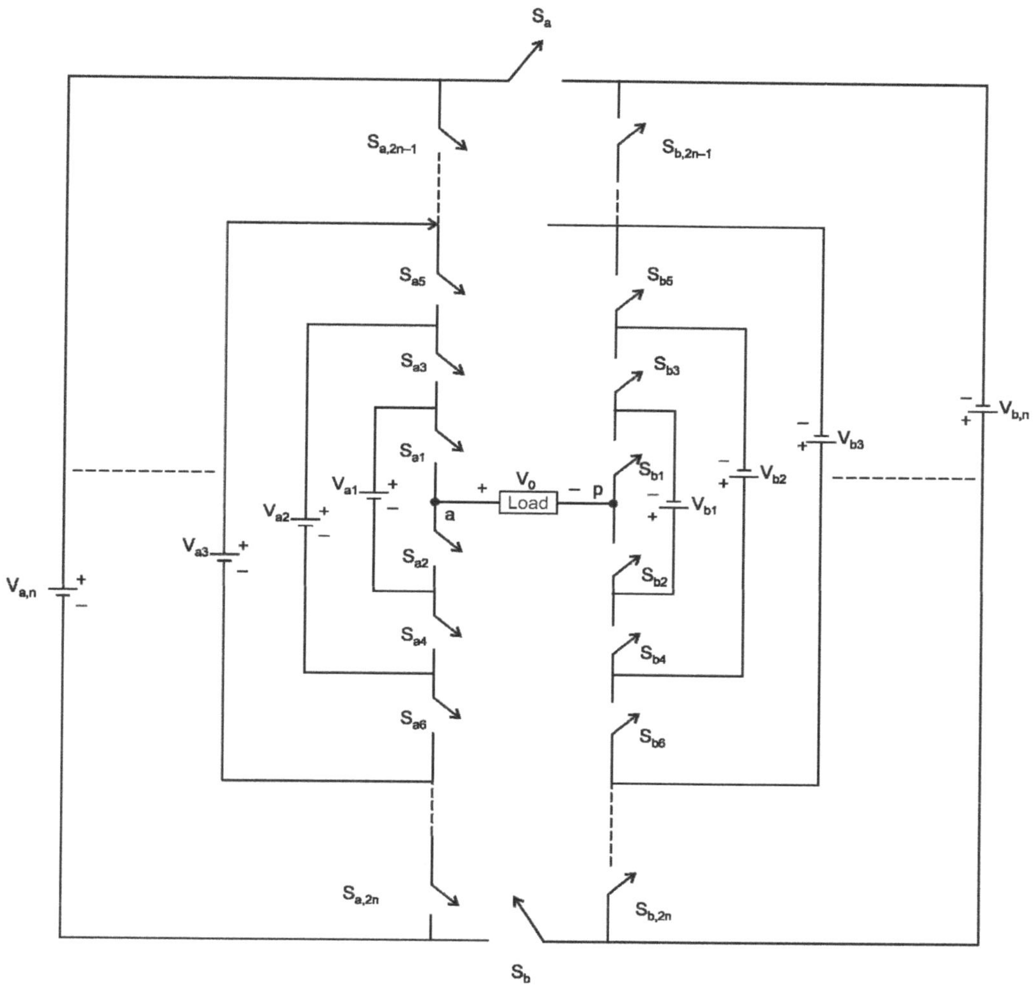

Fig. 6.18 Proposed Generalized Topology for n-level Inverter.

The generalized topology consists of 2n dc voltage sources (n is the number of the dc voltage sources on each leg) and 4n + 2 unidirectional power switches. In the proposed generalized topology, the number of output voltage levels (N_{step}), number of switches (N_{switch}), number of dc voltage sources (N_{source}), and the maximum magnitude of the generated voltage ($V_{o,max}$) are

$$N_{step} = 2^{2n+1} - 1 \quad .....(6.2)$$

$$N_{switch} = 4n + 2 \quad .....(6.3)$$

$$N_{source} = 2n \quad .....(6.4)$$

$$V_{o,max} = V_{a,n} + V_{b,n} \quad .....(6.5)$$

The other important parameters of the total cost of a multilevel inverter for evaluation are the variety of the values of dc voltage sources and the value of the blocking voltage of the switches. As the variety of dc voltage sources and the value of the blocking voltage of the switches are low, the inverter's total cost decreases [20]. The number of varieties of the values of dc voltage sources $N_{variety}$ is given by

$$N_{variety} = 2n \quad .....(6.6)$$

The following pattern is utilized to calculate the maximum magnitude of the blocking voltage of the power switches. As shown in Fig. 6.15(b), the blocking voltage of switches S_{b1}, and S_{b2} are calculated as follows:

$$V_{Sb1} = V_{Sb2} = V_{b1} \quad .....(6.7)$$

Where V_{Sb1} and V_{Sb2} indicate the maximum blocking voltages of V_{Sa1} and V_{Sa2}, respectively. The blocking voltage of V_{Sa1} and V_{Sa2} are as follows:

$$V_{Sa1} = V_{Sa2} = V_{a1} \quad .....(6.8)$$

Where V_{Sb1} and V_{Sb2} indicate the maximum blocking voltages of S_{a1} and S_{a2}, respectively. Therefore, the maximum blocking voltage of all switches in the proposed seven-level inverter ($V_{block,1}$) is calculated as follows:

$$V_{block,1} = V_{Sb1} + V_{Sb2} + V_{Sa1} + V_{Sa2} + V_{sa} + V_{sb} = 4(V_{b1} + V_{a1}) \quad .....(6.9)$$

Considering Fig. 6.16, the maximum blocking voltage of the switches of 31-level inverter is as follows:

$$V_{Sb1} = V_{Sb2} = V_{b1} \quad .....(6.10)$$

$$V_{Sb3} = V_{Sb4} = V_{b2} - V_{b1} \quad .....(6.11)$$

$$V_{Sa1} = V_{Sa2} = V_{a1} \quad .....(6.12)$$

$$V_{Sa3} = V_{Sa4} = V_{a2} - V_{a1} \quad .....(6.13)$$

$$V_{Sa} = V_{Sb} = V_{b2} + V_{a2} \quad .....(6.14)$$

Therefore, the maximum blocking voltage of all switches of the proposed 31-level inverter ($V_{block,2}$) is as follows:

$$V_{block,2} = V_{Sb1} + V_{Sb2} + V_{Sb3} + V_{Sb4} + V_{Sa1} + V_{Sa2} + V_{Sa3} + V_{Sa4} + V_{Sa} + V_{Sb} = 4\,(V_{b2} + V_{a2}) \quad .....(6.15)$$

Similarly, the maximum blocking voltage of all switches of the 127-level inverter is calculated as follows:

$$V_{block,3} = 4\,(V_{b3} + V_{a3}) \quad .....(6.16)$$

Finally, the maximum blocking voltage of all the switches of the generalized topology ($V_{block,n}$)is calculated as follows:

$$V_{block,n} = 4(V_{b,n} + V_{a,n}) \quad .....(6.17)$$

However, the proposed topology has a higher number of varieties of dc voltage sources in comparison with others.

A 31-level inverter based on the topology shown in Fig. 6.16 has been used for the simulation. Table 6.8 shows the switching states of the 31-level inverter. The simulation is done by using PSCAD software. According to 31-level inverter, the maximum output voltage of this inverter will be 225 V. The fundamental frequency switching control method has been used. In this method, the sinusoidal reference voltage is compared with the available dc voltage levels and the level that is nearest to the reference voltage is chosen. The main advantage of this control method is related to its low switching frequency, which leads to reduction of switching losses.

Table. 6.8 Switching pattern and output voltage for 31-level Inverter.

Sl.no.	S_{a1}	S_{a2}	S_{a3}	S_{a4}	S_{b1}	S_{b2}	S_{b3}	S_{b4}	S_a	S_b	V_o
1.	1	0	1	0	1	0	1	0	0	1	$V_{a2} + V_{b2}$
2.	0	1	1	0	1	0	1	0	0	1	$V_{a2} + V_{b2} - V_{a1}$
3.	1	0	1	0	0	1	1	0	0	1	$V_{b2} + V_{a2} - V_{b1}$
4.	0	1	1	0	0	1	1	0	0	1	$V_{a2} + V_{b2} - V_{a1} - V_{b1}$
5.	1	0	0	1	1	0	1	0	0	1	$V_{a1} + V_{b2}$
6.	0	1	0	1	1	0	1	0	0	1	V_{b2}
7.	1	0	0	1	0	1	1	0	0	1	$V_{a1} - V_{b1} + V_{b2}$
8.	0	1	0	1	0	1	1	0	0	1	$V_{b2} - V_{b1}$
9.	1	0	1	0	1	0	0	1	0	1	$V_{a2} + V_{b1}$
10.	0	1	1	0	1	0	0	1	0	1	$V_{a2} + V_{b1} - V_{a1}$
11.	1	0	1	0	0	1	0	1	0	1	V_{a2}
12.	0	1	1	0	0	1	0	1	0	1	$V_{a2} - V_{a1}$
13.	1	0	0	1	1	0	0	1	0	1	$V_{a1} + V_{b1}$
14.	0	1	0	1	1	0	0	1	0	1	V_{b1}
15.	1	0	0	1	0	1	0	1	0	1	V_{a1}
16.	1	0	1	0	1	0	1	0	1	0	0
17.	0	1	1	0	1	0	1	0	1	0	$-V_{a1}$
18.	1	0	1	0	0	1	1	0	1	0	$-V_{b1}$
19.	0	1	1	0	0	1	1	0	1	0	$-(V_{a1} + V_{b1})$
20.	1	0	0	1	1	0	1	0	1	0	$-(V_{a2} - V_{a1})$
21.	0	1	0	1	1	0	1	0	1	0	$-V_{a2}$

Contd...

Sl.no.	S_{a1}	S_{a2}	S_{a3}	S_{a4}	S_{b1}	S_{b2}	S_{b3}	S_{b4}	S_a	S_b	V_o
22.	1	0	0	1	0	1	1	0	1	0	$-(V_{a2} + V_{b1} - V_{a1})$
23.	0	1	0	1	0	1	1	0	1	0	$-(V_{a2} + V_{b1})$
24.	1	0	1	0	1	0	0	1	1	0	$-(V_{b2} - V_{b1})$
25.	0	1	1	0	1	0	0	1	1	0	$-(V_{a1} - V_{b1} + V_{b2})$
26.	1	0	1	0	0	1	0	1	1	0	$-V_{b2}$
27.	0	1	1	0	0	1	0	1	1	0	$-(V_{a1} + V_{b2})$
28.	1	0	0	1	1	0	0	1	1	0	$-(V_{a2} + V_{b2} - V_{a1} - V_{b1})$
29.	0	1	0	1	1	0	0	1	1	0	$-(V_{b2} + V_{a2} - V_{a1})$
30.	1	0	0	1	0	1	0	1	1	0	$-(V_{a2} + V_{b2} - V_{a1})$
31.	0	1	0	1	0	1	0	1	1	0	$-(V_{a2} + V_{b2})$

Modes of Operation:

The proposed 31-level inverter circuit is comprising of ten unidirectional switches (S_a, S_b, S_{a1}, S_{a2}, S_{a3}, S_{a4}, S_{b1}, S_{b2}, S_{b3}, S_{b4}) and four dc voltage sources (V_{a1}, V_{a2}, V_{b1}, V_{b2}). The 31- Level Cascaded Multilevel Inverter has 31 modes of operation and they are as follows: First 15 modes are with positive output voltages load positive load currents, 16th mode is for zero output voltage and zero load current, and last 15 modes (mode 17th to mode 31st) are with negative output voltages and negative load currents.

Mode 1: In this mode, the output voltage is ($V_{a2} + V_{b2}$) to generate this output voltage level, switches S_{a1}, S_{b1}, S_{a3}, S_{b3} and S_b are turned on. Switches S_{a2}, S_{b2}, S_{a4}, S_{b4} and S_a are turned off. The conduction takes place from dc source V_{a2}, switches S_{a3}, S_{a1}, load terminals '*a*' to '*b*' switches S_{b1}, S_{b3}, dc source V_{b2} and central bottom switch S_b. The load current I_o is positive.

Mode 2: To obtain the output voltage level of ($V_{a2} + V_{b2} - V_{a1}$), the switches S_{a1}, S_{b2}, S_{a3}, S_{b3} and S_b are turned on. Switches S_{a2}, S_{b1}, S_{a4}, S_{b4} and S_a are turned off. The conduction takes place from dc source V_{a2}, switches S_{a3}, S_{a1}, load terminals '*a*' to '*b*' switches S_{b2}, dc source voltage V_{b1}, S_{b3}, dc source V_{b2} and central bottom switch S_b. The load current I_o is positive.

Mode 3: In this mode, the output voltage is ($V_{b2} + V_{a2} - V_{b1}$) to generate this output voltage level, switches S_{a1}, S_{b2}, S_{a3}, S_{b3} and S_b are turned on. Switches S_{a2}, S_{b1}, S_{a4}, S_{b4} and S_a are turned off. The conduction takes place from dc source V_{a2}, switches S_{a3}, S_{a1}, load terminals '*a*' to '*b*' switches S_{b2}, dc source voltage V_{b1}, S_{b3}, dc source V_{b2} and central bottom switch S_b. The load current I_o is positive.

Mode 4: To obtain the output voltage level of ($V_{a2} + V_{b2} - V_{a1} - V_{b1}$) switches S_{a2}, S_{b2}, S_{a3}, S_{b3} and S_b are turned on. Switches S_{a1}, S_{b1}, S_{a4}, S_{b4} and S_a are turned off. The conduction takes place from dc source V_{a2}, switch S_{a3}, dc source V_{a1}, switch S_{a2}, load terminals '*a*' to '*b*' switches S_{b2}, dc source voltage V_{b1}, S_{b3}, dc source V_{b2} and central bottom switch S_b. The load current I_o is positive.

Mode 5: In this mode, the output voltage is (V_{a1} + V_{b2}) to generate this output voltage level, switches S_{a1}, S_{b1}, S_{a4}, S_{b3} and S_b are turned on. Switches S_{a2}, S_{b2}, S_{a3}, S_{b4} and S_a are turned off. The conduction takes place from dc source V_{a1}, switch S_{a1}, load terminals '*a*' to '*b*' switches S_{b1}, S_{b3}, dc source V_{b2}, central bottom switch S_b and switch S_{a4}. The load current I_o is positive.

Mode 6: To obtain the output voltage level of (V_{b2}), the switches S_{b1}, S_{a2}, S_{b3}, S_{a4} and S_b are turned on. Switches S_{b2}, S_{a1}, S_{b4}, S_{a3} and S_a are turned off. The conduction takes place from dc source V_{b2}, central bottom switch S_b, switches S_{a4}, S_{a2}, load terminals '*a*' to '*b*' switches S_{a1}, S_{a3}, dc source V_{b2}. The load current I_o is positive.

Mode 7: In this mode, the output voltage is (V_{a1} - V_{b1} + V_{b2}) to generate this output voltage level, switches S_{a1}, S_{b2}, S_{b3}, S_{a4} and S_b are turned on. Switches S_{a2}, S_{b1}, S_{b4}, S_{a3} and S_a are turned off. The conduction takes place from dc source V_{a1}, switch S_{a1}, load terminals '*a*' to '*b*' switch S_{b2}, dc source voltage V_{b1}, switch S_{b3}, dc source V_{b2}, central bottom switch S_b and switch S_{a4}. The load current I_o is positive.

Mode 8: In this mode, the output voltage is (V_{b2} - V_{b1}) to generate this output voltage level, switches S_{b2}, S_{a2}, S_{b3}, S_{a4} and S_b are turned on. Switches S_{b1}, S_{a1}, S_{b4}, S_{a3} and S_a are turned off. The conduction takes place from dc source V_{b2}, central bottom switch S_b, switches S_{a4}, S_{a2}, load terminals '*a*' to '*b*' switch S_{b2}, dc source voltage V_{b1}, switch S_{b3}, and back to dc source V_{b2}. The load current I_o is positive.

Mode 9: In this mode, to generate output voltage level (V_{a2} + V_{b1}), switches S_{a1}, S_{a3}, S_{b1}, S_{b4}, and Sb are turned on. Switches S_{a2}, S_{a4}, S_{b2}, S_{b3}, and S_a are turned off. The conduction takes place from dc source V_{a2}, switches S_{a3}, S_{a1}, load terminals '*a*' to '*b*' switches S_{b1}, dc source V_{b1}, switch S_{a4}, and central bottom switch S_b. The load current I_o is positive.

Mode 10: In this mode, to generate output voltage level (V_{a2} + V_{b1} - V_{a1}), switches S_{a2}, S_{a3}, S_{b1}, S_{b4}, and S_b are turned on. Switches S_{a1}, S_{a4}, S_{b2}, S_{b3}, and S_a are turned off. The conduction takes place from dc source V_{a2}, switch S_{a3}, dc source V_{a1}, switch S_{a2}, load terminals '*a*' to '*b*' switch S_{b1}, dc source voltage V_{b1}, switch S_{b4}, central bottom switch S_b and back to dc source V_{a2}.

Mode 11: In this mode, the output voltage (V_{a2}) to generate this output voltage level, switches S_{a1}, S_{a3}, S_{b2}, S_{b4}, and S_b are turned on. Switches S_{a2}, S_{a4}, S_{b1}, S_{b3}, and S_a are turned off. The conduction takes place from dc source V_{a2}, switches S_{a3}, S_{a1}, load terminals '*a*' to '*b*' switches S_{b2}, S_{b4}, central bottom switch S_b and back to dc source V_{a2}. The load current I_o is positive.

Mode 12: In this mode, the output voltage is (V_{a2} - V_{a1}) to generate this output voltage level, switches S_{a2}, S_{a3}, S_{b2}, S_{b4} and S_b are turned on. Switches S_{a1}, S_{a4}, S_{b1}, S_{b3} and S_a are turned off. The conduction takes place from dc source V_{a2}, switches S_{a3}, dc source V_{a1}, switch S_{a2}, load terminals '*a*' to '*b*' switches S_{b2}, S_{b4}, central bottom switch S_b and back to dc source V_{a2}. The load current I_o is positive.

Mode 13: In this mode, the output voltage is (V_{a1} + V_{b1}) to generate this output voltage level, switches S_{a1}, S_{a4}, S_{b1}, S_{b4} and S_b are turned on. Switches S_{a2}, S_{a3}, S_{b2}, S_{b3} and S_a are turned off. The conduction takes place from dc source V_{a1}, switch S_{a1}, load terminals '*a*' to '*b*', switch S_{b1}, dc source V_{b1}, switch S_{b4}, central bottom switch S_b, switch S_{a4} and back to dc source V_{a1}. The load current I_o is positive.

Mode 14: In this mode, the output voltage is (V_{b1}) to generate this output voltage level, switches S_{a2}, S_{a4}, S_{b1}, S_{b4} and S_b are turned on. Switches S_{a1}, S_{b2}, S_{a3}, S_{b3} and S_a are turned off. The conduction takes place from dc source V_{b1}, switch S_{b4}, central bottom switch S_b, switches S_{a4}, S_{a2}, load terminals '*a*' to '*b*' switch S_{b1} and back to dc source V_{b1}. The load current I_o is positive.

Mode 15: In this mode, the output voltage is (V_{a1}), to generate this output voltage level, switches S_{a1}, S_{b2}, S_{a4}, S_{b4} and S_b are turned on. Switches S_{a2}, S_{b1}, S_{a3}, S_{b3} and S_a are turned off. The conduction takes place from dc source V_{a1}, switch S_{a1}, load terminals '*a*' to '*b*', switches S_{b2}, S_{b4}, central bottom switch S_b, switch S_{a4} and back to dc source V_{a1}. The load current I_o is positive.

Mode 16: In this mode, the output voltage is Zero ($V_o = V_{ab} = V_{ba} = 0$) to generate this output voltage level, switches S_{a1}, S_{b1}, S_{a3}, S_{b3} and S_a are turned on. Switches S_{a2}, S_{b2}, S_{a4}, S_{b4} and S_b are turned off or vice versa. The conduction takes place from switches S_{a3}, S_{a1}, load terminals '*a*' to '*b*', switches S_{b1}, S_{b3} and central top switch S_a, or switches S_{a2}, S_{a4}, central bottom switch S_b, switches S_{b4}, S_{b2}, load terminals '*b*' to '*a*'. The load current is also Zero.

Mode 17: In this mode, the output voltage is ($-V_{a1}$) to generate this output voltage level, switches S_{a2}, S_{b1}, S_{a3}, S_{b3} and S_a are turned on. Switches S_{a2}, S_{b2}, S_{a4}, S_{b4} and S_b are turned off. The conduction takes place from dc source V_{a1}, switch S_{a3}, central top switch S_a, switches S_{b3}, S_{b1}, load terminals '*b*' to '*a*', switch S_{a2} and back to dc source V_{a1}. The load current I_o is negative.

Mode 18: In this mode, to generate output voltage level ($-V_{b1}$), switches S_{a1}, S_{b1}, S_{a3}, S_{b3} and S_a are turned on. Switches S_{a2}, S_{b2}, S_{a4}, S_{b4} and S_b are turned off. The conduction takes place from dc source V_{b1}, switch S_{b2}, load terminals '*b*' to '*a*', switches S_{a1}, S_{a3}, central top switch S_a, switch S_{b3} and back to dc source V_{b1}. The load current I_o is negative.

Mode 19: In this mode, $-(V_{a1} + V_{b1})$ to generate this output voltage level, switches S_{a2}, S_{b2}, S_{a3}, S_{b3} and S_a are turned on. Switches S_{a1}, S_{b1}, S_{a4}, S_{b4} and S_b are turned off. The conduction takes place from dc source V_{b1}, switch S_{b2}, load terminals '*b*' to '*a*', switch S_{a2}, dc source V_{a1}, switch S_{a3}, central top switch S_a, switch S_{b3}, back to dc source V_{b1}. The load current I_o is negative.

Mode 20: In this mode, the output voltage is $-(V_{a2} - V_{a1})$, to generate this output voltage level, switches S_{a1}, S_{b1}, S_{a4}, S_{b3} and S_a are turned on. Switches S_{a2}, S_{b2}, S_{a3}, S_{b4} and S_b are turned off. The conduction takes place from dc source V_{a2}, central top switch S_a, switches S_{b3}, S_{b1}, load terminals '*b*' to '*a*', Switch S_{a1}, dc source V_{a1}, switch S_{a4} and back to dc source V_{a2}. The load current I_o is negative.

Mode 21: In this mode, to generate the output voltage level ($-V_{a2}$), switches S_{a2}, S_{b1}, S_{a4}, S_{b3} and S_a are turned on. Switches S_{a1}, S_{b2}, S_{a3}, S_{b4} and S_b are turned off. The conduction takes place from dc source V_{a2}, central top switch S_a, switches S_{b3}, S_{b1}, load terminals '*b*' to '*a*', switches S_{a2}, S_{a4} and back to dc source V_{a2}. The load current I_o is negative.

Mode 22: In this mode, the output voltage is $-(V_{a2} + V_{b1} - V_{a1})$ to generate this output voltage level, switches S_{a1}, S_{b2}, S_{a4}, S_{b3} and S_a are turned on. Switches S_{a2}, S_{b1}, S_{a3}, S_{b4} and S_b are turned off. The conduction takes place from dc source V_{b1}, switch S_{b2}, load terminals '*b*' to '*a*', switch S_{a1}, dc source V_{a1}, switch S_{a4}, dc source V_{a2}, central top switch S_a, switch S_{b3} and back to dc source V_{b1}. The load current I_o is negative.

Mode 23: In this mode, the output voltage is $-(V_{a2} + V_{b1})$ to generate this output voltage level, switches S_{a2}, S_{b2}, S_{a4}, S_{b3} and S_a are turned on. Switches S_{a1}, S_{b1}, S_{a3}, S_{b4} and S_b are turned off. The conduction takes place from dc source V_{b1}, switch S_{b2}, load terminals '*a*' to '*b*', switches S_{a2}, S_{a4}, dc source V_{a2}, central top switch S_a, switch S_{b3} and back to dc source V_{b1}. The load current I_o is negative.

Mode 24: In this mode, the output voltage is $-(V_{b2} - V_{b1})$ to generate this output voltage level, switches S_{a1}, S_{b1}, S_{a3}, S_{b4} and S_a are turned on. Switches S_{a2}, S_{b2}, S_{a4}, S_{b3} and S_b are turned off. The conduction takes place from dc source V_{b2}, switch S_{b4}, dc source V_{b1}, switch S_{b1}, load terminals '*b*' to '*a*', switches S_{a1}, S_{a3}, central top switch S_a and back to dc source V_{b2}. The load current I_o is negative.

Mode 25: In this mode, the output voltage is $-(V_{a1} - V_{b1} + V_{b2})$ to generate this output voltage level, switches S_{a2}, S_{b1}, S_{a3}, S_{b4} and S_a are turned on. Switches S_{a1}, S_{b2}, S_{a4}, S_{b3} and S_b are turned off. The conduction takes place from dc source V_{b2}, switch S_{b4}, dc source V_{b1}, switch S_{b1}, load terminals '*b*' to '*a*', switch S_{a2}, dc source V_{a1}, switch S_{a3}, central top switch S_a and back to dc source V_{b2}. The load current I_o is negative.

Mode 26: In this mode, the output voltage is ($-V_{b2}$) to generate this output voltage level, switches S_{b2}, S_{a1}, S_{b4}, S_{a3} and S_a are turned on. Switches S_{b1}, S_{a2}, S_{b3}, S_{a4} and S_b are turned off. The conduction takes place from dc source V_{b2}, switches S_{b4}, S_{b2}, load terminals '*b*' to '*a*', switches S_{a1}, S_{a3}, central top switch Sa and back to dc source V_{b2}. The load current I_o is negative.

Mode 27: In this mode, to generate the output voltage level $-(V_{a1} + V_{b2})$, switches S_{a2}, S_{b2}, S_{b4}, S_{a3} and S_a are turned on. Switches S_{a1}, S_{b1}, S_{b3}, S_{a4} and S_b are turned off. The conduction takes place from dc source V_{b2}, switches S_{b4}, S_{b2}, load terminals '*b*' to '*a*', switch S_{a2}, dc source V_{a1}, Switch S_{a3}, central top switch S_a and back to dc source V_{b2}. The load current I_o is negative.

Mode 28: In this mode, to generate output voltage $[-(V_{a2} + V_{b2} - V_{a1} - V_{b1})]$, switches S_{a1}, S_{b1}, S_{a4}, S_{b4} & S_a are turned on. S_{a2}, S_{b2}, S_{a3}, S_{b3} and S_b are turned off. The conduction takes place from dc source V_{b2}, switch S_{b4}, dc source V_{b1}, switch S_{b1}, load terminals '*b*' to '*a*', switch S_{a1}, dc source V_{a1}, switch S_{a4}, dc source V_{a2}, central top switch S_a and back to dc source V_{b2}. The load current I_o is negative.

Mode 29: In this mode, the output voltage is $[-(V_{b2} + V_{a2} - V_{b1})]$ to generate this output voltage level, switches S_{b1}, S_{a2}, S_{a4}, S_{b4} and S_a are turned on. Switches S_{b2}, S_{a1}, S_{b3}, S_{a3} and S_b are turned off. The conduction takes place from dc source V_{b2}, switch S_{b4}, dc source V_{b1}, switch S_{b1}, load terminals '*b*' to '*a*', switches S_{a2}, S_{a4}, dc source V_{a2}, central top switch S_a and back to dc source V_{b2}. The load current I_o is negative.

Mode 30: In this mode, the output voltage is $-(V_{a2} + V_{b2} - V_{a1})$ to generate this output voltage level, switches S_{a1}, S_{b2}, S_{a4}, S_{b4} and S_a are turned on. Switches S_{a2}, S_{b1}, S_{a3}, S_{b3} and S_b are turned off. The conduction takes place from dc source V_{b2}, switches S_{b4}, S_{b2}, load terminals '*b*' to '*a*', switch S_{a1}, dc source V_{a1}, switch S_{a4}, dc source V_{a2}, central top switch S_a and back to dc source V_{b2}. The load current I_o is negative.

Mode 31: In this mode, the output voltage is $-(V_{a2} + V_{b2})$ to generate this output voltage level, switches S_{a2}, S_{b2}, S_{a4}, S_{b4} and S_a are turned on. Switches S_{a1}, S_{b1}, S_{a3}, S_{b3} and S_b are turned off. The conduction takes place from V_{b2}, switches S_{b4}, S_{b2}, load terminals '*b*' to '*a*', switches S_{a2}, S_{a4}, dc source V_{a2}, central top switch S_a back to dc source V_{b2}. The load current I_o is negative.

The simulated output voltage and current waveforms are shown in Fig. 6.18. As Fig. 6.18 (a) shows, the proposed topology is able to generate 31-levels (15 positive levels, 15 negative levels, and 1 zero level) with the maximum voltage of 225 V. Comparing the output voltage and current waveforms indicates that the output current waveform is more similar to the ideal sinusoidal waveform than the output voltage, because the R–L load acts as a lowpass filter. In addition, there is a phase difference between the output voltage and current waveforms, which is caused by the inductive feature of the load. The total harmonic distortions of the output voltage and current are equal to 0.94% and 0.19%, respectively.

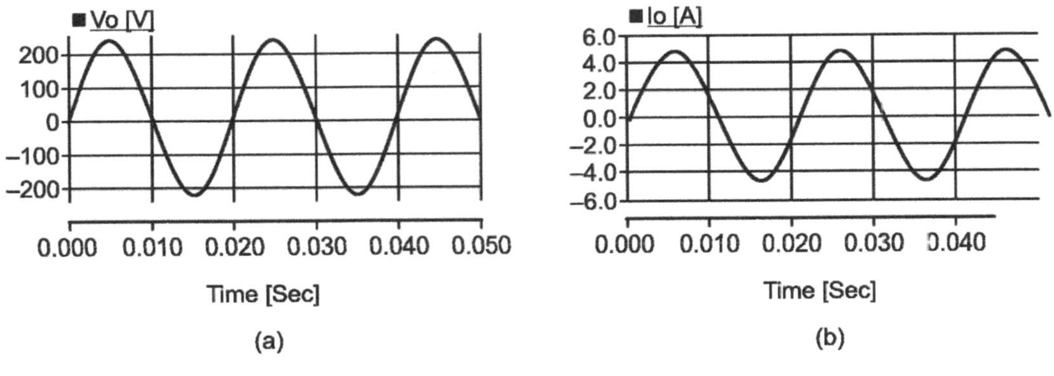

Fig. 6.19 Proposed 31-level inverter, (a) Output voltage waveform and (b) Output current waveform.

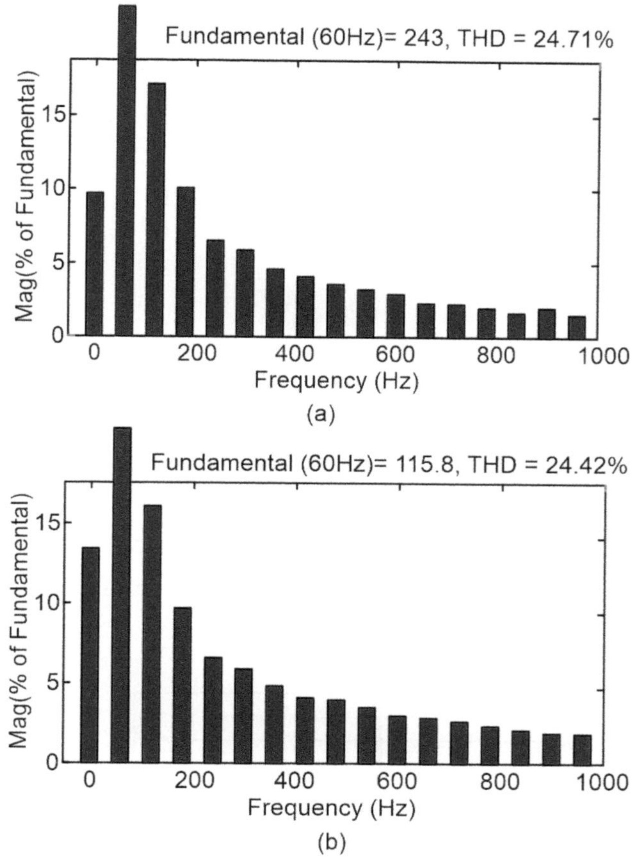

Fig. 6.20 Harmonic spectrum of (a) output voltage and (b) output current for proposed 31-level inverter.

Fig. 6.20 (a) and (b) shows the harmonic spectrum of the output voltage and current, respectively. The figure shows that the magnitudes of harmonics of both voltage and current waveforms are low. However, the harmonics of the current waveform are lower than the voltage waveform. It is important to note that in Fig. 6.20, in order to show the magnitudes of the fundamental and high-order harmonics, the scale of the vertical axis is considered nonlinear. As mentioned before, the power switches in the proposed topology are unidirectional from the voltage viewpoint. In order to prove this issue, the voltages on the switches of a single leg of the inverter (S_{a1}, S_{a2}, S_{a3}, S_{a4}, and S_a) are shown in Fig. 6.21.

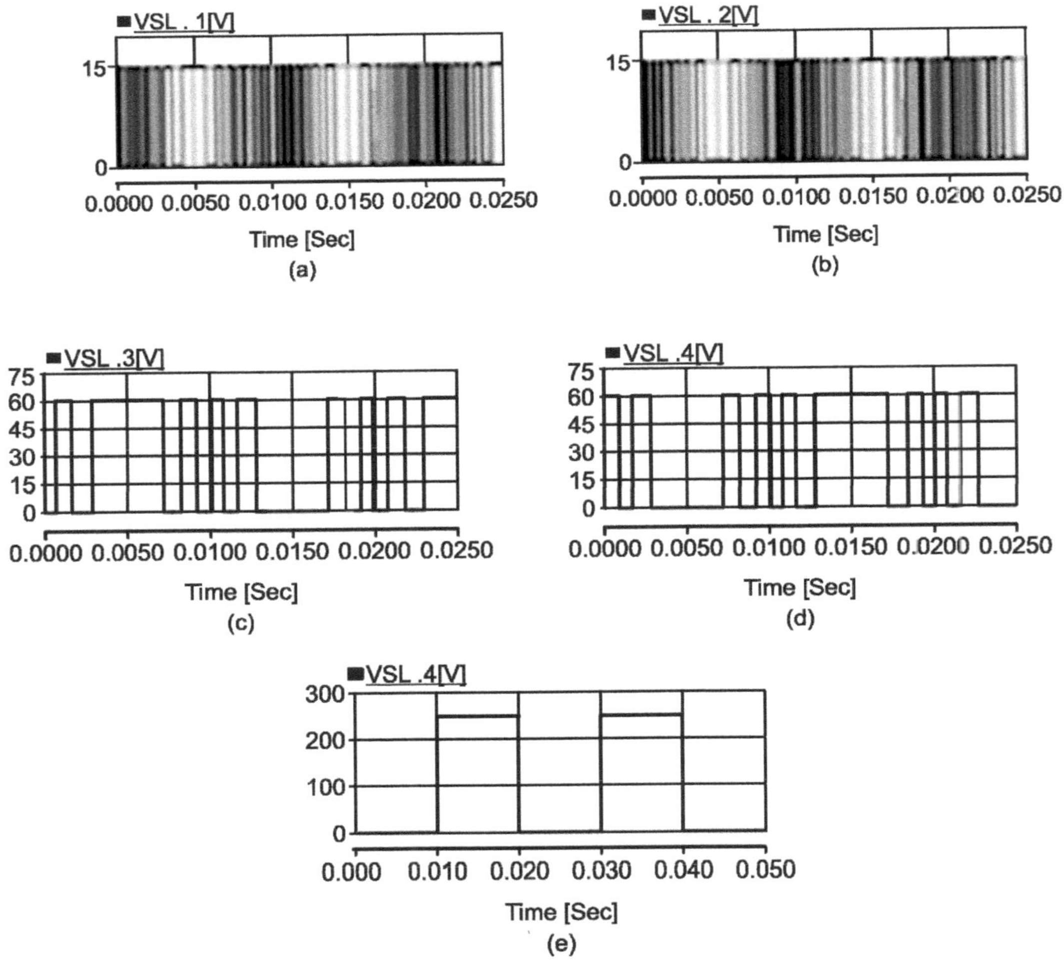

Fig. 6.21 Voltage of switches (a) S_{a1}, (b) S_{a2}, (c) S_{a3}, (d) S_{a4}, and (e) S_a.

As can be seen, the maximum blocking voltage by switches S_{a1}, S_{a2}, S_{a3}, S_{a4} and S_a are equal to 15, 15, 60, 60, and 225 V, respectively. Obviously, the voltage values are zero or equal to the positive ones, which is well in accordance to the unidirectional feature of the switches from the voltage view point. Considering the magnitude of the blocking voltage of the switches, the relations associated to the maximum voltage drop of the switches are well confirmed.

6.3.2 49-Level Inverter from Two-cascaded Seven-level Inverters

The 49-level inverter has the same number of voltage sources as that of 31-level inverter presented here. However, it has two more switches. So, in this case the increase in the number of voltage levels is by increasing the number of switches, without increasing the number of voltage sources. The number of switches for the 49-level inverter is 12 compared to 10 for 31-level inverters.

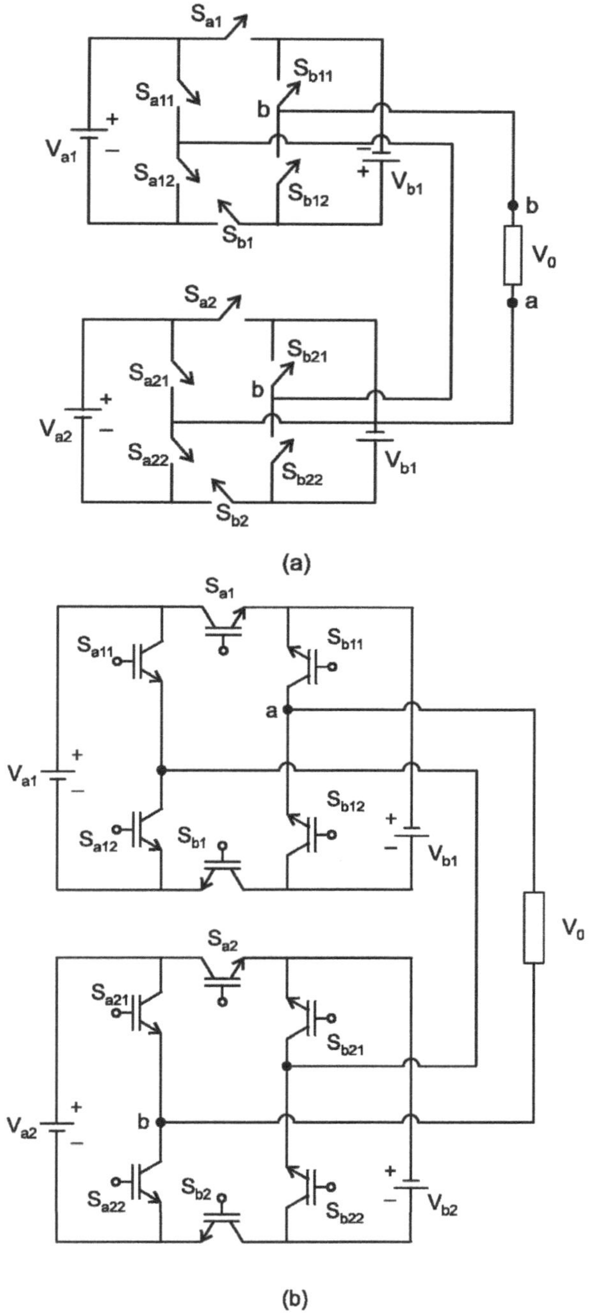

Fig. 6.22 49-level inverter developed from two-cascaded 7-level inverters with reduced number of switches (a) Basic Circuit and (b) Realization with IGBTs.

Two seven-level inverters are connected in cascade to form a 49-level inverter. Total twelve unidirectional switches are required, namely, S_{a11}, S_{a12}, S_{b11}, S_{b12}, S_{a1} and S_{b1} for first bridge and S_{a21}, S_{a22}, S_{b21}, S_{b22}, S_{a2} and S_{b2} for second bridge circuits. Four dc sources V_{a1}, V_{a2} are with positive polarity and V_{b1}, V_{b2} are with negative polarity. The load is connected between the bridges as shown in the Fig. 6.22 and its output is the sum of the two cascaded inverters. Let us take that V_{o1} is the output voltage of first seven level inverter and V_{o2} is the output voltage of second seven level inverter, then the output of the 49-level inverter will be $V_{o1} + V_{o2} = V_o$.

6.4 Advanced Hybrid Multilevel Inverter Classification

The most popular Multilevel converter classification is given in Fig. 6.23 including few combinations of various multilevel inverter topologies. Advanced Hybrid Multilevel Voltage Source Inverters classification is presented in Fig.6.24. The classification on Hybrid Multilevel Converter (HMC) are made based on combinations of any two variety of Multilevel Inverters as Hybrid Clamped Multilevel Converter, Hybrid Cascaded Multilevel Converter, Hybrid Flying Capacitor based Multilevel Converter, Hybrid Flying Capacitor based Multicell Converters and Hybrid Sub-module Cascaded Converters. The Voltage levels can be from three-level to m-level depending upon the requirement of the loads in three-phase, or higher-order phases. To implement these advanced Hybrid Multilevel Converters, various options of power semiconductor devices can be used based on the voltage, current, power, and thermal ratings, turn-on and turn-off times, switching frequency, etc. The judiciously select a suitable power semiconductor device among Thyristors, GTOs, IGBTs, IGCTs, and ETOs, MTOs, MCTs, etc.

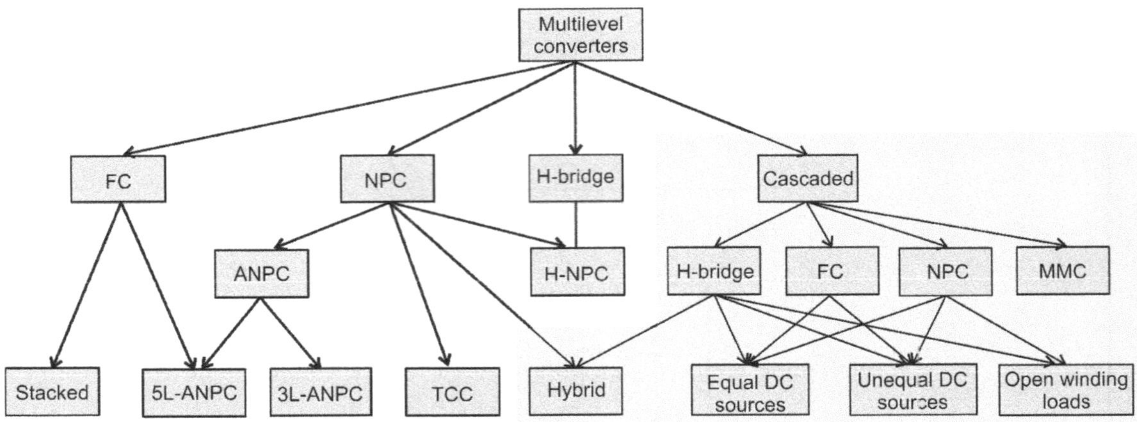

Fig. 6.23 Classification of most popular Multilevel Converters.

FC: Flying Capacitor converter, NPC: Neutral Point Clamped converter, ANPC: Active Neutral Point Clamped converter, TCC: Transistor Clamped Converter, H-NPC: H-Bridge-Neutral Point Clamped converter, MMC: Modular Multilevel Converters.

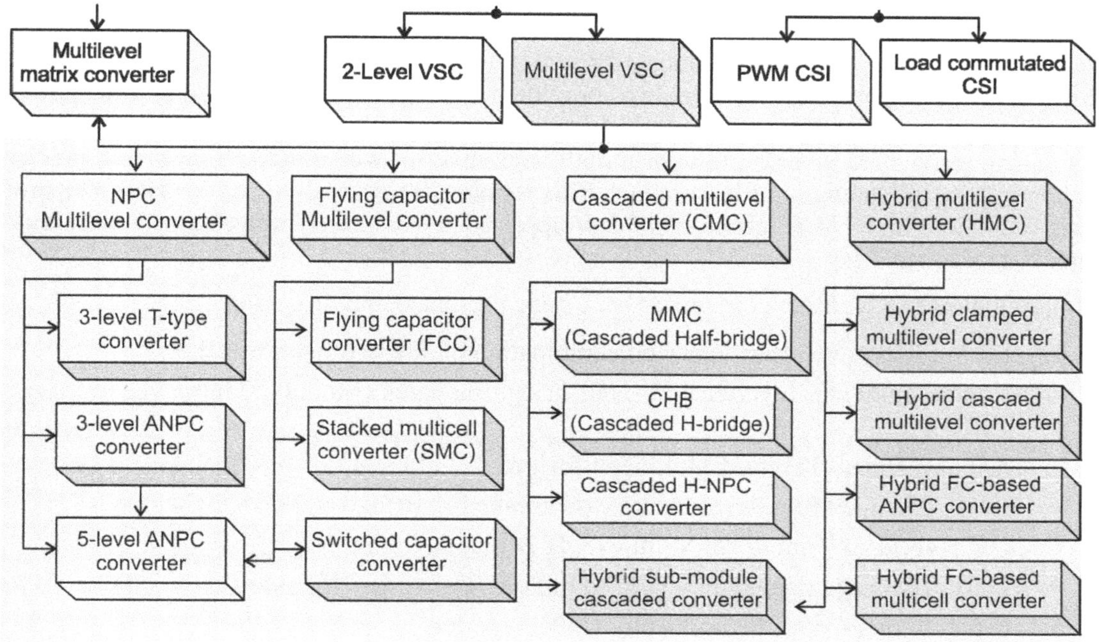

Fig. 6.24 Classification of Advanced Hybrid Multilevel Converters
Ref: ("Hybrid Multilevel Converters: Topologies, Evolutions and Verifications", Jianzhong Zhang, Shuai Xu, Zakuid Din and Xing Hu).

6.5 Chapter Summary

The different varieties of Hybrid-multilevel inverters are presented and described in this chapter. It can be concluded that every topology has its own advantages and disadvantages. Thus, as per the application, topology can be chosen and implemented. As asymmetrical multilevel inverters are alternative to minimize the harmonic distortion of the output voltages without increasing the number of power devices. The use of different DC voltage values naturally leads to hybrid multilevel topologies, which employ distinct types of semiconductors and modulation strategies, in an effort to optimize the power processing of the overall system. On the other hand, these features increase significantly the flexibility and complexity of hybrid multilevel inverter design. Distinct series connected cells hybrid multilevel inverter reduces the complexity for hybrid topology for distinct applications, thus minimizing the number of switching devices and reducing the circulating energy among the series-connected cells. Compared with an H-bridge cascaded multilevel inverter, the number of overall isolated DC sources is reduced in the single-phase hybrid symmetrical multilevel inverter while the number of semiconductors is kept the same. Thus, this concept appears as a useful and suitable solution for medium voltage applications where input-side isolation is required along with high efficiency and modularity. New hybrid asymmetrical multilevel H-bridge inverter reduces the harmonic components of output voltage. Hybrid multilevel inverter employing half-bridge modules and a three-phase inverter is able to provide better losses distribution among the power semiconductors and to limit the maximum device loss to a lower level when compared to a fully high frequency switched inverter. In hybrid multilevel inverter with single DC source topology capacitor voltage balancing is of importance. This topology is generally used for HEV (Hybrid Electric Vehicles) and EV (Electric Vehicles) applications, while hybrid clamped multilevel inverter reduces filter size.

The following Hybrid Multilevel Inverters have been presented and discussed:

- Hybrid Multilevel Inverters based on combination of any two types of MLIs.
  - Hybrid Multilevel inverter with Half-bridge modules (Modular Multilevel Inverters: MMI).
  - New Symmetrical Hybrid Multilevel Inverter.
  - Hybrid Clamped Five-level Inverter (Combined H-Bridge and Diode-Clamped).
  - New Hybrid Asymmetrical Multilevel H-Bridge Inverter.
  - Hybrid H-Bridge type diode-clamped multilevel inverter (H-bridge based NPC).
  - Hybrid Flying-Capacitor and Active NPC Inverter.
- Hybrid Inverter with reduced power semiconductor switches:

Two basic topologies have been discussed for multilevel inverters to generate seven voltage levels at the output. The basic topologies can be developed to any number of levels at the output where the 31-level, 49-level, 127-level, and generalized topologies are consequently presented. In addition, a new algorithm to determine the magnitude of the dc voltage sources has been explained. The generalized topology was compared with the different kinds of presented topologies in literature from different points of view. According to the comparison results, the proposed topology requires a lesser number of IGBTs, power diodes, driver circuits, and dc voltage sources. Moreover, the magnitude of the blocking voltage of the switches is lower than that of conventional topologies. However, the proposed topology has a higher number of varieties of dc voltage sources in comparison with the others. The performance of a 31-level inverter is obtained by using PSCAD tools and presented the results. Also, a 49-level inverter with 12 switches is proposed by using two seven-level bridges in series, which has the same number of voltage sources as that of 31-level inverter to show that the number of levels can be increased without increasing the number of voltage sources.

At the end the classification of advanced Hybrid multilevel inverters is presented for the benefit of interested readers.

6.6 Overall Summary

The multilevel inverters can immediately replace the existing systems that use traditional multi-pulse inverters without the need for transformers. For a Three-phase system, the relationship between the number of levels 'm', and the number of pulses, p, can be formulated by p = (m - 1) x 6.

The multilevel inverters are used to produce from three-level output voltages to higher-level output voltages. The multilevel inverters are classified into four categories namely, diode-clamped multilevel inverter, flying capacitor multilevel inverter, cascaded H-bridge multilevel inverter, and hybrid multilevel inverter.

The diode-clamped multilevel inverters are presented and described from basic construction of the circuit, three-level, five-level, seven-level and m-level inverters with detailed circuits and their operations in different modes, switching tables, and output load voltage waveforms.

The flying capacitor multilevel inverters are described with detailed circuit operations in various modes, switching tables, and output load voltages for three-level, five-level, seven-level and m-level inverters.

The cascaded H-bridge inverters are further classified as conventional cascaded H-bridge inverter and transistor-clamped cascaded H-bridge inverter, they are further categorized as symmetrical and asymmetrical values of dc source voltages. when dc source voltages of each H-bridge cell are equal, it is called symmetrical configuration and when dc source voltages of each H-bridges are unequal, then it is called asymmetrical configuration. The transistor-clamped cascaded H-bridge inverters have advantages of using reduced number of switching devices when compared to conventional cascaded H-bridge inverters.

The symmetrical configuration conventional cascaded H-bridge inverters are explained with detailed circuits with various modes of operation, switching tables, and output load voltages, for three-level with one-H-bridge, five-level with two-H-bridges and seven-level inverters with three-H-bridges in cascade. The asymmetrical configuration conventional cascaded H-bridge inverters are also presented and explained with detailed circuits with various modes of operation, switching tables, and output load voltages, for seven-level with two-H-bridges and thirteen-level inverters with three-H-bridges.

The transistor-clamped cascaded H-bridge inverters have advantages of using reduced number of switching devices and produces higher-level output load voltages. To produce five-level output load voltage, a transistor-clamped single H-bridge with equal dc sources voltages in symmetrical configuration is used and to produce nine-level output load voltage, transistor-clamped two H-bridges with equal dc source voltages in symmetrical configuration is used. They are presented and described with the detailed circuits and various modes of operations, switching tables, and output load voltage waveforms.

Similarly, asymmetrical configuration of transistor-clamped cascaded H-bridge inverters with unequal dc source voltages are explained by using one-H-Bridge it can produce seven-level output load voltage and whereas by using two-H bridges in cascade it can produce twenty-one-level output load voltage. These are explained in detailed by considering circuit diagrams for various modes of operation, switching tables, and output load voltages.

Three-phase circuit configurations are presented for all varieties of multilevel inverters for star-connected RL loads. The load current depends on the nature of the loads. The m-level configurations are also presented. The advantages, disadvantages and applications of all the above discussed multilevel inverters are also given.

Finally, introduced Hybrid-multilevel inverters which are popularly used in the literature are presented and described for the completeness of the multilevel inverters. Then most of existing Hybrid Multilevel inverters can be derived, and further new Hybrid Multilevel inverter topologies might be inspired that are applicable to specific emerging areas in the future years. In view of this, the classification of advanced Hybrid-multilevel inverters is introduced for carrying out future research activities.

Generalized Multilevel Cells: The existing multilevel inverters such as diode-clamped and capacitor-clamped multilevel inverters can be derived from the generalized inverter topology. Moreover, the generalized multilevel inverter topology can balance each voltage level by itself regardless of load characteristics. Therefore, the generalized multilevel inverter topology provides a true multilevel structure that can balance each DC voltage level automatically at any number of levels, regardless of active or reactive power conversion, and without any assistance from other circuits. Thus, in principle, it provides a complete multilevel topology that embraces the existing multilevel inverters.

The case studies and exercise problems are given in all varieties of multilevel inverters for practice by modeling and simulation.

Table 6.9 Comparison of power component requirements per phase leg among three multilevel inverters.

Inverter type	Diode-clamped	Flying-capacitor	Cascaded H-bridge
Main switching devices	(m-1) x 2	(m-1) x 2	(m-1) x 2
Main diodes	(m-1) x 2	(m-1) x 2	(m-1) x 2
Clamping diodes	(m-1) x (m-2)	0	0
DC bus capacitors	(m-1)	(m-1)	(m-1)/2
Balancing capacitors	0	(m-1) x (m-2)/2	0

m-level Inverters: Any m-level inverter six-sectors and m^3 switching states in space vector diagram, while each sector comprises $(m-1)^2$ triangles. There are $1 + S\Sigma^{m-1}{}_{i=1}\, i$, where S is total number of sectors and m shows number of levels. The switching states with equal line to line voltages are known as redundant states in space vector diagram.

The space vector diagrams for 2-level inverter has 6-triangles and for 3-level inverter has 24-triangles and for 5-level inverter has 96-triangles and for 7-level inverter has 216-triangles and for 9-level inverter has 384-triangles. A single space vector diagram is shown in Fig.6.25 for demonstrating 2-level, 3-level,5-level,7-level and 9-level inverters with their respective number of triangles. The innermost six sectors are shown for 2-level inverter.

Table 6.10 Number of Switching states and triangles in each sector for m-level inverters.

m -Level	Switching States m^3	Each Sector $(m-1)^2$ triangles
2 - level	2^3 = 8 switching states	$(2-1)^2$ = 1 triangle in each sector 1 triangle x 6 sectors = total 6 triangles
3 - level	3^3 = 27 switching states	$(3-1)^2$ = 4 triangles in each sector 4 triangles X 6 sectors = 24 triangles
5 - level	5^3 = 125 switching states	$(5-1)^2$ = 16 triangles in each sector 16 triangles X 6 sectors = 96 triangles
7 - level	7^3 = 343 switching states	$(7-1)^2$ = 36 triangles in each sector 36 triangles X 6 sectors = 216 triangles
9 - level	9^3 = 729 switching states	$(9-1)^2$ = 64 triangles in each sector 64 triangles X 6 sectors = 384 triangles

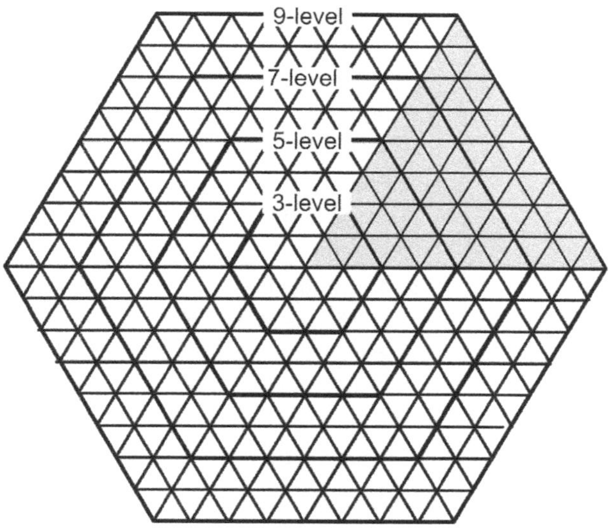

Fig. 6.25 The space vector diagram for 2-level (6-triangles), 3-level (24-triangles), 5-level (96 - triangles), 7-level (216-triangles) and 9-level (384-triangles) inverters.

Performance Summary of Multilevel Inverters: The performance summary of the three conventional multilevel inverters and two hybrid multilevel inverters are tabulated based on advantages, disadvantages, and preferred modulation techniques and presented in Table 6.11.

Table 6.11 Performance Summary of Multilevel Inverters.

Topology	Advantages	Disadvantages	Modulation Technique
Diode-clamped Inverter (NPC)	1. Simple structure 2. Equalized blocking voltage of power switches	1. Unequal distribution of power losses. 2. DC-link voltage balance limits the converter to 3-level topology.	1. CBPWM 2. SVPWM
Flying-Capacitor Inverter (FCI)	1. Modular structure 2. Owning a high number of redundant states.	1. Poor dynamic response of dc voltage balancing. 2. Large amounts of flying capacitors reduce the system reliability.	1. Phase-shifted carrier PWM. (achieves neutral balancing of flying capacitors.
Cascaded H-bridge Inverter (CHB)	1. Modular structure. 2. Owning fault tolerant capability.	1. Require independent dc sources. 2. Isolated Transformers increase the system volume.	1. Phase-shifted carrier PWM. (achieves equalization of power losses)
Multi-Modular Inverters (MMI)	1. Modular structure. 2. Easy to achieve fault tolerant operation	1. Low frequency voltage oscillation of floating capacitors. 2. Complex data acquisition and communication for each power cells.	1. Nearest level modulation. (achieves equalization of power losses and dc voltage balance)
Active Neutral Point Clamped Inverters (ANPC)	1. Simple structure. 2. Easily extendable to higher level by stacking flying capacitor cells.	1. Unequal usage of switches flying capacitors. 2. Requires series switches to handle high voltage.	1. Hybrid phase-shifted PWM (low frequency for high voltage switches and high frequency for flying capacitor cells)

Applications of various topologies of Multilevel inverters.

In general, multilevel inverters are used with a focus on Industrial medium voltage motor drives, Utility interface for renewable energy systems, FACTS and Traction drive systems. In particular, as given in the Table.6.12.

Table 6.12 Applications of four types of Multilevel inverters.

Sl.no.	Type of Multilevel Inverter	Applications
1.	Diode-clamed Inverter	Static Var Compensation, Variable speed motor drives, HVDC & AC Transmission Lines, HV System Interconnections.
2.	Flying-capacitor Inverter	Static Var Generation, Flexible AC Transmission Systems Induction motor control using Direct Torque Control.
3.	Cascaded H-bridge Inverter	AC motor drives, Active filters, Electric Vehicle drives, DC power source utilization, Power factor compensators, Back-to-back frequency link systems, Interfacing with renewable energy sources.
4.	Hybrid Multilevel Inverters	High performance AC motor drives Renewable Power Generation Systems Electric Aircrafts

Learning Outcomes

- Acquire knowledge about various topologies of hybrid multilevel inverters from basic three-level to higher order levels operation including with reduced switching devices: CO1 :: (PO1)
- Ability to analyze the operation of various topologies of hybrid multilevel inverters of different levels of output voltages: CO4 :: (PO2)
- Ability to analyze the operation of each type of hybrid inverters including(i) half-bridges (MMC), (ii) hybrid-clamped, (iii) hybrid symmetrical and asymmetrical H-bridge, (iv) hybrid H-bridge diode-clamped, (v) hybrid H-bridge flying-capacitor and Active neutral point clamped inverter topologies. CO4 :: (PO2)
- To be able to apply suitable pulse width modulation techniques for various hybrid inverters. : CO3 :: (PO1)

- To be able to develop the simulation model using software tools for various types of hybrid H-bridge inverters. CO4 :: (PO5)
- To be able to evaluate the performance of the various types of hybrid H-bridge inverters for various loads by simulation. CO5 :: (PO3)
- To be able to investigate the performance of higher order levels of combinational clamping hybrid inverters by simulation. CO6 ::(PO4)

Questions

6.1 Analyze the hybrid multilevel inverter using single-leg for phase 'a' of three half-bridge cascaded inverter and explain the various modes of operation for positive load current?

6.2 Analyze the hybrid multilevel inverter using single-leg for phase 'a' of three half-bridge cascaded inverter and explain the various modes of operation for negative load current?

6.3 Analyze the new symmetrical hybrid multilevel inverter shown in Fig.6.3 in the chapter 6, and explain the operation?

6.4 Analyze the hybrid-clamped (both diode-clamped and flying-capacitor) five-level inverter as shown in Fig.6.5 (b) and explain the various modes of operation for positive load current?

6.5 Analyze the hybrid-clamped (both diode-clamped and flying-capacitor) five-level inverter as shown in Fig.6.5 (b) and explain the various modes of operation for negative load current?

6.6 Analyze the new hybrid asymmetrical multilevel H-bridge inverter shown in Fig.6.7, and explain the various modes of operation for 5-level output voltage for positive load current?

6.7 Analyze the new hybrid asymmetrical multilevel H-bridge inverter shown in Fig.6.7, and explain the various modes of operation for 5-level output voltage for negative load current?

6.8 Analyze the Hybrid H-bridge type diode-clamped Multilevel Inverter (H-bridge based NPC) topology shown in Fig.6.8 for 5-level output voltage and explain its functionality?

6.9 Analyze the configuration of the seven-level flying-capacitor based active NPC inverter shown in Fig.6.11 and Fig.6.12 and discuss its functionality?

6.10 Realize the equivalent circuit of Hybrid inverter with reduced power semiconductor switches shown in Fig.6.15 for seven-level output voltage?

6.11 Analyze the realized configuration of seven-level hybrid inverter with reduced power semiconductor switches shown in Fig.6.15 (b) and explain various modes of operation for positive load current?

6.12 Analyze the realized configuration of seven-level hybrid inverter with reduced power semiconductor switches shown in Fig.6.15 (b) and explain various modes of operation for negative load current?

6.13 Analyze the configuration of thirty-one-level hybrid inverter with reduced power semiconductor switches shown in Fig.6.16 and explain various modes of operation for positive load current?

6.14 Analyze the configuration of thirty-one-level hybrid inverter with reduced power semiconductor switches shown in Fig.6.16 and explain various modes of operation for negative load current?

6.15 Draw the topology for 49-Level Inverter by realizing from two-cascaded seven-level inverters using unidirectional IGBTs and explain the various modes of operation for positives load current.

6.16 Mention the advantages and disadvantages of diode-clamped inverter (DCI)?

6.17 Mention the advantages and disadvantages of flying-capacitor inverter (FCI)?

6.18 Mention the advantages and disadvantages of cascaded H-bridge inverter (CHB)?

6.19 Mention the advantages and disadvantages of multi-modular inverter (MMI)?

6.20 Mention the advantages and disadvantages of Active neutral point clamped (ANPC) inverter?

Multiple Choice Questions

6.1 In a Hybrid multilevel inverter using three-half-bridge modules in series for producing 'a' phase output voltage, how many dc sources are used?

(a) two (b) three
(c) four (d) five

6.2 In a Hybrid multilevel inverter using three-half-bridge modules in series for producing 'a' phase output voltage, how many bidirectional power semiconductor switches are used?

(a) four (b) six
(c) eight (d) twelve

6.3 The new symmetrical cascaded H-bridge inverter can have a number of levels in its output Voltage given by 2N+1, where N is,

(a) number of switches (b) number of dc sources
(c) number of levels (d) number of H-bridges

6.4 In hybrid-clamped five-level inverter of single-leg arrangement, how many dc source capacitors are required?

(a) two (b) three
(c) four (d) five

6.5 In hybrid-clamped five-level inverter of single-leg arrangement, how many main switching devices are used?

(a) six (b) eight
(c) four (d) twelve

6.6 In hybrid-clamped five-level inverter of single-leg arrangement, how many clamping-switching devices are employed?

(a) eight (b) four
(c) six (d) three

6.7 In hybrid-clamped five-level inverter of single-leg arrangement, how many clamping diodes are used?

(a) eight (b) four
(c) five (d) six

6.8 In hybrid-clamped five-level inverter of single-leg arrangement, how many flying-capacitors are required?

(a) three (b) two
(c) four (d) no flying-capacitors are required

6.9 In m-level hybrid-clamped inverter topology, of single-leg arrangement, how many dc link source capacitors are needed?

(a) 2(m-1) (b) (m-1)
(c) 2(m-2) (d) (m-2)

6.10 In m-level hybrid-clamped inverter topology, of single-leg arrangement, how many main switching devices are required?

(a) (m-1) (b) 2(m-2)
(c) 2(m-1) (d) (m-2)

6.11 In m-level hybrid-clamped inverter topology, of single-leg arrangement, how many clamping switching devices are desired?

(a) 2(m-1) (b) (m-2)
(c) (m-1) (d) 2(m-2)

6.12 In m-level hybrid-clamped inverter topology, of single-leg arrangement, how many clamping diodes are required?

(a) (m-1) x (m-2) (b) (m-1) – (m-2)
(c) 2(m-1) x (m-2) (d) (m-1) + (m-2)

6.13 In m-level hybrid-clamped inverter topology, of single-leg arrangement, how many flying-capacitors are essential?

(a) 2(m-2) (b) (m-2)
(c) (m-1) (d) 2(m-1)

6.14 In new hybrid asymmetrical (transistor-clamped) five-level H-bridge inverter, how many dc link source capacitors are needed?

(a) four (b) three
(c) two (d) five

6.15 In new hybrid asymmetrical five-level H-bridge inverter, how many transistors are essential?
(a) five (b) six
(c) four (d) seven

6.16 Hybrid H-bridge type diode-clamped five level (single phase) inverter, how many dc link source capacitors are required?
(a) four (b) two
(c) three (d) five

6.17 Hybrid H-bridge type diode-clamped five level (single phase) inverter, how many clamped diodes are essential for entire bridge circuit topology?
(a) two (b) six
(c) four (d) eight

6.18 Hybrid H-bridge type diode-clamped five level (single phase) inverter, how many switching devices are required in total for entire bridge circuit topology?
(a) two (b) four
(c) six (d) eight

6.19 In the seven-level flying capacitor based active neutral point clamped (NPC) inverter topology, how many dc link source capacitors are needed?
(a) two (b) six
(c) five (d) seven

6.20 In the seven-level flying capacitor based active neutral point clamped (NPC) inverter topology, how many flying-capacitors are essential?
(a) four (b) two
(c) six (d) three

6.21 In the seven-level flying capacitor based active neutral point clamped (NPC) inverter topology, how many main forward direction switching devices are employed?
(a) three (b) five
(c) four (d) two

6.22 In the seven-level flying capacitor based active neutral point clamped (NPC) inverter topology, how many main reverse direction switching devices are needed?
(a) three (b) five
(c) two (d) four

6.23 In the seven-level flying capacitor based active neutral point clamped (NPC) inverter topology, how many forward direction clamping switching devices are needed?
(a) three (b) five
(c) six (d) eight

6.24 In the seven-level flying capacitor based active neutral point clamped (NPC) inverter topology, how many reverse direction clamping switching devices are needed?
(a) eight (b) three
(c) five (d) six

6.25 In a seven-level new hybrid bridge inverter topology with reduced power semiconductor switches, how many dc sources are required?
(a) three (b) one
(c) two (d) four

6.26 In a seven-level new hybrid bridge inverter with reduced power semiconductor switches, how many main unidirectional switching devices are needed?
(a) four (b) six
(c) eight (d) five

6.27 In a seven-level new hybrid bridge inverter with reduced power semiconductor switches, how many centrally (or commonly) connected unidirectional switching devices are used?
(a) one (b) two
(c) four (d) six

6.28 In a thirty-one-level new hybrid bridge inverter topology with reduced power semiconductor switches, how many dc sources are required?
(a) three (b) one
(c) two (d) four

6.29 In a thirty-one-level new hybrid bridge inverter with reduced power semiconductor switches, how many main unidirectional switching devices are needed?
(a) four (b) six
(c) eight (d) five

6.30 In a thirty-one-level new hybrid bridge inverter with reduced power semiconductor switches, how many centrally (or commonly) connected unidirectional switching devices are used?
(a) one (b) two
(c) four (d) six

6.31 In a 49-Level new hybrid bridge Inverter constructed from two-cascaded hybrid seven-level inverters, how many dc sources are required?
(a) two (b) three
(c) six (d) four

6.32 In a 49-Level new hybrid bridge Inverter constructed from two-cascaded hybrid seven-level inverters, how many unidirectional main switching devices are used?
(a) eight (b) four
(c) six (d) twelve

6.33 In a 49-Level new hybrid bridge Inverter constructed from two-cascaded hybrid seven-level inverters, how many centrally connected unidirectional switching devices are required?
(a) eight (b) two
(c) four (d) six

Exercises

Ex.6.1: Simulate the performance of Hybrid multilevel inverter based on half-bridge modules by considering single phase and single-leg arrangement topology for phase 'a' of RL load for four-level output voltage with positive load current using any of the technical software tools? (Refer Fig. 6.1 for circuit topology)

Ex.6.2: Evaluate the performance of Hybrid multilevel inverter based on half-bridge modules by considering single phase and single-leg arrangement topology for phase 'a' of RL load for four-level output voltage with negative load current by simulation using any of the technical software tools. (Refer Fig. 6.1)

Ex.6.3: Investigate the performance of Hybrid-clamped five-level inverter topology by considering single phase and single-leg arrangement topology shown in Fig.6.5, for phase 'a' of RL load for five-level output voltage with positive load current using any of the technical software tools?

Ex.6.4: Apply any PWM scheme and simulate the performance of new Hybrid asymmetrical five-level H-Bridge Inverter topology proposed in Fig.6.7 for single H-bridge (transistor-clamped) for RL load and obtain load voltage and load current waveforms by simulation.

Ex.6.5: Evaluate the performance of Hybrid H-bridge type diode-clamped five-level Inverter (H-bridge based NPC) shown in Fig.6.8 for RL load by applying Sinusoidal PWM scheme and obtain the load voltage and load current waveforms and also its % total harmonic distortions. Use any available simulation tools.

Ex.6.6: Investigate the performance of Hybrid Flying-Capacitor and Active NPC seven-level Inverter configuration shown in Fig. 6.11 for both positive half-period and negative half-period by simulation and obtain the load voltage and current waveform. Use any available simulation tools.

Ex.6.7: Simulate the performance of seven-level hybrid inverter with reduced power semiconductor devices shown in Fig.6.15 by applying Sinusoidal PWM scheme to obtain the load voltage and load current by using any simulation tools.

Ex.6.8: Evaluate the performance of thirty-one-level hybrid inverter with reduced power semiconductor devices shown in Fig.6.16 by applying Sinusoidal PWM scheme to obtain the load voltage and load current by using any simulation tools.

Ex.6.9: Simulate the performance of forty-nine-level hybrid inverter with reduced power semiconductor devices shown in Fig.6.21 by applying Sinusoidal PWM scheme to obtain the load voltage and load current by using any simulation tools.

A. Course Outcomes (COs) as per Bloom's Taxonomy:

Sl. nos.	Parameters	Activities	COs
1.	Remembering/Knowledge	Recognize, listing, describe, finding	CO 1
2.	Understanding/Comprehension	Explain, classify, summarize, interpret	CO 2
3.	Applying	Implement, use, execute, carryout	CO 3
4.	Analyzing	Demonstrate, find, interrogate	CO 4
5.	Evaluating	Checking, experimenting, judging	CO 5
6.	Creating	Design, Construct, Planning, Producing, Inventing	CO 6

B. Mapping Course outcomes to Program Outcomes

Program Outcomes (POs)	Course Outcomes (COs)					
	CO1	CO2	CO3	CO4	CO5	CO6
PO 1: (Engineering Knowledge)	•	•	•			
PO 2: (Problem Analysis)				•		
PO 3: (Design/Development of solutions)					•	
PO 4: (Investigation)						•
PO 5: (Modern tools usage)				•		
PO 6: (The Engineer and Society)						
PO 7: (Environment and Sustainability)						
PO 8: (Ethics)						
PO 9: (Individual and Team Work)						
PO 10: (Communication)						
PO 11: (Project Management)						
PO 12: (Lifelong learning)						

Answers to Multiple Choice Questions

6.1	(c)	6.2	(b)	6.3.	(b)	6.4	(c)	6.5	(b)
6.6	(c)	6.7	(d)	6.8	(a)	6.9	(b)	6.10	(c)
6.11	(d)	6.12	(a)	6.13	(b)	6.14	(c)	6.15	(a)
6.16	(b)	6.17	(c)	6.18	(d)	6.19	(a)	6.20	(b)
6.21	(d)	6.22	(c)	6.23	(a)	6.24	(b)	6.25	(c)
6.26	(a)	6.27	(b)	6.28	(d)	6.29	(c)	6.30	(b)
6.31	(d)	6.32	(a)	6.33	(c)				

References

1. J. Rodriguez, J.S. Lai and F.Z. Peng, “Multilevel inverters: survey of topologies, controls, and applications,” IEEE Trans Ind. Appl., vol. 49-4, pp. 724–738, Aug. 2002.
2. J. S. Lai and F. Z. Peng, “Multilevel converters–A new breed of power converters,” IEEE Transactions Industry Application, vol. 32, pp. 509–517, May/June 1996.
3. 07_chapter 2.pdf : (Introduction to Multilevel Inverters)
4. J. Rodriguez, S. Bernet, P. Steimer, and I. Lizama, “A survey on neutral point clamped inverters,” IEEE Transactions on Industrial Electronics, vol. 57, no. 7, pp. 2219–2230, Jul. 2010.
5. 08_chapter 3.pdf : (Cascaded Multilevel Inverters)
6. M. Malinowski, K. Gopakumar, J. Rodriguez, and M. A. Pérez, “A survey on cascaded multilevel inverters,” IEEE Transactions on Industrial Electronics, Vol. 57, No. 7, pp. 2197–2206, July 2010.
7. M. R. Banaei, E. Salary, "Analysis of Generalized Symmetrical Multilevel Inverter" Journal of Circuits Systems Computations, Vol. 20, No. 2, pp. 299- 311, 2011.
8. Pallavi Appaso Arbune and Dr. Asha Gaikwad, “Comparative Study of Three level and Five -level Inverter”, International Journal of Advanced Research in Electrical, Electronics and Instrumentation Engineering (An ISO 3297: 2007 Certified Organization) Vol. 5, Issue 2, February 2016, pp.681-686.
9. S. Khomfoi and L.M. Tolbert, Power Electronics Handbook, 2. Edition, Chapter 31 – “Multilevel Power Converters”, Amsterdam: Elsevier, 2007, pp. 31/1-31/50.
10. “Power Electronics Handbook” by Muhammad H. Rashid, Third Edition, Elsevier, 2011.
11. “Power Electronics Circuits Devices and Applications”, By Muhammad. H. Rashid, Third Edition, Prentice Hall, Englewood Cliffs, New Jersey.
12. R. Teodorescu, F. Beaabjerg, J. K. Pedersen, E. Cengelci, S. Sulistijo,B. Woo, and P. Enjeti, “Multilevel converters - A survey,” in Proceedings European Power Electronics Conference EPE’99), Lausanne, Switzerland, 1999.
13. Meynard, T.A., Fadel, M. and Aouda, N., “Modeling of multilevel converters”, IEEE Transactions on Industrial Electronics, Vol. 44, No. 3, June 1997, pp. 356 – 364.

14. M. Malinowski, "Cascaded multilevel converters in recent research and applications", Bulletin of The Polish Academy of Sciences Technical Sciences, Vol. 65, No. 5, 2017, pp.567-578.

15. Ebrahim Babaei, Somayeh Alilu, and Sara Laali, "A New General Topology for Cascaded Multilevel Inverters with Reduced Number of Components Based on Developed H-Bridge", IEEE Transactions on Industrial Electronics, Vol. 61, No. 8, August 2014, pp.3932-3939.

16. Ebrahim Babaei *, Mohammad Farhadi Kangarlu, Farshid Najaty Mazgar, "Symmetric and asymmetric multilevel inverter topologies with reduced switching devices", Electric Power Systems Research Vol. 86, May 2012, Pages 122-130.

17. E. Babaei and S.H. Hosseini, "New cascaded multilevel inverter topology with minimum number of switches," Energy Conversion and Management, vol. 50, pp. 2761–2767, 2009.

18. R.A. Ahmed, S. Mekhilef and H.W. Ping, "New multilevel inverter topology with reduced number of switches," in Proceedings 2010, 14th International Middle East Power Systems Conference, pp. 565-570.

19. Hulusi Karaca, "A Novel Topology for Multilevel Inverter with Reduced Number of Switches", Proceedings of the World Congress on Engineering and Computer Science 2013 Vol. I, WCECS 2013, 23-25 October, 2013, San Francisco, USA.

20. Moirangthem Samson and Mohammed Saleem, "A New General Topology for Cascaded Multilevel Inverters with Reduced Number of Components Based On Developed H-Bridge", International Journal of Research in Advanced Engineering Technologies, Vol.4, Issue.4, July 2015, PP.1-12.

21. Ebrahim Babaei, Concettina Buccella, and Maryam Saeedifard, "Recent Advances in Multilevel Inverters and Their Applications—Part. I", IEEE Transactions on Industrial Electronics, Vol. 63, No. 11, November 2016, pp.7145-7147.

22. Ebrahim Babaei, Concettina Buccella, and Maryam Saeedifard, "Recent Advances in Multilevel Inverters and Their Applications—Part. II", IEEE Transactions on Industrial Electronics, Vol. 63, No. 12, November 2016, pp.7777-7779.

23. S. Kouro, M. Malinowski, K. Gopakumar, J. Pou, L. G. Franquelo, B. Wu, J. Rodriguez, M. A. Perez, and J. I. Leon, "Recent advances and industrial applications of multilevel converters," IEEE Transactions on Power Electronics, Vol. 57, No. 8, pp. 2553–2580, Aug. 2010.

24. Shadi H. Shehadeh, Hamed H. Aly, and M. E. El-Hawary, "Photovoltaic Multi-level Inverters Technology", Proceeding of the IEEE 28th Canadian Conference on Electrical and Computer Engineering Halifax, Canada, May 3-6, 2015, pp.648-655.

25. K. K. Gupta, and S. Jain, "A Novel Multilevel Inverter Based on Switched DC Sources," IEEE Transaction on Industrial Electronics, Vol. 61, No. 7, pp. 3269 - 3278, July 2014.

26. A. Ajami, M.R.J. Oskuee, and A. Mokhberdoran, "Developed cascaded multilevel inverter topology to minimize the number of circuit devices and voltage stresses of switches," IET Power Electronics, Vol. 7, No. 2, pp. 459 – 466, Feb. 2014.
27. Babaei, Sara Laali, and Zahra Bayat, "A Single-Phase Cascaded Multilevel Inverter Based on a New Basic Unit with Reduced Number of Power Switches", IEEE Transactions on Industrial Electronics Vol. 62, No.2, February 2015, pp. 922-929.
28. M. Manjrekar and T. A. Lipo, "A hybrid multilevel inverter topology for drive application," in Proceedings APEC, 1998, pp. 523–529.
29. N. Abd Rahim, M. F. Mohamad Elias, and W. P. Hew, "Transistor-clamped H-bridge based cascaded multilevel inverter with new method of capacitor voltage balancing," IEEE Transactions Industrial Electronics, vol. 60, no. 8, pp. 2943–2956, Aug. 2013.
30. Pauline Jothi Kiruba G. J, Mercy P, Parveen Nisha M. G, "Cascaded H-Bridge Multilevel Inverter with Reduced Number of Power Components", International Journal of Science and Research (IJSR) ISSN (Online): 2319-7064, pp.1167-1172.
31. J. Liu, K. W. E. Cheng, and Y. Ye, "A Cascaded Multilevel Inverter Based on Switched-Capacitor for High-Frequency AC Power Distribution System," IEEE Transactions on Power Electronics, Vol. 29, No. 8, pp. 4219 - 4230, Aug. 2014.
32. Y.Suresh: "Investigation on Cascade Multilevel Inverter for Medium and High-Power Applications", a Ph.D. Thesis, Prof. Anup Kumar Panda (Supervisor), Department of Electrical Engineering National Institute of Technology, Rourkela July 2012.
33. S.R.Nandhini, G.Akalya, K.Aswini, K.Nandhini, M.Valan Rajkumar, and B.Rajeshkumar, "A New Topology of H-bridge based Multilevel Inverter for PV System with Reduced Switches", International Journal of Emerging Technologies in Engineering Research (IJETER) Vol. 5, Issue- 4, April (2017), pp.60-68.
34. E. Babaei, and S. S. Gowgani, "Hybrid Multilevel Inverter Using Switched Capacitor Units," IEEE Transactions on Industrial Electronics, Vol. 61, No. 9, pp. 4614 - 4621, Sept. 2014.
35. Holtz, J, "Pulse width modulation–A survey", IEEE Transactions Industrial Electronics, 1992, 30, (5), pp. 410–420.
36. K.A. Corzine, M.W. Wielebski, F.Z. Peng and J. Wang, "Control of cascaded multi-level inverters," IEEE Transactions Power Electronics, vol. 19-3, pp. 732–738, 2004.
37. P.Chaithanya Deepak and S. Nagaraja Rao, "Cascaded H-Bridge Multilevel Inverter Using Inverted Sine Wave PWM Technique", International Journal of Emerging Trends in Electrical and Electronics (IJETEE – ISSN: 2320-9569) Vol. 6, Issue. 1, Aug-2013, pp.39-44.
38. Satish Kumar Peddapelli, "Recent Advances in Pulse Width Modulation Techniques and Multilevel Inverters", World Academy of Science, Engineering and Technology International Journal of Electrical and Computer Engineering Vol.8, No.3, 2014, pp.593-601.

39. L-38(DP)(PE) ((EE)NPTEL).pdf (Lesson 38 Other Popular PWM Techniques) Version 2 EE IIT, Kharagpur, pp.1-13.
40. Jani Rushiraj G. and P.N. Kapil, "Analysis of Different Modulation Techniques for Multilevel Inverters", 1st IEEE International Conference on Power Electronics, Intelligent Control and Energy Systems (ICPEICES-2016), pp.1-6.
41. B.P. McGrath, and D.G. Holmes, "Multicarrier PWM strategies for multilevel ınverters," IEEE Transactions Industrial Electronics, vol. 49-4, Aug. 2002.
42. Jagdish Kumar, Biswarup Das, and Pramod Agarwal, "Selective Harmonic Elimination Technique for a Multilevel Inverter", Fifteenth National Power Systems Conference (NPSC), IIT Bombay, December 2008, pp.608-613.
43. R.Saravana kumar, R.M.Anusuya, V.Kavitha, and A.Gopi, "Selective Harmonic Elimination in Seven Level Cascaded Inverter", 2013 International Conference on Renewable Energy and Sustainable Energy [ICRESE'13], pp.51-57.
44. Sourabh Rathore, Mukesh Kumar Kirar and S. K Bhardwaj, "Simulation of Cascaded H- Bridge Multilevel Inverter using PD, POD, APOD Techniques", Electrical & Computer Engineering: An International Journal (ECIJ) Volume 4, Number 3, September 2015, pp.27-41.
45. J. I. Leon, S. Kouro, L. G. Franquelo, J. Rodriguez, and Bin Wu, "The Essential Role and the Continuous Evolution of Modulation Techniques for Voltage Source Inverters in Past, Present and Future Power Electronics", IEEE Transactions on Industrial Electronics, 2015.
46. 09_chapter 4.pdf (Space Vector Pulse Width Modulation)
47. Irfan Ahmed, Vijay B. Borghate, Amarendra Matsa, Prafullachandra M. Meshram, H. M. Suryawanshi, and Madhuri A. Chaudhari, "Simplified Space Vector Modulation Techniques for Multilevel Inverters", IEEE Transactions on Power Electronics, 2015.
48. Dorin O. Neacsu, "Space Vector Modulation –An Introduction" = Tutorial at IECON2001: The 27th Annual Conference of the IEEE Industrial Electronics Society, pp.1583-1592.
49. Bengi Tolunay, "Space Vector Pulse Width Modulation for Three-Level Converters - a LabVIEW Implementation", pp.1-61.
50. Qamar Muhammad Attique, Yongdong Li, and Kui Wang, "A Survey on Space-Vector Pulse Width Modulation for Multilevel Inverters", CPSS Transactions on Power Electronics and Applications, Vol. 2, No. 3, September 2017, pp.226-236.
51. Soumitra Das, G. Narayanan, and M. Pandey, "Space-Vector-Based Hybrid Pulse Width Modulation Techniques for a Three-Level Inverter", IEEE Transactions on Power Electronics, Vol. 29, No. 9, September 2014, pp.4580-4591.
52. N. Ghiasnezhad, M. Tavakoli Bina, and M. A. Golkar, "Three-Dimensional Space-Vector Modulation Algorithm for All Types of Multilevel Converters using ABC Coordinate", 5th International Advanced Technologies Symposium (IATS'09), May 13-15, 2009, Karabuk, Turkey.

53. M.M.Prats, L.G.Franquelo, J.I.Leon, R.Portillo, E.Galvan and J.M. Carrasco, "A SVM-3D Generalized Algorithm for Multilevel Converters", pp.24-29.

54. Felix Rojas, Roberto Cardenas, Ralph Kennel, Jon Clare, and Matias Diaz, "A Simplified Space Vector Modulation Algorithm for Four-Leg NPC Converters", IEEE Transactions on Power Electronics, 32 (11). pp. 8371-8380.

55. Nikola Celanovic, and Dushan Boroyevich, "A Fast Space-Vector Modulation Algorithm for Multilevel Three-Phase Converters", IEEE Transactions on Industry Applications, Vol. 37, No. 2, March/April 2001, pp.637-641.

56. S. M. Padmaja, and G. Tulasi Ram Das, "SVPWM Based 3-Level STATCOM for Reactive Power Management under Line-Line Fault in a Transmission Line", International Journal of Electrical Engineering & Technology (IJEET), Volume 4, Issue 3, May - June (2013), pp. 167-187.

57. G. Annapurna and G. Tulasi Ram Das, "Multicarrier Modulated Multi Level Inverter Based DVR and DSTATCOM for Mitigating Swell", ARPN Journal of Engineering and Applied Sciences, Vol. 10, No. 18, October 2015, pp.7969-7976.

58. G. Sree Lakshmi, S. Kamakshaiah and G. Tulasi Ram Das, "Closed Loop Control of Three-Level Diode Clamped Inverter Fed IPMSM with Different Modulation Techniques", Global Journal of Researches in Engineering Electrical and Electronics Engineering Volume 13 Issue 9 Version 1.0 Year 2013, pp.9-17.

59. G. Sree Lakshmi S. Kamakshaiah and G. Tulasi Ram Das, "Three-level DCML fed IPMSM using SPWM and CBSVPWM varying Modulation Index", 2013, International Conference on Control, Automation, Robotics and Embedded Systems (CARE).

60. R.W. Menzies, Y. Zhuang, "Advanced Static Compensation using a Multilevel GTO Hytistor inverter", IEEE Trans. On Power Delivery, Vol. 10, No. 2, PP. 732-738, Apr. 1995.

61. F.Z. Peng, J.S. Lai, J.W. McKeever, J. VanCoe versing, "A Multilevel Voltage. Source inverter with separate dc sources for static Var generation", IEEE Trans. On Indi. Electrons., Vol. 32, No.5, pp.1130-1138, Sept. 1996.

62. F.Z. Peng, J.S. Lai, "Dynamic Performance and Control of a Static Var generation using cascade multilevel inverter", IEEE Trans. on Ind. Electron., Vol. 33, No. 3, pp.478-755, May 1997.

63. A. Nabae, I. Takashashi, H. Akagi, "A New Neutral-point clamped PWM inverter", IEEE Trans. on Ind. Appli., Vol. IA-17, PP518-523, Sept.-Oct. 1981.

64. Jiom Zhong Zhang, Shai Xu, Zakuid Din and Xing Hu, in Hybrid Multilevel Converters: Topologies Revolutions and verification", Journals. Energies, Vol.12, Issue 4, Feb. 2019.

Recommended Books/E-books for further reading:

65. Krishna Kumar Gupta, Pallavee Bhatnagar, "**Multilevel Inverters** - Conventional and Emerging Topologies and Their Control", Academic Press, 2017.
66. "Multilevel Inverters – Introduction and Emergent Topologies", 1st edition, Feb.2021, edited by Ersan Kabalci.
67. "Multilevel Converters: Analysis, Modulation, Topologies, and Applications" Edited by Gabriele Grandi and Alex Ruderman, Published in Energies, MDPI.
68. "Single-DC-Source Multilevel Inverters" by Hani Vahedi, Mohamed Trabelsi, Springer, E-book, 2019.
69. "Advanced Multilevel Converters and Applications in Grid Integration", by Ali Iftekhar Maswood (Editor) and Hossein Dehghani Tafti (Editor), Wiley Press, E-book, 2019.
70. "Multilevel Inverter" by Subhashini M, E-book, 2020.
71. "Modular Multilevel Converters: Analysis, Control, and Applications", *Wiley-IEEE Press*, January 2018.

Index

I

L

M

N

O

P

R

S

T

www.ingramcontent.com/pod-product-compliance
Ingram Content Group UK Ltd.
Pitfield, Milton Keynes, MK11 3LW, UK
UKHW061534190726
13845UKWH00005B/98

9 789348 73470